Ergebnisse der Mathematik und ihrer Grenzgebiete. 3. Folge / A Series of Modern Surveys in Mathematics

Volume 70

More information about this series at http://www.springer.com/series/728

Matthias Schütt · Tetsuji Shioda

Mordell–Weil Lattices

Springer

Matthias Schütt
Institut für Algebraische Geometrie
Leibniz Universität Hannover
Hannover, Niedersachsen, Germany

Tetsuji Shioda
Department of Mathematics
Rikkyo University
Tokyo, Japan

ISSN 0071-1136 ISSN 2197-5655 (electronic)
Ergebnisse der Mathematik und ihrer Grenzgebiete. 3. Folge / A Series of Modern Surveys in Mathematics
ISBN 978-981-32-9300-7 ISBN 978-981-32-9301-4 (eBook)
https://doi.org/10.1007/978-981-32-9301-4

Mathematics Subject Classification (2010): 14J27, 14J20, 14J26, 14J28, 11G05, 11G07, 11G25, 11R09, 11G50, 11H31, 06B15

This Springer imprint is published by the registered company Springer Nature Singapore Pte Ltd.
The registered company address is: 152 Beach Road, #21-01/04 Gateway East, Singapore 189721, Singapore

Dedicated to Yukiko

Preface

In this book, we will study the Mordell–Weil lattice of an elliptic curve over a function field and discuss various applications.

The idea for the book arose in the early 1990s. Yet it took more than 20 years to finish it. At least, this delay came with the incidental advantage that we were able to cover several recent developments building on Mordell–Weil lattices. Let us highlight here only the notion of (multiplicative) excellent families and the impact of Mordell–Weil lattices on the theory of K3 surfaces in recent years. Once again, these applications display the powerful interactions which Mordell–Weil lattices foster in a seemingly elementary way between foundational topics such as algebraic equations, elliptic curves and lattices.

One characteristic of the Mordell–Weil lattice is the fact it is equipped with a clear-cut height pairing, based on the intersection theory of the associated elliptic surface. As a consequence, we are naturally led to some beautiful lattices such as E_8 (the positive-definite unique even unimodular lattice of rank eight) among the Mordell–Weil lattices.

When writing the present book, we had a fairly broad readership in mind, from novice to advanced researchers ranging through different fields such as classical and algebraic geometry and number theory, reflecting the diverse input into the theory of Mordell–Weil lattices. To keep the manuscript as self-contained as possible, we decided to include concise reviews of all the background material needed (notably lattices, elliptic curves, and algebraic surfaces). Necessarily, this means that some readers will be familiar with parts of the early chapters, and that the level of the material presented rises steadily as the book advances. Numerous examples are to be found throughout the book, some building explicitly on each other so the reader can see the ideas of the subjects under consideration unfold one after the other until the full impact and connections of Mordell–Weil lattices have been revealed.

Hannover, Germany
Tokyo, Japan
June 2019

Matthias Schütt
Tetsuji Shioda

Acknowledgements

This book, and some of the developments described in it, would have been impossible without Jean-Pierre Serre and Friedrich Hirzebruch, whose advice and encouragement were invaluable to us. We are no less indebted to our colleagues and co-authors throughout the years, of whom we should like to mention Noboru Aoki, Bert van Geemen, Torsten Ekedahl, Noam D. Elkies, Klaus Hulek, Toshiyuki Katsura, Remke Kloosterman, Shigeyuki Kondo, Masato Kuwata, Abhinav Kumar, Keiji Oguiso, Ichiro Shimada, Tomohide Terasoma, Jaap Top, and Hisashi Usui. Special thanks go to Simon Brandhorst for many valuable comments which helped us improve the manuscript.

We are indebted to the referees for their numerous suggestions which helped streamline the exposition and for providing many additions and corrections. Our thanks go to everyone at Springer who helped to make this book possible, especially to Reinhold Remmert.

We thank the institutions which have supported, or even influenced, the genesis of this book in the last 25 years, notably Rikkyo University, MPIM Bonn, Leibniz Universität Hannover, University of Tokyo, University of Copenhagen, and RIMS Kyoto.

Funding from various sources is gratefully acknowledged. Among others we received generous support from JSPS under Grant-in-Aid for Scientific Research No. 17540044, (C) 20540051, (C) 25400052, from ERC under StG 279723 (SURFARI), and from DFG in the framework of RTG 1463.

Most importantly, we would like to thank our families for their patience and support throughout all these years.

Hannover, Germany
Tokyo, Japan
June 2019

Matthias Schütt
Tetsuji Shioda

Contents

Chapter 1
Introduction

The purpose of this book is to define the Mordell–Weil lattice of an elliptic curve over a function field, and explore its numerous applications.

The key idea of this notion is rather simple: to endow an interesting and important object (the Mordell–Weil group of an elliptic curve, which is known to be a finitely generated abelian group) with an additional structure (a natural bilinear pairing) to make it into a lattice (the Mordell–Weil lattice). We carry out this idea in the case of an elliptic curve defined over a function field, by using the intersection theory on the associated elliptic surface. This allows us to gain an understanding far better than before and to develop many intriguing applications. Even more remarkably, all the ingredients themselves will not be too hard to understand (and they will be laid out in detail throughout the first five chapters of this book), yet all together the outcome will produce a very powerful tool—the Mordell–Weil lattice. A bit picturesque, admittedly, we like to think of the Mordell–Weil lattices as the center of a circle pointing to many important areas of modern mathematics and even physics (Fig. 1.1).

The starting point for all our considerations will be elliptic curves, one of the most classical, most fundamental and most fascinating objects of study in mathematics. The decisive feature of elliptic curves is the group law which their rational points are equipped with, and which despite its classical nature is still not completely understood.

The structure of the group of rational points on an elliptic curve heavily depends on the base field. For instance, over $\mathbb{C}$ the curve is isomorphic to a complex torus $\mathbb{C}/\Lambda$. At once, this gives rise to the group structure while also showing that the points on the curve form a continuum. In comparison, over $\mathbb{R}$, the points on a real elliptic curve decompose into one or two connected components. The case of $\mathbb{Q}$, the field of rational numbers (or more generally global fields), has caught most attention over the years. A celebrated theorem of Mordell (later generalized by Weil) states that the rational points of an elliptic curve over $\mathbb{Q}$ (or any number field) form a finitely generated abelian group. Let us briefly comment on the importance of this fundamental object, paralleling similar groups in algebra, number theory and algebraic geometry. First, we recall Abel's theorem

M. Schütt and T. Shioda, *Mordell–Weil Lattices*, Ergebnisse der Mathematik und ihrer Grenzgebiete. 3. Folge / A Series of Modern Surveys in Mathematics 70,
https://doi.org/10.1007/978-981-32-9301-4_1

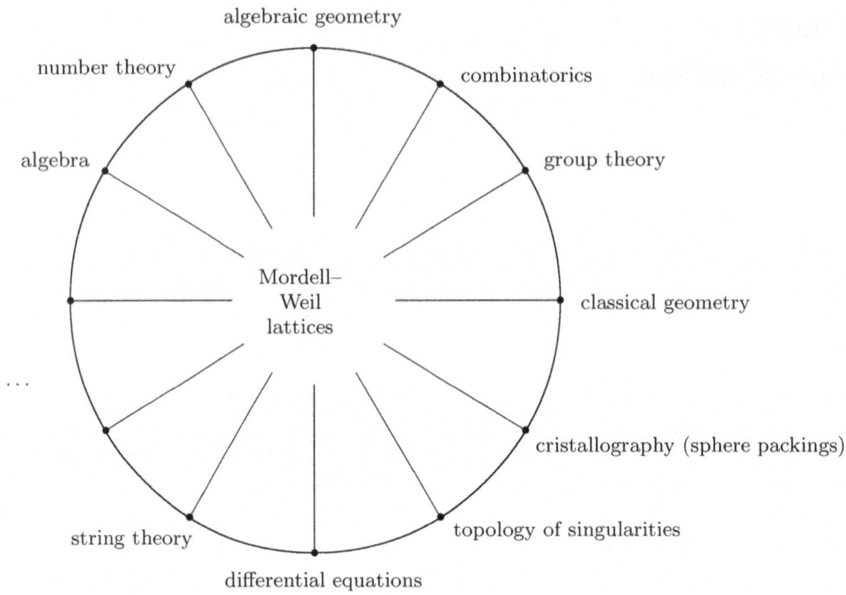

Fig. 1.1 Mordell–Weil lattices

$$E(K) \cong \mathrm{Pic}^0(E/K),$$

stating that the group of K-rational points $E(K)$ is isomorphic to the group of linear equivalence classes of degree zero K-rational divisors on the elliptic curve E defined over a field K. The importance of the Mordell–Weil group in this sense compares to the importance of the ideal class group of a number field. Secondly, in another aspect, it is similar to the unit group of a number field, both being finitely generated (cf. Dirichlet's theorem). Thirdly, the Mordell–Weil group can be regarded as the simplest non-trivial example of a group of algebraic cycles on some algebraic variety. In fact, we would almost go so far as to say that the Mordell–Weil group has the rare ability to visualize the algebraic cycles in question.

Despite this clear abstract picture, it remains to this very day a non-trivial task to determine the Mordell–Weil group of an elliptic curve explicitly, or already its rank, although the situation improved substantially after the introduction of the canonical height for the rational points on elliptic curves due to Néron, Tate, Manin and others. In retrospect, this development clearly set the scene for the discovery of Mordell–Weil lattices, giving rise to pairings, regulators etc., but to introduce it properly we have to switch the scenery slightly.

Consider a one-dimensional function field K which for simplicity the reader can think of as the field of rational functions in one variable, $k(t)$, for any field k, algebraically closed or not, say $k = \mathbb{Q}$ or $\mathbb{C}$. Then Mordell's theorem extends to elliptic curves over K under a mild condition which essentially requires that the

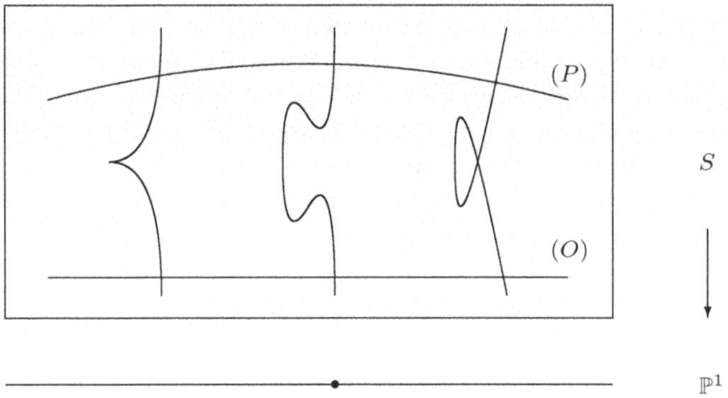

Fig. 1.2 Elliptic surface with section and singular fibres

curve be non-constant when thought of as varying over the base. Indeed, this last description conveys exactly the right impression of a non-constant elliptic curve over a function field: over the base, it fills out a whole algebraic surface over k as depicted below in Fig. 1.2—an elliptic surface S!

With this picture in mind (which can and will be made precise in this book), a K-rational point P, as well as the zero O, on the original elliptic curve E over K (called the generic curve) spread out to horizontal curves on the surface S (called sections) which thus lend themselves to the application of classical surface theory. This is exactly where intersection theory enters into the study of elliptic curves over function fields, a seemingly elementary, yet far reaching tool as we shall explore. Naturally, intersection theory defines a pairing on curves and their formal linear combinations (the so-called divisors), and thus an equivalence relation (numerical equivalence) whose quotient is a finitely generated free abelian group. Automatically, this group is equipped with a non-degenerate integral pairing which thus makes it into what we call a lattice (the so-called Néron–Severi lattice $\mathrm{Num}(S)$ which will play an important role throughout this book). The reader may simply think of a lattice as the $\mathbb{Z}$-span of some vectors in $\mathbb{R}^n$ which are $\mathbb{R}$-linear independent, such as the natural lattice $\mathbb{Z}^n \subset \mathbb{R}^n$. We should add, though, that some extra care is required since the lattice will depend in an essential way on the chosen pairing, i.e. the non-degenerate bilinear form which may be definite or indefinite. Below we depict the hexagonal lattice A_2 in $\mathbb{R}^2$ which will continue to reappear throughout this book.

Tying everything together, it is exactly this kind of (definite) lattice structure which we want to put on the group of rational points of an elliptic curve (non-constant over a one-dimensional function field). We shall now give a rough idea how this leads to the notion of Mordell–Weil lattices which is the central theme of this book.

The elliptic fibration, as depicted above, literally divides the irreducible curves on S into two kinds: horizontal and vertical curves. While the latter simply comprise all fibre components, the former consist not only of sections, but also of multisections.

Here, Abel's theorem will enable us to express any multisection in terms of sections and vertical curves up to linear equivalence. What prevents us from defining the intersection pairing directly on $E(K)$ is that a section usually does not have square zero when considered as a curve on S, but this would, of course, be required for the chosen neutral element O. One can work around this by considering the so-called trivial lattice $\mathrm{Triv}(S)$ generated by fibre components together with the zero section O. The next step is to derive an isomorphism

$$E(K) \cong \mathrm{Num}(S)/\mathrm{Triv}(S)$$

as finitely generated abelian groups. It then remains to define a suitable pairing on $E(K)$. This is achieved by means of the orthogonal projection from $\mathrm{Num}(S)$ with respect to $\mathrm{Triv}(S)$. Note, however, that the orthogonal projection is a priori only defined on the $\mathbb{Q}$-vector spaces obtained by extending scalars, so the resulting pairing on $E(K)$ need not be integral. In fact, the pairing may even be degenerate, but after dividing out by the torsion in $E(K)$ (which can thus be understood completely in terms of $\mathrm{Triv}(S)$), we obtain the desired lattice structure. Finally, changing sign gives the announced positive-definite *Mordell–Weil lattice* $\mathrm{MWL}(S)$ which forms the centerpiece of this book. Its precise definition will be given in Chap. 6 after explaining all the necessary ingredients.

Without further ado, let us dive into an explicit example of an elliptic curve given affinely as a plane cubic over $k(t)$, the field of rational functions in one variable t over some field k:

$$E: \quad y^2 + ty = (x-\alpha_1)(x-\alpha_2)(x-\alpha_3), \quad \alpha_i \in k \text{ pairwise distinct.} \tag{1.1}$$

Other than the invisible point O at infinity, i.e. O lies in a different affine chart of the projective plane over $k(t)$, this curve has six obvious rational points at

$$P_i = (\alpha_i, 0), \quad Q_i = (\alpha_i, -t) \quad i = 1, 2, 3.$$

The viewpoint of Mordell–Weil lattices will allow us to prove directly that the group of rational points on E (with O serving as zero element) is generated by any two of the six points (with different indices). Indeed, we will have

$$P_i = -Q_i \quad \text{and} \quad P_1 + P_2 + P_3 = O \tag{1.2}$$

with respect to the group law. More precisely, the Mordell–Weil lattice of E will be isomorphic to A_2 up to scaling (in fact, it will equal the dual $A_2^\vee$ of A_2), with the six minimal vectors as depicted in Fig. 1.3 exactly given by the above six rational points. That the Mordell–Weil lattice arises so naturally for the above elliptic curve is by no means accidental or exceptional. Rather this should be taken as a first illustration of the intriguing interaction of many different concepts which forms the core of our exposition.

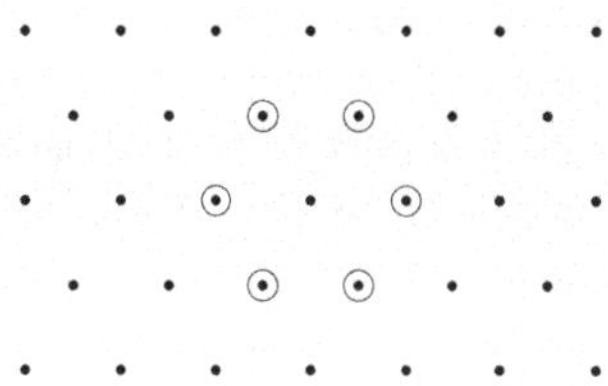

Fig. 1.3 Hexagonal lattice

Mordell–Weil lattices were introduced simultaneously by Noam D. Elkies and one of us at the end of the 1980s. This discovery was a central step for many subsequent developments. Not only did it provide a very natural framework for the previous considerations on elliptic curves as alluded to above, but it also featured many applications, for example:

- concrete techniques, especially to determine the group of rational points itself or geared towards algebraic equations and Galois representations,
- constructive methods, such as for elliptic curves of high rank (following Néron's method),
- classification techniques, such as for rational elliptic surfaces and K3 surfaces,
- a strong bridge between arithmetic and geometry, facilitating results on modular forms and moduli spaces,
- connections to deformation theory for singularities,
- new records for dense sphere packings.

We conclude this introduction with a brief outline of the contents of the book. Throughout we have tried to include instructive examples, whenever possible, and put the results into both the historical context and the logical context of this book. To this end, we have supplemented each chapter with some historical comments, but, of course, we cannot claim these to be anywhere close to complete. Each chapter is preceded by a short summary to prepare the reader for what is to come.

Chapter 2 reviews all the basics about lattices needed for this book. We lay special emphasis on sphere packings, root lattices, and Weyl groups.

The third chapter introduces a minimum of background on elliptic curves necessary for the understanding of Mordell–Weil lattices. In detail, we define the group law and review the relation to elliptic functions. Then we state the Mordell–Weil theorem for elliptic curves over a number field which is the starting point for any modern treatment of the rational points on an elliptic curve.

The third basic ingredient of Mordell–Weil lattices is the theory of algebraic surfaces as reviewed in Chap. 4. We will mostly be concerned with intersection theory on algebraic surfaces; these form a classical instance of a deep structure (given by lattices) put on a certain abelian group (given by divisors up to a certain equivalence). After discussing blow-ups and minimal models, we then continue to study invariants of surfaces and highlight the Enriques–Kodaira classification of algebraic surfaces.

With these tools at our disposal, we will be ready to introduce and investigate elliptic surfaces in thorough detail in Chap. 5. Since we will eventually aim for

Mordell–Weil lattices, we will right away restrict our considerations to jacobian elliptic surfaces, i.e. those admitting a section, since this endows the generic fibre with a rational point and thus makes it an elliptic curve in the strict sense (as opposed to a genus one curve). We discuss the singular fibres and their relation with Dynkin diagrams following Kodaira and Tate, and the interplay of the Kodaira–Néron model and the Weierstrass model. For classification purposes as well as extensive later use, we then explain how the canonical divisor and other invariants of an elliptic surface may be computed explicitly.

Chapter 6 brings together the concepts from the preceding sections in order to introduce Mordell–Weil lattices. From an overly simplifying viewpoint, this theory will essentially give us a better understanding of the Néron–Severi lattice of an elliptic surface (which, for instance, also lends itself to explicit computations, a notoriously difficult problem for curves on algebraic surfaces). The proper definition of Mordell–Weil lattices, involving trivial lattices, essential lattices and frames, will take almost all of Chap. 6. It concludes with important special considerations regarding torsion sections, the unimodular case and functorial properties of Mordell–Weil lattices.

The remainder of the book is devoted to a diversity of applications of the theory of Mordell–Weil lattices as indicated at the beginning of the introduction.

As the first application, we study rational elliptic surfaces in Chaps. 7 and 8. We will develop a complete hierarchy formed by the Mordell–Weil lattices of rational elliptic surfaces, dominated by the root lattice E_8, the unique positive-definite even unimodular lattice of rank 8. The details will be discussed in full throughout Chap. 8, providing a comprehensive understanding, for instance, of torsion sections and integral sections. Meanwhile Chap. 7 will set the scene by detailing more basic properties of rational elliptic surfaces. Here the reader should especially think of cubic pencils and elliptic curves of high rank (i.e. rank 6, 7 or 8).

Chapter 9 highlights Galois representations and algebraic equations which arise naturally from Mordell–Weil lattices. As pointed out before, these subjects have been amidst the central themes in developing the theory of Mordell–Weil lattices. We review the main ideas from the beginning of the development and then formulate the notion of excellent families as it arose in the additive setting. This will unfold a deep connection between Mordell–Weil lattices of rational elliptic surfaces, algebraic equations and the invariant theory of Weyl groups.

Subsequently, Chap. 10 proceeds to discuss more recent developments about the multiplicative excellent families. This is followed by various important applications of Mordell–Weil lattices. Notably, we will shed new light on the 27 lines on a cubic surface, one of the most classical topics in algebraic geometry, and closely related to the algebraic equation of degree 27 with Galois group $W(E_6)$, discovered by [Cay49, Salm49] for the former and first studied in Jordan's Treatise [Jor70] for the latter. The theory of Mordell–Weil lattices allows us to assume a new point of view which turns out to be quite beneficial. As an illustration, we systematically construct new examples where all the 27 lines on the cubic surfaces have $\mathbb{Q}$-coefficients.

The next two Chaps. 11 and 12, deal with K3 surfaces, one of the central objects of study in algebraic geometry and many neighbouring areas such as differential geometry, number theory and mathematical physics, in particular string theory, for the last half century. Even more than abelian surfaces, K3 surfaces are singled out by the way in which lattices govern their moduli. Together with other basics on elliptic fibrations on K3 surfaces, including many instructive examples, this will be reviewed in Chap. 11 with a strong focus on geometry.

Chapter 12 then provides details on more specific topics in the area of K3 surfaces related to, or in fact, relying on Mordell–Weil lattices. Here the focus shifts notably to the arithmetic side, demonstrating once again the versatility of Mordell–Weil lattices. Next to isogeny structures, we discuss the problem of classifying all elliptic fibrations on a given K3 surface, and in particular on supersingular K3 surfaces.

The book concludes in Chap. 13 with another circle of considerations and applications which played a central role for the discovery of Mordell–Weil lattices: the rank problem for elliptic curves over a given field and the sphere packing problem. We discuss the rank problem over various fields such as $\mathbb{Q}$, $\mathbb{Q}(t)$ and $\mathbb{F}_q(t)$, reviewing work ranging from classical by Tate and Shafarevich to fairly recent by Ulmer and Elkies. Finally, we highlight the discovery, simultaneously due to Elkies and one of us in the late 1980s, that Mordell–Weil lattices of special elliptic curves may provide new record densities for the sphere packing problem which in fact, to the best of our knowledge, are still standing.

Chapter 2
Lattices

In this chapter, we fix the terminology concerning lattices used in this book. Moreover, we review some elementary facts and properties, augmented by more specific results and constructions which will be relevant in later chapters. Special emphasis is laid on even definite lattices (especially root lattices of ADE-type), and on the relation to the sphere packing problem. For details, we refer to standard monographs such as [Bou68, CS99, Ebe13] or [Kne56], although we will also retrieve important input from selected papers, notably from [Nik80] and [Morr84].

2.1 Generalities on Lattices

Throughout this book, we employ the following definition of a lattice. In short, a lattice is a free $\mathbb{Z}$-module of finite rank with a non-degenerate symmetric bilinear pairing. Here we set up the notation.

Let L be a free $\mathbb{Z}$-module of finite rank, given with a symmetric bilinear pairing with values in $\mathbb{R}$:

$$\langle \cdot , \cdot \rangle : L \times L \longrightarrow \mathbb{R}, \quad (x, y) \longmapsto \langle x, y \rangle. \tag{2.1}$$

Thus we have $\langle x, y\rangle = \langle y, x\rangle$ for all $x, y \in L$. The $\mathbb{R}$-linear extension of the pairing to the finite-dimensional $\mathbb{R}$-vector space $L_\mathbb{R} = L \otimes \mathbb{R}$ is denoted by $\langle \cdot , \cdot \rangle_\mathbb{R}$. We say that the pairing $\langle \cdot , \cdot \rangle$ is *non-degenerate* if and only if $\langle \cdot , \cdot \rangle_\mathbb{R}$ is non-degenerate, i.e. $\langle x, y\rangle_\mathbb{R} = 0$ for all $y \in L_\mathbb{R}$ implies $x = 0$ in $L_\mathbb{R}$.

Definition 2.1 The $\mathbb{Z}$-module L together with the non-degenerate pairing $\langle \cdot , \cdot \rangle$ is called a *lattice*.

Note that in some papers the non-degeneracy condition is not always assumed in the definition of a lattice, though for our purposes it is really essential (compare

M. Schütt and T. Shioda, *Mordell–Weil Lattices*, Ergebnisse der Mathematik und ihrer Grenzgebiete. 3. Folge / A Series of Modern Surveys in Mathematics 70,
https://doi.org/10.1007/978-981-32-9301-4_2

Example 2.3). In many cases, we speak of a lattice L without explicit mention of the bilinear form $\langle \cdot\,,\cdot \rangle$, but whenever this is done, the bilinear form should be clear from the context.

Example 2.2 The most familiar example of a lattice is $\mathbb{Z}^r \subset \mathbb{R}^r$, the lattice of all integral vectors in Euclidean space, with the pairing induced by the usual inner product in the r-dimensional Euclidean space $\mathbb{R}^r$:

$$\langle x, y \rangle = \sum_{i=1}^{r} x_i y_i \quad \text{for} \quad x = (x_i), y = (y_i) \in \mathbb{R}^r. \tag{2.2}$$

Note that, for instance, we can scale the coefficient of each summand $x_i y_i$ in (2.2) by a non-zero real number $a_i \in \mathbb{R}^\times$ to obtain another lattice in $\mathbb{R}^r$. On the other hand, we can apply the pairing (2.2) to any other $\mathbb{Z}$-module $L \subset \mathbb{R}^r$ (and check whether it is non-degenerate on $L \otimes \mathbb{R}$ to determine whether L is a lattice). Here comes an instructive case where the pairing indeed degenerates:

Example 2.3 Let $L = \mathbb{Z}^2$ with pairing given by

$$\langle (x_1, x_2), (y_1, y_2) \rangle = (x_1 + \gamma x_2)(y_1 + \gamma y_2)$$

for some $\gamma \in \mathbb{R}^\times$. Then the pairing is non-degenerate on L for $\gamma \in \mathbb{R} \setminus \mathbb{Q}$, but it always degenerates on $L_\mathbb{R}$.

We point out that this kind of problem does not persist when the initial pairing (2.1) is assumed to be $\mathbb{Q}$-valued.

The following two lattices play a key role in the theory of elliptic curves (and thus also for elliptic surfaces). In particular, the second one (coined "hexagonal") will keep on reappearing throughout this book.

Example 2.4 (*Square and hexagonal lattice*) In the two-dimensional case, consider:

(i) the square lattice $\mathbb{Z}^2 \subset \mathbb{R}^2$ and
(ii) the hexagonal lattice (see Fig. 2.1).

When $\mathbb{R}^2$ is identified with the complex plane $\mathbb{C}$, they are given respectively by:

(i) $\{a + b\sqrt{-1} \mid a, b \in \mathbb{Z}\}$ (the Gaussian integers) and
(ii) $\{a + b\omega \mid a, b \in \mathbb{Z}\}, \omega = \frac{-1+\sqrt{-3}}{2}$ (the Eisenstein ring of integers).

One easily checks that the pairing (2.2) indeed defines a lattice structure on either $\mathbb{Z}$-module.

Two lattices $\{L, \langle \cdot\,,\cdot \rangle\}$ and $\{L', \langle \cdot\,,\cdot \rangle'\}$ are said to be *isomorphic* (or isometric) if there is a module isomorphism of L onto L' preserving the pairings, i.e. if there is an isomorphism $\varphi : L \to L'$ such that

$$\langle \varphi(x), \varphi(y) \rangle' = \langle x, y \rangle \quad \forall x, y \in L.$$

Isomorphic lattices will often be identified, as is common to many mathematical notions.

Let L be a lattice of rank r and take a $\mathbb{Z}$-basis $\{\xi_1, ..., \xi_r\}$ of L. Given $x, y \in L$, we can write

$$x = \sum_{i=1}^{r} x_i \xi_i, \quad y = \sum_{j=1}^{r} y_j \xi_j \quad (x_i, y_j \in \mathbb{Z})$$

and hence we get a bilinear form

$$\langle x, y \rangle = \sum_{i,j=1}^{r} a_{ij} x_i y_j, \quad a_{ij} = \langle \xi_i, \xi_j \rangle$$

and a quadratic form

$$\langle x, x \rangle = \sum_{i,j=1}^{r} a_{ij} x_i x_j \quad (a_{ij} = a_{ji}).$$

The matrix

$$I = (a_{ij}) = (\langle \xi_i, \xi_j \rangle) \quad (1 \leq i, j \leq r)$$

is called the *Gram matrix* of L with respect to the basis $\{\xi_i\}$. It is a real symmetric invertible matrix, which transforms under a change of bases with a unimodular matrix U as

$$I \longrightarrow I' = {}^t U I U.$$

Here a matrix $U \in \mathrm{Mat}(n, \mathbb{Z})$ is called *unimodular* if $U \in \mathrm{GL}_r(\mathbb{Z})$, so that $\det U = \pm 1$.

The Gram matrix I determines the lattice L as follows: for fixed r, the isomorphism classes of lattices of rank r are in one-to-one correspondence with the equivalence classes of non-degenerate real quadratic forms in r variables, or with the equivalence classes of $r \times r$ real symmetric invertible matrices with respect to the equivalence relation $I \sim I'$. Various properties or invariants of a lattice can be defined in terms of its Gram matrices. For example, the *determinant* of L is defined by

$$\det L = \det I,$$

the value of $\det I$ being independent of the choice of a basis of L, since $\det I' = (\det U)^2 \det I = \det I$.

Example 2.5 (*Square and hexagonal lattice cont'd*) Recall the square lattice and the hexagonal lattice in the two-dimensional case from Example 2.4. With respect to the pairing (2.2), and with basis vectors for $(a, b) = (1, 0), (0, 1)$, the Gram matrices are

$$\begin{pmatrix} 1 & 0 \\ 0 & 1 \end{pmatrix} \quad \text{resp.} \quad \begin{pmatrix} 1 & -\frac{1}{2} \\ -\frac{1}{2} & 1 \end{pmatrix}.$$

The determinants are 1 and 3/4.

A lattice L is called a *positive-definite* lattice if

$$\langle x, x \rangle > 0 \quad \text{for every } x \in L, \quad x \neq 0.$$

This is the case if and only if the Gram matrix I is a positive-definite matrix, and similarly for *negative-definite* lattices. More generally, the *signature* of L is defined as that of I as a real symmetric matrix.

The symbol L^- denotes the *opposite lattice* of L, which is the module L with the pairing $-\langle \cdot, \cdot \rangle$. For instance, L is a negative-definite lattice if and only if L^- is positive-definite. For $n \in \mathbb{Q}$, the notation $L(n)$ denotes the $\mathbb{Z}$-module L with intersection form scaled by n, i.e. with Gram matrix multiplied by n. Hence $L^- = L(-1)$.

A *sublattice* of L, say T, is a submodule of L such that the restriction of the pairing $\langle \cdot, \cdot \rangle$ to $T \times T$ is non-degenerate (i.e. such that T is also a lattice in our terminology). In case T is a sublattice of finite index in L, then we have

$$\det T = \det L \cdot [L : T]^2. \tag{2.3}$$

This is a special case of the following elementary result:

Lemma 2.6 *Let A be a finitely generated abelian group having a symmetric bilinear pairing $\langle \cdot, \cdot \rangle$ such that the induced pairing on the free $\mathbb{Z}$-module $L = A/A_{tor}$ is non-degenerate. For any free subgroup B of finite index in A, we have*

$$\frac{\det B}{[A : B]^2} = \frac{\det L}{|A_{tor}|^2}. \tag{2.4}$$

Proof This is an immediate consequence of elementary divisor theory. Namely, it tells us that, given a submodule $B \subset A$, there exist some generators $x_1, \ldots, x_n$ of A such that $e_1 x_1, \ldots, e_r x_r$ generate B where $r = \operatorname{rk} B = \operatorname{rk} L$ and $e_1, \ldots, e_r$ are positive integers with $e_i \mid e_{i+1}$. Then

$$\det B = \det(\langle e_i x_i, e_j x_j \rangle) = (e_1 \cdots e_r)^2 \det(\langle x_i, x_j \rangle).$$

On the other hand, A_{tor} is generated by $x_{r+1}, \ldots, x_n$ since B is a free submodule of finite index in A. It follows that $[A : B] = e_1 \cdots e_r |A_{tor}|$ and that $L = A/A_{tor}$ is generated by $x_1, \ldots, x_r$. Thus we have $\det L = \det(\langle x_i, x_j \rangle)$ and (2.4) follows. □

Remark 2.7 By formulating Lemma 2.6 in terms of finitely generated abelian groups as opposed to $\mathbb{Z}$-modules, it points already towards the distinction between Mordell–Weil groups and lattices, to be exhibited in Chap. 6.

An *integral* lattice is a lattice with a $\mathbb{Z}$-valued pairing. An integral lattice L is called *even* if $\langle x, x \rangle \in 2\mathbb{Z}$ for all $x \in L$, *odd* otherwise, and *unimodular* if $\det L = \pm 1$.

Example 2.8 (*Square and hexagonal lattice normalized*) The square lattice from Example 2.5 is integral, odd and unimodular, while the hexagonal lattice is not integral. Of course, both are positive-definite.

Note that rescaling the pairing (2.2) by a factor of 2, both the square and the hexagonal lattice become even integral lattices, of determinant 4 resp. 3. In the classical notation (to be introduced below), they are usually denoted by $A_1^{\oplus 2}$ resp. A_2.

It may be surprising at first how big a difference the parity of a lattice makes. As an illustration, we point out that unimodular odd lattices can be directly constructed with any given signature (take a diagonal Gram matrix with entries ± 1), while for even lattices a severe restriction applies. This is expressed in terms of the *index* of a lattice L, defined as the difference of the number of positive and negative eigenvalues of the Gram matrix associated with L. That is, if L has signature (s_+, s_-), then its index is $s_+ - s_-$.

Theorem 2.9 *Let L be an even unimodular lattice. Then the index of L is divisible by* 8.

For a proof of the theorem, see [Ser70] or [CS99]. Note the immediate consequence that any even unimodular lattices has even rank.

Clearly Theorem 2.9 does not hold for odd lattices regardless of being definite or indefinite, but it has a strong implication for definite *even* lattices:

Corollary 2.10 *The rank of a definite even unimodular lattice is divisible by* 8. *In particular, there is no such lattice of rank less than* 8.

It is known that for any given rank, the number of definite unimodular even lattices of this rank up to isometry is finite. We will briefly touch upon the first few cases around Remark 2.33 in Sect. 2.3.4 and then again in Sect. 12.3.2.

This finiteness can be viewed as the first instance of a *class number*; here we count all integral (often even) lattices with given invariants (usually determinant and signature) up to isometry. The most classical case consists of even binary positive-definite quadratic forms (i.e. of signature (2, 0)) of fixed determinant d, with class number $h(-d)$ and has famous connections to class groups, number fields and elliptic curves (see (13.5) for class number one, for instance). For instance, for $d = 3, 4$, we have $h(-d) = 1$ with quadratic form represented exactly by the hexagonal resp. square lattice scaled as in Example 2.8.

The *dual lattice* $L^\vee$ (sometimes also denoted by L^*) of an integral lattice L is defined by

$$L^\vee = \{x \in L \otimes \mathbb{Q} \mid \langle x, y \rangle \in \mathbb{Z} \text{ for all } y \in L\},$$

with pairing naturally extended from L to $L \otimes \mathbb{Q}$. As the name suggests, the dual lattice can also be defined as a natural lattice structure on the dual module $L^\vee = \mathrm{Hom}(L, \mathbb{Z})$. Though this viewpoint is useful in many contexts, we will always regard $L^\vee$ as a lattice containing L as a sublattice of finite index. We have

$$[L^\vee : L] = |\det L|, \quad \det L^\vee = 1/(\det L). \tag{2.5}$$

In fact, if we take a basis $\{\xi_i\}$ of L and identify the underlying module L with $\mathbb{Z}^r$, then $L^\vee$ is identified with $I^{-1}\mathbb{Z}^r$ in terms of the Gram matrix I. Hence $L^\vee/L$ is isomorphic to $\mathbb{Z}^r/I\mathbb{Z}^r$, and elementary divisor theory gives the first equality. The second then follows from (2.3).

Note that if L is an integral lattice, then $L^\vee$ is not integral unless L is unimodular. In the latter case, we have in fact $L = L^\vee$.

Example 2.11 (*Hexagonal Mordell–Weil lattice*) We have already pointed out in the introduction that the hexagonal elliptic curve from (1.1) will have Mordell–Weil lattice $A_2^\vee$, the dual of the hexagonal lattice A_2, scaled to be even as in Example 2.8. This example will form a common thread throughout this book.

A sublattice T of a lattice L is called *primitive* if the quotient L/T is torsion-free. The *primitive closure* of a sublattice T of L is:

$$T' = \{x \in L \mid nx \in T \text{ for some positive integer } n\}.$$

The primitive closure is the smallest primitive sublattice of L containing T, i.e. $T' = (T \otimes \mathbb{Q}) \cap L$, and we have $T' = T$ if and only if T is primitive.

Suppose that T is a sublattice of a lattice L. Then the *orthogonal complement* of T, denoted by $T^\perp$, is defined by

$$T^\perp = \{x \in L \mid \langle x, y\rangle = 0 \text{ for all } y \in T\}.$$

By definition, $T^\perp$ is a primitive sublattice of L. It follows that one can obtain the primitive closure as $T' = (T^\perp)^\perp$. Moreover, we have $T \cap T^\perp = \{0\}$, since T is non-degenerate by definition, and the *direct (orthogonal) sum* $T \oplus T^\perp$ is a lattice of finite index in L. In case L is integral, the formulas (2.3) and (2.5) imply

$$\det T \cdot \det T^\perp = \det L \cdot [L : T \oplus T^\perp]^2.$$

Lemma 2.12 *Suppose that T is a primitive sublattice of a unimodular (thus integral) lattice L. Then we have*

$$|\det T| = |\det T^\perp| = [L : T \oplus T^\perp]. \tag{2.6}$$

Proof First note that $L^\vee = L$, since L is unimodular. Next consider the dual map $L^\vee \longrightarrow T^\vee$ induced by the inclusion map $T \longrightarrow L$. The dual map is surjective, since T is primitive in L. Now the kernel of the composed map

$$L = L^\vee \longrightarrow T^\vee \longrightarrow T^\vee/T$$

is easily seen to be $T \oplus T^\perp$. Hence

$$[L : T \oplus T^\perp] = [T^\vee : T] = |\det T|$$

by (2.5). Interchanging the role of T and $T^{\perp}$, we complete the proof. □

Remark 2.13 The quotient groups $T^{\vee}/T$ and $(T^{\perp})^{\vee}/T^{\perp}$, arising in the setting of the above lemma, will be studied under the name of *discriminant group* in Sect. 2.3.5.

Recall that an *automorphism* (or isometry) of a lattice L is an isomorphism $\varphi : L \to L$ such that

$$\langle \varphi(x), \varphi(y) \rangle = \langle x, y \rangle \quad \forall x, y \in L.$$

If φ is represented by a matrix $A \in M_r(\mathbb{Z})$ with respect to a basis ξ_i of L, then the above is equivalent to

$$^{t}AIA = I,$$

where I is the Gram matrix. We denote by $\mathrm{Aut}(L)$ or $O(L)$ the automorphism group of L. It is isomorphic to the group of integer matrices A satisfying the above equation.

The following property is essential for many applications:

Theorem 2.14 *For a positive-definite lattice L, $\mathrm{Aut}(L)$ is a finite group.*

Proof This holds simply because $\mathrm{Aut}(L)$ is contained in a discrete subgroup of a compact group (i.e. the group of integral matrices in the orthogonal group of $L \otimes \mathbb{R}$). □

Example 2.15 (*Square and hexagonal lattice*) For the square lattice $A_1^{\oplus 2}$ and the hexagonal lattice A_2 from Example 2.8, the automorphism groups are the dihedral group of order eight and the product $S_3 \times \{\pm 1\}$.

We will study automorphism groups of root lattices, and in particular their relation to Weyl groups, further in Sect. 2.3.6.

2.2 Sphere Packings

Now suppose that L is a positive-definite lattice of rank r. Then $L \otimes \mathbb{R} = \mathbb{R}^r$ becomes an r-dimensional Euclidean space with the inner product $\langle \cdot, \cdot \rangle$, and L is a discrete subgroup in $\mathbb{R}^r$ with compact quotient $\mathbb{R}^r/L$ (a torus). This is what a lattice usually means in the geometric sense. This viewpoint makes visible the relations with many interesting subjects in mathematics. Thus it becomes clear that lattices have been extensively studied from various angles. To name but a few examples (some of which have been mentioned at the beginning of the introduction):

(a) sphere packings, crystallography;
(b) elliptic functions, complex tori, abelian varieties;
(c) quadratic forms, theta functions, modular forms;
(d) differential geometry and analysis on a torus;
(e) topology, (co)homology.

The interrelations among these subjects are also quite interesting. Although it is not our aim to discuss all of these matters in this book, let us take a glance at some of the relationships between lattices, sphere packings and moduli of complex tori (For more details, see [CS99].).

Given a positive-definite lattice L of rank $r = \mathrm{rk}(L)$, the minimum distance of two distinct points in L is given by

$$d = \sqrt{\mu(L)},$$

where $\mu(L)$ denotes the *minimal norm* of L:

$$\mu(L) = \mathrm{Min}_{x \in L, x \neq 0} \langle x, x \rangle. \tag{2.7}$$

By attaching to each lattice point of L a sphere of radius $d/2$, we obtain a non-overlapping arrangement of spheres in the ambient Euclidean space $L \otimes \mathbb{R} = \mathbb{R}^r$; this is called the *sphere packing* associated with a lattice L or more simply a *lattice packing*. The *density* $\Delta(L)$ of this sphere packing is defined as the ratio of the volume of one sphere to that of the fundamental domain of L, i.e. $\sqrt{\det(L)}$. Thus we have

$$\Delta(L) = \frac{(\frac{1}{2}d)^r V_r}{\sqrt{\det(L)}}$$

where V_r is the volume of an r-dimensional sphere of radius 1, given by $V_r = \pi^{\frac{r}{2}} / \Gamma(\frac{r}{2} + 1)$ expressed in terms of the Γ-function; for r even, this simplifies as

$$V_r = \frac{\pi^{r/2}}{(r/2)!}.$$

The quantity $\delta(L) = \Delta(L)/V_r$ is called the *center density*: we have

$$\delta(L) = \frac{(\frac{1}{2}\sqrt{\mu(L)})^r}{\sqrt{\det(L)}}. \tag{2.8}$$

In other contexts, the center density $\delta(L)$ is sometimes also known as Hermite constant. Obviously both $\Delta(L)$ and $\delta(L)$ are independent of the scaling of L.

Example 2.16 (*Square and hexagonal lattice cont'd*) Recall in the two-dimensional case from Example 2.4:

(i) the square lattice $\mathbb{Z}^2 \subset \mathbb{R}^2$ and
(ii) the hexagonal lattice (see Fig. 2.1).

With unscaled pairing (2.2), the above formula gives $V_2 = \pi$, $d = 1$, and the volume (area) of the fundamental domain is equal to 1 or $\frac{\sqrt{3}}{2}$ according to the case (i) or (ii). Hence the density $\Delta(L)$ is equal to

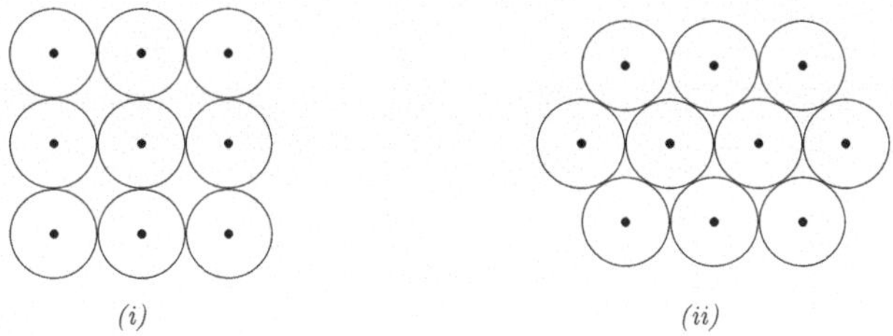

Fig. 2.1 Square lattice and hexagonal lattice

$$(i)\ \pi/4 = 0.785... \quad \text{and} \quad (ii)\ \pi/(2\sqrt{3}) = 0.9069\ldots.$$

Remark 2.17 In general, given a positive-definite lattice, it may be not so easy to control the minimal norm. It will be one of the key features of Mordell–Weil lattices that they have built-in a good lower bound for the minimal norm.

The famous *sphere packing problem* is to find a sphere packing with the highest density in r-dimensional space when r is given. This is trivial for $\boldsymbol{r = 1}$, since it is obvious that the lattice $\mathbb{Z} \subset \mathbb{R}$ has the density $\Delta = 1$. For $\boldsymbol{r = 2}$, the densest lattice packing is given by the hexagonal lattice. A proof of this fact will be given below (Proposition 2.18). Actually, it is known that it has the highest density among all two-dimensional sphere packings (not necessarily coming from lattices) (see [CS99] or the original [Tot43]).

For $\boldsymbol{r = 3}$, we all 'know' the solution to the sphere packing problem from inspecting how oranges, say, are usually piled up. A serious proof of this observation, famously known as Kepler's conjecture, was only exhibited by Hales in 1998 [Hal05] (and recently supplemented by a formal argument which may be verified with proof-checking software).

For any dimension $\boldsymbol{r > 3}$, the sphere packing problem is usually considered separately for lattices alone and for packings in general. Restricted to lattice packings, this problem has been solved for dimension $r \leq 8$ and $r = 24$; the answer is summarized in Table 2.1. The lattices A_r, D_r, E_r in this table are *root lattices* which will be described in the next section. The lattice Λ_{24} is the Leech lattice; its record density and uniqueness among lattices was recently proved by [CK09] (see also [CS99]).

The very latest development only concerns ranks $\boldsymbol{r = 8, 24}$ (quite surprisingly at first, maybe). Namely, Viazovska in the spring of 2016 managed to let go the restriction to lattices in dimension 8, thus proving that the packing provided by the root lattice E_8 is optimal among all sphere packings [Via17]. Subsequently, a collaboration with four colleagues yielded the same result for the Leech lattice in dimension 24 [CK+17].

Table 2.1 Record lattice packings

r	2	3	4	5	6	7	8	24
L	A_2	A_3	D_4	D_5	E_6	E_7	E_8	Λ_{24}
δ	$\frac{1}{2\sqrt{3}}$	$\frac{1}{4\sqrt{2}}$	$\frac{1}{8}$	$\frac{1}{8\sqrt{2}}$	$\frac{1}{8\sqrt{3}}$	$\frac{1}{16}$	$\frac{1}{16}$	1
τ	6	12	24	40	72	126	240	196560

Another important invariant of a lattice L is its *kissing number* $\tau(L)$, which is defined as the number of minimal vectors in L:

$$\tau(L) = \#\{x \in L \mid \langle x, x\rangle = \mu(L)\}. \tag{2.9}$$

In the example above, Fig. 2.1 shows that (i) $\tau(L) = 4$ and (ii) $\tau(L) = 6$.

The *kissing number problem* is to find lattices or sphere packings having the largest kissing number in a given dimension. The answer is known for lattice packings for $r \leq 8$ and $r = 24$; it is given by Table 2.1.

Going back to the case $r = 2$, let us show the following:

Proposition 2.18 *For any lattice $L \subset \mathbb{R}^2$, we have*

$$\delta(L) \leq \frac{1}{2\sqrt{3}},$$

and equality holds if and only if L is proportional to the hexagonal lattice.

Proof Recall that the center density $\delta(L)$ does not change if L is replaced by any lattice proportional to it. We identify $\mathbb{R}^2$ with the complex plane $\mathbb{C}$, so that the induced bilinear pairing becomes

$$\langle x + iy, u + iv\rangle = xu + yv \qquad (x, y, u, v \in \mathbb{R}).$$

Then any lattice in $\mathbb{R}^2$ is proportional to a lattice of the form

$$L_z = \mathbb{Z} + \mathbb{Z}z = \{m + nz \mid m, n \in \mathbb{Z}\} \tag{2.10}$$

for some $z \in \mathcal{H}$ where $\mathcal{H}$ denotes the upper half plane:

$$\mathcal{H} = \{z \in \mathbb{C} \mid \Im(z) > 0\}.$$

As is easily seen, two such lattices L_z and $L_{z'}$ are proportional if and only if z' and z are transformed to each other under a modular transformation:

$$z' = \frac{az + b}{cz + d}, \quad \begin{pmatrix} a & b \\ c & d \end{pmatrix} \in SL_2(\mathbb{Z}).$$

Now the modular group $SL_2(\mathbb{Z})$ acting on $\mathcal{H}$ has a well-known fundamental domain:

$$\mathcal{F} = \left\{ z \in \mathcal{H} \mid |z| \geq 1, -\frac{1}{2} < \Re(z) \leq \frac{1}{2} \right\}$$

(cf. [Ser70, Chap. 7], [Shimu71, Chap. 1]).

Consider any lattice $L = L_z$ with $z \in \mathcal{F}$. It has the minimal norm 1 and the volume of the fundamental parallelotope is equal to $\mathrm{Im}(z)$. Therefore, by the formula (2.8), we have

$$\delta(L) = \frac{1}{4\,\mathrm{Im}(z)}.$$

Note that, for $z \in \mathcal{F}$, we have $\mathrm{Im}(z) \geq \mathrm{Im}(\omega) = \frac{\sqrt{3}}{2}$, with equality only for $z = \omega$ or $z = \omega' = \omega + 1$. It follows that

$$\delta(L) \leq \frac{1}{2\sqrt{3}},$$

and that equality holds if and only if L is proportional to L_ω which is nothing but the hexagonal lattice. □

Remark 2.19 For each $z \in \mathcal{H}$, we can define the complex torus $E_z = \mathbb{C}/L_z$. Then E_z and $E_{z'}$ are isomorphic precisely when L_z and $L_{z'}$ are proportional. Thus the classification of lattices of rank two in $\mathbb{C}$ and that of complex tori are equivalent. On the other hand, complex tori are elliptic curves over $\mathbb{C}$, and from this viewpoint the quotient space $SL_2(\mathbb{Z}) \setminus \mathcal{H}$ gives the most classical moduli space of elliptic curves (the modular curve). See [Shimu71].

Among all complex elliptic curves, E_ω and $E_{\sqrt{-1}}$ are singled out by the property that they admit an automorphism of order six respectively four, while generally there are only identity and inversion. In fact, these two are the first examples of elliptic curves with complex multiplication, and they will recur throughout this book.

Remark 2.20 Sometimes the invariants of a lattice such as packing density or kissing number can be improved by adding lattice points located at the so-called holes of the lattice, i.e. points in the ambient space $\mathbb{R}^n$ which locally maximize the distance to all the lattice points. Often it is beneficial to choose *deep holes*, i.e. global maxima (easily located in Fig. 2.1, for instance). These are also relevant for many aspects of lattice constructions and classifications. Famously, by [CPS82], there are 23 inequivalent holes in the Leech lattice, and they are in one-to-one correspondence with the other 23 Niemeier lattices which we will recall in Theorem 12.2.

2.3 Root Lattices and Their Dual Lattices

In the following paragraphs, we consider the root lattices of type A, D, E:

$$A_r \ (r \geq 1), \quad D_r \ (r \geq 4), \quad E_6, \ E_7 \text{ and } E_8,$$

and their dual lattices. We put emphasis on the description of the dual lattices, compared with standard references such as [Bou68, Chap. 6], [CS99, Chap. 4] or [Man86, Chap. 4]. This is because they arise quite naturally in considering Mordell–Weil lattices, and such a viewpoint suggests a construction given below in which the dual lattice comes to the scene first, and the root lattice can then be defined as a sublattice. This is particularly well-suited for describing the minimal vectors, especially for type E_6, E_7, E_8.

The results in this section will be used in later chapters of this book, but at the moment they can serve as concrete examples of important lattices. The reader could skip some details at first and come back later when the results here are needed.

We note here some common features of the root lattices under consideration, recalling the standard terminology. Any of these root lattices, say L, is a positive-definite even integral lattice of rank indicated by the subscript r. The minimal norm in such a lattice is 2, hence the minimal vectors earn a special name:

Definition 2.21 Let L be a definite even integral lattice.

(i) Elements $x \in L$ with $\langle x, x \rangle = \pm 2$, are called root vectors or more simply *roots*.
(ii) The set of roots in L will be denoted by $\mathcal{R}(L)$.

Let V be the subspace of $L \otimes \mathbb{R}$ spanned by $\mathcal{R}(L)$. Then the set of roots $\mathcal{R}(L)$ forms a *root system* in the sense of Lie algebras (cf. [Bou68, Chap. 6], [CS99, Chap. 4]). It is known that if it is irreducible (i.e. if it is not a union of two orthogonal root systems), it is a root system of type A_r, D_r or E_r, which explains why these ADE lattices are so named (see Theorem 2.25). Generally, we have the following definition:

Definition 2.22 A definite even integral lattice is called a *root lattice* if it is generated by its roots.

More concretely, we make the following definition:

Definition 2.23 A lattice L of rank r is a root lattice of *type* $A_r (r \geq 1)$, $D_r (r \geq 4)$ or $E_r (r = 6, 7, 8)$ if there exists a basis $\{\alpha_1, \ldots, \alpha_r\} \subset \mathcal{R}(L)$ of L such that the following holds: for $1 \leq i < j \leq r$, we have zero pairing $\langle \alpha_i, \alpha_j \rangle = 0$ unless

$$
\begin{aligned}
(A_r) \quad & \langle \alpha_i, \alpha_j \rangle = -1 \Leftrightarrow i + 1 = j. \\
(D_r) \quad & \langle \alpha_i, \alpha_j \rangle = -1 \Leftrightarrow i + 1 = j < r, \quad \text{or } i = r - 2, \ j = r. \\
(E_r) \quad & \langle \alpha_i, \alpha_j \rangle = -1 \Leftrightarrow i + 1 = j < r, \quad \text{or } i = 3, \ j = r.
\end{aligned}
$$

That this definition gives a positive-definite lattice can be directly verified, but we omit the verification here, since it will follow from the elementary construction of A_r, D_r or E_r which we describe in the subsequent Sects. 2.3.2–2.3.4.

In Table 2.2, we list the determinant, the minimal norm μ and the number of minimal vectors τ for each root lattice of type A_r, D_r or E_r and its dual lattice.

Table 2.2 Basic data for root lattices

	A_r	$A_r^\vee$	D_r	$D_r^\vee$	E_6	$E_6^\vee$	E_7	$E_7^\vee$	E_8
det	$r+1$	$1/(r+1)$	4	1/4	3	1/3	2	1/2	1
μ	2	$r/(r+1)$	2	1	2	4/3	2	3/2	2
τ	$r(r+1)$	2 $(r=1)$ $2(r+1)$ $(r\geq 2)$	$2r(r-1)$	24 $(r=4)$ $2r$ $(r\geq 5)$	72	54	126	56	240

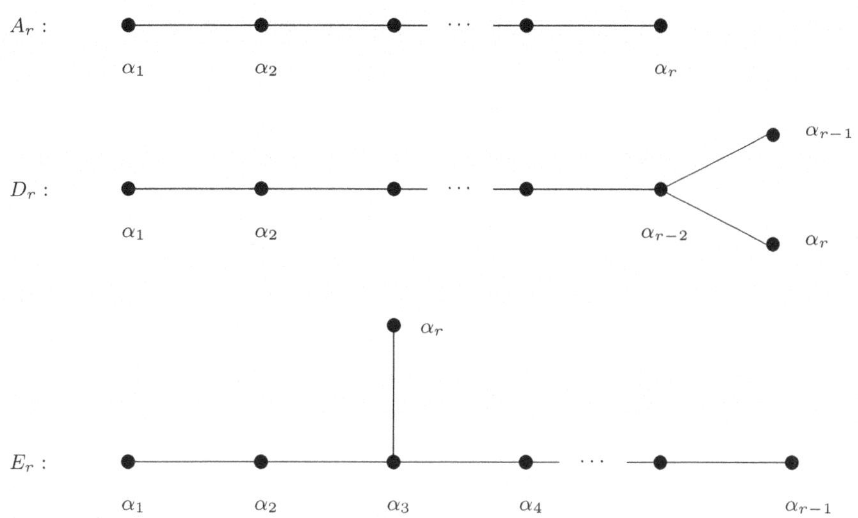

Fig. 2.2 Dynkin diagrams of type A_r, D_r, E_r

Remark 2.24 Some mathematicians, especially geometers, tend to define the ADE root lattices to be negative-definite (and with good reason coming from intersection numbers of curves on algebraic surfaces, as we will see later from Chap. 7 on), but we decided to stick to the historical definition here. The negative-definite root lattices will thus be denoted by A_r^-, D_r^- and E_r^-.

To the basis of a positive-definite even lattice as in Definition 2.23 one attaches a diagram in which the r vertices correspond to r roots α_i (called simple roots) and two vertices α_i, α_j are joined by an edge if and only if $\langle \alpha_i, \alpha_j \rangle = -1$ holds. This diagram is called the *Dynkin diagram* of type A_r, D_r or E_r; see Fig. 2.2.

In other words, the root lattice A_r, D_r or E_r can be defined as the lattice generated by a basis corresponding to the Dynkin diagram of type A_r, D_r or E_r.

Once such a basis is fixed, it is known that every root is either *positive* or *negative* in the sense that it is a linear combination of the simple roots with all nonnegative (or all nonpositive) integer coefficients.

Of course, there are general definitions of simple or positive roots, but setting this up properly would require introducing more machinery than strictly necessary for our purposes. Indeed, we can avoid this since our approach is more algebraic, based on the invariant theory of Weyl groups (in Sects. 2.3.6, 2.3.7), and leading to the results in Chap. 9 about excellent families, motivated by our diophantine viewpoint of Mordell–Weil lattices.

The importance of the root lattices of type A_r, D_r and E_r stems from their ubiquity in many seemingly disconnected areas of mathematics, ranging from Lie theory and combinatorics to singularity theory, and in particular including the Mordell–Weil lattices which are the central object of study in this book. The classification of root lattices is governed by the following fundamental theorem:

Theorem 2.25 *Any positive-definite even integral root lattice is isometric to an orthogonal sum of root lattices of type A_n, D_k and E_l.*

We conclude this section by introducing the Weyl group of a (root) lattice which will be of great relevance for our study of rational elliptic surfaces in Chaps. 9 and 10. To this end, consider a root $\alpha \in L$, i.e. $\alpha^2 = \langle \alpha, \alpha \rangle = \pm 2$. Then the homomorphism

$$s_\alpha : \ x \mapsto x - 2\,\frac{\langle x, \alpha \rangle}{\langle \alpha, \alpha \rangle}\,\alpha \tag{2.11}$$

defines an automorphism of the lattice L, which induces a *reflection* of the ambient space $L \otimes \mathbb{R}$ with respect to the hyperplane

$$\alpha^\perp = \{x \in L \otimes \mathbb{R};\ \langle x, \alpha \rangle = 0\}.$$

The subgroup of $\mathrm{Aut}(L)$ generated by these reflections is called the *Weyl group* of L and denoted by $W(L)$. We will discuss some details about Weyl groups, especially their invariants, in Sects. 2.3.6, 2.3.7.

2.3.1 Auxiliary Lattice $L_{r,s}$

In this and the subsequent subsections, we provide alternative constructions of the root lattices of type A_r, D_r, E_r. In favour of a uniform treatment, we first introduce a series of auxiliary lattices which turn out to be quite useful. For the root lattices of type E_r, this approach has been introduced in [Shio95b].

Definition 2.26 Let r be a positive integer and let s be a real number greater than $-1/r$. Let $L_{r,s}$ be a lattice of rank r generated by r elements $u_1, \ldots, u_r$ such that the symmetric bilinear pairing on $L_{r,s}$ is defined by the rule:

$$\langle u_i, u_j \rangle = \delta_{ij} + s. \tag{2.12}$$

Lemma 2.27 *For any $s > -1/r$, $L_{r,s}$ is a positive-definite lattice of rank r with* $\det L_{r,s} = 1 + rs$.

Proof This is an immediate consequence of the following:

Lemma 2.28 *Suppose that $A = (a_{ij})$ is a real symmetric matrix of size $r \times r$ such that $a_{ij} = \delta_{ij} + s$ for all i, j for a fixed real number s. Then* $\det A = 1 + rs$. *In particular, such a matrix A is positive-definite if $s > -1/r$.*

Proof Noting that each column sums up to $1 + rs$, we have:

$$\det A = \begin{vmatrix} 1+s & s & \cdots & s \\ s & 1+s & \cdots & s \\ \vdots & \vdots & \ddots & \vdots \\ s & s & \ldots & 1+s \end{vmatrix} = \begin{vmatrix} 1+s & s & \cdots & s \\ s & 1+s & \cdots & s \\ \vdots & \vdots & \ddots & \vdots \\ 1+rs & 1+rs & \ldots & 1+rs \end{vmatrix}$$

$$= (1+rs) \begin{vmatrix} 1+s & s & \cdots & s \\ s & 1+s & \cdots & s \\ \vdots & \vdots & \ddots & \vdots \\ 1 & 1 & \ldots & 1 \end{vmatrix} = (1+rs) \begin{vmatrix} 1 & 0 & \cdots & 0 \\ 0 & 1 & \cdots & 0 \\ \vdots & \vdots & \ddots & \vdots \\ 1 & 1 & \ldots & 1 \end{vmatrix} = 1 + rs.$$

This proves the first assertion. Next assume that $s > -1/r$. For any $n = 1, 2, \ldots, r$, the n-th principal minor of A has $\det = 1 + ns > 1 - n/r \geq 0$. Hence by Sylvester's criterion A is a positive-definite matrix. □

Lemma 2.29 *For any $i \neq j$, let $\alpha_{ij} = u_i - u_j \in L_{r,s}$. Then we have*

$$\langle \alpha_{ij}, u_k \rangle = \delta_{ik} - \delta_{jk} \in \mathbb{Z} \tag{2.13}$$

$$\langle \alpha_{ij}, \alpha_{kl} \rangle = \delta_{ik} - \delta_{jk} - \delta_{il} + \delta_{jl} \in \mathbb{Z}. \tag{2.14}$$

In particular, we have $\langle \alpha_{ij}, \alpha_{ij} \rangle = 2$ for any $i \neq j$.

Proof Immediate. □

2.3.2 A_r and $A_r^\vee$

Given r, we take $s = -1/(r+1)$ which satisfies $1 + rs = 1/(r+1) > 0$, and consider the lattice $\tilde{L} = L_{r,s}$. By definition, it is a positive-definite lattice generated by $u_1, \ldots, u_r$ such that

$$\langle u_i, u_j \rangle = \delta_{ij} - \frac{1}{r+1}. \tag{2.15}$$

If we set $u_{r+1} = -u_1 - \cdots - u_r$, this formula holds for all $i, j = 1, \ldots, r+1$. The determinant of $\tilde{L}$ is $1/(r+1)$.

Let L^0 be the sublattice of $\tilde{L}$ generated by $\alpha_i = u_i - u_{i+1}$ $(i = 1, \ldots, r)$. Then we have

$$\langle \alpha_i, \alpha_j \rangle = \begin{cases} 2 & (i = j) \\ -1 & (i - j = \pm 1) \\ 0 & (\text{otherwise}). \end{cases}$$

The lattice L^0 corresponds to the (positive-definite) Dynkin diagram of type A_r. Therefore the lattice L^0 generated by α_i $(i = 1, \ldots, r)$ is isomorphic to the root lattice A_r; let us identify $L^0 = A_r$.

Note that $\tilde{L}/L^0$ is generated by u_1 with $(r+1)u_1 \in L^0$; in fact,

$$u_1 = \frac{1}{r+1}(r\alpha_1 + \cdots + \alpha_r).$$

Hence we have $[\tilde{L} : L^0] = r + 1$ and $\det L^0 = r + 1$ (since $\det \tilde{L} = 1/(r+1)$).

Proposition 2.30 *With the above notation, we have*

(i) $L^0 = A_r$ *and* $\tilde{L} = A_r^\vee$.
(ii) All the roots of A_r *are given by* $\{\alpha_{ij} = u_i - u_j \ (1 \le i \ne j \le r+1)\}$, *and the positive roots with respect to the basis* $\{\alpha_i\}$ *are* α_{ij} $(i < j)$.
(iii) The minimal vectors of $A_r^\vee$ *(of norm* $\frac{r}{r+1}$*) are given by* $\{u_1, \ldots, u_{r+1}\}$ *and* $\{-u_1, \ldots, -u_{r+1}\}$.
(iv) The Weyl group $W(A_r)$ *acts on* $A_r^\vee$ *as the symmetric group* S_{r+1} *on* $r+1$ *letters on the set* $\{u_1, \ldots, u_{r+1}\}$.

Proof By the Eq. (2.13), we have $\langle \alpha_i, u_k \rangle \in \mathbb{Z}$ for all i, k. Hence each u_k is contained in the dual lattice of A_r so that we have $L^0 \subset \tilde{L} \subset A_r^\vee$. Noting that

$$[\tilde{L} : L^0] = r + 1, \ [A_r^\vee : A_r] = \det A_r = r + 1,$$

we see that $\tilde{L} = A_r^\vee$.

Now it follows from (2.15) that $\pm u_i$ are minimal vectors of $A_r^\vee$ and α_{ij} are roots of A_r. Since these form exactly the right number of minimal vectors and roots (i.e. $2(r+1)$ as soon as $r > 1$ respectively $r(r+1)$), they constitute all the minimal vectors in $A_r^\vee$ and all the roots of A_r. (In case $r = 1$, the two minimal vectors are given by u_1 and $u_2 = -u_1$.) The reflection s_α for a root $\alpha = \alpha_{ij}$ is defined by $s_\alpha(u_k) = u_k - \langle \alpha, u_k \rangle \alpha$, so it interchanges u_i, u_j and fixes u_k for any $k \ne i, j$ (i.e. the transposition $(ij) \in S_{r+1}$). Hence the Weyl group $W(A_r)$ generated by s_α acts as the permutation group of $u_1, \ldots, u_{r+1}$; in particular, it has two orbits among the minimal vectors of $A_r^\vee$ if $r > 1$, i.e. $\{u_1, \ldots, u_{r+1}\}$ and $\{-u_1, \ldots, -u_{r+1}\}$.

Note finally that, for $i < j$, $\alpha_{ij} = \alpha_i + \alpha_{i+1} + \cdots + \alpha_{j-1}$ (a positive root). This proves the proposition. □

2.3.3 D_r and $D_r^\vee$

For any $r \geq 4$, we take $s = 0$ and consider the lattice $\tilde{L}' = L_{r,s}$. By definition, it is generated by $u_1, \ldots, u_r$ such that

$$\langle u_i, u_j \rangle = \delta_{ij}.$$

In other words, $\tilde{L}' \simeq \mathbb{Z}^r$, the most familiar lattice in $\mathbb{R}^r$.

Let $\tilde{L}$ be the lattice generated by $\tilde{L}'$ and the following v_0:

$$v_0 = \frac{1}{2}(u_1 + \cdots + u_r).$$

Obviously we have $[\tilde{L} : \tilde{L}'] = 2$ and $\det \tilde{L} = 1/4$. Define

$$\alpha_{ij} = u_i - u_j, \ \beta_{ij} = u_i + u_j \ (i < j).$$

Let L^0 be the sublattice of $\tilde{L}'$ generated by the r elements:

$$\alpha_i = \alpha_{i\ i+1} \ (i = 1, \ldots, r-1), \ \alpha_r = \beta_{r-1\ r}.$$

It is immediate that $\{\alpha_1, \ldots, \alpha_r\}$ corresponds to the Dynkin diagram of type D_r (cf. Fig. 2.2).

Proposition 2.31 *With the above notation, we have:*

- *(i)* $L^0 = D_r$ *and* $\tilde{L} = D_r^\vee$.
- *(ii) All the roots of* D_r *are given by* $\{\pm\alpha_{ij}, \pm\beta_{ij} \ (i < j)\}$; *the positive roots with respect to the basis* $\{\alpha_i\}$ *are* α_{ij} $(i < j)$ *and* β_{ij} $(i < j)$.
- *(iii) The minimal vectors of* $D_r^\vee$ *(of norm 1) are given by* $\{\pm u_1, \ldots, \pm u_r\}$ *provided that* $r > 4$. *In case* $r = 4$, *they are given by* $\{\pm u_i\}$, $\{\pm(v_0 - u_i)\}$ *and* $\{\pm v_0, v_0 - u_i - u_j \ (i < j)\}$.
- *(iv) The Weyl group* $W(D_r)$ *acts on* $D_r^\vee$ *as follows: each reflection* s_α *for* $\alpha = \alpha_{ij}$ *(or* β_{ij}*) interchanges* u_i, u_j *(or maps* $u_i \mapsto -u_j, u_j \mapsto -u_i$*), fixing other* u_k $(k \neq i, j)$. *In particular,* $W(D_r)$ *acts transitively on the set of minimal vectors in* $D_r^\vee$ *if* $r > 4$, *but it has three orbits if* $r = 4$ *which are the three sets listed in (iii).*

The proof is similar to the case of type A_r. We leave it to the reader as an exercise.

2.3.4 E_r and $E_r^\vee$ $(r = 6, 7, 8)$

For $r = 6, 7, 8$, we consider the lattice $L_r = L_{r,s}$, where we take $s = 1/d, d = 9 - r$, i.e. $s = \frac{1}{3}, \frac{1}{2}, 1$ for $r = 6, 7, 8$. Thus L_r is generated by r elements $u_1, \ldots, u_r$ such that

$$\langle u_i, u_j \rangle = \delta_{ij} + \frac{1}{d} = \delta_{ij} + \begin{cases} \frac{1}{3} \ (r = 6), \\ \frac{1}{2} \ (r = 7), \\ 1 \ \ (r = 8). \end{cases} \tag{2.16}$$

Then L_r is a positive-definite lattice of rank r with $\det L_r = 1 + \frac{r}{d} = \frac{9}{d}$, i.e. $\det L_r = 3, \frac{9}{2}, 9$ for $r = 6, 7, 8$.

Let $\tilde{L}_r$ be the lattice in $L_r \otimes \mathbb{Q}$ generated by L_r and v_0:

$$v_0 = \frac{1}{3} \sum_{i=1}^{r} u_i. \tag{2.17}$$

It is a positive-definite lattice such that $[\tilde{L}_r : L_r] = 3$ and

$$\det \tilde{L}_r = \frac{1}{3^2} \det L_r = \begin{cases} \frac{1}{3} \ (r = 6), \\ \frac{1}{2} \ (r = 7), \\ 1 \ \ (r = 8). \end{cases} \tag{2.18}$$

$$\langle v_0, u_j \rangle = \begin{cases} 1 \\ \frac{3}{2} \\ 3 \end{cases}, \quad \langle v_0, v_0 \rangle = \begin{cases} 2 \ (r - 6), \\ \frac{7}{2} \ (r = 7), \\ 8 \ (r = 8). \end{cases} \tag{2.19}$$

Further we set

$$\begin{aligned} \alpha_{ij} &= u_i - u_j \quad (i \neq j) \\ \beta_{ijk} &= v_0 - (u_i + u_j + u_k) \quad (i, j, k \text{ distinct}). \end{aligned}$$

Each element α_{ij} or β_{ijk} has norm 2:

$$\langle \alpha_{ij}, \alpha_{ij} \rangle = 2, \quad \langle \beta_{ijk}, \beta_{ijk} \rangle = 2. \tag{2.20}$$

Let L_r^0 denote the sublattice of $\tilde{L}_r$ generated by the r elements:

$$\alpha_i = \alpha_{i\ i+1} \ (i = 1, \ldots, r-1), \ \beta = \beta_{123}. \tag{2.21}$$

Then we have for $i \neq j$

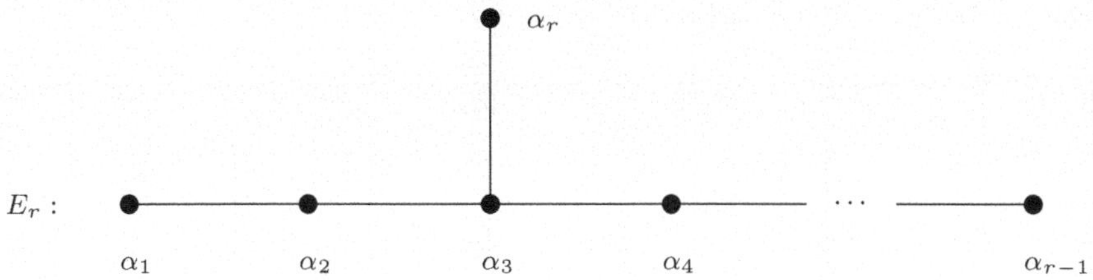

Fig. 2.3 Dynkin diagram of type E_r ($r = 6, 7, 8$)

$$\langle \alpha_i, \alpha_j \rangle = \begin{cases} -1 & (|i-j| = 1) \\ 0 & \text{otherwise,} \end{cases} \quad \langle \beta, \alpha_i \rangle = \begin{cases} -1 & (i = 3) \\ 0 & \text{otherwise.} \end{cases} \tag{2.22}$$

It follows that $\{\alpha_i (i < r), \alpha_r = \beta\}$ forms a basis of the root system of type E_r, associated with the Dynkin diagram (cf. Fig. 2.3). Therefore we can identify L_r^0 with the root lattice E_r (where the dots signify the $(r-6)$ roots connecting α_4 and α_{r-1}).

Proposition 2.32 *For $r = 6, 7, 8$, we have $L_r^0 = E_r$ and $\tilde{L}_r = E_r^\vee$ (the dual lattice of E_r).*

Proof First note that $\det L_r^0 = \det E_r = d = 3, 2$ or 1 for $r = 6, 7$ or 8. This implies that the index of L_r^0 in $\tilde{L}_r$ is d.

On the other hand, $\tilde{L}_r$ is contained in the dual lattice of $L_r^0 = E_r$. Indeed, for the generators $\{u_i, v_0\}$ of $\tilde{L}_r$, we have

$$\langle u_i, \alpha_j \rangle = \delta_{ij} - \delta_{i\ j+1} \in \mathbb{Z}, \ \langle u_i, \beta \rangle = -(\delta_{i1} + \delta_{i2} + \delta_{i3}) \in \mathbb{Z}$$

and

$$\langle v_0, \alpha_j \rangle = 0, \quad \langle v_0, \beta \rangle = -1. \tag{2.23}$$

This shows that $\tilde{L}_r \subset E_r^\vee$. By noting that the index $[L_r^0 : \tilde{L}_r] = d$ is equal to $[E_r : E_r^\vee]$, we conclude that $\tilde{L}_r = E_r^\vee$. □

Note that (2.23) implies the following: for any $\xi \in \tilde{L}_r$,

$$\xi = -\langle \xi, v_0 \rangle \beta + \text{a linear combination of } \alpha_1, \ldots, \alpha_{r-1}. \tag{2.24}$$

In particular, the orthogonal complement $\langle v_0 \rangle^\perp$ of v_0 in $L_r \otimes \mathbb{Q}$ is generated by $\{\alpha_i\ (i < r)\}$.

Remark 2.33 In case $r = 8$, we see directly that $\tilde{L}_8$ is an even integral lattice. Indeed, its generators satisfy $\langle u_i, u_i \rangle = 2$ and $\langle v_0, v_0 \rangle = 8$. Hence $\tilde{L}_8$ is a positive-definite even unimodular lattice of rank 8. It is well known that such a lattice is unique up to isomorphism (cf. [Ser70, Chap. V], [CS99]), and this gives another proof of the fact $\tilde{L}_8 = L_8^0 \simeq E_8$.

Table 2.3 Positive roots of E_r and minimal vectors of $E_r^\vee$

	E_6	E_7	E_8
Basis	$\{\alpha_1, \dots, \alpha_5, \beta\}$	$\{\alpha_1, \dots, \alpha_6, \beta\}$	$\{\alpha_1, \dots, \alpha_7, \beta\}$
pos. roots	$\begin{cases} \alpha_{ij} & (15) \\ \beta_{ijk} & (20) \\ -v_0 & (1) \end{cases}$	$\begin{cases} \alpha_{ij} & (21) \\ \beta_{ijk} & (35) \\ u_i' & (7) \end{cases}$	$\begin{cases} \alpha_{ij} & (28) \\ \beta_{ijk} & (56) \\ -\gamma_{ij} & (28) \\ -u_i & (8) \end{cases}$
min. vectors	$\pm\begin{cases} u_i & (6) \\ u_i' & (6) \quad (2\cdot 27 = 54) \\ \gamma_{ij} & (15) \end{cases}$	$\begin{cases} \pm u_i & (14) \\ \pm\gamma_{ij} & (42) \end{cases}$	$\pm$ pos. roots (240)

The above remark touches upon the finiteness statement from Corollary 2.10: for rank 8, the root lattice E_8 is the only positive-definite even unimodular lattice up to isometry. For rank 16, there is one more lattice in addition to $E_8^{\oplus 2}$; it can be constructed as an index 2 overlattice of the root lattice D_{16} and is thus often denoted by D_{16}^+. For rank 24, there are already 24 definite unimodular even lattices, as classified by [Nie73]. Interesting enough, each is determined by its root lattice (see Theorem 12.1). These lattices are named after Niemeier; they will play an instrumental role when we study elliptic K3 surfaces in Chap. 11, and particularly in our classification of all elliptic fibrations on a given K3 surface in Sect. 12.3.2. Beyond rank 24, the number of definite even unimodular lattices goes into the thousands and millions.

From now on, we make the identification:

$$L_r^0 = E_r, \quad \tilde{L}_r = E_r^\vee.$$

Keeping the same notation as before: $u_i, v_0, \alpha_{ij}, \beta_{ijk}, \dots$, we determine the positive roots of E_r with respect to the chosen basis $\{\alpha_i, \beta\}$, and describe the minimal vectors of $E_r^\vee$ in terms of u_i, v_0.

Theorem 2.34 *Set*

$$\begin{cases} \alpha_{ij} &= u_i - u_j \quad (i < j) \\ \beta_{ijk} &= v_0 - (u_i + u_j + u_k) \quad (i < j < k) \\ \gamma_{ij} &= v_0 - u_i - u_j \quad (i < j) \\ u_i' &= u_i - v_0. \end{cases} \tag{2.25}$$

Then the positive roots of E_r and the minimal vectors of $E_r^\vee$ are given by Table 2.3. (The number in parentheses denotes the number of elements of that type.)

Proof First we note that the elements in the table have the proper norm. We know already that α_{ij}, β_{ijk} have norm 2. Further, by (2.16), (2.19), we have

$$\langle u_i', u_i' \rangle = \begin{cases} \frac{4}{3} \\ 2 \\ 4 \end{cases}, \quad \langle \gamma_{ij}, \gamma_{ij} \rangle = \begin{cases} \frac{4}{3} \ (r = 6), \\ \frac{3}{2} \ (r = 7), \\ 2 \ (r = 8). \end{cases}$$

Next let us check that each element, say α, listed as a "positive root" above, is indeed such. For $\alpha = \alpha_{ij}$, the proof is the same as in the case of A_r. For other α, it is a root of E_r if it belongs to E_r (since it has norm 2); the latter holds because $\langle \alpha, u_i \rangle \in \mathbb{Z}$ and $\langle \alpha, v_0 \rangle \in \mathbb{Z}$. Then we note that $-\langle \alpha, v_0 \rangle > 0$, which will imply by (2.24) that α is a positive root. Explicitly, this gives

$$\begin{cases} \alpha_{ij} &= \alpha_i + \alpha_{i+1} + \ldots + \alpha_{j-1} \quad (i < j) \\ \beta_{ijk} &= \beta + \ldots \\ -v_0 &= 2\beta + \alpha_1 + 2\alpha_2 + 3\alpha_3 + 2\alpha_4 + \alpha_5 \quad (r = 6) \\ u_i' &= 2\beta + \ldots \quad (r = 7) \\ -\gamma_{ij} &= 2\beta + \ldots \quad (r = 8) \\ -u_i &= 3\beta + \ldots \quad (r = 8) \end{cases}$$

where $\ldots$ stands for a linear combination of α_l with coefficients ≥ 0.

The number of positive roots obtained in this way is 36, 63 or 120 for $r = 6, 7, 8$ respectively, which is half of the number of roots, and hence we have found all the positive roots of E_r.

Similarly, all the minimal vectors of the dual lattice $E_r^\vee$ are given as in the table.
□

Now we consider the Weyl group $W(E_r)$. Let us describe its action on $E_r^\vee$, by looking at the effect of the reflections s_α, (2.11), on the basic vectors $u_1, \ldots, u_r$.

Proposition 2.35 *For any root α in E_r, let $s(\alpha) = s_\alpha$ be the reflection (2.11). With the notation in the above theorem, we have for $r = 6, 7, 8$:*

$$s(\alpha_{ij}) : \begin{cases} u_i \mapsto u_j \\ u_j \mapsto u_i \\ u_l \mapsto u_l \ (l \neq i, j) \end{cases} \qquad s(\beta_{ijk}) : \begin{cases} u_i \mapsto \gamma_{jk} \\ u_j \mapsto \gamma_{ik} \\ u_k \mapsto \gamma_{ij} \\ u_l \mapsto u_l \ (l \neq i, j, k). \end{cases}$$

Further we have

$$(r = 6) \ \ s(v_0) : \ \ u_l \mapsto u_l'$$

$$(r = 7) \ \ s(u_i') : \begin{cases} u_i \mapsto u_i \\ u_l \mapsto -\gamma_{li} \ (l \neq i) \end{cases}$$

$$(r = 8) \ \ s(u_i) : \begin{cases} u_i \mapsto -u_i \\ u_l \mapsto \alpha_{li} \ (l \neq i), \end{cases} \qquad s(\gamma_{ij}) : \begin{cases} u_i \mapsto u_i \\ u_j \mapsto u_j \\ u_l \mapsto -\beta_{ijl} \ (l \neq i, j) \end{cases}$$

The verification is straightforward, so it is omitted. As an immediate consequence, we have:

Proposition 2.36 *For any $r = 6, 7, 8$, the Weyl group $W(E_r)$ acts transitively on the set of the roots in E_r. Also it acts transitively on the set of the minimal vectors in $E_r^\vee$, except for the case $r = 6$, where it has the two orbits, Ω and $-\Omega$, each consisting of 27 elements, where*

$$\Omega = \{u_i,\ u_i'\ (i = 1, \ldots, 6),\ \gamma_{ij}\ (i < j)\}.$$

Observe that $\langle \xi, \eta \rangle \equiv \frac{1}{3} \mod \mathbb{Z}$ for any $\xi, \eta \in \Omega$. Therefore the $W(E_6)$-orbit of a minimal vector $\zeta \in E_6^\vee$ is characterized as the set of those vectors ξ satisfying $\langle \xi, \zeta \rangle \equiv \frac{1}{3} \mod \mathbb{Z}$).

There are many more properties of root lattices which we would have liked to include here, but for lack of space decided to omit. For instance, the root lattices E_r cannot be embedded into any A_n, D_m, and similarly, D_m does not embed into A_n for any integers m, n, r (see [Nis96], for instance). We will use this when studying specific elliptic K3 surfaces in Sect. 12.3.5.1.

2.3.5 *Discriminant Groups*

In this section we review results about discriminant groups and discriminant forms that go back to Kneser and Nikulin as announced in Sect. 2.2. We also collect the relevant data for the discriminant groups of root lattices of type A_r, D_r, E_r using the description from the previous paragraphs.

Let L be an integral lattice. Recall the dual lattice $L^\vee$ which can be viewed as a (generally non-integral) overlattice of L of index the absolute value of the determinant of L.

Definition 2.37 The quotient

$$A_L := L^\vee / L$$

is a finite abelian group which is called the *discriminant group* of L.

The bilinear form on L can be extended $\mathbb{Q}$-linearly to $L^\vee$. Thus it induces a bilinear form on A_L with values in $\mathbb{Q}/\mathbb{Z}$. The discriminant group can always be decomposed into a product of cyclic groups (of prime power order, if this helps). Note that direct summands of A_L whose orders are relatively prime are automatically orthogonal with respect to the induced bilinear form. (This can ease both presentation and computations substantially.)

Now consider the case where L is even.

Definition 2.38 The *discriminant form* of an even lattice L is defined as the quadratic form

$$q_L : A_L \to \mathbb{Q}/2\mathbb{Z}$$
$$x \mapsto \langle x, x \rangle \mod 2\mathbb{Z}.$$

The importance of the discriminant form is illustrated by the relation with genus theory of lattices. A genus behaves roughly like an isogeny class, comprising lattices with the same signature and discriminant form. Precisely, two lattices are said to be in the same *genus* if they are isometric over the p-adic integers for all primes p and over $\mathbb{R}$. In many situations (for instance indefinite lattices with a relatively mild condition, see Theorem 2.40 below) there is only one isomorphism class of lattices in a genus, but we will also encounter important cases in the context of K3 surfaces where this does not hold true (as discussed for positive-definite even unimodular lattices, see also Theorem 12.21). Mostly we will be concerned with even (thus integral) lattices:

Theorem 2.39 (Nikulin [Nik80, Corollary. 1.9.4]) *The genus of an even lattice is determined by its signature and discriminant form.*

We remark that there are very powerful results to determine when the genus of an even lattice consists of a single class which is determined uniquely by signature and discriminant form. For instance, this holds in the following situation which we will often appeal to in Chap. 11:

Theorem 2.40 (Kneser [Kne56], Nikulin [Nik80, Corollary 1.13.3]) *Let L be an even indefinite lattice. Assume that the length $l(A_L)$ of A_L (i.e. the minimum number of generators) does not exceed rank$(L) - 2$. Then L is uniquely determined by signature and discriminant form.*

Often we will consider even lattices with a primitive embedding into some unimodular lattice. Here the discriminant form captures the essential information about the orthogonal complement:

Proposition 2.41 (Nikulin [Nik80, Proposition 1.6.1]) *Let $L \hookrightarrow M$ be a primitive embedding of even lattices. Assume that M is unimodular and consider the orthogonal complement $L^\perp$ of L inside M. Then there is a canonical isomorphism*

$$q_L \cong -q_{L^\perp}.$$

In other words, the following diagram commutes:

$$\begin{array}{ccc} L^\vee/L & \cong & (L^\perp)^\vee/L^\perp \\ \downarrow q_L & & \downarrow q_{L^\perp} \\ \mathbb{Q}/2\mathbb{Z} & \xrightarrow{-1} & \mathbb{Q}/2\mathbb{Z} \end{array}$$

Sketch of Proof of Proposition 2.41. We have an inclusion of lattices

$$L \oplus L^\perp \subset M = M^\vee \subset (L \oplus L^\perp)^\vee,$$

where the first and last inclusion have index m, say. This gives an embedding of finite abelian groups

$$H_M = M/(L \oplus L^\perp) \hookrightarrow A_{L\oplus L^\perp} \cong A_L \oplus A_{L^\perp}.$$

The primitivity of L is equivalent to the above embedding composed with projection to A_L being injective, and same for $L^\vee$ (which is automatically primitive). Since the groups $H_M, A_L, A_{L^\perp}$ all have the same size m (cf. Lemma 2.12), the claimed isomorphism

$$A_L \cong H_M \cong A_{L^\perp} \tag{2.26}$$

follows at once. Moreover, H_M is isotropic considered as a subgroup of $A_{L\oplus L^\perp}$ (or simply with induced quadratic form taking values in $\mathbb{Q}/2\mathbb{Z}$). Hence the isomorphism (2.26) has to reverse signs between q_L and $q_{L^\perp}$. □

Conversely, let L, L' be even lattices. Then any isomorphism $q_L \cong -q_{L'}$ encodes sufficient information to glue together L and L' as primitive orthogonal sublattices of some unimodular lattice (which may depend on the chosen isomorphism if the lattices are definite, compare Sect. 12.3.2). All these considerations can easily be extended to non-unimodular (integral) overlattices, see [Nik80, Proposition 1.5.1].

Along similar lines, one can study the embeddings of a given lattice L into some reference lattice M. Indeed, if both lattices are non-degenerate as usual, then the number of embeddings up to isometry is always finite by [Kne02, (30.2)]. As an application, one obtains that the number of representations, up to isometry, of any given non-zero integer $m \in \mathbb{Z}$ is finite (take $L = \langle m \rangle$). (This is a trivial fact for definite lattices, used in Sect. 2.2, even without considering isometries.) For later reference (when studying K3 surfaces in Chap. 11), we note the following special situation where existence and uniqueness can be settled at the same time (assuming primitivity):

Theorem 2.42 (Nikulin [Nik80, Proposition 1.14.4]) *Let L be an even lattice of signature (s_+, s_-) and M even unimodular of signature (t_+, t_-). Assume that*

1. $s_+ < t_+$,
2. $s_- < t_-$,
3. $l(A_L) \leq \operatorname{rank}(M) - \operatorname{rank}(L) - 2$.

Then there exists a unique primitive embedding $L \hookrightarrow M$.

If we allow ourselves to include *degenerate* lattices L, then the above finiteness statements no longer hold. This can already be seen from a single isotropic vector v (taking $L = \mathbb{Z}v$), compare [Kne02, (30.3)]. However, even this can be helped by requiring that the embedding be primitive, as we shall discuss briefly for later use.

Lemma 2.43 *Let v be an isotropic vector and M be any non-degenerate integral lattice. There are only finitely many primitive embeddings of v into M up to isometries.*

Table 2.4 Discriminant groups and discriminant forms of ADE lattices

L	A_{n-1}	D_{2n}	D_{2n+1}	E_6	E_7
A_L	$\mathbb{Z}/n\mathbb{Z}$	$(\mathbb{Z}/2\mathbb{Z})^2$	$\mathbb{Z}/4\mathbb{Z}$	$\mathbb{Z}/3\mathbb{Z}$	$\mathbb{Z}/2\mathbb{Z}$
q_L	$\frac{n-1}{n}$	$\begin{pmatrix} 1 & \frac{1}{2} \\ \frac{1}{2} & \frac{n}{2} \end{pmatrix}$	$\frac{2n+1}{4}$	$\frac{4}{3}$	$\frac{3}{2}$

Proof Given an embedding of v into M, consider the set

$$\{\langle v, w\rangle; \quad w \in M\} \subset \mathbb{Z}.$$

In fact, this is an ideal, so it equals $m\mathbb{Z}$ for some $m \in \mathbb{N}$. By primitivity, v/m has order m in the discriminant group $A_M = M^\vee/M$. In particular, $m \mid \det(M)$, so m is bounded (and better bounds can often be obtained from the precise structure of A_M as a finite abelian group). Fix some $w \in M$ such that $\langle v, w\rangle = m$. This gives the non-degenerate lattice

$$L = \mathbb{Z}v + \mathbb{Z}w = \begin{pmatrix} 0 & m \\ m & u \end{pmatrix}.$$

Using the obvious isometries, the Gram matrix can immediately be reduced to the case $|u| \leq m$. It follows that there are only finitely many non-degenerate lattices L to consider, and to each of them [Kne02, (30.2)] applies as above to prove the claim. □

Note that not only the bilinear form extends from an even lattice L to $L \otimes \mathbb{Q}$ and thus to A_L, but also any isometry of L. Here we see one of the special features of the Weyl group: any reflection of L acts trivially on A_L as one easily verifies.

We conclude this section with a table listing the discriminant groups and discriminant forms (determined on generators of the discriminant group) for the root lattices of type A, D, E (Table. 2.4). All these data can be extracted from the discussion of the auxiliary lattices $L_{r,s}$, see in particular Table 2.2.

2.3.6 *Invariant Theory of the Weyl Groups, I*

We recall some facts about the invariant theory for the Weyl groups (see [Bou68, CS99]).

First suppose that G is a finite group acting on a vector space V over a field F (with coordinates $u_1, \ldots, u_r$), and consider the action of G on the polynomial ring $F[u_1, \ldots, u_r]$. Let G be a finite reflection group, i.e. a discrete group generated by reflections of V (where V is endowed with a definite pairing, cf. [Hum92]). If the order of G is not divisible by the characteristic of F, then the ring of invariants is a graded polynomial ring. Namely, there exist r homogeneous polynomials, say

$f_1, \ldots, f_r$, such that

$$F[u_1, \ldots, u_r]^G = F[f_1, \ldots, f_r]. \tag{2.27}$$

For details and the proof, see [Bou68, Chap. V, Sect. 5] (esp. Theorem 4) or [Hum92, Chap. I.3]. The polynomials $f_1, \ldots, f_r$ are called the *fundamental invariants* of G. The set $\{d_i = \deg f_i\}$ of the degrees of fundamental invariants is uniquely determined by G, and one has $|G| = \prod_{i=1}^r d_i$.

In the case of the Weyl group $G = W(L)$ of a root lattice L, let $V = L \otimes F$ be the vector space over a field F of characteristic not dividing $|G|$, e.g. $F = \mathbb{Q}$. Then we have:

Theorem 2.44 *Suppose $G = W(L)$ is the Weyl group for $L = A_r, D_r, E_r$. Then the degrees of fundamental invariants and the order of the Weyl group are given as follows:*

$$\begin{cases} W(A_r) & \{2, 3, \ldots, r, r+1\} \ (r \geq 1) & (r+1)!, \\ W(D_r) & \{2, 4, \ldots, 2(r-1), r\} \ (r \geq 4) & 2^{r-1} r!, \\ W(E_6) & \{2, 5, 6, 8, 9, 12\} & 2^7 3^4 5, \\ W(E_7) & \{2, 6, 8, 10, 12, 14, 18\} & 2^{10} 3^4 5\, 7, \\ W(E_8) & \{2, 8, 12, 14, 18, 20, 24, 30\} & 2^{14} 3^5 5^2 7. \end{cases}$$

Next a polynomial $f \in F[u_1, \ldots, u_r]$ is called an *anti-invariant* of $G = W(L)$ if

$$f(g(u_1, \ldots, u_r)) = \epsilon(g) f(u_1, \ldots, u_r) \text{ for any } g \in W(L).$$

Here the sign function $\epsilon(g) = \pm 1$ is defined as the determinant of the natural action of g on the lattice L:

$$\epsilon(g) = \det(g) \in \{\pm 1\} \quad (g \in W(L)).$$

Among the anti-invariants, there is a basic anti-invariant δ such that every anti-invariant f is equal to a product $f = \delta \cdot h$ with h an invariant. Explicitly, it is given by (see Proposition 6, [Bou68, Chap. V, Sect. 5]):

$$\delta = \prod_{\alpha > 0} \alpha, \tag{2.28}$$

where the product is taken over the positive roots α of L, expressed as linear forms of u_i, as e.g. in case $L = E_r$ given in Theorem 2.34 and similarly given for other L in the previous sections. Moreover, the basic anti-invariant δ is equal to the Jacobian determinant of the basic invariants $\{f_j\}$ w.r.t. $\{u_i\}$:

$$J = \det\left(\frac{\partial f_j}{\partial u_i}\right) \tag{2.29}$$

up to constants.

As for the structure of the Weyl groups, we saw above that $W(A_r)$ is isomorphic to $\mathcal{S}_{r+1}$, the symmetric group on $r+1$ letters. $W(D_r)$ is a semi-direct product of $\mathcal{S}_r$ with $(\mathbb{Z}/2\mathbb{Z})^{r-1}$. For the Weyl groups $W(E_r)$ $(r = 6, 7, 8)$ we (have to) give a few more details, but we just state the bare facts, as the background, in particular concerning the groups involved, can be found [Con85] from where we also take the notation.

Notably, the Weyl group $W(E_r)$ $(r = 6, 7, 8)$ is almost a simple group. More precisely, $W(E_6)$ contains a simple subgroup of index 2 (consisting of elements with determinant 1), which is isomorphic to $U_4(2) \simeq S_4(3)$. $W(E_7)$ contains -1, and the quotient group $W(E_7)/\{\pm 1\}$ is a simple group isomorphic to $S_6(2) = Sp(6, \mathbb{Z}/2\mathbb{Z})$. Also $W(E_8)$ contains -1 and the quotient group $W(E_8)/\{\pm 1\}$ has a simple subgroup of index 2 which is isomorphic to $O_8^+(2)$.

For later use, we also note that $\mathrm{Aut}(E_r)$ is equal to $W(E_r)$ for $r = 7$ or 8, while $\mathrm{Aut}(E_6) = W(E_6) \cdot \{\pm 1\}$.

2.3.7 *Invariant Theory of the Weyl Groups, II*

Now we consider another kind of action of the Weyl group $G = W(L)$, not on the vector space $L \otimes \mathbb{Q}$ as above, but on the group ring $\mathbb{Z}[L^\vee]$ of the dual lattice $L^\vee$ of the root lattice L, which is also called the *weight lattice* of L in the language of Lie algebras ([Bou68, Chap. VI Sect. 5.3]). This group ring is defined as the free abelian group generated by the elements $\tilde{\alpha}$ corresponding to all $\alpha \in L^\vee$. (Note that this abelian group has infinite rank.) Denoting $\tilde{\alpha} = e^\alpha$ symbolically, the multiplication in the group ring is defined by $\tilde{\alpha} \cdot \tilde{\beta} = e^\alpha \cdot e^\beta = e^{\alpha+\beta}$. Any $w \in G = W(L)$ acts as $e^\alpha \to e^{w(\alpha)}$. Thus for any $u \in L^\vee$, the G-orbit sum

$$\chi(u) := \sum_{w \in G} e^{w(u)} \quad (u \subset L^\vee)$$

is obviously a G-invariant, i.e. $\chi(u)$ belongs to $\mathbb{Z}[L^\vee]^G$.

Suppose that $B = \{\alpha_1, \dots, \alpha_r\}$ is a $\mathbb{Z}$-basis of L corresponding to the Dynkin diagram of type L, and let $\{u_1, \dots, u_r\}$ be its dual basis in $L^\vee$; then $u_1, \dots, u_r$ are called the *fundamental weights* (with respect to the chosen basis B), and $\chi_i = \chi(u_i)$ are called the *fundamental characters*.

Now, if we set $s_i = e^{u_i}$, the elements $\sum_i n_i u_i$ of $L^\vee$ correspond bijectively to the monomials $\prod_i s_i^{n_i} \in \mathbb{Z}[L^\vee]$. Thus $\mathbb{Z}[L^\vee]$ is identified with the ring of Laurent polynomials:

$$\mathbb{Z}[L^\vee] = \mathbb{Z}[s_1^{\pm 1}, \dots, s_r^{\pm 1}].$$

Rewriting $\chi_i = \chi(u_i)$ in terms of s_i (and abusing the notation), we set

$$\chi_i := \chi(s_i) = \sum_{w \in G} w(s_i).$$

We quote the following theorem (Theorem 1 from [Bou68, Chap. VI Sect. 3]):

Theorem 2.45 *The ring of invariants of the Weyl group $W(L)$ in $\mathbb{Z}[L^\vee]$ is a polynomial ring*

$$\mathbb{Z}[s_1^{\pm 1}, \ldots, s_r^{\pm 1}]^{W(L)} = \mathbb{Z}[\chi_1, \ldots, \chi_r] \tag{2.30}$$

generated by the fundamental characters $\chi_i = \chi(s_i)$.

The anti-invariants in the ring of Laurent polynomials are defined in the same way as before. Again there exists a basic anti-invariant

$$\delta' = e^{-\rho} \prod_{\alpha>0} (e^\alpha - 1). \tag{2.31}$$

Here $\rho = 1/2(\sum_i u_i)$ is half of the sum of fundamental weights u_i and α runs through the positive roots of L, while e^α, as before, is the symbol expressing the monomial of x_i corresponding to α. Moreover, the basic anti-invariant δ' is equal to the Jacobian determinant of the invariants $\{\chi_j\}$ w.r.t. $\{s_i\}$,

$$J' = \det(\frac{\partial \chi_j}{\partial s_i}), \tag{2.32}$$

up to units in the Laurent polynomial ring.

The reader will see, later in this book, that the root lattices and their dual lattices play a very important role in the study of Mordell–Weil lattices, first in the description of singular fibres of an elliptic surface (see especially Sect. 5.5), and second in analysing the structure of the Mordell–Weil lattice itself (from Chap. 6 on). For example, for the case of root lattices such as

$$L = A_2, E_6, E_7 \quad \text{or} \quad E_8,$$

we will see some interesting ("excellent" or "multiplicative excellent", as they will be called later) families of elliptic surfaces with Mordell–Weil lattice of type L such that the parameters of each family correspond essentially to the fundamental invariants of the Weyl group $W(L)$ (see Theorems 9.15, 9.21, 10.8).

Historical Notes

Lattice theory nowadays is ubiquitous in many branches of mathematics. For instance, this is reflected by a variety of textbooks such as [Bou68, CS99, Ebe13] or [Mar03], each of them focussing on many different aspects of the theory and its applications. In particular, several of them include historical accounts (e.g. [Bou68, Note historique], [CS99, Sects. I and IV]), so we will not repeat all of this here, but concentrate on a few aspects which will show up later in this book.

Historically, there were many independent developments leading to the current state of knowledge on lattices. To mark one decisive starting point, we highlight Gauss' study of class numbers of imaginary quadratic fields which he phrased in terms of positive-definite binary quadratic forms (i.e. definite even integral lattices of rank two, see Sect. 13.1.3.2, for instance). In this arithmetic direction, the most influential step was Minkowski's pioneering work on the geometry of numbers which shaped a whole subject for years to come. The connection with diophantine equations is maybe best understood through the local-to-global principle using p-adic rational numbers, discovered by Hasse (cf. [Ser70]). The geometric perspective on quadratic forms and lattices which we often employ was pioneered by [Wit37].

From the point of view of Lie groups, we have already discussed Dynkin's classification of root lattices, with many contributions starting from Killing and Cartan through Weyl and van der Waerden to Witt. The fundamental invariants of the Weyl groups give particularly nice cases of Chevalley's theorem for the structures of invariant rings (here again being polynomial rings, with notable dimension and weight formulas). There were classical computations for the root lattices of A_n and D_n type, but for type E_n we will see that our approach through excellent families of rational elliptic surfaces (to be detailed in Chaps. 9 and 10) provides explicit and concise formulae for the basic invariants, which can then be applied to the arithmetic applications.

There are two definite even unimodular lattices which are arguably singled out among all others: on the one hand E_8 of rank 8 as discussed above, on the other hand the Leech lattice of rank 24 (which we only touch upon very briefly in Sect. 12.3.2). They share remarkable properties, for instance measured in terms of their symmetries; in the case of E_8, these will play a guiding role in our study of rational elliptic surfaces and their Mordell–Weil lattices in Chaps. 9 and 10. It was long known that both these lattices, E_8 and Leech lattice, admit very dense packings, and their exceptional status was amplified by the spectacular recent results of Viazovska and her collaborators, confirming their optimality among all lattice and non-lattice packings [Via17, CK+17].

In complex and algebraic geometry, lattices had been taking a prominent role for a long time through the study of complex elliptic curves (as we shall review in the next chapter) and higher-dimensional abelian varieties and complex tori, and maybe a little less prominently through cohomology and lattices of algebraic cycles, such as the Néron–Severi lattice (see Sect. 4.2). Prototypically, this features in the study of the 27 lines on a smooth cubic surface (related to E_6 and its Weyl group); another early use of root lattices (this time E_7) and Weyl group invariants in algebraic geometry may be found in [Bram18]. In comparison, E_8 shows up in the study of del Pezzo surfaces, see [Man86]. Fundamentally, the last-mentioned structures were influenced by algebraic topology. Notably, Poincaré duality gave a perfect reason for the study of unimodular lattices, both even and odd, starting maybe from Milnor. Then distinguishing definite and indefinite lattices led to many classifications (such as Niemeier lattices, see Theorem 12.2) which are shadowed, for instance, in the Néron–Severi lattice and the cycle class map.

The subject rose to its current level of importance in algebraic geometry through the proof of the global Torelli theorem for algebraic K3 surfaces (due to [PSS71], with clarifications and extensions in [BR75, Shio78, LP81]); we will come back to this in Sect. 11.3.3. In turn, this also led to a different view on quadratic forms, geared towards K3 surfaces and largely due to Nikulin (e.g. [Nik80]). Notably the notion of the discriminant group is due to him, although the techniques involved are naturally based on foundational work of [Kne56]. Lately there has been another direction of applications opened up by the rising topic of Hyperkähler manifolds, also referred to as irreducible holomorphic symplectic manifolds (see e.g. the foundational paper by [Huy99]).

Chapter 3
Elliptic Curves

Probably most readers are familiar with elliptic curves. They are treated in most textbooks on algebraic or arithmetic geometry, and further there are excellent textbooks specifically on elliptic curves such as [Cas91, Hus14, Ser97, Sil09] (and [Sil94]) or [Was08].

Hence we will allow ourselves to restrict this chapter to the bare minimum necessary to speak about elliptic curves: (several) definitions of elliptic curves and the group law. Then we state the Mordell–Weil theorem for elliptic curves over a number field (Theorem 3.11); the function field case will need some further preparations (see Chap. 5).

The rest of this chapter serves as an introduction to the main theme of this book which is to view the Mordell–Weil group of elliptic curves over a function field as a lattice. In this vein we discuss degenerations of elliptic curves, as a step to introduce elliptic surfaces in Chap. 5 and then define the Kodaira–Néron model of an elliptic curve over a function field (see Sect. 5.6).

3.1 Elliptic Curves

Let K be a field and $\bar{K}$ its algebraic closure. An elliptic curve E defined over K, often denoted by E/K, is a smooth projective curve of genus 1 defined over K together with a K-rational point O. We denote the set of K-rational points of E by $E(K)$. This set will play a central role in our studies.

In more elementary terms, an elliptic curve over K can be defined as a plane cubic curve without singular points (over $\bar{K}$) and with a K-rational point. The most familiar equation of an elliptic curve, valid outside characteristics 2 and 3, is the following one, called the *Weierstrass normal form*:

$$Y^2Z = X^3 + pXZ^2 + qZ^3 \quad (p, q \in K). \tag{3.1}$$

M. Schütt and T. Shioda, *Mordell–Weil Lattices*, Ergebnisse der Mathematik und ihrer Grenzgebiete. 3. Folge / A Series of Modern Surveys in Mathematics 70,
https://doi.org/10.1007/978-981-32-9301-4_3

The resulting curve in $\mathbb{P}^2$ has no singular points if and only if the *discriminant* Δ (of the right-hand side of (3.1)) is different from 0:

$$\Delta = -2^4(4p^3 + 27q^2) \neq 0. \tag{3.2}$$

Clearly $O = (0 : 1 : 0)$ is a K-rational point of this curve.

Fact 3.1 *Any elliptic curve over a field of characteristic $\neq 2, 3$ can be realized by a Weierstrass normal form (3.1).*

A justification of Fact 3.1 will be given momentarily. Meanwhile we point out that an elliptic curve is often defined by the corresponding affine equation,

$$y^2 = x^3 + px + q, \tag{3.3}$$

in terms of the inhomogeneous coordinates $x = X/Z$, $y = Y/Z$. The point O, as chosen above, then is the point at infinity (i.e. the unique point on the line at infinity $Z = 0$), and we have

$$E(K) = \{O\} \cup \{(x, y) \in K^2 \mid y^2 = x^3 + px + q\}.$$

In the case of arbitrary characteristic (including 2 and 3), (3.1) and (3.3) should be replaced by the *generalized Weierstrass form*,

$$Y^2Z + a_1XYZ + a_3YZ^2 = X^3 + a_2X^2Z + a_4XZ^2 + a_6Z^3, \quad (a_i \in K) \tag{3.4}$$

and the corresponding affine equation,

$$y^2 + a_1xy + a_3y = x^3 + a_2x^2 + a_4x + a_6. \tag{3.5}$$

Again we take $O = (0 : 1 : 0)$, and the smoothness condition is given by $\Delta \neq 0$, where the discriminant Δ now is a certain polynomial in a_i with integral coefficients, which reduces to (3.2) if $a_1 = a_2 = a_3 = 0$, $a_4 = p$, $a_6 = q$ (see (3.6)). Though we will only rarely need the general formula of the discriminant Δ in this book (when we deal with or include characteristics 2 and 3), we copy it from Tate [Tat75] for the sake of completeness:

$$\Delta = -b_2^2b_8 - 8b_4^3 - 27b_6^2 + 9b_2b_4b_6, \tag{3.6}$$

where

$$\begin{aligned} b_2 &= a_1^2 + 4a_2, \\ b_4 &= a_1a_3 + 2a_4, \\ b_6 &= a_3^2 + 4a_6, \\ b_8 &= a_1^2a_6 - a_1a_3a_4 + 4a_2a_6 + a_2a_3^2 - a_4^2. \end{aligned}$$

Further, by setting

$$c_4 = b_2^2 - 24b_4, \; c_6 = -b_2^3 + 36b_2b_4 - 216b_6,$$

we have also

$$\Delta = (c_4^3 - c_6^2)/1728.$$

It is actually easy how to simplify a generalized Weierstrass form (3.4) to the Weierstrass normal form (3.1). First one completes the square on the left-hand side of (3.4) (outside characteristic 2), and then, if necessary, one completes the cube on the resulting right-hand side (outside characteristic 3). It thus remains to show that every elliptic curve can be transformed to the generalized Weierstrass form, or to a plane cubic with a rational point as claimed in the beginning lines of this section.

Theorem 3.2 *The three given definitions of an elliptic curve are equivalent to each other:*

(i) a genus 1 curve with a rational point,
(ii) a plane cubic with a rational point, or
(iii) a Weierstrass cubic.

Proof The implications (iii) $\Rightarrow$ (ii) $\Rightarrow$ (i) are elementary and easily verified (for instance, the non-zero regular 1-form on an elliptic curve given in generalized Weierstrass form can be given affinely by $dx/(2y + a_1x + a_3)$, up to scaling, and similarly for a smooth cubic (regardless of the K-rational point)). To go from (i) to (iii), the standard argument uses the Riemann–Roch theorem for curves (see e.g. [Har77, IV.1]). □

Note that to go directly from (ii) to (iii) by means of a linear transformation, only one ingredient is required: the existence of a point of inflection on any smooth plane cubic, i.e. a point on the curve whose tangent meets the curve with multiplicity three at the given time. Thus the tangent line does not intersect the curve in any other point. This observation plays a central role in the group law of the elliptic curves as exploited below. On any smooth plane cubic, there is a point of inflection over $\bar{K}$, but there may be none over K (non-algebraically closed). However, given a smooth cubic with a K-rational point, there are standard means (circumventing Riemann–Roch) to derive an isomorphic cubic curve where this point becomes a point of inflection, the isomorphism in fact being defined over K (see for instance [Cas91] or [Mord69, p. 139]).

To exhibit the step from (ii) to (iii) one simply applies a linear transformation that takes the point of inflection to O and its tangent line to the line $(Z = 0)$. One easily checks that the inflection condition ensures that the transformed plane cubic takes exactly the shape of the generalized Weierstrass form (again after normalizing).

Implicitly, we have used in the above the notion of a *K-isomorphism*, i.e. an isomorphism between two curves defined by polynomials (or rational functions) with coefficients in K. Here one may intuitively think of a linear transformation over

K, but, of course, isomorphisms derived from Riemann–Roch cannot be expected to be linear. In this context, it is useful to introduce the absolute invariant (or *j-invariant*) of E/K defined by

$$j = c_4^3/\Delta. \tag{3.7}$$

Remark 3.3 Here we have used the standard normalization of the j-invariant such that the elliptic curve with Weierstrass form

$$y^2 = x^3 - x$$

has j-invariant 12^3. This has several advantages on the arithmetic side, for instance j-invariants of elliptic curves with complex multiplication will always be algebraic integers. The reader should be aware that the normalization for the above curve to $j = 1$ also persists in the literature.

One of the decisive obvious features of the j-invariant is that it remains constant under *admissible coordinate changes*, i.e. those affine transformations which preserve the given shape of Weierstrass form. For instance, for the short Weierstrass form (3.3), the admissible coordinate changes over the base field K amount to scalings

$$(x, y) \mapsto (u^2 x, u^3 y) \quad \text{for } u \in K^\times. \tag{3.8}$$

In fact, the j-invariant captures much more information than this, at least over an algebraically closed ground field:

Theorem 3.4 *Two elliptic curves E, E' are isomophic over an algebraically closed field K if and only if $j = j'$.*

To see that the condition in Theorem 3.4 that K be algebraically closed is necessary, take a non-square $u \in K$ (given it exists), for instance, and apply the coordinate change

$$(x, y) \mapsto (ux, u^{3/2} y). \tag{3.9}$$

This so-called *quadratic twist* results in two $K(\sqrt{u})$-isomorphic elliptic curves which are both defined over K, but which are in general not K-isomorphic. Anticipating the group structure explained in the next section, we would like to point out the intriguing conjectures about how the rank of an elliptic curve over a number field might vary when averaging over quadratic twists (see e.g. [GM91]). For elliptic surfaces, quadratic twists play a similarly important role (see Sect. 5.9.1).

3.2 The Group Law

A special feature of an elliptic curve among all smooth projective curves is that it has a natural structure of a (commutative) group such that the group operations are morphisms in the sense of algebraic geometry; in other words, it has the structure of an abelian variety of dimension one.

The group law can be based on the notion of *divisors* on a smooth curve C (a concept which we will also adapt for surfaces in Chap. 4). Here we consider formal finite sums of points on C, i.e. in $C(\bar{K})$ (although you might, of course, restrict to points in $C(K)$). Thus we obtain an abelian group

$$\mathcal{D}(C) = \left\{ \sum_{i=1}^{r} n_i P_i; \quad r \in \mathbb{N},\ n_i \in \mathbb{Z},\ P_i \in C(\bar{K}) \right\}.$$

A divisor $D = \sum_{i=1}^{r} n_i P_i$ is called *effective* (notation $D \geq 0$) if all n_i are non-negative, and its *degree* is naturally defined as

$$\deg(D) = \sum_{i=1}^{r} n_i.$$

Of course, we have $D \geq 0 \Rightarrow \deg(D) \geq 0$. We obtain a subgroup $\mathcal{D}^0(E)$ consisting of divisors of degree zero. A rational function $h \in \bar{K}(C)$ gives rise to a divisor (h) of degree zero through zeroes minus poles (with multiplicity). Such divisors are called *principal*.

Usually these concepts are employed over algebraically closed fields (as we have done above), but one can easily extend the theory to arbitrary fields K. In detail, let us write a divisor $D \in \mathcal{D}(C)$ uniquely as

$$D = \sum_{P \in C(\bar{K})} n_P P \quad \text{with} \quad n_P \in \mathbb{Z}, \quad \text{where} \quad n_P = 0 \quad \text{for almost all } P \in C(\bar{K}).$$

Then D is said to be K-rational if

(i) $n_P = n_{P^\sigma}$ whenever $\sigma \in \mathrm{Aut}(\bar{K}/K)$ and
(ii) each n_P is divisible by the degree of inseparability of the extension $K(P)/K$.

Here the field extension $K(P)$ is obtained from K by adjoining the quotients of the non-zero homogeneous coordinates of $P \in C(\bar{K})$ (or, equivalently, the affine coordinates in any given affine chart over K containing P).

The condition (ii) can be discarded in case K is a perfect field (for example if K has characteristic 0). Note that for a K-rational divisor D, the vector space

$$L(D) = \{h \in \bar{K}(C) \setminus \{0\} \mid (h) + D \geq 0\} \cup \{0\}$$

has a basis consisting of functions in $K(C)$ (for K perfect, a proof can be found in Silverman [Sil09, Proposition II.5.8]). We shall denote the groups of K-rational divisors (of degree zero) by $\mathcal{D}(C)_K$ resp. $\mathcal{D}^0(C)_K$.

All these groups are endowed with an additional structure by introducing *linear equivalence*. Here two divisors are called linear equivalent ($D \sim D'$) over K if their difference is a principal divisor defined over K:

$$D \sim D' \iff D - D' = (h) \text{ for some } h \in K(C)^\times.$$

We now specialize to the situation where $C = E$ is an elliptic curve over a field K. The group law of E is based on the divisor class group

$$J(K) = \mathcal{D}^0(E)_K / \sim .$$

To avoid confusion, we denote by $[P]$ the divisor of degree 1 corresponding to a point $P \in E(K)$. Here $J(K)$ naturally comes equipped with the structure of an abelian group. By the following lemma, this group structure carries over to $E(K)$.

Lemma 3.5 *For an elliptic curve E/K, the map*

$$\alpha : E(K) \longrightarrow J(K)$$

sending $P \in E(K)$ to the linear equivalence class of $[P] - [O]$ is bijective.

The bijection established in the lemma can serve two purposes:

1. to give the structure of an abelian group on the set $E(K)$;
2. to define the structure of a variety (in fact, of an abelian variety of dimension 1) on the divisor class group J; in general, this makes it into the *Jacobian variety* of E, $J = \mathrm{Jac}(E)$, which presently means $J = E$.

To explain this, first note that the lemma holds for any extension K' of K; in other words, we have a bijection:

$$\alpha_{K'} : E(K') \longrightarrow J(K').$$

Moreover, for any $K' \subset K''$, the maps $\alpha_{K'}$ and $\alpha_{K''}$ are compatible with the natural maps $E(K') \to E(K'')$ and $J(K') \to J(K'')$. Thus we may write $P \in E$ for $P \in E(K')$ (some K') by omitting K'.

In what follows, let us denote by $P + Q \in E$ the sum of $P, Q \in E$ in the group law, transported via the map α from the divisor class group. (Warning: Do not confuse the notation $P + Q$ with that of the degree two divisor $[P] + [Q]$. For instance, $P + Q = R$ if and only if the divisor classes of $[P] + [Q]$ and $[R] + [O]$ are the same.)

Theorem 3.6 *The map $(P, Q) \mapsto P + Q$ defines a morphism of $E \times E \to E$, and the inversion $P \mapsto -P$ is likewise a morphism of E to itself.*

To get an idea of how Theorem 3.6 works, we may assume without loss of generality by Theorem 3.2 that E is a Weierstrass cubic and O is the unique point at infinity. Then any three collinear points P, Q, R add up to zero, since $[P]+[Q]+[R] \sim 3[O]$, and the converse statement can also be seen:

$$P+Q+R=O \Longleftrightarrow P, Q, R \text{ are collinear.} \tag{3.10}$$

For the sake of simplicity, assume that E has the defining affine Eq. (3.3):

$$E: \quad y^2 = x^3 + px + q.$$

Let $P=(x_1, y_1)$, $Q=(x_2, y_2)$, $R=(x_3, y_3)$. Then the inversion map is given by $P=(x_1, y_1) \mapsto P'=(x_1, -y_1)$; indeed, the three points P, P', O lie on the projective line $X - x_1 Z = 0$, so $P + P' = O$ by (3.10) as required.

Further, $P + Q = R'$ is given by first taking the third intersection point R of the line $\overline{PQ}$ connecting P and Q with the cubic curve E and then taking the inverse $R' = -R$. If $\mathbf{x_1} \neq \mathbf{x_2}$, the line $\overline{PQ}$ is defined by

$$y - y_1 = \mu(x - x_1), \quad \mu = (y_2 - y_1)/(x_2 - x_1),$$

so the three roots of the cubic equation in x

$$x^3 + px + q - \{\mu(x - x_1) + y_1\}^2 = 0$$

must be x_1, x_2 and x_3. By the relation of the roots and the coefficients of the equation, we have

$$x_3 = -(x_1 + x_2) + \mu^2 = -(x_1 + x_2) + \{(y_2 - y_1)/(x_2 - x_1)\}^2$$

and, of course,

$$y_3 = \mu(x_3 - x_1) + y_1.$$

Thus the sum of two points P and Q, $P + Q = (x, y)$, is given by the *addition formula*:

$$\begin{cases} x = -(x_1 + x_2) + \left(\dfrac{y_2 - y_1}{x_2 - x_1}\right)^2 \\ y = -\dfrac{y_2 - y_1}{x_2 - x_1}(x - x_1) - y_1. \end{cases} \tag{3.11}$$

This proves that $(P, Q) \mapsto P + Q$ is a morphism of $E \times E \to E$ if $x_1 \neq x_2$. On the other hand, $\mathbf{x_1} = \mathbf{x_2}$ holds only when either $P = Q$ or $-P = Q$. In case $P = Q$, the line connecting P, Q should be understood as the tangent line to the cubic curve E at P (e.g. the line in yellow tangent at $(-3/4, -13/8)$ in Fig. 3.1 to follow), and we obtain the following *duplication formula* for $2P = (x, y)$ as the limit of the above addition formula as $x_2 \to x_1$:

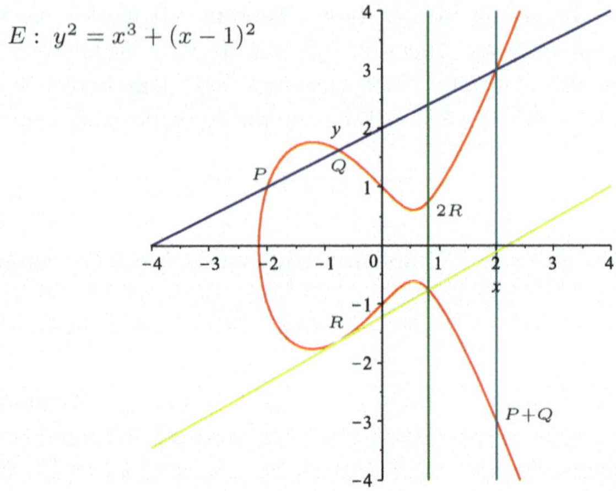

Fig. 3.1 Elliptic curve with chord, tangent and vertical lines

$$\begin{cases} x = -2x_1 + (y_1')^2, \quad y_1' = \dfrac{3x_1^2 + p}{2y_1} \\ y = -y_1'(x - x_1) - y_1. \end{cases} \tag{3.12}$$

In this way, one can prove (by some non-elementary computations) that $(P, Q) \mapsto P + Q$ is a morphism.

Geometrically, one often refers to the chord-tangent construction when defining the group law on a Weierstrass cubic (or more generally, on any plane cubic with a rational point). This is visualized in Fig. 3.1 where, as usual, the point O at ∞ serves as zero element for the group law. Here the chord (between $P = (-2, 1)$ and $Q = (-3/4, 13/8)$) is printed in dark blue, while the tangent (at $R = (-3/4, -13/8)$) is printed in yellow.

Example 3.7 In Fig. 3.1, one can compute

$$P + Q = (2, -3) \quad \text{and} \quad 2R = (2193/2704, 106079/140608).$$

(This already illustrates that the expressions tend to become complicated rather fast, so the reader might prefer using the assistance of a computer algebra system.)

Example 3.8 (*Addition on the hexagonal elliptic curve*) Consider the hexagonal elliptic curve E over $K(t)$ from (1.1), equipped with rational points P_i, Q_i. For $\{i, j, k\} = \{1, 2, 3\}$, we calculate (under the assumption that the $\alpha_i \neq \alpha_j$ for all $i \neq j$)

$$P_i + P_j = Q_k = -P_k$$

as required by (1.2), and $P_i - P_j = P_i + Q_j$ in terms of x- and y-coordinate as

$$\mathrm{x}(P_i - P_j) = \frac{t^2 + \alpha_k(\alpha_i - \alpha_j)^2}{(\alpha_i - \alpha_j)^2}, \quad \mathrm{y}(P_i - P_j) = \frac{t((\alpha_i - \alpha_j)^2(\alpha_j - \alpha_k) - t^2)}{(\alpha_i - \alpha_j)^3}.$$

Along the same lines we compute the coordinates of $2P_i$:

$$x(2P_i) = \frac{(\alpha_j + \alpha_k - \alpha_i)t^2 + (\alpha_i - \alpha_j)^2(\alpha_i - \alpha_k)^2}{t^2}$$

$$y(2P_i) = -\frac{(t^2 - (\alpha_i - \alpha_j)(\alpha_i - \alpha_k)^2)(t^2 - (\alpha_i - \alpha_k)(\alpha_i - \alpha_j)^2)}{t^3}.$$

We have seen that an elliptic curve E/K has the structure of a *commutative algebraic group*. We remark that there are purely geometric ways to see that the addition of points is associative, but we will not touch on the details here.

In summary, we obtain the following famous result:

Theorem 3.9 (Abel's theorem for an elliptic curve) *An elliptic curve E/K is isomorphic to its Jacobian variety* $\mathrm{Jac}(E)$ *defined over K via the natural correspondence*

$$P \leftrightarrow class([P] - [O])$$

associating to a point $P \in E(K')$ over any extension K'/K the class of the K'-rational divisor $[P] - [O]$ and vice versa.

Remark 3.10 In the complex case, an elliptic curve is a torus $\mathbb{C}/L$ (as in Remark 2.19). This gives an obvious group structure, including the shape of torsion subgroups $E[N] \cong (\mathbb{Z}/N\mathbb{Z})^2$ (which stays valid in characteristic p as long as $p \nmid N$).

Over $\mathbb{C}$, the theory of elliptic curves is almost identical to that of elliptic functions (in particular, the Weierstrass $\wp$-function). For example, the addition formula (3.11) is a geometric counterpart of the classical addition theorem for elliptic functions (cf. [Ahl78, Chap. 7], [Shimu71, Chap. 4.2]).

With the group structure at our disposal, we introduce one further important invariant of an elliptic curve E: the *endomorphism ring* $\mathrm{End}(E)$. Of course, one always has the inclusion

$$\mathbb{Z} \subseteq \mathrm{End}(E).$$

In characteristic zero, the only alternative is that $\mathrm{End}(E) = \mathcal{O}$, an order in an imaginary quadratic field K. This is immediate from the analytic description of E as a complex torus $\mathbb{C}/(\mathbb{Z} + \mathbb{Z}\tau)$ as this admits extra endomorphisms if and only if τ is imaginary quadratic (so that $K = \mathbb{Q}(\tau)$). In this case, one says that E has *complex multiplication (CM)* by $\mathcal{O}$. The easiest examples are the two elliptic curves with extra automorphisms from Remark 2.19 with j-invariants 0 and 12^3, but obviously, there are countably infinitely many more elliptic curves with CM (see [Sil94, II]).

CM elliptic curves will play an important role in our study of elliptic K3 surfaces and their arithmetic, see Chap. 12. There we will also touch upon their connection to class field theory (see [Lan87, Shimu71]).

In positive characteristic $p > 0$, the picture changes drastically. In particular, this holds over finite fields where the Frobenius morphism endows any elliptic curve with extra endomorphisms, accounting for CM in the above terminology. If this gives the full endomorphism ring of E, then one says that E is *ordinary*. Otherwise, one can show that the endomorphism algebra is a quaternion algebra over $\mathbb{Q}$ ramified at p and ∞. In particular, End(E) is a $\mathbb{Z}$-module of rank 4, and one says that E is *supersingular*, following Hasse [Has36] and Deuring [Deu41] (cf. [Sil09, V.3]).

Of course, we can also interpret End(E) as Hom(E, E), and there are similar structure results for Hom(E, E'), the ring of homomophisms $E \to E'$ (or isogenies). Namely, Hom(E, E') is zero if E and E' are not isogenous, while otherwise it is a $\mathbb{Z}$-module of the same rank as End(E) (or End(E')).

3.3 Mordell–Weil Theorem

In Sect. 3.2 we have seen that the rational points $E(K)$ of an elliptic curve E/K form an abelian group. But the structure of this abelian group depends very much on the field of definition K.

For example, if $K = \mathbb{C}$ is the field of complex numbers, then $E(\mathbb{C})$ is isomorphic to a complex torus $\mathbb{C}/L$, as we saw above. In particular, the group $E(\mathbb{C})$ is a continuum, which is far from being finitely generated.

In contrast, we shall now assume K to be a number field, i.e. an extension of $\mathbb{Q}$ of finite degree (not necessarily Galois). In this setting, a completely different situation persists compared to $\mathbb{C}$. The following result must be viewed as one of the cornerstones marking the interplay between number theory and algebraic geometry:

Theorem 3.11 (Mordell–Weil theorem) *Suppose that K is a number field. Then, for any elliptic curve E/K, $E(K)$ is a finitely generated abelian group.*

This theorem was first proved for $K = \mathbb{Q}$, the rational number field, by Mordell [Mord22] in 1922. Relatively elementary proofs can be found in many textbooks on elliptic curves. Then it was extended by Weil [Wei28] in 1928 as follows. Let Γ be an algebraic curve (of positive genus) defined over an algebraic number field K, and let J be its Jacobian variety. Then $J(K)$ (= the group of K-rational divisor classes of degree 0 on Γ) is a finitely generated abelian group. In particular, if $\Gamma = E$ is an elliptic curve over K, then we have $J \simeq E$ (see Sect. 3.2) and $E(K)$ is finitely generated.

By now, the Mordell–Weil theorem has been generalized in many ways, in particular to abelian varieties over number fields (by Néron, see below).

Definition 3.12 The group of rational points $E(K)$ is often called the *Mordell–Weil group* in view of the Mordell–Weil theorem.

In this book, we will be mainly concerned with the case of elliptic curves E/K where K is a function field of dimension 1, i.e. $K = k(C)$ is the function field of a curve C defined over some field k. The simplest example for such a field K is

$K = k(t)$, the rational function field of one variable t over k, which corresponds to the case $C = \mathbb{P}^1$ (the projective line over k).

In the function field case, the analogue of the Mordell–Weil theorem holds, in general, only under some additional conditions. However, these conditions should be viewed as mild and not too restrictive. To get an idea of the cases to be excluded, consider the following situation: if $E/\mathbb{C}$ is given, then one can consider E over the function field $\mathbb{C}(t)$, but the group $E(\mathbb{C}(t))$ obviously contains $E(\mathbb{C})$ and hence cannot be finitely generated. As a consequence, we are led to consider elliptic curves E over $\mathbb{C}(t)$ which are not constant. We will make this notion precise in Convention 5.10.

In Chap. 6, we shall prove the Mordell–Weil theorem for elliptic curves over function fields under a fairly mild condition (Convention 5.10 again). The condition requires that the associated elliptic surface (the Kodaira–Néron model of E/K) has at least one singular fibre. In particular this will rule out constant elliptic curves as sketched above. These new notions will play a central role in our treatment. Before explaining them we make some preparation in the next section.

As the abstract theory of abelian varieties was founded in the late 1940s by Weil and greatly developed in the next decade by Weil and others, the Mordell–Weil theorem was extended to more general situations. In particular, Néron proved that, if A/K is an abelian variety and K is a finitely generated extension of a prime field, then the group of rational points $A(K)$ is finitely generated (and also usually called Mordell–Weil group). For these results and related topics, we refer to [Lan83, Mum74, Ser97, Sil09].

3.4 Degenerate Plane Cubics

It is a classical and very instructive problem to study how plane cubics may degenerate to singular or even reducible curves. First let us consider the degeneration of a Weierstrass cubic over a field of characteristic $\neq 2, 3$:

$$\Gamma: \quad y^2 = x^3 + px + q.$$

By (3.2), this cubic curve has a singularity if and only if

$$4p^3 + 27q^2 = 0, \text{ i.e. } \left(\frac{q}{2}\right)^2 = \left(\frac{-p}{3}\right)^3$$

Then two cases are possible:

(a) either $p = q = 0$ or
(b) $pq \neq 0$.

Both are sketched in Fig. 3.2.

Fig. 3.2 Nodal cubic and cuspidal cubic

3.4.1 *Case (a)*

The curve Γ is a *cuspidal cubic*

$$y^2 = x^3 \quad \text{i.e.} \quad Y^2 Z = X^3 \tag{3.13}$$

which has a cusp at the point $A = (X : Y : Z) = (0 : 0 : 1)$, but no other singular points. We shall now endow the smooth locus $\Gamma^{\#} = \Gamma \setminus \{A\}$ of the singular curve with a *group structure* (extending the group law from smooth cubics). If we set $u = x/y = X/Y$, then we have $x = 1/u^2$, $y = 1/u^3$. Thus, by letting

$$P(u) = \left(\frac{1}{u^2} : \frac{1}{u^3} : 1\right) = (u : 1 : u^3) \in \Gamma^{\#},$$

the map $u \mapsto P(u)$ defines an isomorphism of $\mathbb{A}^1 \to \Gamma^{\#}$ as (open) curves. In comparison with the previous relation (3.10), note that three distinct points $P(u_i)$ of $\Gamma^{\#}$ are collinear if and only if $u_1 + u_2 + u_3 = 0$, because

$$\begin{vmatrix} u_1 & 1 & u_1^3 \\ u_2 & 1 & u_2^3 \\ u_3 & 1 & u_3^3 \end{vmatrix} = (u_1 - u_2)(u_1 - u_3)(u_2 - u_3)(u_1 + u_2 + u_3).$$

Therefore the correspondence $P(u) \leftrightarrow u$ gives an isomorphism of $\Gamma^{\#}$ with the additive group $\mathbb{G}_a$.

3.4.2 *Case (b)*

We can set $q/2 = \alpha^3$, $-p/3 = \alpha^2$ with $\alpha = -3q/2p \neq 0$, and the curve Γ becomes a *nodal cubic*

$$y^2 = (x - \alpha)^2 (x + 2\alpha) \tag{3.14}$$

with the node at $A = (x, y) = (\alpha, 0)$. The nodal cubic has two distinct tangents at the point A, given by $y = \pm\sqrt{3\alpha}(x - \alpha)$. In this case, the smooth part $\Gamma^{\#} = \Gamma - \{A\}$ is isomorphic to the multiplicative group $\mathbb{G}_m$ having the parameter

$$u = \frac{y - \sqrt{3\alpha}(x - \alpha)}{y + \sqrt{3\alpha}(x - \alpha)}$$

over $k(\sqrt{3\alpha})$ if k is a field of definition of Γ. We leave it to the reader to verify this as an exercise (cf. [SiT15, Chap. 3, Sect. 7]).

We remark that there are other forms to write a nodal cubic—more convenient in avoiding irrationalities such as $\sqrt{3\alpha}$ above. A typical example arises from the generalized Weierstrass form as

$$y^2 + \beta xy = x^3.$$

This has the benefit of working in any characteristic, and it will be important for the set-up of multiplicative excellent families in Sect. 10.1, but only at the cost of letting go the geometric symmetry $y \mapsto -y$ immediately visible in Fig. 3.2.

3.4.3 Further Degenerations

If we consider more general plane cubic curves, there are further possibilities for their degenerations other than just a nodal cubic or a cuspidal cubic. In fact, a plane cubic can be reducible: it may admit a decomposition into a union of a conic and a line, or even into a union of three lines (not necessarily distinct). The following example will reappear on several occasions in this book:

Example 3.13 (*Hesse pencil*) Consider the following plane cubic curve depending on the parameter $t \in k$:

$$E = E_t : \quad X^3 + Y^3 + Z^3 = tXYZ = 0. \tag{3.15}$$

If char$(k) \neq 3$, then this curve splits into a union of three distinct lines exactly for the following special values of the parameter t: $t = 3, 3\omega, 3\omega^2$ (where ω is a primitive cube root of 1). For instance, for $t = 3$, we have

$$X^3 + Y^3 + Z^3 - 3XYZ = (X + Y + Z)(X + \omega Y + \omega^2 Z)(X + \omega^2 Y + \omega Z).$$

In characteristic 3, meanwhile, the given special values come together, and the cubic degenerates to a triple line at $t = 0$. In any case, there is also a triangle XYZ of $\mathbb{P}^1$'s at $t = \infty$ (understood using the notion of elliptic surfaces as sketched in Sect. 3.4.4).

In this last example, it is far from accidental that the given special values of t at the same time describe all values where the cubic curve attains singularities. That is, for all other t, the curve defined by Eq. (3.15) is elliptic since we have rational points such as $(X : Y : Z) = (1 : -1 : 0)$. In this situation, it is natural to consider the family of cubic curves parametrized by $C_0 = \mathbb{A}^1$ (with affine parameter t). Its total space can be compactified in a natural way (compare Sect. 3.4.4). This will produce the first example of what we will call an *elliptic surface* over $C = \mathbb{P}^1$. On the other hand, we could, of course, view E as an elliptic curve over the function field $k(t)$ (with the same rational point as above). This interplay of elliptic curves over function fields and elliptic surfaces, as sketchy as it may be for the time being, will be instrumental throughout the following chapters where we shall also take up the above example continuously.

Example 3.14 As an example for a different degeneration (outside characteristic 3), consider another plane cubic curve with parameter $t \in k$:

$$E: \quad (X + Y + Z)^3 - tXYZ = 0. \tag{3.16}$$

We observe three different degenerations: At $t = \infty$ (i.e. where only the right-most monomial remains, cf. Sect. 3.4.4), the curve splits into a union of three distinct lines as in Example 3.13. Meanwhile at $t = 0$, the curve consists of the triple line $X + Y + Z = 0$. Finally (outside characteristic 3), there is a nodal singular curve at $t = 27$ with node at $(1 : 1 : 1)$. For all other values of $t \in k$, one can check that the curve defined by (3.16) is smooth.

One of the main subjects in this book is elliptic curves over a function field $K = k(C)$ and their rational points. For their understanding, it is essential to consider an *elliptic surface* which is naturally associated with E/K. By an elliptic surface, we roughly mean a smooth algebraic surface having a fibration by elliptic curves. This short section is a preparation for this important notion.

In the above, we examined some examples of plane cubic curves with singularities. In contrast to this "static" viewpoint, a more "dynamic" one is to regard a given plane cubic to have variable coefficients (such as p, q or t) and to consider the mode of degeneration as the coefficients or parameters take some special value. That is, instead of considering the elliptic curves E_t in (3.15) or (3.16) for fixed values of $t \in k$, we consider the whole cubic as an elliptic curve over the function field $k(t)$. Setting $K = k(t)$, we can pursue the same concepts as before such as $k(t)$-rational points, Mordell–Weil group etc.

In the next two chapters, we will explain how to make an elliptic curve over a function field into an elliptic surface. In particular, surface theory will allow us to derive structure results for the Mordell–Weil group of the elliptic curve over the function field K.

Generally, there is an abstract approach through Kodaira–Néron models and an explicit approach through Weierstrass forms. As an illustration, we shall only discuss two examples here that follow naturally from what we have seen before.

3.4.4 The First Elliptic Surface

Consider the elliptic curve E_t from Example 3.13. Interpreting t as an affine coordinate of $\mathbb{P}^1$, there is a canonical way to projectivize the algebraic surface given by (3.15) in $\mathbb{P}^2 \times \mathbb{P}^1$. Let $(t_0 : t_1)$ denote homogeneous coordinates in $\mathbb{P}^1$ such that $t = t_1/t_0$. Then we find the surface

$$S: \quad t_0(X^3 + Y^3 + Z^3) - t_1 XYZ = 0.$$

Projection onto $\mathbb{P}^1$ defines a morphism $f : S \to \mathbb{P}^1$. This morphism induces an elliptic fibration, i.e. the fibres are exactly the cubic curves E_t which are almost always smooth and thus elliptic. In particular, the generic fibre is the elliptic curve E defined by (3.15) over the rational function field $K = k(\mathbb{P}^1) = k(t)$. Equivalently, in the language of schemes (see [Har77, Mum66] or [Liu06]), it is the fibre over the generic point t of the base curve $\mathbb{P}^1$.

It is easy to check that the surface S is smooth over k if the characteristic is different from 3 (although the same does not hold for all fibres E_t). The surface S (together with the morphism $f : S \to C$) gives the first example within this book of what we call the *elliptic surface over* $C = \mathbb{P}^1$ *associated with the elliptic curve* E/K (cf. Sect. 5.1).

Note that E has some obvious points over K. For any $\omega \in k$ such that $\omega^3 = 1$, $(1 : -\omega : 0)$ and all its coordinate permutations are points on the elliptic curve E_t. In fact, these are the points of inflection on E_t (which also leads to a moduli theoretic interpretation of E_t). Later we shall show using surface theory that E_t has no further K-rational points (even when k is assumed to be algebraically closed, and regardless of the characteristic). On the opposite end of the scale, we will also study the same type of question for general cubic pencils and determine in Sect. 7.6 whether their base points generate the Mordell–Weil group.

3.4.5 An Elliptic Surface with Constant Moduli

Generally, we let E be an elliptic curve over K where $K = k(C)$ is the function field of a curve C defined over a base field k. Assume that C is a smooth projective curve and k is an algebraically closed field. Later we shall also consider non-closed fields.

If E is given by a (generalized) Weierstrass equation (3.1) or (3.4), of course the coefficients p, q (or a_i) belong to the field K. This means that these coefficients are then functions on the curve C. It follows that the absolute invariant j, defined in (3.7), is also a function on the curve C. Usually the j-invariant will not be constant, but if it is (i.e. $j \in k$), then E/K is said to have *constant moduli*; otherwise, E/K is said to have *variable moduli*.

Example 3.15 Let char$(k) \neq 2, 3$ and $K = k(t)$ be the rational function field of one variable t, i.e. the function field of $C = \mathbb{P}^1$, a projective line. Fix a positive integer m and consider the elliptic curve

$$E: \quad y^2 = x^3 + t^m + 1.$$

The j-invariant is equal to 0, so this is an elliptic curve with constant moduli.

Note that despite having constant moduli, the previous elliptic curve E is not constant over K. That is, any two given fibres are isomorphic over k (since $k = \bar{k}$), but $E \not\cong E_{t_0}$ over K for any value $t_0 \in k$.

We will see later that this example alone (for various values of m and for the base field k with various characteristics) leads to pretty interesting results from the viewpoint of Mordell–Weil lattices such as neat lattices (for instance E_8 for $m = 5, 6$) and actually lattices of very high rank and with large density (as we shall explore in Chap. 13). Indeed, for many values of m, the elliptic curve E has some obvious rational points; for instance $(-\omega, \pm t^{m/2})$ for even m and $(-\omega t^{m/3}, \pm 1)$ if $3 \mid m$ where $\omega^3 = 1$. In general, however, there can be many non-trivial points, and it is a remarkable circumstance that we will in fact be able to determine all of them (see Sect. 13.2.2 for a discussion of the characteristic zero case).

We conclude this chapter with a brief explanation of how to make the elliptic curve E over $k(t)$ from Example 3.15 into an elliptic surface. The obvious idea is to projectivize the fibres as cubics in $\mathbb{P}^2$ by adding the point O at ∞:

$$S: \quad Y^2Z = X^3 + (t^m + 1)Z^3.$$

Thus we obtain an elliptic fibration over the affine t-line. Here the fibres are singular exactly at the roots of $t^m + 1$, giving cuspidal cubics. In contrast, by the Jacobian criterion, the surface S is smooth in $\mathbb{P}^2 \times \mathbb{A}^1$ unless char(k) divides m.

It remains to compactify the base curve $\mathbb{A}^1$ to $\mathbb{P}^1$, i.e. we have to add a fibre at ∞ while extending the morphism $S \to \mathbb{A}^1$ to $\mathbb{P}^1$. For this, we do *not* projectivize the above equation in $\mathbb{P}^2 \times \mathbb{P}^1$ (as in Sect. 3.4.4), but rather extend the morphism $S \to \mathbb{A}^1$ to $\mathbb{P}^1$. In order to compactify $\mathbb{A}^1$, we write $t = 1/s$. Then one can use admissible coordinate changes as in (3.8) to ensure that all coefficients in the Weierstrass form are polynomial in s. In detail, let

$$m = 6l - i \quad \text{for } i \in \mathbb{N} \text{ and } 0 \leq i \leq 5.$$

Affinely, the admissible transformation $(x', y') = (s^{2l}x, s^{3l}y)$ leads to the Weierstrass form

$$S': \quad y'^2 = x'^3 + s^i(1 + s^m).$$

Through this transformation, we can then glue together the two quasi-projective polynomial models S, S' with compatible elliptic fibration to $\mathbb{A}^1$ to obtain a projective surface with an elliptic fibration to $\mathbb{P}^1$. This is our first candidate for the elliptic surface associated with E (which could also be derived in terms of certain vector bundles or in weighted projective space).

Note that the fibre of S' at $s = 0$ depends on i: S'_0 is smooth if and only if $i = 0$, and cuspidal otherwise. Another problem is the smoothness of S'. Namely, S' has a surface singularity at $(0, 0, 0)$ if and only if $i > 1$ (again by the Jacobian criterion). In that case, we will have to resolve the singularity first in order to obtain the elliptic surface associated with E. In a general framework, this is established by the Kodaira–Néron model that we will study in Chap. 5 (see also Sect. 4.10 for a brief discussion of surface singularities).

Remark 3.16 A similar problem arises when we compactify the surface from (3.16) in $\mathbb{P}^2 \times \mathbb{P}^1$. Then there are three surface singularities (type A_2, cf. Sect. 4.10) on the triple line at ∞. The resolution gives rise to a fibre of Kodaira type IV^* (to be defined in Chap. 5), with central triple component given by (the strict transform of) the line $X + Y + Z = 0$.

Remark 3.17 Along the same lines, we can construct the elliptic surface arising from the elliptic curve (1.1) over $k(t)$ with Mordell–Weil lattice $A_2^\vee$ to which we alluded in the introduction (see Example 5.9).

Historical Notes

Historically, the general theory of elliptic curves (and hence of algebraic curves) started with elliptic functions as studied by Gauss, Abel, Jacobi, Riemann and Weierstrass (see e.g. [Scho74, Lan87]).

Over the years, elliptic curves have been a driving force for algebraic and arithmetic geometry, see e.g. the landmark papers by Cassels [Cas66] and Tate [Tat74]. In fact, there have been numerous important developments over the last 50 or 60 years in this once classical area, so we shall mention just three of the absolute highlights:

- The Shimura–Taniyama–Weil conjecture on the modularity of elliptic curves over $\mathbb{Q}$ and the proof of Fermat's Last Theorem by Wiles, Taylor and others [Wil95, TW95, BCDT01]. We will touch upon modularity (for certain K3 surfaces) briefly in Sect. 13.1.4.
- The Birch–Swinnerton-Dyer conjecture relating the rank of the Mordell–Weil group of an elliptic curve over a number field to its analytic rank, i.e. the vanishing order of the L-series $L(E, s)$ at $s = 1$ [BS-D65]. This is known for elliptic curves over $\mathbb{Q}$ up to rank 1 by now (by [Kol89, GZ86] using modularity), and thus for a positive proportion of all elliptic curves over $\mathbb{Q}$, cf. [BS15]. We will see an analogue of the Birch–Swinnerton-Dyer conjecture for elliptic surfaces over finite fields, the Artin–Tate conjecture, in Sect. 13.3.
- The manifold applications of elliptic curves in cryptography and coding theory, of growing importance even to everyday life (cf. [Was08]).

As regards the Mordell–Weil theorem, it might be worth emphasizing that at the time of Weil's proof for Jacobians, there was no algebraic theory of abelian varieties developed in a satisfactory way. This was one of the main motivations for Weil to build exactly this theory, cf. Weil [Wei46, Wei48a, Wei48b], not much later taken on in [Lan59], for instance. On this topic, one may want to compare the work of Igusa, Matsusaka, Chow, Grothendieck and others on the Picard variety (resp. Picard scheme), for instance (see [Kle14] and the references therein). The impact of the Mordell–Weil theorem in special (arithmetic) settings is seen prototypically in Mazur's classical work [Maz77] determining all possible torsion subgroups of $E(\mathbb{Q})$ for any elliptic curve E over $\mathbb{Q}$.

As we have seen in Sect. 3.4, degenerations of plane cubic curves are easy to understand, especially for cubics in (generalized) Weierstrass form. On elliptic surfaces, we will have to take a few more possibilities into account; the classification, due to Kodaira (over $\mathbb{C}$) and Tate (over any perfect field) will be reviewed in Sect. 5.4. In higher genus, considering certain degenerate curves (the so called stable curves) played an essential role in the formation of a good compactification of the moduli space (or moduli stack) of curves of given genus, due to Deligne and Mumford [DM69].

Chapter 4
Algebraic Surfaces

In this chapter we will review the basics from surface theory that we will use to introduce and study elliptic surfaces. Standard references on the subject include: Badescu [Bad01], Barth–Hulek–Peters–van de Ven [BHPV04], Beauville [Bea96], Hartshorne [Har77], Kodaira [Kod63, Kod64+], Mumford [Mum66], Shafarevich [Sha94].

Throughout this chapter we will be exclusively concerned with projective surfaces (also called algebraic surfaces), i.e. geometrically irreducible algebraic varieties of dimension two admitting a closed embedding into some projective space $\mathbb{P}^n$. Note that for most practical issues we can always break down the information and calculations into local affine pieces. Until the final section of this chapter, we will also assume the surfaces to be smooth.

Throughout this book we will try to keep the exposition independent of the characteristic of the base field, whenever possible. It is our strong belief that this will give a better view of the geometric world with harmony when everything is explained in a clear-cut way. Nonetheless, any reader who feels more comfortable with the picture in characteristic zero should feel free, at least at first reading, to skip some parts, especially those parts dealing with small characteristics (mostly 2 and 3), supersingular surfaces, or what some might call pathologies. We point out, though, that in later chapters base fields of positive characteristic will play a prominent independent role; in this direction, let us only highlight the framework of sphere packings in Chap. 13, where the new contribution of Mordell–Weil lattices will in fact be achieved by the use of supersingular surfaces over fields of positive characteristic.

4.1 Divisors and Picard Group

In order to gain detailed knowledge about algebraic surfaces, it is essential to understand the curves on a given surface S. Here S will always be assumed to be an irreducible smooth projective surface over an algebraically closed field k. In particular, we will fix an embedding

M. Schütt and T. Shioda, *Mordell–Weil Lattices*, Ergebnisse der Mathematik und ihrer Grenzgebiete. 3. Folge / A Series of Modern Surveys in Mathematics 70,
https://doi.org/10.1007/978-981-32-9301-4_4

$$S \hookrightarrow \mathbb{P}^n \tag{4.1}$$

which also determines the degree of S (defined formally through the leading coefficient of the Hilbert polynomial, or alternatively by Example 4.13).

Many notions and constructions considered in this section carry over more or less directly from the case of curves. We start by mimicking our previous considerations for divisors on curves. To this end, consider the *divisor group* consisting of formal sums of irreducible subvarieties of codimension one, i.e. curves on S:

$$\mathcal{D}(S) = \left\{ \sum_{i=1}^{r} n_i C_i;\ r \in \mathbb{N}, n_i \in \mathbb{Z},\ C_i \subset S \text{ irreducible curve} \right\}.$$

Just like on a curve, a divisor is called *effective* if all $n_i \geq 0$ (assuming the curves C_i to be pairwise distinct). The notion of the *degree* of a *divisor* carries over as well, but we have to take into account that contrary to a point on a curve, an irreducible curve on S will usually not have degree one, but the degree inherited from the embedding (4.1). Thus a divisor $D = \sum_{i=1}^{r} n_i C_i$ has degree

$$\deg(D) = \sum_{i=1}^{r} n_i \deg(C_i).$$

As before, a non zero rational function $g \in k(S)^\times$ gives rise to a divisor $(g) \in \mathcal{D}(S)$ of degree zero (independent of the chosen embedding (4.1)), the difference of the zero divisor and the pole divisor of g. Such divisors are called *principal.*

We continue to call two divisors *linearly equivalent* ($D \sim D'$) if their difference is a principal divisor:

$$D \sim D' \iff D - D' = (g) \text{ for some } g \in k(S).$$

Here is one of the most important examples: for any embedding $S \hookrightarrow \mathbb{P}^n$, any two hyperplane sections (not containing S) are linearly equivalent (take for g the quotient of their linear forms). For this reason, one usually refers to the *hyperplane section* H without mentioning the equivalence class if the ambient projective space is clear (as given by (4.1)).

By definition, the hyperplane sections for all possible embeddings (4.1) give exactly (the classes of) the very ample divisors on S. Conversely, a very ample divisor H gives rise to an embedding (4.1) by way of its linear system $|H|$ after choosing a basis (so that the embedding, usually denoted by $\phi_{|H|}$, is uniquely determined up to linear transformations).

The quotient of the divisor group $\mathcal{D}(S)$ by linear equivalence is called the *Picard group* of S:

$$\text{Pic}(S) = \mathcal{D}(S)/\sim.$$

The equivalence classes in $\mathrm{Pic}(S)$ are called divisor classes, although we will sometimes allow ourselves to just refer to divisors. Amplifying this abuse of terminology, we will often call a divisor effective if it is linearly equivalent to an effective divisor.

Equivalently, we can consider invertible sheaves on S. Here we have a canonical bijection that associates a divisor class, represented by some divisor D, with the invertible sheaf $\mathcal{O}_S(D)$ consisting of functions in $k(S)$ which may have poles along D (with multiplicity), but which are regular outside D (with zeroes of prescribed multiplicities if D is not effective). Then we can define the Picard group as the group of invertible sheaves up to isomorphism with the structure sheaf $\mathcal{O}_S$ as identity, tensor product $\otimes$ as multiplication and the dual invertible sheaf as inverse.

One benefit of this equivalence is that we can extract information about a divisor D from the sheaf cohomology of $\mathcal{O}_S(D)$. We will later see this in more detail when discussing and using Riemann–Roch from (4.9) on. Here we content ourselves with two important pointers. First, D being effective is equivalent to $h^0(\mathcal{O}_S(D)) > 0$. Secondly, there is an additional symmetry coming from *Serre duality* [Har77, III, Corollary 7.7]:

$$h^i(\mathcal{O}_S(D)) = h^{2-i}(\mathcal{O}_S(K_S - D)). \tag{4.2}$$

Here K_S denotes the canonical divisor of S, an invariant of great importance defined as follows. Consider the sheaf Ω_S^1 of differential 1-forms on S. Since S has dimension two, the wedge product of Ω_S^1 defines an invertible sheaf,

$$\omega_S := \wedge^2 \Omega_S^1,$$

which is called the *canonical bundle* of S. The identification of divisors and invertible sheaves associates with ω_S the *canonical divisor class* K_S (which we just call the *canonical divisor*).

Example 4.1 The canonical divisor can be obtained from the difference of the zero divisor and the pole divisor of any rational 2-form on the given surface S. For instance, for a smooth surface $S \subset \mathbb{P}^3$ of degree d, one directly finds that $K_S = (d-4)H, \quad \omega_S = \mathcal{O}_S(d-4)$.

The last formulas are special cases of the adjunction formula [Har77, II, Proposition 8.20] for a smooth codimension one subvariety Y of a smooth projective variety X:

$$\omega_Y = (\omega_X \otimes \mathcal{O}_X(Y))|_Y. \tag{4.3}$$

A special case of it will feature in this chapter as Theorem 4.11, but we wanted to make a point here that they are equally well computable in an explicit and elementary way (just like many invariants of elliptic surfaces will be, cf. e.g. Sect. 5.13). To see this, consider an affine chart U of the smooth projective surface

$$S = Z(f) \subset \mathbb{P}^3, \quad f \text{ irreducible of degree } d,$$

with coordinates x, y, z. Then $dx \wedge dy$ clearly defines a regular 2-form on U, but it has zero divisor $(\partial f/\partial z)$. To see this, note that $(dx \wedge dy)$ is supported on those $P \in U$ where the differentials dx and dy fail to generate the maximal ideal $\mathfrak{m}_P$ in the local ring $\mathcal{O}_{U,P}$. Since dx, dy and dz together trivially generate $\mathfrak{m}_P$ for any $P \in U$, the relation

$$0 = df = \frac{\partial f}{\partial x} dx + \frac{\partial f}{\partial y} dy + \frac{\partial f}{\partial z} dz \tag{4.4}$$

shows that this failure occurs exactly when $\partial f/\partial z(P) = 0$. But at such a point, by smoothness, we have $\partial f/\partial x(P) \neq 0$ or $\partial f/\partial y(P) \neq 0$. In the former case, for instance, we can use (4.4) to switch to the 2-form on the RHS in

$$dx \wedge dy = \left(\frac{\partial f}{\partial z} dy \wedge dz\right) \Big/ \frac{\partial f}{\partial x}.$$

Since dy, dz generate $\mathfrak{m}_P$ by assumption, we infer that the multiplicity of $(dx \wedge dy)$ at P equals the vanishing order of $\partial f/\partial z$ as claimed.

By choosing a suitable U, we may assume that $\partial f/\partial z$ has degree $d - 1$. Gluing in the hyperplane H_∞ at ∞ by one of the standard variable transformations, one sees that $dx \wedge dy$ attains a pole of order 3 along H_∞, i.e.

$$(dx \wedge dy) = (\partial f/\partial z) - 3H_\infty \sim (d-4)H,$$

where the linear equivalence (giving the claimed canonical divisor K_S) follows since we are only dealing with hypersurfaces in $\mathbb{P}^3$ intersected with S. From this we can also compute the *geometric genus* $p_g(S) = h^0(\omega_S)$ of S:

$$p_g(S) = \dim_k V = \binom{d-1}{3} \quad \text{where } V = \{g \in k[x, y, x];\ \deg(g) \leq d-4\}. \tag{4.5}$$

Accordingly, the regular 2-forms on S are affinely represented as follows:

$$H^0(\omega_S) = \left\{ g \frac{dx \wedge dy}{\partial f/\partial z};\ g \in V \right\}. \tag{4.6}$$

4.2 Néron–Severi Group

For the next few paragraphs we shall consider the complex case as a motivation. Following [Ser55/56], GAGA allows us to switch freely between the algebraic and the analytic category (with holomorphic functions, holomorphic line bundles etc.). In particular, we obtain the isomorphism

$$\mathrm{Pic}(S) \cong H^1(S, \mathcal{O}_S^*).$$

The exponential sequence

$$0 \longrightarrow \mathbb{Z} \longrightarrow \mathcal{O}_S \longrightarrow \mathcal{O}_S^* \longrightarrow 0$$

then induces a long cohomology sequence

$$0 \longrightarrow H^1(S,\mathbb{Z}) \longrightarrow H^1(S,\mathcal{O}_S) \longrightarrow \mathrm{Pic}(S) \stackrel{c_1}{\longrightarrow} H^2(S,\mathbb{Z}) \longrightarrow \cdots$$

Here we break down $\mathrm{Pic}(S)$ into the kernel and the image of c_1 which amount to the continuous and the discrete constituents of $\mathrm{Pic}(S)$.

The connection morphism c_1 is called the first Chern map. The image of $\mathrm{Pic}(S)$ under c_1 is called the *Néron–Severi group* and denoted by $\mathrm{NS}(S)$. As a subgroup of $H^2(S,\mathbb{Z})$, $\mathrm{NS}(S)$ is a finitely generated abelian group. Its rank is called the *Picard number* and denoted by $\rho(S)$. In essence, $\mathrm{NS}(S)$ encodes the discrete part of $\mathrm{Pic}(S)$.

On the other hand, the kernel of c_1 is isomorphic to the quotient

$$\mathrm{Pic}^0(S) := H^1(S,\mathcal{O}_S)/H^1(S,\mathbb{Z}).$$

This is a complex torus that has the structure of an abelian variety called the Picard variety of S. Its dimension $q(S) = h^1(S,\mathcal{O}_S)$ is also referred to as the *irregularity* of S.

Example 4.2 A smooth irreducible hypersurface $X \subset \mathbb{P}^n$ of dimension at least two ($n \geq 3$) has irregularity zero. This follows from the standard short exact sequence

$$0 \to \mathcal{O}_{\mathbb{P}^n}(-X) \to \mathcal{O}_{\mathbb{P}^n} \to \mathcal{O}_X \to 0$$

since all cohomology groups $H^i(\mathbb{P}^n, \mathcal{O}_{\mathbb{P}^n}(d))$ vanish in the range $0 < i < n$ by Hartshorne [Har77, III, Theorem 5.1].

A characteristic-free definition of the Néron–Severi group relies on the notion of algebraic equivalence. Roughly speaking, two divisors are called *algebraically equivalent* ($D \approx D'$) if they belong to the same family of divisors on X. Most relevant for our issues is the case of a fibration $f : S \to C$: here two fibres are linearly equivalent if and only if either they are the same or the base curve C is rational (i.e. has genus zero). Independent of the base curve C, any two fibres are algebraically equivalent.

Here's the precise definition. A divisor D on X is said to be *algebraically equivalent* to zero ($D \approx 0$) if the following holds: there is a connected scheme W and an effective divisor $\bar{D}$ on $X \times W$, such that $\bar{D}$ is flat over W and

$$D = \bar{D}_{w_1} - \bar{D}_{w_2}$$

for two fibres $\bar{D}_{w_1}, \bar{D}_{w_2}$ of $\bar{D}$ at some $w_1, w_2 \in W$. (W should be connected, but it may be non-complete; moreover, it can always be assumed to be a curve.) Two divisors D_1, D_2 are said to be algebraically equivalent ($D_1 \approx D_2$) if $D_1 - D_2 \approx 0$. This is known to be an equivalence relation (see [Lan59, III, Sect. 1]).

With algebraic equivalence at hand, one can define the Néron–Severi group of the surface S independently of the characteristic as

$$\mathrm{NS}(S) := \mathcal{D}(S)/\approx .$$

The following result is an important instance of the theorem of the base [Lan83, Chap. 5] valid for smooth projective varieties in general:

Theorem 4.3 *The Néron–Severi group* $\mathrm{NS}(X)$ *of a smooth projective variety* X *is a finitely generated abelian group.*

The two equivalence relations are related as follows:

Lemma 4.4 *Linear equivalence implies algebraic equivalence:*

$$D_1 \sim D_2 \Rightarrow D_1 \approx D_2.$$

Remark 4.5 On a smooth projective variety X over a field of positive characteristic p, the first Chern map can be replaced by the cycle class map

$$\gamma : \mathrm{NS}(X) \to H^2(X) \tag{4.7}$$

to ℓ-adic étale or crystalline cohomology ($\ell \neq p$).

4.3 Intersection Theory

Our aim is to endow the Picard group $\mathrm{Pic}(S)$ of a smooth algebraic surface S with an intersection form. We start by pursuing the intuitive geometric approach on the level of irreducible curves. The standard general reference is Fulton's book on intersection theory [Ful98, Chap. X], going far beyond what will be needed here, but for our purposes Mumford's book on curves on algebraic surfaces [Mum66] would be very suitable or likewise [Bea96, I], [Har77, V], [Sha94].

Given two distinct irreducible curves C, C' on S, we can define the intersection multiplicity of C and C' in a point P geometrically using local equations f resp. g by

$$\mathrm{mult}_P(C, C') := \dim_k \mathcal{O}_{S,P}/(f, g).$$

The intersection multiplicity at P returns a well-defined finite number which is in fact

$$\mathrm{mult}_P(C, C') = \begin{cases} 0 & \text{if } P \notin C \cap C'; \\ 1 & \text{if } C, C' \text{ are smooth and intersect transversally at } P; \\ > 1 & \text{otherwise.} \end{cases}$$

Globally, the intersection number of C and C' is defined as

$$(C.C') = \sum_P \mathrm{mult}_P(C, C').$$

Alternatively, one can consider the length of the intersection scheme $C \cap C'$. As an illustration, we consider the easiest case where $S = \mathbb{P}^2$.

Theorem 4.6 (Bézout's theorem) *Let $C, C' \subset \mathbb{P}^2$ be irreducible curves of degrees d, d' without common components. Then $(C.C') = dd'$.*

This classical result will be of importance soon when we consider rational elliptic surfaces as cubic pencils (see Sect. 7.5).

We aim for a bilinear symmetric pairing on $\mathrm{Pic}(S)$ (descending to $\mathrm{NS}(S)$) which is compatible with the above intersection numbers. To this end, we employ the *Euler characteristic* $\chi(\mathcal{L})$ of an invertible sheaf $\mathcal{L}$ on S:

$$\chi(\mathcal{L}) = h^0(\mathcal{L}) - h^1(\mathcal{L}) + h^2(\mathcal{L}). \tag{4.8}$$

Definition 4.7 Define an intersection form on $\mathrm{Pic}(S)$ by

$$\begin{array}{rcl} (\cdot,\cdot): \ \mathrm{Pic}(S) \times \mathrm{Pic}(S) & \to & \mathbb{Z} \\ (\mathcal{L}, \mathcal{L}') & \mapsto & \chi(\mathcal{O}_S) - \chi(\mathcal{L}^{-1}) - \chi(\mathcal{L}'^{-1}) + \chi(\mathcal{L}^{-1} \otimes \mathcal{L}'^{-1}). \end{array}$$

Theorem 4.8 *The form $(\cdot,\cdot)$ on $\mathrm{Pic}(S)$ is symmetric and bilinear. Moreover, for any two distinct irreducible curves $C, C' \subset S$, one has*

$$(C.C') = (\mathcal{O}_S(C), \mathcal{O}_S(C')).$$

There are two key steps in proving Theorem 4.8 (see [Bea96, Theorem I.4], for instance): first, the natural property that for any smooth irreducible curve $C \subset S$ one has

$$(\mathcal{O}_S(C), \mathcal{L}) = \deg(\mathcal{L}\,|_C) \quad \forall\, \mathcal{L} \in \mathrm{Pic}(S)$$

(where, of course, the degree function is linear); second, the consequence of Bertini's theorem that any divisor $D \in \mathrm{Div}(S)$ can be written as

$$D \sim C - C'$$

for smooth irreducible curves C, C'. Additionally, one may arrange for transverse intersections with each other or with any given finite number of curves, and for C, C'

to avoid any given finite number of points on S. The above properties explain why this result often files under the name of *moving lemma*.

Corollary 4.9 *Consider a fibration $f : S \to C$. Let F denote any fibre of f. Then $F^2 = 0$.*

The symmetric bilinear form $(\cdot, \cdot)$ on $\mathrm{Pic}(S)$ gives rise to the Riemann–Roch formula by inserting $\mathcal{L}^{-1}$, $\mathcal{L} \otimes \omega_S$ and using Serre duality (4.2):

Theorem 4.10 (Riemann–Roch) *Let $\mathcal{L}$ be an invertible sheaf on the surface S. Then*

$$\chi(\mathcal{L}) = \chi(\mathcal{O}_S) + \frac{1}{2}(\mathcal{L}, (\mathcal{L} - \omega_S)). \tag{4.9}$$

In view of (4.8), Riemann–Roch can often be used to infer that $h^0(\mathcal{L}) + h^2(\mathcal{L}) > 0$. Then Serre duality (4.2) implies that $\mathcal{L}$ or $\omega_S \otimes \mathcal{L}^{-1}$ is effective. (This will be particularly useful for K3 surfaces where $\omega_S = \mathcal{O}_S$, see especially Sect. 11.3.6.)

The Riemann–Roch formula (4.9) can be used to derive the adjunction formula for an irreducible curve $C \subset S$. To cover singular curves, we need the arithmetic genus $p_a(C) = 1 - \chi(\mathcal{O}_C)$; for smooth curves, this agrees with the genus $g(C)$.

Theorem 4.11 (Adjunction formula) *Let $C \subset S$ be an irreducible curve. Then*

$$2p_a(C) - 2 = C^2 + (C.K_S).$$

The adjunction formula provides a convenient way to compute the self-intersection number of an irreducible smooth curve $C \subset S$ in terms of the arithmetic genus of C and the canonical divisor of S.

Example 4.12 We will later discuss cubic surfaces in $\mathbb{P}^3$ in detail, see Example 4.18 and Sect. 10.5. One classical fact about them is that each smooth cubic surface S contains exactly 27 lines ℓ; this is known as the Cayley–Salmon Theorem (see [Cay49, Salm49]) and is covered in most textbooks on algebraic geometry (e.g. [Har77, V. 4]). By Example 4.1 we have $K_S = -H$, with H a hyperplane section, and $(\ell.H) = 1$. Thus the adjunction formula gives $(\ell)^2 = -1$.

Later we will use adjunction to compute the self-intersection number of a section P of an elliptic surface $S \to C$ as $(P)^2 = -\chi(\mathcal{O}_S)$ (Corollary 5.45).

Alternatively, one can use the compatibility of the intersection form with linear equivalence to compute self-intersection numbers (and then, in turn, obtain the arithmetic genus from Theorem 4.11).

Example 4.13 If $S \subset \mathbb{P}^3$ of degree d, then replacing a given hyperplane section H by another one, say H' (linearly equivalent), leads to $(H)^2 = (H.H') = d$ on S. By definition, this self-intersection number equals the degree of S with respect to the given embedding (4.1). Using Example 4.1, we also find $(K_S)^2 = d(d-4)^2$.

The intersection pairing introduced above is compatible with algebraic equivalence. Hence we obtain an induced pairing on $\mathrm{NS}(S)$ which we will use throughout this book. Cohomologically, the pairing commutes with the first Chern map in the following sense: $H^2(S, \mathbb{Z})$ and $H^2_{\text{ét}}(S, \mathbb{Q}_\ell)$ come equipped with cup-product $\cup$, a symmetric pairing whose image sits inside $H^4(S, \mathbb{Z}) \cong \mathbb{Z}$ resp. $H^4_{\text{ét}}(S, \mathbb{Q}_\ell) \cong \mathbb{Q}_\ell$. For divisor classes D, D', one obtains

$$(D.D') = c_1(D) \cup c_1(D')$$

by applying the above canonical isomorphism.

We conclude this section by defining an important class of divisors based on the above intersection pairing. A divisor $D \in \mathrm{Div}(S)$ is called *numerically effective* (in short, *nef*) if it meets any effective divisor on S non-negatively:

$$(D.D') \geq 0 \quad \forall\, D' \in \mathrm{Div}(S),\ \ D \geq 0.$$

Equivalently, $(D.C) \geq 0$ for any curve $C \subset S$. This notion plays a central role in the surface classification (Theorem 4.26), which highly depends on the canonical divisor K_S being nef or not.

4.4 Hodge Index Theorem

We shall now look further into the intersection pairing on $\mathrm{NS}(S)$. First we eliminate torsion in $\mathrm{NS}(S)$ (if any) by tensoring with $\mathbb{R}$:

$$\mathrm{N}_{\mathbb{R}}(S) := \mathrm{NS}(S) \otimes \mathbb{R}.$$

This gives an $\mathbb{R}$-vector space of dimension $\rho(S)$, the Picard number of S. The intersection pairing extends linearly from $\mathrm{NS}(S)$ to $\mathrm{N}_{\mathbb{R}}(S)$. As a bilinear form over $\mathbb{R}$, we have to discuss whether it may be degenerate and what its signature is.

Theorem 4.14 (Hodge index theorem) *The induced symmetric bilinear form*

$$(\cdot, \cdot) : \mathrm{N}_{\mathbb{R}}(S) \times \mathrm{N}_{\mathbb{R}}(S) \to \mathbb{R}$$

is non-degenerate of signature $(1, \rho(S) - 1)$.

We are now ready to explore a connection with lattice theory. Namely, consider the quotient

$$\mathrm{Num}(S) := \mathrm{NS}(S)/\text{torsion}. \tag{4.10}$$

By the above considerations, this defines a non-degenerate integral lattice of signature $(1, \rho(S) - 1)$. Often one can also decide whether this lattice is odd or even. For instance, for surfaces $S \subset \mathbb{P}^3$, $\mathrm{Num}(S)$ is odd if the degree d of S is odd by Example 4.13. On the other hand, evenness can often be deduced from the adjunction formula if K_S is two-divisible (or torsion) in $\mathrm{NS}(S)$. As an example, consider smooth surfaces $S \subset \mathbb{P}^3$ of even degree d where $K_S = (d-4)H$ by Example 4.1. Here any irreducible curve has self-intersection

$$C^2 \equiv (d-4)(C.H) \equiv 0 \mod 2$$

by Theorem 4.11. For an arbitrary divisor $D = \sum_{i=1}^r C_i$ the claim thus follows from

$$D^2 = \left(\sum_{i=1}^r C_i\right)^2 \equiv \sum_{i=1}^r C_i^2 \equiv 0 \mod 2.$$

Another way to define $\mathrm{Num}(S)$ is by *numerical equivalence*. Here we identify two divisors D, D' if and only if they have the same intersection numbers with any curve on S:

$$D \equiv D' \iff (D.C) = (D'.C) \quad \forall \text{ curves } C \subset S.$$

We have the following implication (cf. [Mum66, Lecture 12]):

Lemma 4.15 *Algebraic equivalence implies numerical equivalence. That is,*

$$D_1 \approx D_2 \implies D_1 \equiv D_2.$$

For $\mathrm{Num}(S)$, the lattice defined in (4.10), Lemma 4.15 gives way to the following interpretation (which also explains the abbreviation):

$$\mathrm{Num}(S) = \mathcal{D}(S)/\equiv .$$

There are several standard instances where some of these notions of equivalence of divisors coincide. On a smooth hypersurface $S \subset \mathbb{P}^3$, linear equivalence equals algebraic equivalence equals numerical equivalence. Essentially this holds because S is (algebraically) simply connected (see [SGA2, Exp. XII, Corollary 3.5], for instance). Hence $\mathrm{Pic}(S) = \mathrm{NS}(S) = \mathrm{Num}(S)$, and we attain complete control over the curves on S by computing intersection numbers.

Elliptic surfaces with section will turn out to be almost as convenient to work with because algebraic and numerical equivalence coincide. That is, $\mathrm{NS}(S)$ will be torsion-free and can be studied through Gram matrices collecting intersection numbers. Note, however, that $\mathrm{Pic}(S)$ will contain a continuous part depending on (the genus of) the base curve C. Thus $\mathrm{Pic}(S) \neq \mathrm{NS}(S)$ unless C has genus zero.

4.5 Blow-Ups

We are now in the position to start looking into the geometry of algebraic surfaces. As a motivation, we consider one important property of algebraic surfaces that shows the substantial difference to the curve case right away. Namely, recall Zariski's theorem for curves:

Theorem 4.16 (Zariski's theorem) *Any birational map between any two smooth algebraic curves extends to an isomorphism.*

Remark 4.17 In a more general framework, Zariski's theorem can be interpreted as a consequence of the valuative criterion of properness [Har77, II, Theorem 4.7]. For a rational map between two smooth surfaces, this implies that the locus of indeterminacy has dimension zero.

The analogous statement of Theorem 4.16 does not hold for algebraic surfaces. In essence this is due to the existence of blow-ups as we shall now explain.

The *blow-up* of a surface S in a smooth point P consists of a surface $\tilde{S}$ together with a morphism $\pi : \tilde{S} \to S$ such that

1. $\pi^{-1}(P)$ is a smooth rational curve and
2. π gives an isomorphism when restricted to the open subsets $\tilde{S} \setminus \pi^{-1}(P)$ and $S \setminus \{P\}$.

The smooth rational curve $\pi^{-1}(P)$ is called the *exceptional divisor* and is usually denoted by E. Independent of the original surface S, one finds that $E^2 = -1$. Generally, any smooth rational curve E on S with $E^2 = -1$ is called a (-1)-curve. The canonical divisor transforms via pull-back of divisor classes through π^* as

$$K_{\tilde{S}} = \pi^* K_S + E.$$

In fact, the moving lemma tells us generally for any divisors D, D' on S that $\pi^* D$ is orthogonal to E on $\tilde{S}$, and $(\pi^* D.\pi^* D') = (D.D')$. Hence π^* induces the decomposition

$$\operatorname{Pic}(\tilde{S}) \cong \operatorname{Pic}(S) \oplus \mathbb{Z},$$

and we also deduce

$$\rho(\tilde{S}) = \rho(S) + 1. \tag{4.11}$$

There is an important converse to this situation: by Castelnuovo's theorem (cf. [Har77, Chap. V]) any (-1)-curve on a smooth surface can be contracted to a smooth point. This process is often also called blowing down.

Example 4.18 On a smooth cubic surface S, we can contract any line. It is an instructive task to determine how this changes the intersection numbers of the other lines, and in particular, which lines result in (-1)-curves after the contraction.

In fact, for any two skew lines there are exactly five lines meeting both. Projection from S to the two skew lines exhibits a birational morphism

$$S \to \mathbb{P}^1 \times \mathbb{P}^1.$$

Since the inverse birational map fails to be well-defined exactly on the five image points of the connecting lines, this shows that S is isomorphic to $\mathbb{P}^1 \times \mathbb{P}^1$ blown up in five points (which itself is isomorphic to $\mathbb{P}^2$ blown up in six points, compare Example 4.20, but see also Sect. 10.5.3).

Blow-ups play an important role in algebraic geometry for several reasons. Notably the concept of blowing up a point can be generalized to blowing up along subvarieties (which can, of course, also be phrased in terms of ideals, or rather ideal sheaves). Here are two key properties:

(1) Any rational map between projective varieties can be made into a morphism by blowing up along the locus of indeterminacy. This property follows directly from the definition. For instructive examples, consider the projection from a point on a surface, or the standard Cremona transformation on $\mathbb{P}^2$ given by inverting the homogeneous coordinates:

$$\begin{array}{ccc} \mathbb{P}^2 & \dashrightarrow & \mathbb{P}^2 \\ [x_0, x_1, x_2] & \mapsto & \left[\frac{1}{x_0}, \frac{1}{x_2}, \frac{1}{x_2}\right]. \end{array}$$

(2) In characteristic zero, any singularity can be resolved by a succession of blow-ups. This property, in contrast to the previous, is an extremely deep and important result in algebraic geometry. It was originally proved by Hironaka in 1964 and subsequently extended in positive characteristic up to dimension 3 (see [CP08, CP09]), but despite recent progress still lacks a proven counterpart in positive characteristic in dimensions greater than 3.

In the theory of algebraic surfaces, the concept of blowing up plays an even more central role. This is due to the fact that any birational map between two algebraic surfaces can be factorized into a succession of blow-ups and blow-downs (and isomorphisms).

4.6 Minimal Models

In this section, we discuss the notion of minimal models as a means to overcome the failure of Zariski's theorem (Theorem 4.16) for surfaces. An algebraic surface is called *minimal* if it does not contain any (-1)-curves. Here's the main result:

Theorem 4.19 *Any smooth projective surface possesses a minimal model.*

The strategy to produce a minimal model consists in successively contracting (-1)-curves. To prove the theorem, one has to show that the blowing down process terminates eventually, i.e. that after a finite number of contractions, we reach a minimal surface. In particular, the termination does not depend on the (-1)-curves that we decided to contract (but the number of blow-downs may do, see Example 4.20).

The termination can be seen with the Picard group. Namely, it follows from (4.11) that the Picard number drops by one upon blowing down a (-1)-curve. Since the Picard number is always positive and finite, the minimalization process thus has to terminate.

A word of warning is in place: a minimal model of a projective surface need not be unique. As an illustration, we give a simple example:

Example 4.20 It is easy to work out an explicit isomorphism between $\mathbb{P}^2$ blown up in two points and $\mathbb{P}^1 \times \mathbb{P}^1$ blown up in one point. For instance, it suffices to interpret $\mathbb{P}^1 \times \mathbb{P}^1$ as a quadric S in $\mathbb{P}^3$ and then project from a point on S to $\mathbb{P}^2$. Here $\mathbb{P}^2$ and $\mathbb{P}^1 \times \mathbb{P}^1$ are both minimal and not isomorphic. Both claims follow directly from the respective shape of the lattice $\mathrm{Num}(S)$ (which equals $\mathrm{Pic}(S)$ in the present cases):

- $\mathrm{Pic}(\mathbb{P}^2) = \mathbb{Z}h$ with $h^2 = 1$ by Bézout's theorem;
- $\mathrm{Pic}(\mathbb{P}^1 \times \mathbb{P}^1) \cong \mathbb{Z}^2$, the hyperbolic plane with intersection form $\begin{pmatrix} 0 & 1 \\ 1 & 0 \end{pmatrix}$, a basis being given by fibres of the two projections.

In fact, one could continue this line of examples with the so-called Hirzebruch surfaces (cf. [BHPV04, V. 4]) to exhibit all minimal smooth rational surfaces (but this statement relies on a deep result).

In contrast, for surfaces of non-negative Kodaira dimension (see Sect. 4.8), there is an analogue of Zariski's theorem (Theorem 4.16) for surfaces. Namely, for a minimal such surface S, K_S is always nef, which means that $(C.K_S) \geq 0$ for any curve C on S. This can be seen to imply that any birational map (i.e. a priori not only morphisms!) from a minimal surface of non-negative Kodaira dimension to a smooth surface is in fact an isomorphism; in particular, the minimal model is *unique* up to isomorphism.

Remark 4.21 The above concepts can be extended to higher dimensions. Contrary to the surface case, there need not be unique minimal models even in the case of non-negative Kodaira dimension. This can be seen already in dimension three from the presence of flops connecting minimal models (see e.g. [KM98, Kaw08]).

4.7 Invariants of Surfaces

In this section we work out the most important invariants of algebraic surfaces S. The first few concepts hold generally for algebraic varieties, but soon we encounter results exclusive to the theory of algebraic surfaces.

Recall the *Betti numbers*

$$b_i(S) = \dim H^i(S).$$

Here we can work with singular cohomology over $\mathbb{C}$ in the case of characteristic zero or generally with ℓ-adic étale cohomology ($\ell \neq \mathrm{char}(k)$, with the usual comparison theorems asserting that the Betti numbers do not depend on the chosen cohomology theory and even extending to equivariance as Galois modules etc. [Har77, Appendix C].

The *topological Euler number* (or Euler–Poincaré characteristic) is then defined as

$$e(S) := \sum_{i=0}^{4} (-1)^i b_i(S). \tag{4.12}$$

Over $\mathbb{C}$ we can further decompose the complex cohomology according to the sheaf of regular p-forms Ω_S^p on the surface (a vector bundle of rank $\binom{2}{p}$). Let $H^{p,q}(S) = H^q(\Omega_S^p)$ and define the Hodge numbers $h^{p,q}(S) = \dim H^{p,q}(S)$. Then the *Hodge decomposition* reads

$$H^n(S, \mathbb{C}) = \oplus_{p+q=n} H^{p,q}(S).$$

Here we have the following symmetries:

- $h^{p,q} = h^{q,p}$ since by complex conjugation $H^{p,q}(S) = \overline{H^{q,p}(S)}$.
- $h^{p,q} = h^{n-p,n-q}$ by Serre duality (combined with standard properties of wedge products).

The irregularity $q(S)$ occurs as $h^{0,1}$. Another important invariant is the *geometric genus* $p_g(S) = h^{2,0}(S) = h^0(\omega_S)$. By the above symmetries, these invariants determine the Hodge numbers of S up to $h^{1,1}$. We collect them conveniently in the Hodge diamond:

$$\begin{array}{ccccc} & & 1 & & \\ & q & & q & \\ p_g & & h^{1,1} & & p_g \\ & q & & q & \\ & & 1 & & \end{array}$$

Regardless of the characteristic, the *Euler characteristic* of S is defined as $\chi(S) = \chi(\mathcal{O}_S)$. By Serre duality, we find

$$\chi(S) = 1 - q(S) + p_g(S).$$

Finally we recall *Noether's formula*, which describes an important relation between the invariants studied so far:

$$12\chi(S) = e(S) + K_S^2. \tag{4.13}$$

In practice, it is often relatively easy to compute $q(S)$, $p_g(S)$ and K_S^2. Then $e(S)$ is obtained from Noether's formula, and $h^{1,1}(S)$ follows from (4.12).

Example 4.22 For a smooth surface $S \subset \mathbb{P}^3$ of degree d, we found $q(S) = 0$ in Example 4.2. Since $K_S = (d-4)H$, one has $K_S^2 = d(d-4)^2$ and $p_g(S) = \binom{d-1}{3}$ as in (4.5). This gives $e(S)$, $b_2(S)$, and, in characteristic zero, $h^{1,1}(S)$. It is sometimes useful that the space of regular 2-forms was also worked out explicitly in (4.6) (much like what we will do for elliptic surfaces in Sect. 5.13).

Similar formulas hold for complete intersections and in weighted projective space.

4.7.1 Relation with Picard Number

Recall the Picard group with the first Chern map over $\mathbb{C}$ or alternatively the cycle class map for ℓ-adic cohomology. It is originally due to Igusa [Igu60] that this induces the following bound for the Picard number:

$$\rho(S) \leq b_2(S). \tag{4.14}$$

In positive characteristic, this bound can in fact be attained; such surfaces are often called *supersingular*. In characteristic zero, there is a stronger bound for surfaces with $h^{2,0}(S) \neq 0$. It goes back to Lefschetz' theorem, which states that

$$\mathrm{NS}(S) = c_1(\mathrm{Pic}(S)) = H^2(S, \mathbb{Z}) \cap H^{1,1}(S).$$

Here the intersection is taken in $H^2(S, \mathbb{C})$ into which we can embed $H^2(S, \mathbb{Z})$ by tensoring with $\mathbb{C}$. In conclusion, we derive Lefschetz' bound

$$\rho(S) \leq h^{1,1}(S) = b_2(S) - 2h^{2,0}(S). \tag{4.15}$$

Surfaces attaining this bound are often called *singular* (in the sense of exceptional, but not meaning non-smooth, cf. Definition 11.3). This terminology was adapted from the classical theory of elliptic curves where, following German tradition, elliptic curves with complex multiplication were said to have "singular moduli".

Example 4.23 If S is a complex surface with $p_g(S) = 0$ (e.g. if S is rational), then $H^{1,1}(S) = H^2(S, \mathbb{C})$. Hence $\rho(S) = h^{1,1}(S)$ and S is singular. These are the "trivial singular surfaces"; usually one is interested in singular surfaces with $p_g > 0$ such as K3 surfaces and abelian surfaces, see Chap. 11.

4.8 Enriques–Kodaira Classification

The Enriques–Kodaira classification of algebraic surfaces is greatly inspired by the classification of algebraic curves. For curves, the discrete invariant of the genus g suffices to classify them. In detail, as $\deg(K_C) = 2g - 2$, the genus predicts the positivity of K_C. This motivates the following classification:

$$\begin{array}{ll} g = 0 & \text{rational,} \\ g = 1 & \text{elliptic,} \\ g > 1 & \text{general type.} \end{array}$$

It is noteworthy that there is a deep interplay with arithmetic over number fields according to Faltings' Theorem [Fal83]: While curves from the first two classes always attain a dense set of rational points over a finite extension of the base field, curves of genus $g > 1$ will always have a finite set of rational points over any number field. The Bombieri–Lang conjecture states that a similar situation ought to persist in higher dimensions as well, but so far we have only partial evidence of this.

Let us now turn to the surface case. Here positivity of K_S alone is not sufficient for classification purposes. Instead we introduce the *higher plurigenera*:

$$P_n(S) = h^0(nK_S).$$

Note that for $n = 1$ we recover the geometric genus. It is a non-trivial result that these invariants are suitable for classification purposes:

Theorem 4.24 *For algebraic surfaces, the irregularity $q(S)$ and the plurigenera $P_n(S)$ $(n \in \mathbb{N})$ are birational invariants.*

Hence we can study algebraic surfaces up to birational equivalence through irregularity and plurigenera. This should be compared with, for instance, the Picard number $\rho(S)$, which is not a birational invariant because it is affected by blow-ups by (4.11).

Based on the plurigenera, we introduce another birational invariant that roughly measures the positivity of the canonical divisor: the *Kodaira dimension* $\kappa(S)$. There are several equivalent definitions, each with advantages and disadvantages. The most intuitive definition may be the maximum of the dimensions of the images of the pluricanonical maps:

$$\kappa(S) = \max\{\dim(\mathrm{im}\varphi_{|nK_S|});\ n \in \mathbb{N}\} \tag{4.16}$$

where, by convention, $\dim(\emptyset) = -\infty$. At once this shows that there are only four possible values for the Kodaira dimension of an algebraic surface:

$$\kappa(S) = -\infty, 0, 1, 2.$$

Alternatively, one may analyse the asymptotic behaviour of $P_n(S)$ as n tends to ∞. We say that S has Kodaira dimension $-\infty$ if $P_n(S) = 0$ for all $n \in \mathbb{N}$. Otherwise there is a minimum integer $r \in \mathbb{N}_0$ such that $P_n(S)/n^r$ is bounded. This integer r is defined to be the Kodaira dimension of S. More precisely, $\kappa(S) = 0$ if and only if the plurigenera $P_n(S)$ are bounded; since $\mathrm{Sym}^l H^0(nK_S) \hookrightarrow H^0(nlK_S)$, this assumption actually implies the very restrictive condition that $P_n(S) \in \{0, 1\}$ for all $n \in \mathbb{N}$ without being identically zero (which, of course, agrees with the definition from (4.16)). In fact, for minimal surfaces S this precisely corresponds to the case where K_S is torsion. Finally, if $\kappa(S) > 0$, then there are positive numbers $c_1, c_2 \in \mathbb{R}^+$ such that

$$c_1 n^\kappa \leq P_n(S) \leq c_2 n^\kappa \quad \forall n \gg 0$$

(and again, $P_n(S)/n^\kappa$ is bounded).

Example 4.25 Let $S \subset \mathbb{P}^3$ be smooth of degree d. Then $K_S = (d-4)H$. If $d < 4$, then we have seen that S is rational, and $\kappa(S) = -\infty$. If $d = 4$, then $K_S = 0$, and S is a so-called K3 surface (see Chap. 11); of course, this has $\kappa(S) = 0$. If $d > 4$, then one obtains $\kappa(S) = 2$ since H is very ample.

In analogy with algebraic curves, surfaces with $\kappa(S) = 2$ are said to be of *general type*. The following classification is the cornerstone of algebraic surface theory; it is due to Enriques originally [Enr14, Enr49] and was completed (and extended to analytic surfaces) by Kodaira over $\mathbb{C}$ (see [Kod63, Kod64+]); the positive characteristic counterpart was established by Bombieri–Mumford [BM76/77, Mum69] (with the only essential addition that in characteristics 2, 3, the notions elliptic and bielliptic ought to encompass quasi-elliptic and quasi-bielliptic as well, compare Definition 7.21 and in particular Convention 11.22). For a quick introduction explaining all the surface types occurring, see e.g. Barth et al. [BHPV04, VI].

Theorem 4.26 (Enriques–Kodaira classification) *Given an algebraic surface S, then a minimal model S′ of S falls into the following cases:*

$\kappa(S)$	*Minimal model S′*	$K_{S'}$
$-\infty$	*Rational, ruled*	*Not nef*
0	*abelian, K3, Enriques, bielliptic*	*Torsion:* $12K_{S'} = 0$
1	*(Honestly) elliptic*	*Nef,* $K_{S'}^2 = 0$
2	*General type*	*Nef,* $K_{S'}^2 > 0$

In the light of the Enriques–Kodaira classification, it is the intermediate case of Kodaira dimension $\kappa = 1$ that attracts our particular attention: elliptic surfaces which we shall introduce and discuss in detail in the next chapter, but with the additional convention that our fibrations ought to admit a section (see Remark 5.3). The terminology *honestly elliptic* pays tribute to the fact that surfaces with $\kappa < 1$ may also admit elliptic fibrations (e.g. cubic pencils, product abelian surfaces $E \times E'$ or many K3 surfaces). However, those elliptic fibrations are somewhat limited; for instance, the base curve is always rational or elliptic.

4.9 Castelnuovo's Criterion

We seek a cohomological criterion when a surface S is *rational*, i.e. birational to $\mathbb{P}^2$. The most obvious idea would be to require

$$q(S) = p_g(S) = 0,$$

but this fails since it would include Enriques surfaces and Godeaux surfaces, both of which are not rational (see [Bad01, BHPV04]). Over $\mathbb{C}$, Castelnuovo derived the following criterion replacing the geometric genus p_g (i.e. the first plurigenus) by the second plurigenus:

Theorem 4.27 (Castelnuovo's criterion) *A smooth projective surface S is rational if and only if*

$$q(S) = P_2(S) = 0.$$

A key point in the original proof was the fact that a minimal surface S with K_S *not nef* is either $\mathbb{P}^2$ or a geometrically ruled surface. With the conditions of the theorem above, one can indeed infer, independently of the characteristic, that S is either $\mathbb{P}^2$ or a Hirzebruch surface, i.e. geometrically ruled over $\mathbb{P}^1$. The first characteristic-free proof of Castelnuovo's criterion was due to [Zar58] (see also [Bad01, Sect. 13]) for a proof and for references to simplified proofs due to Artin and Kurke).

We close this section with a word towards unirationality. Here we are concerned with surfaces which are dominated by $\mathbb{P}^2$:

$$\mathbb{P}^2 \dashrightarrow S. \tag{4.17}$$

Over $\mathbb{C}$, one can use Castelnuovo's criterion to deduce that unirationality implies rationality. To this end, one uses that S is automatically simply connected. Moreover, after resolving the indeterminacies of (4.17) by a rational surface Y which is a blow-up of $\mathbb{P}^2$, pluricanonical forms pull back from S to Y for any $n \in \mathbb{N}$ (and they can only pull back identically to zero if the morphism $Y \to S$ is inseparable). Since $P_n(Y) = P_n(\mathbb{P}^2) = 0$ for all $n \in \mathbb{N}$ by Theorem 4.24, we obtain the claim by applying Castelnuovo's criterion to the case $n = 2$.

The above reasoning stays valid in positive characteristic as long as the dominant rational map (4.17) is separable. If it becomes inseparable, though, then unirationality need not imply rationality anymore. One intrinsic way to produce counterexamples is to apply purely inseparable base change to rational elliptic surfaces, see Sect. 12.4.3 for two extended examples.

On the other hand, among unirational surfaces (mostly) of general type, a familiar example is given by Fermat surfaces. Precisely, the Fermat surface of degree m in characteristic p is unirational (and supersingular) if and only if $p^\nu \equiv -1 \mod m$ for some $\nu \in \mathbb{N}$ by [KatS79]. This will appear in a disguised form in the last chapter of this book, see Sect. 13.4.

4.10 Non-smooth Surfaces

In algebraic geometry, it is impossible to advance without eventually considering non-smooth varieties. In characteristic zero, singularities can be resolved by successive blow-ups—an important property that is still unknown in positive characteristic in higher dimensions, but that holds for surfaces (and threefolds). In this section, we shall only study a particular class of singularities (which are nice enough so that also their higher-dimensional analogues admit resolutions unconditionally).

We want to allow the surface S to admit isolated rational double points as singularities (also called ADE-singularities). This is a generalization of simple curve singularities (see for instance [DuV33]). Over $\mathbb{C}$, we can give them in local coordinates uniquely up to analytic isomorphism as follows:

$$\begin{array}{ll} A_n \; (n \geq 1) & y^2 + x^2 + t^{n+1} = 0, \\ D_n \; (n \geq 4) & y^2 + t(x^2 + t^{n-2}) = 0, \\ E_6 & y^2 + x^3 + t^4 = 0, \\ E_7 & y^2 + x(x^2 + t^3) = 0, \\ E_8 & y^2 + x^3 + t^5 = 0. \end{array}$$

In small positive characteristic, slightly different equations may have to be allowed which are not necessarily unique anymore (cf. [Art77]). For an equivalent definition which is characteristic free, consider a minimal resolution

$$\pi : \tilde{S} \to S.$$

Note that due to the singularities, the components of the exceptional divisor will not have self-intersection -1 anymore, but smaller self-intersection numbers. For a rational double point, we require that the exceptional locus above that point consists only of (-2)-curves, i.e. smooth rational curves C with self-intersection $C^2 = -2$. Moreover, the graph of these components forms a Dynkin diagram of the given type (as sketched in Fig. 2.2). In other words, the exceptional curves span a root lattice of ADE type inside $\mathrm{NS}(\tilde{S})$; note that the resulting root lattice is negative-definite, i.e. it equals one out of

$$A_n^- \; (n \geq 1), \quad D_l^- \; (l \geq 4), \quad E_6^-, \; E_7^-, \; E_8^-$$

in the notation from Chap. 2.

Rational double points will play an important role for elliptic surfaces in the context of the Kodaira–Néron model and Weierstrass forms, and we will discuss them in the framework of Tate's algorithm in Sect. 5.8. We conclude this section by giving two important properties of rational double points:

(1) Let ω_S° denote the dualizing sheaf of the surface S with isolated rational double points. Then $\omega_{\tilde{S}} = \pi^* \omega_S^\circ$ (unlike for blow-ups in smooth points). (Depending on one's point of view, this may be regarded either as the definition of rational double points, or as a local computation.)

(2) Let S_0 be a smooth surface and degenerate it to a surface S with isolated rational double points (for instance, think of S_0 as a smooth hypersurface in $\mathbb{P}^3$ and vary the coefficients of the equation). Then the minimal resolution $\tilde{S}$ of S lies in the same deformation class as S_0 (see [Tju70] and also [Art74b] for a discussion in the context of simultaneous resolution of singularities following Brieskorn). In particular, the minimal resolution $\tilde{S}$ shares several invariants with the original surface S_0 (Betti numbers, Euler number and characteristic, Hodge numbers (over $\mathbb{C}$)).

This explains why it is usually considered a natural concession to allow surfaces with isolated rational double points. This point of view will be employed in the theory of elliptic surfaces when thinking of them in terms of Weierstrass models, see Sect. 5.7.

Historical Notes

Once curves (or compact Riemann surfaces) were fairly well understood in complex geometry, the central objects of study for the development of algebraic geometry came to be algebraic surfaces. Starting from the work of Max Noether, with important contributions from several mathematicians such as Picard and Poincaré, this was taken to another level by the Italian school of geometers led by Castelnuovo, Enriques, Severi and Segre, a definite cornerstone being Enriques classical monograph [Enr49]. Later this led many mathematicians to jointly build the rigorous foundations of what the Italian school had invented—what was eventually to become modern algebraic geometry.

Much of the algebraic theory was shaped by Zariski and Weil, as documented in a series of papers and monographs, e.g. Weil [Wei46, Wei48a, Wei48b, Zar71] (the first edition of which appeared in 1935). A brief account of their ideas, including applications to number theory, can be found in the ICM notes [Wei50, Wei54, Zar50].

Of similar importance was the input algebraic geometry received from neighbouring areas. We highlight but a few, such as harmonic integrals leading to Hodge theory [Hod52] and Serre's theory of coherent algebraic sheaves [Ser55] and GAGA, the comparison theorems between algebraic and analytic geometry [Ser55/56]. Input from algebraic topology can be seen in Hirzebruch's monograph [Hir78] and in his proof of the Riemann–Roch theorem for smooth projective varieties.

All of this predates the revolution of algebraic geometry that goes back to Grothendieck and his monumental work. This opened up a whole world of new techniques which led to resolution of singularities [Hir64], the study of moduli spaces (as schemes or stacks, see [DM69, MFK94]), the theory of étale cohomology (due to Artin and Grothendieck, cf. [Mil80]) and eventually the proof of the Weil conjectures by [Del74] (which will play a modest role in later chapters).

For the theory of algebraic surfaces, Zariski's work was most instrumental, both through his fundamental research (also leading to one of the classical textbooks [Zar71]) and through his constitution of a whole group of young algebraic geometers working on surfaces at Harvard those days, including names such as Artin, Hironaka, and Mumford.

We would like to point out two other big developments in the field. One consisted in Kodaira's work on the structure and classification of compact complex surfaces, thus extending the area of research widely in a series of classical papers such as Kodaira [Kod63, Kod64+]. Incidentally, these are also of great relevance to the subsequent chapters of this book since they advanced the study and knowledge of elliptic surfaces, if they did not invent them, to a completely new level. It may be fair to say that Kodaira's insights reflect to a large extent what led to the modern standard reference on complex compact surfaces [BHPV04], encompassing both algebraic and analytic theory.

Another stream of research, almost completely independent from the others for political reasons, was initiated by Shafarevich's seminar in Moscow which set out to build the whole theory of algebraic surfaces anew from scratch (see [Sha+65]). This culminated in the proof of the global

Torelli theorem for K3 surfaces mostly due to [PSS71] which we will discuss in more detail in Sect. 11.3.3. Shafarevich and his seminar were also the first to consider the Kodaira dimension of an algebraic surface systematically [Sha+65], although it was named after Kodaira for varieties of arbitrary dimension only later by [Iit71].

There are many other important developments in the field, notably the extension of the theory to fields of positive characteristic which was already initiated by Zariski, a notable peak being the thesis of Artin, and then culminated in Bombieri–Mumford's classification results [Mum69, BM76/77].

The story of isolated rational singularities on algebraic surfaces largely builds on the classification of simple curve singularities; a systematic treatment can be found in a series of papers by [Art62, Art66, Art77] and Brieskorn [Bri68]. Isolated rational double points relate directly to the topics from Chap. 2. Namely, the configuration of exceptional divisors in a minimal desingularization corresponds to a (negative-definite) Dynkin diagram. A general discussion of resolution of singularities on algebraic surfaces can be found in [Lip75]. Remarkably, for elliptic surfaces, we will not need this general machinery at all, since the construction of minimal Weierstrass models will reduce all considerations to (the resolution of) isolated rational double points (see Sect. 5.10).

We shall not go into any more details here, but at the same time not miss the opportunity to reiterate that one of the key lines of research on algebraic surfaces, starting at least from [Enr49] and then systematically from [Kod63], was and continues to be elliptic surfaces (as we shall start exploring in the next chapter).

Chapter 5
Elliptic Surfaces

In this chapter we are going to define the notion of an *elliptic surface*, assuming the basic terminology in algebraic geometry reviewed in Chap. 4. Then we will describe a few examples to see what elliptic surfaces are really like, in this and the next sections. The key reference for much of the theory still seems to be the foundational paper by Kodaira [Kod63], written in the language of complex manifolds. Many important results, especially for the algebraic setting, are spread over the literature (see for instance [CD89, Mir89, Sil94, SSh10]).

For geometric considerations, we take the ground field k to be an algebraically closed field of arbitrary characteristic. From the arithmetic viewpoint, however, it will be important to consider also varieties over non-closed fields k. Then the geometric conditions have to be fulfilled for the base extension to the algebraic closure $\bar{k}$. Scheme theory (as in [Har77, Liu06]) only enters for the definition and geometric interpretation of the generic fibre, although the precise arguments can also be carried out without assuming this abstract point of view.

5.1 Definition of an Elliptic Surface

We still need some warming-up towards the definition of an elliptic surface. Recall that for a variety X over k and any field extension $K \supset k$, $X(K)$ denotes the set of K-rational points of X.

Let C be a smooth projective curve over an algebraically closed field k. Suppose that S is a smooth projective algebraic surface and

$$f : S \to C$$

is a surjective morphism with connected fibres, both defined over k. We call $f : S \to C$ a *fibration* on the surface S with the base curve C.

M. Schütt and T. Shioda, *Mordell–Weil Lattices*, Ergebnisse der Mathematik und ihrer Grenzgebiete. 3. Folge / A Series of Modern Surveys in Mathematics 70,
https://doi.org/10.1007/978-981-32-9301-4_5

For each point $v \in C(k)$, let $F_v = f^{-1}(v)$ be the *fibre* over v. The subset

$$|F_v| = \{x \in S(k) \,|\, f(x) = v\} \subset S(k)$$

is called the support of F_v. The fibre F_v itself is the divisor of S (with multiplicities) defined by the local equation $u \circ f = 0$ where u is a local parameter at $v \in C$ (i.e. u is a generator of the maximal ideal $m_{C,v}$ of the local ring $\mathcal{O}_{C,v}$ of C at v).

We call $f : S \to C$ a *genus g fibration* if all fibres F_v except for finitely many ones are smooth curves of genus g. In this case, $\{F_v | v \in C\}$ forms a family of curves of genus g, parametrized by C, with finitely many exceptions. The *generic fibre* of f is the fibre over the generic point of the base curve C; thus the generic fibre is a curve of genus g defined over the function field $k(C)$ of C/k.

We are mainly interested in the case $g = 1$. Recall that an elliptic curve is a curve of genus 1 given with a rational point (see Chap. 3). Thus when we speak of a family of elliptic curves $\{F_v | v \in C\}$, it will be natural to ask that the chosen points $O_v \in F_v$ vary rationally with $v \in C$. This leads to the following definition.

Definition 5.1 Given a fibration $f : S \to C$ as above, a *(global) section* of f is a morphism $\sigma : C \to S$ such that $f \circ \sigma : C \to C$ is the identity map of C.

A section thus marks a point $\sigma(v)$ on each fibre F_v $(v \in C)$. Now we make a provisional definition: an *elliptic surface* is a genus 1 fibration $f : S \to C$ with a section $\sigma_0 : C \to S$. All but finitely many fibres F_v are elliptic curves with the origin $\sigma_0(v)$. We call σ_0 the *zero section* (cf. Proposition 5.4). The generic fibre is an elliptic curve over the function field $k(C)$ with the origin O corresponding to σ_0.

The above definition is provisional because it leaves some unnecessary freedom for the shape of fibres. This is due to existence of blow-ups for surfaces. As explained in Sect. 4.5, a blow-up is a morphism $\tilde{S} \to S$ that replaces a point P on a smooth surface S by a $\mathbb{P}^1$. In the present situation, assume that S admits a fibration $f : S \to C$ to some curve C. Then naturally this induces a fibration $\tilde{S} \to C$. All fibres stay isomorphic except at $v = f(P)$ as $\tilde{F}_v$ contains the exceptional divisor. In order to avoid this situation, we invoke a minimality condition in the definition of elliptic surface that resembles the general surface situation.

Definition 5.2 An elliptic surface is a genus 1 fibration

$$f : S \to C$$

from a smooth projective surface S to a smooth projective curve C with a section $\sigma_0 : C \to S$ which is *relatively minimal* in the sense that no fibre contains a (-1)-curve.

We note that relative minimality in general does not imply minimality. For an instructive counterexample, see the discussion about the base points of a cubic pencil in Example 5.7.

In summary, an elliptic surface is defined by a smooth projective surface S having a surjective morphism $f : S \to C$ and a section $\sigma_0 : C \to S$ such that

(1) all but finitely many fibres of f are elliptic curves and
(2) no fibre contains an exceptional curve of the first kind.

We often speak of an elliptic surface S over C, or more simply an elliptic surface S, by suppressing the morphism f or the zero section σ_0 or even the base curve C, when the choice of f (or σ_0 or C) is clear from the context. This simplification is justified first of all since we can obtain essential information about the curve C a priori from the surface. Moreover, many surfaces only admit a single elliptic fibration (if any); more precisely, only surfaces of Kodaira dimension zero may admit distinct elliptic fibrations (largely due to the assumption that the fibration be relatively minimal). Indeed, this is an important fact in the study of K3 surfaces (and, for genus one fibrations, of Enriques surfaces) that we will come back to in Chap. 11.

Any fibre which is not an elliptic curve is called a *singular fibre*. We study the singular fibres in detail in Sect. 5.4.

Remark 5.3 In the literature, the word "elliptic surface" or "genus one fibration" is sometimes used in a wider sense. We list some instances:

(a) the provisional definition above, where minimality is not required;
(b) a genus 1 fibration, where the existence of a section is not assumed (for instance on Enriques surfaces);
(c) in characteristic $p = 2, 3$, we may have a smooth surface even if the general fibre is not smooth, but a cuspidal curve of arithmetic genus one. Often these surfaces are called quasi-elliptic (see Definition 7.21);
(d) in the complex analytic category, one considers also genus 1 fibration on smooth compact complex surfaces which may not even be algebraic (the best reference for this is [Kod63]).

Our definition that requires all these decisive properties (existence of a section, relative minimality and projectivity) will suit our purpose best, to deal with the Mordell–Weil group of E/K, an elliptic curve over a function field. That said, we will soon impose one further convention in order to exclude the trivial case of products (Convention 5.10).

5.2 Sections Versus Rational Points

Before looking into explicit examples, we investigate the relationship of an elliptic surface $f : S \to C$ and its generic fibre E/K (an elliptic curve over the function field $K = k(C)$). The central connection is covered in the following proposition (which holds true in much greater generality, see Liu ([Liu06, Sect. 8.3]) and compare also Theorem 6.5 and its proof in Sect. 6.3).

Proposition 5.4 *The global sections of $f : S \longrightarrow C$ are in a natural one-to-one correspondence with the K-rational points of E.*

Proof Any section $\sigma : C \to S$ of f defines a curve $D = \sigma(C) \cong C$ inside S which meets every fibre transversally in a single point. The curve D extends naturally to the underlying scheme $\mathcal{S}$ by taking the Zariski closure; thus it meets the generic fibre in a single K-rational point. We sketch this geometric picture in the following figure (essentially borrowed from the introduction, but enhanced with generic point and generic fibre on the left-hand side).

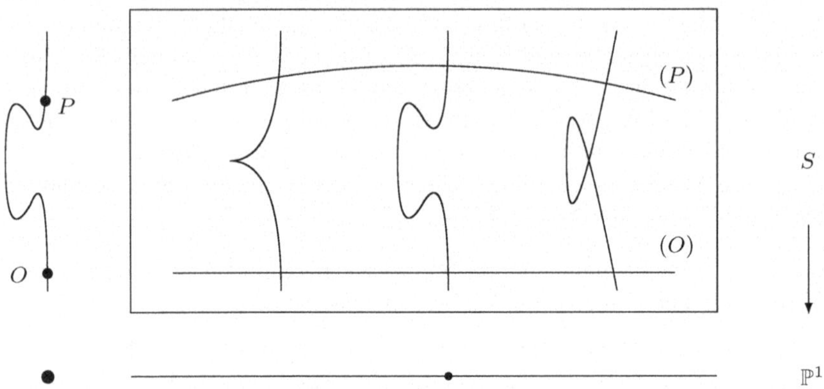

Conversely, given any $P \in E(K)$, the closure of P in $\mathcal{S}$, say $\mathcal{D}$, is a curve in $\mathcal{S}$. Let D denote the restriction to the elliptic surface S. By construction, D is endowed with a finite, birational morphism

$$f_1 = f|_D : D \to C.$$

Since C is smooth (and hence normal), f_1 must be an isomorphism of D onto C. (Indeed, for any point $w \in D$, let $B = \mathcal{O}_{D,w}$ be the local ring of D at w, and similarly, set $A = \mathcal{O}_{C,v}$ with $v = f_1(w) \in C$. Then via f_1^*, A becomes a subring of B, and B is integral over A. Since both A and B have K as their fraction field and A is integrally closed, we conclude that $B = A$. This shows that $f_1 : D \to C$ is locally isomorphic at w, and hence isomorphic everywhere.)

Thus the inverse of f_1 gives a unique section $\sigma : C \to S$ with $\mathrm{Im}(\sigma) = D$. The correspondence $\sigma \leftrightarrow P$ is clearly bijective. □

In view of this fact, we can identify the sections of an elliptic surface S with the K-rational points of its generic fibre E/K. In the following, we identify $\sigma = P$ in the above situation, and use the same notation $E(K)$ to denote also the group of sections of $f : S \to C$. Likewise we will use the phrases *section* and *rational point* interchangeably unless a precise distinction seems strictly necessary.

Notation 5.5 For $P \in E(K)$, we denote by (P) the curve on S which is the image of the section $P : C \to S$ (i.e. $(P) = D$ in the notation of the above proof).

By abuse of terminology, such a curve $(P) \subset S$ will also be called a section. In particular, we have $\sigma_0 = O$, and (O) denotes the image of the zero section (as a curve in S). The reason for this twist in notation lies in the different group structures that we are presently dealing with: we can add rational points P, Q in $E(K)$ as well as the corresponding curves (P), (Q) in $\mathrm{NS}(S)$, but

$$(P+Q) \neq (P) + (Q)$$

because when intersecting with a fibre F, the curve $(P+Q)$ obtained from the section $P+Q$ on the LHS returns 1, while the two curves $(P), (Q)$ on the RHS return 2 in total. Indeed, this subtlety will play a central role in the definition of Mordell–Weil lattices, compare Theorem 6.5.

5.3 Examples

Let us give a few concrete examples of elliptic surfaces. Some of them have appeared before, mostly as families of elliptic curves, but here we will give them a concise treatment as elliptic surfaces.

Example 5.6 Suppose E_0/k is an elliptic curve over the field of constants k. Let $S = E_0 \times C$ be the product surface of E_0 with C, and let $f = p_2$ be the projection from S to the second factor C. Then S is an elliptic surface over C such that all the fibres are isomorphic to E_0.

We point out that this is *not* a typical example of elliptic surfaces that we will treat in this book. Rather it will soon be excluded when we impose Convention 5.10. In this example, the generic fibre E is the elliptic curve E_0 viewed as a curve over $K = k(C)$, i.e. $E = E_0 \otimes_k K$. Note that the group $E(K)$ of K-rational points contains $E_0(k)$ as a subgroup, since $k \subset K$. In particular, the cardinality of $E(K)$ is at least as large as that of $E_0(k)$. For example, if $k = \mathbb{C}$ is the field of complex numbers, then $E(K)$ has the cardinality of a continuum, far too large to be finitely generated. Thus any Mordell–Weil theorem is bound to break down for such surfaces.

The interested reader is advised to compare Example 5.6 to Example 3.15 (see also Example 5.8). There we studied isotrivial elliptic surfaces, i.e. all smooth fibres are isomorphic over $\bar{k}$. However, due to the presence of singular fibres, the fibration was not trivial.

Example 5.7 (*Hesse pencil revisited*) Consider the algebraic surface $S \subset \mathbb{P}^2 \times \mathbb{P}^1$ from Sect. 3.4.4 (see also Example 3.13), defined by the equation

$$t_0(X^3 + Y^3 + Z^3) - t_1 XYZ = 0,$$

where $(X : Y : Z)$ and $(t_0 : t_1)$ denote the homogeneous coordinates in $\mathbb{P}^2$ respectively $\mathbb{P}^1$. Take $C = \mathbb{P}^1$ so that $K = k(t)$ is the rational function field generated by $t = t_1/t_0$ over k. Define the morphism

$$f : S \to C$$

as the composite of the inclusion map of S into $\mathbb{P}^2 \times \mathbb{P}^1$ followed by the projection to the second factor. Then (S, f) is an elliptic surface over $C = \mathbb{P}^1$, defined over k, and the generic fibre E is the plane cubic over K defined by the very same equation as S,

$$X^3 + Y^3 + Z^3 - tXYZ = 0,$$

but now interpreted in $\mathbb{P}^2$ over K.

For clarity, we shall now take some time to check all the conditions on an elliptic surface for the above example.

Assume that $\text{char}(k) \neq 3$. We claim that S is a smooth surface and that f satisfies the two conditions (1) and (2). To see that S is smooth, take any point on it. Assume, for instance, that it has $Z \neq 0$ and $t_0 \neq 0$. Then S is defined locally by the affine equation $F = x^3 + y^3 + 1 - txy$ in the (x, y, t)-space. At a singular point, we should have $\partial F/\partial x = 3x^2 - ty = 0$, $\partial F/\partial y = 3y^2 - tx = 0$, $\partial F/\partial t = -xy = 0$, which would imply $x = y = 0$, a contradiction. Similarly, we see that S is smooth everywhere.

For any point $t = \tau \in \mathbb{P}^1$, the fibre $f^{-1}(\tau)$ over τ is the plane cubic defined by the above Hessian equation with $t = \tau$. As we saw in Example 3.13, it is not smooth if and only if $\tau = \infty$ or $\tau^3 = 27$. Hence f satisfies the condition (1). Next, if $\tau = \infty$ (i.e. $t_0 = 0, t_1 \neq 0$), then the singular fibre over it is defined by $XYZ = 0$, which decomposes into three lines $X = 0$, $Y = 0$ or $Z = 0$. Similarly, if τ is a cube root of 27, then the singular fibre $f^{-1}(\tau)$ is a reducible curve which is a union of three lines. Thus any irreducible component Θ of such a reducible fibre is a smooth rational curve. However, it is not an exceptional curve of the first kind since its self-intersection number is $\Theta^2 = -2$. Thus f also satisfies the condition (2).

The intersection number $\Theta^2 = -2$ can be computed in several ways. For instance, it follows in general from the canonical bundle formula (Theorem 5.44) combined with adjunction (Theorem 4.11). Presently, of course, the canonical divisor can also be computed directly by the adjunction formula (4.3) as $K_S = -F$ for any fibre F. Alternatively, write $F_\tau = \Theta + \Theta' + \Theta''$ and use that F_τ is algebraically equivalent to any other fibre F_υ. Hence we obtain

$$0 = (\Theta.F_\upsilon) = (\Theta.F_\tau) = \Theta^2 + \underbrace{(\Theta.\Theta')}_{=1} + \underbrace{(\Theta.\Theta'')}_{=1}.$$

In the same way, the self-intersection number of any other irreducible component of a reducible fibre is -2.

Note that the projection $S \to \mathbb{P}^2$ enables us to view S as the blow-up of $\mathbb{P}^2$ in nine points (six of which may require the extension $k(\sqrt{-3})$ in case k is not algebraically closed). These are the base points of the cubic pencil and serve as sections for the elliptic fibration. Thus the surface S does contain nine exceptional curves (self-intersection -1), but none is contained in a fibre. That is, S is relatively minimal with respect to the elliptic fibration $S \to \mathbb{P}^1$, but not absolutely minimal.

In comparison, consider characteristic 3. Here the cubic at $t = 0$ degenerates into a triple line, and the surface S attains three singularities in that specific fibre at $(0 : 1 : -1), (1 : 0 : -1), (1 : -1 : 0)$. It is easily verified that these singularities have type A_2. The minimal desingularization gives the desired elliptic surface, i.e. it is smooth without exceptional curves contained in the fibres.

In the next example, we start from an elliptic curve E/K where K is a function field over k, and construct an elliptic surface having E/K as its generic fibre. As a matter of fact, this is a special case of the general construction of the Kodaira–Néron model indicated in the proof of Theorem 5.19.

Example 5.8 Consider the elliptic curve E defined by the Weierstrass form

$$y^2 = x^3 + t^5 + 1$$

over the rational function field $K = k(t)$ with $\operatorname{char}(k) \neq 2, 3, 5$. (This is the case $m = 5$ in Example 3.15.) Let S' denote the affine surface defined by the same equation in the affine space $\mathbb{A}^3$ with coordinates x, y, t. Then there is an elliptic surface S over $C = \mathbb{P}^1$ whose generic fibre is E and which contains S' as an open subset.

The surface S is constructed following Example 3.15. First notice that S' is smooth (by the Jacobi criterion since the above defining equation and its partial derivatives have no common solutions). By dividing the given equation by t^6 and rewriting it in terms of

$$\bar{t} = \frac{1}{t}, \quad \bar{x} = \frac{x}{t^2}, \quad \bar{y} = \frac{y}{t^3},$$

we obtain

$$\bar{y}^2 = \bar{x}^3 + \bar{t} + \bar{t}^6.$$

Let S'' be the affine surface defined by this equation in the $(\bar{x}, \bar{y}, \bar{t})$-space $\mathbb{A}^3$; this is again a smooth surface. By patching together S' and S'' (i.e. by identifying $(x, y, t) \in S'$ with $(\bar{x}, \bar{y}, 1/\bar{t}) \in S''$ in the open subset $\{t\bar{t} \neq 0\}$ of the union $S' \cup S''$), we obtain a smooth (open) surface S^*.

What is instrumental to our purposes, the projection $(x, y, t) \mapsto t$ defines a morphism from S' to $\mathbb{A}^1$, which extends naturally to a morphism $f' : S^* \to \mathbb{P}^1$.

Finally, the desired elliptic surface S is obtained from S^* by adding the zero section. More precisely, replace the affine surface S' by its partial projectivization

$$S_1 \subset \mathbb{P}^2 \times \mathbb{A}^1$$

which is defined by

$$Y^2 Z = X^3 + (t^5 + 1) Z^3,$$

and similarly replace S'' by

$$S_2: \quad \bar{Y}^2\bar{Z} = \bar{X}^3 + (\bar{t} + \bar{t}^6)\bar{Z}^3.$$

Both S_1 and S_2 are smooth, and S' is identified with the open subset $Z \neq 0$ of S_1, and similarly for $S'' \subset S_2$. Then the surface S is obtained by patching together S_1 with S_2 (by an obvious identification) which contains S^* as an open subset, and the morphism $f: S \to \mathbb{P}^1$ by naturally extending $f': S^* \to \mathbb{P}^1$. Note that the complement of S^* in S is given by $Z = 0$ or $\bar{Z} = 0$ which defines the zero section of f.

There are six singular fibres on this elliptic surface, namely over $t^5 = -1$ and $t = \infty$. Each of them is a cuspidal cubic $Y^2Z = X^3$. Thus all fibres are irreducible. We will see later (see Chaps. 7 and 8) that the Mordell–Weil group $E(K)$ has rank 8 and it admits a natural structure of the root lattice E_8!

In comparison, in characteristics 2 and 3, the general fibre is not smooth, but a cuspidal rational curve. The underlying surface, however, does only have a single isolated singularity (in the fibre at $t = 0$). We obtain an example of a *quasi-elliptic fibration* (which will reappear in Example 7.19).

Finally, let us consider base fields k of characteristic 5. In this case, the general fibre is still smooth, but the surface attains an isolated singularity in the fibre at $t = -1$ of type E_8 (as in Sect. 4.10). The minimal desingularization results in a configuration of (-2)-curves that will show up in the next section as Kodaira type II^*.

Example 5.9 (*Hexagonal example*) Along the same lines, we can treat the elliptic curve (1.1) from the introduction related to the hexagonal lattice. Compactifying as above, the only difference consists in resolving an E_6 singularity in the fibre at ∞.

Alternatively, we can interpret the elliptic curve as a cubic pencil. After homogenizing, we find

$$tYZ^2 = (X - \alpha_1 Z)(X - \alpha_2 Z)(X - \alpha_3 Z) - Y^2 Z. \tag{5.1}$$

Here the three sections P_i are base points of the cubic pencil, and so is the point $O = [0, 1, 0]$ which was taken to serve as zero for the group law. Note, however, that O has multiplicity 6 as a base point, accounting for an A_5 singularity of the fibre at $t = \infty$ on the double fibre component $Z = 0$. To see this, compute the resultant of the LHS and the RHS of (5.1) with respect to x, which returns y^3z^6. Upon substituting, the factor y^3 leads to P_1, P_2, P_3, while z^6 corresponds to O with multiplicity 6.

In the last example, we considered two different non-smooth models of an elliptic surface with different ADE-singularities. Yet one easily verifies that the resolution results in the same singular fibre: a certain configuration of seven smooth rational curves which the next section will declare as Kodaira type IV^* (and which accidentally has also appeared in Example 5.7 in the exceptional characteristic 3).

5.4 Singular Fibres

Let $f : S \to C$ be an elliptic surface, as defined in Sect. 5.1; in particular, it is assumed that the zero section O is given. For the remainder of this book, we will always exclude the trivial product case (as in Example 5.6) by the following assumption:

Convention 5.10 *Any elliptic surface $f : S \to C$ has a singular fibre.*

In this section we will discuss the possible singular fibres of elliptic surfaces as classified by Kodaira [Kod63], Neŕon[Ner64] and Tate [Tat75]. Recall that we are working over an algebraically closed base field k. (Over non-perfect base fields there can be other fibre types, see the recent work by Szydlo [Szy04].)

Suppose that $F_v = f^{-1}(v)$ is a singular fibre ($v \in C(k)$). We write it as a divisor on S with multiplicities:

$$F_v = \sum_{i=0}^{m_v - 1} \mu_{v,i} \Theta_{v,i} \tag{5.2}$$

where

m_v: the number of the (distinct) irreducible components in F_v,
$\Theta_{v,i}$ $(0 \leq i \leq m_v - 1)$: irreducible components,
$\mu_{v,i}$: the multiplicity of $\Theta_{v,i}$ in F_v (a positive integer).
$m_v^{(1)}$: the number of *simple components* (i.e. with multiplicity $\mu_{v,i} = 1$).

In the sequel we arrange the ordering of the irreducible fibre components $\Theta_{v,i}$ following [Kod63] as stated in the following theorems.

Theorem 5.11 *(i) There exists a unique component of F_v which intersects the zero section (O); it is called the* identity component *and denoted by $\Theta_{v,0}$. The coefficient $\mu_{v,0}$ is equal to one: $\mu_{v,0} = 1$.*

*(ii) If F_v is an irreducible singular fibre (i.e. if $m_v = 1$ and $F_v = \Theta_{v,0}$), then $\Theta_{v,0}$ is either a rational curve with a node (*type I_1*) or a rational curve with a cusp (*type II*) (see Fig. 3.2).*

(iii) If F_v is a reducible singular fibre ($m_v > 1$), then every component $\Theta_{v,i}$ of F_v is a smooth rational curve which has self-intersection number $(\Theta_{v,i})^2 = -2$.

The first two statements of the theorem should be clear; the third often files under the title of Zariski's lemma ([BHPV04, III. (8.2))] (not to be confused with the theorem that a dominant rational map from some curve onto a smooth curve is a morphism). It can be proven easily using the canonical bundle formula (which will be the subject of Theorem 5.44)—exactly along the lines of our argument for Example 5.7.

In the next theorem, as usual, the symbol $(\Theta.\Theta')$ denotes the intersection number of two irreducible curves Θ, Θ' on S. See Sect. 4.3, where we have reviewed the intersection theory on an algebraic surface. We will sketch the different fibre types in Figs. 5.1 and 5.2.

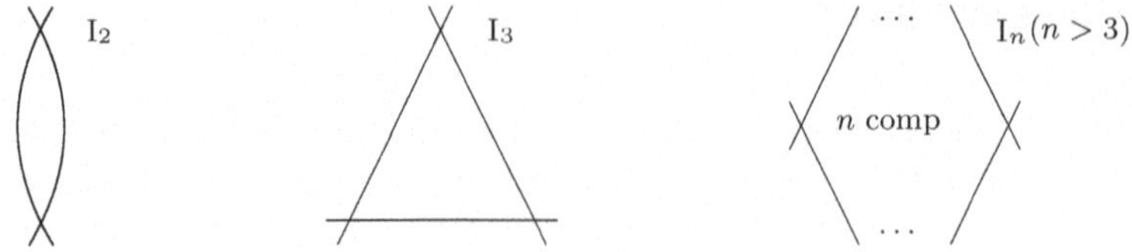

Fig. 5.1 Reducible multiplicative fibres (type I_n $(n > 1)$)

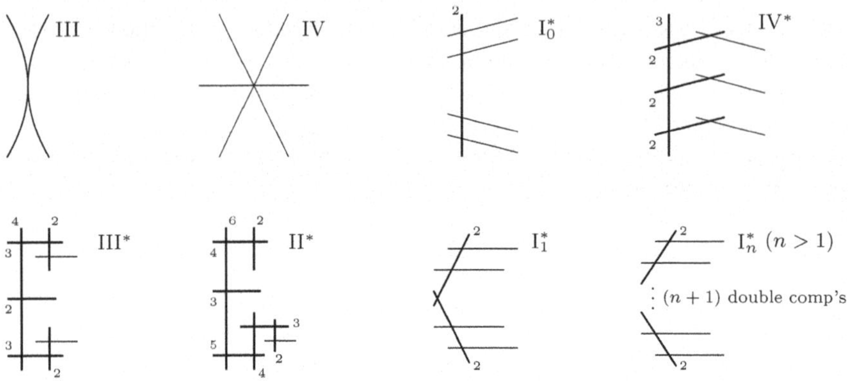

Fig. 5.2 Reducible additive singular fibres

Theorem 5.12 (Kodaira [Kod63], Néron [Ner64],Tate [Tat75]) *All possible types of reducible singular fibres are classified into the following types with $m > 1$ and $b \geq 0$:*

$$\mathrm{I}_m,\ \mathrm{I}_b^*,\ \mathrm{III},\ \mathrm{IV},\ \mathrm{II}^*,\ \mathrm{III}^*,\ \mathrm{IV}^*.$$

(For simplicity, we write $m = m_v(> 1)$, $\Theta_i = \Theta_{v,i}$, dropping the subscript v.)

I_m : $F_v = \Theta_0 + \cdots + \Theta_{m-1}$ *where, for* $m \geq 3$, $(\Theta_i.\Theta_{i+1}) = 1$ *for all* $i = 0, 1, \ldots, m-1$ *cyclically, i.e. the last one meaning* $(\Theta_{m-1}.\Theta_0) = 1$. *For* $m = 2$, *the two components intersect transversally at two points so that* $(\Theta_0.\Theta_1) = 2$.

I_b^* : $F_v = \Theta_0 + \Theta_1 + \Theta_2 + \Theta_3 + 2\Theta_4 + \cdots + 2\Theta_{b+4}$, $m = b + 5$, $b \geq 0$. *Here* $(\Theta_0.\Theta_4) = (\Theta_1.\Theta_4) = 1$, $(\Theta_2.\Theta_{b+4}) = (\Theta_3.\Theta_{b+4}) = 1$, *and* $(\Theta_4.\Theta_5) = \cdots = (\Theta_{b+3}.\Theta_{b+4}) = 1$.

III : $F_v = \Theta_0 + \Theta_1$, $m = 2$, *where the two components intersect at a single point with* $(\Theta_0.\Theta_1) = 2$.

IV : $F_v = \Theta_0 + \Theta_1 + \Theta_2$, $m = 3$, *where all three components meet at a single point and* $(\Theta_0.\Theta_1) = (\Theta_0.\Theta_2) = (\Theta_1.\Theta_2) = 1$.

II^* : $F_v = \Theta_0 + 2\Theta_7 + 3\Theta_6 + 4\Theta_5 + 5\Theta_4 + 6\Theta_3 + 4\Theta_2 + 2\Theta_1 + 3\Theta_8$, $m = 9$, *where* $(\Theta_0.\Theta_7) = (\Theta_7.\Theta_6) = (\Theta_6.\Theta_5) = (\Theta_5.\Theta_4) = (\Theta_4.\Theta_3) = (\Theta_3.\Theta_2) = (\Theta_2.\Theta_1) = (\Theta_3.\Theta_8) = 1$.

III* : $F_v = \Theta_0 + 2\Theta_1 + 3\Theta_2 + 4\Theta_3 + 3\Theta_4 + 2\Theta_5 + \Theta_6 + 2\Theta_7, m = 8,$ *where* $(\Theta_0.\Theta_1) = (\Theta_1.\Theta_2) = (\Theta_2.\Theta_3) = (\Theta_3.\Theta_4) = (\Theta_4.\Theta_5) = (\Theta_5.\Theta_6) = (\Theta_3.\Theta_7) = 1.$

IV* : $F_v = \Theta_0 + \Theta_1 + 2\Theta_2 + 3\Theta_3 + 2\Theta_4 + \Theta_5 + 2\Theta_6, m = 7,$ *where* $(\Theta_1.\Theta_2) = (\Theta_2.\Theta_3) = (\Theta_3.\Theta_4) = (\Theta_4.\Theta_5) = (\Theta_3.\Theta_6) = (\Theta_6.\Theta_0) = 1.$

For any pair $i < j$ *with* $(\Theta_i.\Theta_j)$ *not given explicitly above, we have* $(\Theta_i.\Theta_j) = 0$, *i.e.* Θ_i *and* Θ_j *are disjoint.*

We sketch the singular fibres in the following two figures. We start with fibres of type $I_n (n \geq 1)$ which are often also called *multiplicative*. Type I_1 corresponds to the irreducible nodal curve (Fig. 3.2). Here we draw the reducible multiplicative fibres.

For instance, in Example 5.7, we saw four singular fibres (outside characteristic 3) each consisting of three lines; they are of type I_3 in the above list.

All other singular fibres are called *additive*. As an irreducible fibre, we have the cuspidal cubic (type II, also displayed in Fig. 3.2) that has also appeared six times in Example 5.8 (outside characteristics 2, 3, 5). Some of the reducible additive fibres are non-reduced, i.e. they have multiple components. These appear in bold print in the following figure, with labels indicating the multiplicity.

Implicitly, some of the non-reduced singular fibres have appeared in the previous section. Namely, the Hesse pencil from Example 5.7 degenerates in characteristic 3 to contain a triple line with three surface singularities of type A_2. The desingularization results precisely in a fibre of Kodaira type IV*. The same fibre type has come up in Example 5.9 (in different ways, in fact). Similarly, the elliptic curve from Example 5.8 gave rise immediately to a smooth surface except for characteristic 5 where one fibre contained an E_8 surface singularity. The desingularization leads to a singular fibre of type II*.

The proof that our list is complete generally amounts to Tate's algorithm over perfect fields. This will be outlined in Sect. 5.8.

For later use and easy reference, we introduce the following terminology extending on the notion of identity component defined in Theorem 5.11, which we record here again for completeness. In view of Figs. 5.1 and 5.2, the notions should be self-explanatory, but the relevant information can also be retrieved from the numbering in Theorem 5.12.

Notation 5.13 (i) Among the simple components of the fibre F_v, we call the component Θ_0 meeting the zero section O the *identity component*. All other simple fibre components (Θ_i, $i > 0$) are called *non-identity components*.

(ii) On a fibre of Kodaira type I_{2n} $(n \geq 1)$, the component Θ_n is called the *opposite component*.

(iii) On a fibre of Kodaira type I_n^* $(n \geq 1)$, the component Θ_1 is called the *near (simple) component*, while the components Θ_2, Θ_3 are called *far (simple) components*.

Remark 5.14 Over a non-algebraically closed field k_0, one distinguishes *split singular fibres* F_v, where all components are defined over the base field $k_0(v)$, and *non-split singular fibres*. Of course, the possible Galois actions are rather limited, as visible from Figs. 5.1 and 5.2. Note that Θ_0 is automatically $k_0(v)$-rational.

5.5 Connection with Dynkin Diagrams

Now we consider the dual graph of the reducible singular fibres and notice its close connection with the Dynkin diagram.

The *dual graph* of a reducible fibre F_v with m irreducible components Θ_i ($i = 0, \dots, m-1$) consists of m vertices and several edges connecting them. Throughout we ignore the coefficients μ_i of Θ_i in F_v in forming the dual graph.

The dual graph is defined as follows. Take m vertices, say θ_i, corresponding to the components Θ_i. For $i \neq j$, draw an edge joining θ_i and θ_j for each intersection point of Θ_i and Θ_j (with multiplicity), i.e. θ_i and θ_j are joined by a line if and only if the two components Θ_i and Θ_j intersect each other. By inspection of the previous figures, the intersection consists of a single point and is always transversal except for fibres of type I_2 and III. There we have two vertices joined by two lines.

By the *restricted dual graph* of a reducible fibre F_v ($v \in C(k)$), we mean the graph obtained in the same way as above, starting from the $m-1$ non-identity components Θ_i ($i = 1, \dots, m-1$). In other words, it is the subgraph of the dual graph obtained by deleting the vertex θ_0 (and the edges starting from it). For instance, for a singular fibre of type II^*, we obtain the graph given in Fig. 5.3. It looks exactly the same as the Dynkin diagram for the root lattice E_8 (see Fig. 2.3 in Chap. 2).

Proposition 5.15 *The restricted dual graph of a reducible fibre is the same as the Dynkin diagram for the root lattice of type A, D, E. More precisely,*

Type of F_v	I_m ($m > 1$)	I_b^* ($b \geq 0$)	III	IV	II^*	III^*	IV^*
Dynkin type	A_{m-1}	D_{b+4}	A_1	A_2	E_8	E_7	E_6

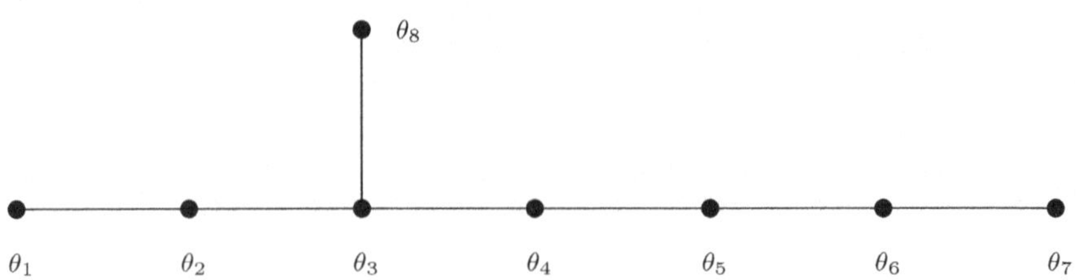

Fig. 5.3 Restricted dual graph for II^*

Proof This is a straightforward consequence of Theorems 5.11 and 5.12. □

For later use, we set-up the following notation:

Definition 5.16 Let T_v denote the lattice of rank $m-1$ generated by θ_i $(i = 1, \ldots, m-1)$, $(m = m_v)$ with the pairing defined by

$$\langle \theta_i, \theta_j \rangle = -(\Theta_i.\Theta_j).$$

Corollary 5.17 *The lattice T_v is a positive-definite lattice isomorphic to the root lattice of type A, D, E associated with F_v as above. In particular, we have*

Type of F_v	I_m $(m > 0)$	I_b^* $(b \geq 0)$	III	IV	II*	III*	IV*
$\det(T_v)$	m	4	2	3	1	2	3

Proof We have a bijection of the sets of the vertices $\{\theta_i\}$ and $\{\alpha_i\}$ (defined in Sect. 2.3) which is compatible with the two graphs. For example, $\theta_i \leftrightarrow \alpha_i$ for type II* as above; for simplicity, assume the same by renumbering α_i if necessary.

As for the value of $\langle \theta_i, \theta_j \rangle$, we have either $(\Theta_i.\Theta_j) = 1$ or 0 for $i \neq j$, which can be read off from the dual graph. Further, we have $(\Theta_i)^2 = -2$ for $i = j$. On the other hand, for the simple roots α_i of a root lattice, we have either $\langle \alpha_i, \alpha_j \rangle = -1$ or 0 for $i \neq j$, which can be read off from the Dynkin diagram (see Sect. 2.3). Of course, we have $\langle \alpha_i, \alpha_i \rangle = 2$ for a root α_i.

Hence we have $\langle \theta_i, \theta_j \rangle = -\langle \alpha_i, \alpha_j \rangle$ for all $i, j \geq 1$. This proves the first assertion. Then the rest follows from 2.3. □

Comparing Proposition 5.15 and Corollary 5.17, we record the following useful observation (involving the discriminant group A_{T_v} of the lattice T_v, compare Sect. 2.3.5, and the number $m_v^{(1)}$ of simple fibre components):

$$|A_{T_v}| = \det T_v = m_v^{(1)}. \tag{5.3}$$

5.6 The Kodaira–Néron Model

We have studied how the generic fibre associates an elliptic curve E over a function field K with an elliptic surface S over a curve C. Here $K = k(C)$. In this section, we discuss how to invert this process. This leads to the notion of the Kodaira–Néron model.

Definition 5.18 Let E be an elliptic curve over the function field K of a curve C. Suppose that there exists an elliptic surface S over C whose generic fibre is isomorphic to E/K. In this case, we say that $f : S \to C$ (or simply S) is the *Kodaira–Néron model* of E/K. We also call S the *elliptic surface associated with* E/K.

In Theorem 5.19 we will show existence and uniqueness of the Kodaira–Néron model. We will use the following universal property that underlies the notion of relative minimality:

If S' is also a smooth projective surface and if the morphism $f' : S' \to C$ has generic fibre E/K, then f' factors through f, i.e. $f' = f \circ g$ for some morphism $g : S' \to S$. This is because the universal property can be equivalently phrased as relative minimality: no fibre of f contains a (-1)-curve (compare the analogue of Zariski's theorem for curves (Theorem 4.16) which holds for minimal surfaces as explained in Sect. 4.6).

With regard to the group structure, there is another universal property of the Kodaira–Néron model, see Theorem 5.22.

Theorem 5.19 *Given an elliptic curve over a function field E/K, the Kodaira–Néron model exists and is unique up to isomorphisms.*

Remark 5.20 Formally, Theorem 5.19 also encompasses the case of a constant elliptic curve E/K, where the Kodaira–Néron model is seen to be a product $E \times C$ even though this does not meet the requirements of an elliptic surface from Convention 5.10. Indeed, the arguments below (and especially Tate's algorithm in Sect. 5.8) cover also the product case.

We give an outline of the proof, assuming some fundamental results in surface theory as sketched in Chap. 4.

Proof We may assume without loss of generality that E/K is given by a *generalized Weierstrass equation* in $\mathbb{P}^2$:

$$y^2 + a_1xy + a_3y = x^3 + a_2x^2 + a_4x + a_6. \tag{5.4}$$

First we show the existence of the Kodaira–Néron model. Let $\Sigma \subset C$ be a finite subset containing all the poles of the coefficients $a_i \in K = k(C)$ and the zeroes of the discriminant Δ of E/K. Let $C' = C - \Sigma$ and let $S' \subset \mathbb{P}^2 \times C'$ denote the (open) surface defined by the above equation. The projection to C' gives a surjective morphism $f' : S' \to C'$ which is a smooth elliptic fibration: for any point $\tau \in C'$, the fibre $f'^{-1}(\tau)$ is an elliptic curve. Let $W \subset \mathbb{P}^2 \times C$ denote the Zariski closure of S'. Then W is a projective surface (possibly with singular points) which admits a morphism $\bar{f} : W \to C$ extending f'. The fibre $\bar{f}^{-1}(v)$ over a point $v \in \Sigma$ is either an elliptic curve or a nodal or cuspidal cubicas a degeneration of a Weierstrass cubic (cf. Sect. 3.4). The singular points of W are contained in the set of the nodes and cusps of non-elliptic fibres, and thus are finite in number. By resolving these singular points, we obtain a smooth projective surface S with a birational morphism $\beta : S \to W$, and $f = \bar{f} \circ \beta$ defines an elliptic fibration $f : S \to C$. If this is a relatively minimal fibration, we are done. Otherwise S contains an exceptional curve of the first kind in some fibre, and we can contract this curve to a point by Castelnuovo's theorem, obtaining a smooth surface S_1 with elliptic fibration $f_1 : S_1 \to C$. By repeating this process a finite number of times (if necessary), we arrive at a relatively minimal fibration with the generic fibre E/K, which shows the existence of the Kodaira–Néron model.

As for the uniqueness of the Kodaira–Néron model (up to isomorphisms), it is clear from the universal property. □

5.6.1 Group Structure

What is intrinsic to the Kodaira–Néron model is that the group operations commute with specialization to a smooth fibre F_v. That is, we can equally well first add two points $P, Q \in E(K)$ on the generic fibre and then specialize to F_v, or first specialize to $P_v, Q_v \in F_v(k(v))$ and then add:

$$(P+Q)_v = P_v + Q_v.$$

The additional feature of Néron's theory, developed in more generality for abelian varieties (see [BLR90]), is the important fact that this compatibility even works at the singular fibres when suitably restricted as follows.

Definition 5.21 The *smooth locus* of a fibre F is denoted by $F^\#$.

The smooth locus $F^\#$ of a singular fibre F is obtained by removing all multiple components and all nodes (i.e. intersection points of fibre components).

The Kodaira–Néron model has the following universal property:

Theorem 5.22 *The group structure extends from the smooth fibres of an elliptic surface over the smooth locus $F^\#$ of any singular fibre F.*

As any algebraic group, $F^\#$ has a normal subgroup $F_0^\#$ with finite quotient group $G(F)$. However, the corresponding exact sequence need not split, as we shall see as a by-product in Example 9.11. Although we did not find any extensive coverage of this problem in the literature, there are, of course, cases where the exact sequence does split (so that the group $F^\#$ often admits a product structure), see Lemma 9.10 for a result in this direction in the case of additive reduction.

In general, one may take for the normal subgroup $F_0^\#$ the smooth locus on the identity component Θ_0 of F (which we already touched upon in Sects. 3.4.1 and 3.4.2, including the structure as multiplicative group $\mathbb{G}_m$ or additive group $\mathbb{G}_a$). One obtains the following relations which, in the complex case, may be extracted from Kodaira ([Kod63, Thm. 9.1]) where every type including the multiplicative one is discussed (compare also [Ner64, Tat75, Table p. 46]):

Additive	$F_0^\# \cong \mathbb{G}_a$
	$F^\#/F_0^\# \cong G(F)$;
Multiplicative	$F_0^\# \cong \mathbb{G}_m$,
(type I_n)	$F^\#/F_0^\# \cong G(F) \cong \mathbb{Z}/n\mathbb{Z}$.

The elements of the group $G(F)$ can be identified with the simple fibre components; $G(F)$ is an abelian group of order equalling the determinant of the restricted Dynkin diagram of the singular fibre because $G(F)$ coincides with the discriminant group of the Dynkin diagram (compare (5.3)). For (split) additive fibres the only possibilities are thus easily determined as follows:

Type	II, II*	III, III*	IV, IV*	I_n^* (n even)	I_n^* (n odd)
$G(F)$	$\{0\}$	$\mathbb{Z}/2\mathbb{Z}$	$\mathbb{Z}/3\mathbb{Z}$	$(\mathbb{Z}/2\mathbb{Z})^2$	$\mathbb{Z}/4\mathbb{Z}$

(5.5)

For later use, we record the following easy application. Many more results in the same spirit can be found in Propositions 6.31 and 6.33.

Lemma 5.23 *Let $S \to C$ be an elliptic surface over an algebraically closed field of characteristic p, equipped with a p-torsion section P (i.e. of order exactly p). At a multiplicative fibre, P intersects either a non-identity component or O.*

Proof If P meets the identity component F_0 of a multiplicative fibre, then it induces a point of order dividing p on $F_0^\# \cong \mathbb{G}_m$. But the only such point is 1, the intersection point of O and F_0. □

Lemma 5.23 has surprising consequences, see for instance Corollary 8.32.

5.7 The Weierstrass Model

In the proof of Theorem 5.19, we have derived the existence of the Kodaira–Néron model in a rather abstract manner. This section, together with the following ones, describes in more detail how one can determine the Kodaira–Néron model of an elliptic surface with section explicitly. At the same time, we will explain how to find the singular fibres of a given elliptic surface. The approach is based on the affine Weierstrass model (5.4) of the generic fibre,

$$E: \quad y^2 + a_1xy + a_3y = x^3 + a_2x^2 + a_4x + a_6. \tag{5.6}$$

Homogenizing the cubic equation in $\mathbb{P}^2$ adds the point $O = (0 : 1 : 0)$ to each fibre which serves as zero of the group law. The coefficients a_i are elements of the function field $K = k(C)$ of the base curve C, not all of them constant by Convention 5.10, so necessarily they have some poles. At a given point $v \in C(k)$, however, we can eliminate the poles of the a_i by some admissible coordinate transformation

$$(x, y) \mapsto (u^2x, u^3y) \quad \text{for } u \in K^\times \tag{5.7}$$

as in (3.8). We will also call the resulting Weierstrass form *integral* at v. This way, we can insert fibres in the Weierstrass model W above points v of C where the a_i, and consequently the discriminant Δ, originally had poles. It thus remains to check whether Δ has a zero at v. If not, then the Weierstrass model W has a smooth fibre at v. At the zeroes of Δ, the Weierstrass W model has a nodal or cuspidal curve as fibre (Fig. 3.2). In this situation, the central question is whether the singular point of the fibre is actually a *surface singularity*. If it is a smooth point on the surface W given by (5.6), then the Kodaira–Néron model S agrees with W at v, i.e. we simply fill in the singular fibre at v (a nodal or cuspical cubic). If the point is a surface singularity, however, then we have to go through the process of desingularization

$$\tilde{S} \to W.$$

By construction, $\tilde{S}$ is endowed with a morphism to C whose general fibre is an elliptic curve. However, $\tilde{S}$ need not be relatively minimal. More precisely, relative minimality turns out to be equivalent to the singularities being rational double points (as in 4.10) as we shall see in the framework of Tate's algorithm in the next section. This will lead to the concept of a minimal Weierstrass form in Sect. 5.10. Blowing down successively (-1)-curves in the fibres of $\tilde{S}$, we arrive at the Kodaira–Néron model S. In particular, if W has only rational double points as singularities, then $\tilde{S} = S$, and the morphism

$$S \longrightarrow W$$

blows down all exceptional curves arising in the process of desingularization. The non-identity components of reducible fibres of $S \to C$ are thus contracted to the surface singularities (nodes or cusps of the corresponding fibre) in the Weierstrass model W.

Example 5.24 Recall the Hesse pencil from Examples 3.13 and 5.7 over some given field k of characteristic different from 3. There are standard ways to convert any given plane cubic (or any given elliptic curve) with a rational point to Weierstrass form, see e.g. [Cas91]. In the present situation, this yields the following generalized Weierstrass form in the affine parameter t:

$$E: \quad y^2 + 9xy - (t^3 - 27)y = x^3. \tag{5.8}$$

The discriminant is $\Delta = (t^3 - 27)^3$, so there can at worst be singular fibres at the cube roots of 27 and at ∞. We compute explicitly the resolution of singularities of the surface S given by (5.8) at $(x, y, t) = (0, 0, 3)$ (the node of the fibre at $t = 3$). The resolution amounts to blowing up the ambient space $\mathbb{A}^3$ and computing the strict transform $\tilde{S}$ of S. This takes us to $\mathbb{A}^3 \times \mathbb{P}^2$ with homogeneous coordinates x', y', t' of $\mathbb{P}^2$ and relations

$$y'x = x'y, \quad x'(t-3) = t'x, \quad y'(t-3) = t'y. \tag{5.9}$$

First consider the affine chart $U = \{t' = 1\}$ such that our affine coordinates x', y', t are obtained from

$$x = (t-3)x', \quad y = (t-3)y'.$$

The strict transform $\tilde{S}$ of S in $\subset \mathbb{A}^3 \times U$ is given by

$$\tilde{S} \cap U: \; y'^2 + 9x'y' - (t^2+3t+9)y' = (t-3)x'^3.$$

This is smooth at $t = 3$ by the Jacobi criterion. Substituting for $t = 3$, we obtain two smooth rational fibre components $\Theta_1 = Z(y')$ and $\Theta_2 = Z(y' + 9x' - 27)$ whose intersection gives a smooth surface point at $(x', y', t) = (0, 3, 3)$.

To see that the singular fibre has Kodaira type I_3 as claimed, it remains to consider the other affine charts of $\mathbb{A}^3 \times \mathbb{P}^2$. In $V = \{x' = 1\}$, we have

$$y = xy', \quad t = 3 + xt'$$

with strict transform

$$\tilde{S} \cap V: \; y'^2 - (t'^3x^2 + 9t'^2x + 27t' - 9)y' = x.$$

The fibre at $t = 3$ visibly has three components:

$$Z(x, y') \cup Z(x, y' + 9 - 27t') \cup Z(t', y'^2 + 9y' - x).$$

The first two correspond to Θ_1, Θ_2 (lying fully inside the exceptional locus of the blow-up of $\mathbb{A}^3$) while the third component Θ_0 is the strict transform of the original nodal cubic. We find that $\tilde{S}$ is smooth on $V \setminus U$ (i.e. at $t' = 0$), and that the components intersect at three distinct points with the following affine coordinates (x, y', t'):

$$\Theta_0 \cap \Theta_1 = \{(0,0,0)\}, \quad \Theta_0 \cap \Theta_2 = \{(0,-9,0)\}, \quad \Theta_1 \cap \Theta_2 = \{(0,0,1/3)\}.$$

The calculations in the affine chart $\{y' = 1\}$ are completely analogous except that the component Θ_1 fails to be visible there. Hence the fibre at $t = 3$ has Kodaira type I_3 as claimed, and in the Weierstrass model W, the components Θ_1, Θ_2 are contracted to the node of the fibre.

At the other third roots of 27, analogous computations go through. Alternatively, one may argue with the symmetry given by multiplying t by a primitive third root of unity. Similar arguments apply to $t = \infty$ after the admissible coordinate transformation $(x, y, t) \mapsto (x/s^2, y/s^3, 1/s)$.

Before taking a closer look at Tate's algorithm, we make a few practical comments. Instead of working with rational functions on the whole of C as entries a_i of the Weierstrass form (5.6), it suffices to consider them on an affine open patch C_0 of C. This has the advantage that the admissible transformation (5.7) can be chosen in such a way that all a_i become polynomial, i.e. they lie in the affine coordinate ring $k[C_0]$, and the same holds for Δ. This is the most convenient and common set-up which we pursued in (5.8) in Example 5.24 (and also already in Sects. 3.4.4 and 3.4.5).

Conversely, the rational functions on C are recovered from the polynomials by homogenizing, but often it is more instructive to homogenize the a_i as elements of the homogeneous coordinate ring $S[C]$ of suitable degree. Conceptually, this relates to the Weierstrass model W being a divisor defined by (5.6) in the $\mathbb{P}^2$-bundle $\mathbb{P}(\mathcal{L}^2 \oplus \mathcal{L}^3 \oplus \mathcal{O}_Y)$ over Y for a suitable invertible sheaf $\mathcal{L}$ on Y, see Miranda ([Mir81, Thm. (2.1)]) (which draws on Mumford's geometric invariant theory). Here $a_i \in H^0(Y, \mathcal{L}^{-i})$ (which also predicts the degrees). All of this will be made explicit for $C = \mathbb{P}^1$ in Sect. 5.10 (see Lemma 5.41).

5.8 Tate's Algorithm

In the sequel, we shall discuss Tate's algorithm to some extent. The interested reader will surely be able to fill in the missing details. One might also consult Tate's original exposition in [Tat75] or [Sil94]. We will only consider the case of perfect fields; the imperfect case was later treated in [Szy04] (see also Example 9.49, where this issue implicitly appears).

For the sake of simplicity, we shall limit ourselves to perfect fields of characteristic different from two. This restriction enables us to work with an *extended Weierstrass form* (after completing the square in the generalized Weierstrass form):

$$y^2 = x^3 + a_2 x^2 + a_4 x + a_6, \tag{5.10}$$

which we may assume to be integral at a given point of C, or polynomial in an affine chart of C following Sect. 5.7. Here the discriminant is given by

$$\Delta = -27\, a_6^2 + 18\, a_2 a_4 a_6 + a_2^2 a_4^2 - 4\, a_2^3 a_6 - 4\, a_4^3. \tag{5.11}$$

In order for a fibre to be singular, the discriminant Δ has to vanish. Very much to our advantage, we can work locally, so we fix a local parameter t on C with normalized valuation v. (This set-up generalizes directly to the number field case where one studies the reduction modulo some prime ideal and picks a uniformizer of the prime.)

Assume that there is (or rather there could be) a singular fibre at $t = 0$, that is, the vanishing order of Δ at $t = 0$ satisfies $v(\Delta) > 0$. By a translation in x, we can move the singularity to $(0, 0)$. Then the extended Weierstrass form transforms to

$$y^2 = x^3 + a_2' x^2 + t\, a_4' x + t\, a_6'. \tag{5.12}$$

If $t \nmid a_2'$, then the above equation at $t = 0$ describes a nodal rational curve. We call the reduction *multiplicative*. If $t \mid a_2'$, then (5.12) defines a cuspidal rational curve at $t = 0$. We refer to *additive reduction*. The terminology is motivated by the group structure of the rational points on the special fibre of the Weierstrass model, as the smooth points of the fibre in (5.12) support the multiplicative resp. additive group over k (see Sect. 3.4). In either case, the special point $(0, 0)$ is a surface singularity if and only if $t|a_6'$.

5.8.1 Multiplicative Reduction

Let $t \nmid a_2'$ in (5.12). From the summand $a_2'^3\, t a_6'$ of Δ in (5.11), it is immediate that $(0, 0)$ is a surface singularity if and only if

$$v(\Delta) > 1.$$

If $v(\Delta) = 1$, then $(0, 0)$ is only a singularity of the fibre, but not of the surface. Hence the singular fibre at $t = 0$ is the irreducible nodal rational curve given by (5.12), cf. Fig. 3.2. Kodaira introduced the notation $\mathbf{I}_1$ for this fibre type. Here the subscript 1 stands for the number of components.

Let us now assume that $v(\Delta) = n > 1$. We have to resolve the singularity at $(0, 0)$ of the surface given by (5.12). This problem can be solved as follows: Let $m = \lfloor \frac{n}{2} \rfloor$ be the greatest integer not exceeding $n/2$. Then translate x by t times a polynomial of degree $m - 1$ such that not only $t \mid a_4$, but $t^{m+1} \mid a_4$:

$$y^2 = x^3 + a_2''\, x^2 + t^{m+1}\, a_4''\, x + a_6''. \tag{5.13}$$

Then $v(\Delta) = n$ is equivalent to $v(a_6'') = n$. Now we blow up the surface m times successively at the point $(0, 0)$. The first $(m - 1)$ blow-ups introduce two $\mathbb{P}^1$'s each. In the chart $x' = t^j x$, $y' = t^j y$ for the jth blow-up, the exceptional divisors are locally given by

$$y'^2 = a_2(0)\, x'^2.$$

In particular, the exceptional divisors come in pairs of rational curves which are conjugate over $k(\sqrt{a_2(0)})/k$. On the nodal curve in (5.12), this extension appears as the splitting field of the two special tangents to the singular point. Depending on whether $\sqrt{a_2(0)} \in k$ or not, one distinguishes *split* and *non-split multiplicative reduction.*

After each blow-up $(j = 1, \ldots, m - 2)$, we continue with another blow-up at $(0, 0)$, the intersection point of the two latest exceptional divisors. After the final blow-up, the local equation of the special fibre is

$$y^2 = a_2(0)\, x^2 + (a_6''/t^{2m})(0).$$

One easily checks that this encodes a single rational component, if $n = 2m$ is even, or again two components if $n = 2m + 1$ is odd. In either case, the surface blown up $(m - 1)$ times is smooth locally around $t = 0$, so we have reached the resolution of the surface singularity in the special fibre. In summary, the process of desingularization has added $(n-1)$ rational curves of self-intersection -2. Hence the singular fibre consists of a cycle of n rational curves, meeting transversally. Following Kodaira, we introduce for this fibre type the symbol $\mathbf{I}_n$.

Remark 5.25 From the local shape of Eq. (5.13), it follows that the surface singularity has exactly the type A_{n-1} in the notation of 4.10.

Example 5.26 (*Hexagonal example cont'd*) Consider the elliptic curve E over $k(t)$ from (1.1) with discriminant Δ. If $\mathrm{char}(k) \neq 2, 3$, then Δ is a degree 4 polynomial in t, $\Delta = 27t^4 + \ldots$, which has multiple roots if and only if

- either $\alpha_i = \alpha_j$ for some $i \neq j$
- or $\sum_i \alpha_i^2 - \sum_{i<j} \alpha_i \alpha_j = 0$.

Outside these two cases, there are thus four singular fibres of type I_1 at the roots of Δ. On the other hand, if $\alpha_i = \alpha_j$, then Δ attains a double root at $t = 0$; resolving the surface singularity at $(x, y, t) = (\alpha_i, 0, 0)$ yields a fibre of Kodaira type I_2. For the second alternative (and further degenerations), see Example 5.28.

In characteristic 3, the discriminant degenerates to a degree 2 polynomial in t, but the same cases as above continue to persist. Meanwhile in characteristic 2, the discriminant degenerates to $\Delta = t^4$, the corresponding singular fibre visibly being additive (see Example 5.28).

At ∞, the Kodaira–Néron model always has a fibre of type IV^*.

5.8.2 Additive Reduction

Consider the case $t \mid a_2'$ in (5.12). We have to determine whether $(0, 0)$ is a surface singularity, i.e. whether $t \mid a_6'$. If the characteristic is different from 2 and 3, then this is equivalent to

$$v(\Delta) > 2.$$

In characteristics 2 and 3, however, the picture is not as uniform. It is visible from (5.11) that $v(\Delta) > 0$ implies $v(\Delta) \geq 3$ if $\mathrm{char}(k) = 3$. In characteristic 2, one even deduces $v(\Delta) \geq 4$ in case of additive reduction. This higher vanishing order is accounted for by *wild ramification* which we will very briefly comment on in Sect. 5.9. Independent of the characteristic, if $(0, 0)$ is a smooth surface point, then the singular fibre is a cuspidal rational curve. We denote it by the Kodaira type II, cf. Fig. 3.2.

If $(0, 0)$ is a surface singularity, then Fig. 5.2 gives the possible singular fibres resulting from the desingularization process. We briefly comment on the single steps.

There are three possibilities for the exceptional divisor of the first blow-up:

1. a rational curve of degree two, meeting the strict transform of the cuspidal curve tangentially in one point;

2. Two lines, possibly conjugate in a quadratic extension of k, meeting the strict transform of the cuspidal curve in one point;
3. a double line.

In the first two cases, we have reached the desingularization and refer to type **III** resp. **IV**. The third case requires further blow-ups, each introducing lines of multiplicity up to six and self-intersection -2. The resolution process branches into three cases here depending on the singularities on the double line (i.e. roots of a certain degree three polynomial):

1. three rational double points (type A_1) $\Rightarrow$ type I_0^*;
2. two singularities $\Rightarrow$ type I_n^* $(n > 0)$;
3. one singularity $\Rightarrow$ types II^*, III^*, IV^* or non-minimal.

We comment on the third case. It is equivalent to the following vanishing orders after a suitable coordinate transformation:

$$v(a_1) \geq 1, \quad v(a_2), v(a_3) \geq 2, \quad v(a_4) \geq 3, \quad v(a_6) \geq 5.$$

Then the above-mentioned cases persist according to the following divisibility conditions:

IV^*	III^*	II^*	Non-minimal
$v(a_3) = 2$	$v(a_3) \geq 3$	$v(a_3) \geq 3$	$v(a_3) \geq 3$
	$v(a_4) = 3$	$v(a_4) \geq 4$	$v(a_4) \geq 4$
		$v(a_6) = 5$	$v(a_6) \geq 6$

In the *non-minimal case*, we have $v(a_i) \geq i$ for each i. This means that the singularity of the Weierstrass model is not a rational double point (as laid out in Sect. 4.10), and we can simplify the Weierstrass form locally as follows: Pick a local coordinate t of the base curve such that the special fibre sits at $t = 0$. Then rescale by the admissible transformation

$$(x, y) \mapsto (t^2 x, t^3 y).$$

The new Weierstrass model W' has coefficients $a_i' = a_i/t^i$, so $v(a_i') = v(a_i) - i \geq 0$, and the new Weierstrass model is still integral at $t = 0$. Thus we can resume the desingularization process for the new Weierstrass model W' at $t = 0$. This process, called *minimalization*, terminates since Δ has finite vanishing order everywhere (and the vanishing order at a given point drops by 12 at each reduction step). This concludes our description of Tate's algorithm.

Remark 5.27 While the above argument applies everywhere, we stress that the procedure is local in nature. Hence we cannot expect to be able to derive a Weierstrass form which is globally minimal (unless the base curve is isomorphic to $\mathbb{P}^1$, see 5.10).

Example 5.28 (*Hexagonal example cont'd*) We continue our analysis of the singular fibres of the Kodaira–Néron model of the elliptic curve $E/k(t)$ from (1.1). Assume first that $\mathrm{char}(k) \neq 2, 3$ and that the second degeneration from Example 5.26 persists.

This makes Δ a perfect square, still of degree 4, while the j-invariant of E turns out to be zero identically. The singular fibres at the two double roots of Δ have Kodaira type II unless the first alternative also persists, that is $\alpha_1 = \alpha_2 = \alpha_3$, which implies $\Delta = 27t^4$ and Kodaira type IV at $t = 0$.

If char(k) = 3, then recall that Δ has only degree 2. In fact, the second degeneration from Example 5.26 makes it a constant, so there are no singular fibres except at ∞. That is, unless all α_i come together as this endows the fibre at t with a singularity at $(\alpha_i + 2^{1/3}t^{2/3}, t)$ and thus leads to a quasi-elliptic surface, with singular fibre of type IV at $t = 0$ as above. (See Sect. 7.5 for a brief discussion of quasi-elliptic surfaces, especially around Definition 7.21.)

If char(k) = 2, then we have already pointed out in Example 5.26 that the discriminant turns out to be a perfect fourth power: $\Delta = t^4$, regardless of the configuration of the α_i. Nonetheless the α_i determine the Kodaira type as follows. As long as all α_i stay distinct, the corresponding singular fibre has type II (regardless of the second degeneration from Example 5.26!), but with two α_i coinciding, it can be promoted to type III, and with all α_i coinciding even to type IV.

5.9 Singular Fibres Versus Discriminant and j-Invariant

If the characteristic differs from 2 and 3, then we can derive the type of the singular fibre directly from few information: the discriminant and the j-invariant plus a small bit of extra information. In essence, this is due to the short Weierstrass form (3.3) to which we can always reduce in these characteristics. Here we only have to consider the vanishing orders of a_4 and a_6 next to Δ.

Assume that the Weierstrass form is minimal and that $v(\Delta) = n > 0$. Then the type of the singular fibre can be read off from the following table. A single entry means equality, while otherwise we will indicate the possible range of valuations. As an addition, we note that Kodaira types III, III* imply $v(j - 12^3) \geq 1$.

The following relation is worth noticing for a singular fibre F_v:

$$m_v = \#\begin{pmatrix}\text{components of} \\ \text{singular fibre } F_v\end{pmatrix} = \begin{cases} v(\Delta) & \text{in the multiplicative case;} \\ v(\Delta) - 1 & \text{in the additive case.}\end{cases} \tag{5.14}$$

In characteristics 2, 3, however, there can be a discrepancy in the additive case in (5.14). To account for this, the notion of wild ramification was introduced. Following Ogg's paper [Ogg67] (see also [Sai88]), we denote the index of wild ramification at an additive fibre by δ_v, defined as follows:

$$\delta_v = v(\Delta) - 1 - m_v. \tag{5.15}$$

In characteristics 2 and 3, some additive fibre types imply higher vanishing order of Δ than predicted by the number of components. We have seen this happen for the cuspidal rational curve (type II) in Sect. 5.8.2. In order to determine the fibre type, one thus has to go through Tate's algorithm—as opposed to the other characteristics where we read off the fibre type from discriminant and Weierstrass form. A case-by-case analysis reveals slightly different bounds for the vanishing orders of the discriminant (see [MT05, SSc13, Prop. 16]).

Example 5.29 (*Hexagonal example cont'd*) Consider the elliptic curve E from (1.1) over $k(t)$ for some field k of characteristic 2 or 3. Then the fibre at $t = \infty$ always has Kodaira type IV* regardless of the possible degenerations of Δ depending on the α_i. However, in characteristic 3, we have seen in Examples 5.26 and 5.28 that Δ has degree 2 or 0; in consequence, there must be wild ramification of index 2 or 4 at ∞.

Similarly, in characteristic 2, we have $\Delta = t^4$ generally corresponding to a singular fibre of type II, thus with wild ramification of index 2. The degenerations from Example 5.28 cause the fibre type to "go up" to III or IV—accordingly the index of wild ramification has to go down by (5.15).

5.9.1 Quadratic Twists

Recall from Theorem 3.4 that two elliptic curves with the same j-invariant $j \in K$ are isomorphic over $\bar{K}$. Here we relate this result for elliptic curves with j-invariant $j \neq 0, 12^3$ with quadratic twists as in (3.9); in particular, this will apply to any elliptic curve over the function field $K = k(C)$ of an algebraic curve C with non-constant j-invariant, i.e. with $j \in K \setminus k$. The case with $j = 0, 12^3$ is a little more delicate due to the presence of quartic and sextic twists (see e.g. [IR82, 9, 18]).

Lemma 5.30 *Any two elliptic curves over any field K with the same j-invariant $j \neq 0, 12^3$ are either isomorphic or quadratic twists of each other.*

Proof For simplicity, we only give a short proof valid in characteristic different from 2, 3 (see [Sil09, App. A] for characteristics 2, 3). By Theorem 3.2, we can convert the two elliptic curves E, E' to Weierstrass normal form (3.3) with coefficients $p, q \in K$ resp. $p', q' \in K$. Then $j = 0$ is equivalent to $p = p' = 0$, and similarly $j = 12^3$ if and only if $q = q' = 0$. By Theorem 3.4, E and E' are isomorphic over $\bar{K}$, so there is an admissible coordinate transformation (3.8) for some $u \in \bar{K}^\times$. That is,

$$p' = u^4 p, \quad q' = u^6 q.$$

But then, since all entries above are non-zero by assumption, and since they all lie in K, we infer that $u^2 = u^6/u^4 \in K$, which exactly encodes the quadratic twist by u (if $u \notin K$) as claimed. □

Corollary 5.31 *Any two elliptic surfaces with the same j-invariant $\neq 0, 12^3$ have the same singular fibres up to some quadratic twist (to be related below in* (5.16)*). In particular, the smooth fibres over some point $v \in C(k)$ are isomorphic as long as we assume that k is algebraically closed.*

If $\mathrm{char}(k) \neq 2$, then any elliptic surface with section admits an extended Weierstrass form (5.10). Hence quadratic twisting can be understood in analogy with (3.9):

$$S_u: \quad y^2 = x^3 + ua_2x^2 + u^2a_4x + u^3a_6, \quad u \in K.$$

Here we point out the effect of a quadratic twist on the fibres, both smooth and singular:

- the fibre type at $v \in C(k)$ stays the same if and only if u has even vanishing order (or pole order) at v; in particular, this applies to all points on C where u neither has a zero nor a pole;
- the fibre type at $v \in C(k)$ changes if and only if u has odd vanishing or pole order at v; in this case, the fibre types satisfy the following pattern:

$$\mathrm{I}_n \longleftrightarrow \mathrm{I}_n^* \ (n \geq 0), \quad \mathrm{II} \longleftrightarrow \mathrm{IV}^*, \quad \mathrm{III} \longleftrightarrow \mathrm{III}^*, \quad \mathrm{IV} \longleftrightarrow \mathrm{II}^*. \quad (5.16)$$

Both cases can be read off directly from Tate's algorithm 5.8, or from Table 5.1 if $\mathrm{char}(k) \neq 3$ as well.

Remark 5.32 In characteristic 2, there is a different notion of quadratic twisting, related to Artin–Schreier extensions (see, for instance, [SSc13, p. 703] with a view towards elliptic surfaces). This will appear briefly in the proof of Proposition 12.32 and also in (13.45).

5.9.2 Base Change

Closely related to the notion of quadratic twists, there is the notion of base change. On the level of elliptic curves, this is easily understood: one just considers the given curve over some extension K' of the given field K. Hence there is no essential change, except that, of course, the group of rational points may grow. We illustrate this with a few examples.

Example 5.33 The constant elliptic curves over the function field of an algebraic curve C which we ruled out with Convention 5.10 can be interpreted as base change from the field of scalars k to the function field $K = k(C)$.

Example 5.34 If two elliptic curves over K are quadratic twists by some non-square $d \in K$, then their base changes to $K(\sqrt{d})$ are isomorphic.

Example 5.35 (*Hesse pencil revisited*) Consider the Hesse pencil from Example 3.13. Apply the base change from $K = k(t)$ to $K' = k(s)$ with $s^2 = -t$. Then $E(K')$ contains the additional point $P = [s, 1, -1]$.

On the level of elliptic surfaces, the notion of base change is more subtle since it usually does not refer to an extension of the base field k (which we usually assume to be algebraically closed anyway). Rather the base change of the elliptic surface ought to be the Kodaira–Néron model of the underlying elliptic curve after applying the base change there. Here a finite extension $K'/k(C)$ corresponds to a finite morphism

$$C' \to C \tag{5.17}$$

defined over k which exhibits the extension with $K' = k(C')$. Hence the base change S' of the elliptic surface S associated with $E/k(C)$ is birational to a fibre product which makes the following diagram commute:

$$\begin{array}{ccc} S' \sim S \times_C C' & \longrightarrow & S \\ \downarrow & & \downarrow \\ C' & \longrightarrow & C \end{array}$$

In practice, we simply pull back the Weierstrass form of S over C via the morphism (5.17), and we find the base change S' of S as the Kodaira–Néron model over C'. Two words of caution might be in place:

- the base change S' need not be an elliptic surface in our strict sense (meeting Convention 5.10); in particular, S' may decompose into the product $E \times C'$;
- usually one will have $p_g(S') \geq p_g(S)$, for instance, but this need not hold if (5.17) is not separable (compare, e.g., Sects. 12.4.3.1 and 12.4.3.2).

We now discuss a more geometric viewpoint of base change of great relevance. To this end, consider a multisection of the fibration $S \to C$, i.e. a smooth irreducible curve C' which meets every fibre in d points, say. Then this induces a morphism of curves (5.17) of degree d such that the base change naturally attains an additional section (usually increasing the Mordell–Weil rank). This construction is very useful when constructing elliptic curves of high rank, as we will use in Chap. 13. For instance, in Example 5.35, the additional rational point P is induced by the (rational) bisection

$$C' = \{Y + Z = X^2 + tY^2 = 0\} \tag{5.18}$$

on the Hesse pencil from Example 3.13. There is also an interesting instance of base change by a very special trisection C' on a K3 surface in [RS15a] which splits into 3-torsion sections on the Galois closure of (5.17), thus allowing us to impose indirectly the restrictive properties of torsion sections which we will discuss in Sect. 6.6.

It remains to discuss the impact of a base change on the fibres of the elliptic surfaces. Clearly, smooth fibres are replaced by smooth fibres, their precise number depending on the ramification (and degree of inseparability) of (5.17). At a singular fibre, however, we have to look closely at the ramification of (5.17), as this indeed affects the fibre type. This is already visible from Table 5.1 as the vanishing orders of the coefficients of the Weierstrass form and of discriminant and j-invariant get multiplied by the ramification index upon base change. For a *multiplicative fibre*

Table 5.1 Vanishing orders of Weierstrass form coefficients, Δ and j

Fibre type	$v(a_4)$	$v(a_6)$	$v(\Delta)$	$v(j)$	Fibre type	$v(a_4)$	$v(a_6)$	$v(\Delta)$	$v(j)$
I_0	$\{ 0$	≥ 0	0	≥ 0	I_0^*	$\{ 2$	≥ 3	6	≥ 0
	≥ 0	0				≥ 2	3		
I_n $(n > 0)$	0	0	n	$-n$	I_{n-6}^* $(n > 6)$	2	3	n	$-n$
II	≥ 1	1	2	≥ 1	IV*	≥ 3	4	8	≥ 1
III	1	≥ 2	3	0	III*	3	≥ 5	9	0
IV	≥ 2	2	4	≥ 1	II*	≥ 4	5	10	≥ 1

Table 5.2 Singular fibres without wild ramification under ramified base change

II	d mod 6	fibre
	0	I_0
	1	II
	2	IV
	3	I_0^*
	4	IV*
	5	II*

I_n^*	d mod 2	fibre
	0	I_{nd}
	1	I_{nd}^*

II*	d mod 6	fibre
	0	I_0
	1	II*
	2	IV*
	3	I_0^*
	4	IV
	5	II

III	d mod 4	fibre
	0	I_0
	1	III
	2	I_0^*
	3	III*

III*	d mod 4	fibre
	0	I_0
	1	III*
	2	I_0^*
	3	III

IV	d mod 3	fibre
	0	I_0
	1	IV
	2	IV*

IV*	d mod 3	fibre
	0	I_0
	1	IV*
	2	IV

(type I_n $(n \geq 0)$), we thus read off from Table 5.1 that ramification of index d causes it to be replaced by a fibre of type I_{nd} (regardless of the characteristic). For *additive fibres*, however, the pull-back of the Weierstrass form might become non-minimal, as is immediate from Tate's algorithm 5.8. Yet, in the absence of wild ramification, the singular fibre resulting from the base change can be predicted exactly. Most directly, this is obtained from a normal Weierstrass form (outside characteristics 2, 3) through the vanishing orders of a_4, a_6 and Δ (which encode how to minimalize). Depending on the equivalence class of the ramification index d modulo a given integer, Table 5.2 collects the behaviour of additive singular fibres without wild ramification under base change.

In the presence of wild ramification, the local analysis can be much more complicated; confer [MT05] for a general account of that situation.

Remark 5.36 Table 5.2 shows that additive fibres (without wild ramification) are *potentially semi-stable*, that is, applying a suitable base change, they are replaced by *semi-stable fibres*, i.e. multiplicative fibres of type I_n $(n \geq 0)$.

Base change will be used in several instances in the next chapters—for instance, for the functoriality of the height pairing in Sect. 6.9, in the appendix 9.6 on special issues over non-perfect fields and in the discussion of ranks of elliptic K3 surfaces in Sect. 12.2 and for elliptic surfaces over $\mathbb{P}^1$ throughout Chap. 13.

Remark 5.37 The interplay of quadratic twist and base change pointed out in Example 5.34 has particular relevance to the determination of Mordell–Weil groups, since in cases of interest (such as around (12.15)) the Mordell–Weil groups of two elliptic curves which are twists of each other generate the Mordell–Weil group of the base changed curve up to finite index.

Example 5.38 (*Hesse pencil cont'd*) Continuing on the base change from Example 5.35, we compute the singular fibres of the Kodaira–Néron model S' associated with E/K'.

First, assume that $\mathrm{char}(k) \neq 2$. Then the base change is separable of degree 2 and ramifies exactly at 0 and ∞. At ∞, this results in a fibre of Kodaira type I_6. At 0, we distinguish whether $\mathrm{char}(k) \neq 3$ where the fibre is and stays smooth (with j-invariant zero), or $\mathrm{char}(k) = 3$ where the original fibre has Kodaira type IV^* with wild ramification of index 1. On S' we calculate type IV, after minimalizing, with wild ramification index 2.

Secondly, assume that $\mathrm{char}(k) = 2$, so the extension K'/K is purely inseparable. Hence all singular fibres are promoted from type I_3 on S to I_6 on S'. Note that all these elliptic surfaces, regardless of the characteristic, share the additional section P from Example 5.35 (which plays a special role as a peculiar torsion section in characteristic 2, see Sect. 12.4.4.2).

5.9.3 Dokchitsers' Refinement

Recently, there was a refinement of the outcome of Tate's algorithm due to [DD13]. Namely, they show that the Kodaira types of the additive singular fibres of an elliptic curve in generalized Weierstrass form (5.4),

$$y^2 + a_1xy + a_3y = x^3 + a_2x^2 + a_4x + a_6 \qquad (a_i \in K),$$

over a perfect field k can be deduced from the minimum of $v(a_i)/i$:

Theorem 5.39 *An elliptic curve $E/k(t)$ with additive reduction at $t = 0$ has a minimal Weierstrass model over k which depends on its Kodaira type as follows:*

	II	III	IV	I_0^*	I_n^* $(n > 0)$	IV^*	III^*	II^*
$\min \frac{v(a_i)}{i} =$	$\frac{1}{6}$	$\frac{1}{4}$	$\frac{1}{3}$	$\frac{1}{2}$	$\frac{1}{2}$	$\frac{2}{3}$	$\frac{3}{4}$	$\frac{5}{6}$
Extra condition			$v(b_6) = 2$	$v(d) = 6$	$v(d) > 6$ $v(a_2^2\text{-}3a_4) = 2$	$v(b_6) = 4$		

Here

$$b_6 = a_3^2 + 4a_6 = \mathrm{disc}(y^2 + a_3 y - a_6), \qquad d = \mathrm{disc}(x^3 + a_2x^2 + a_4x + a_6).$$

Conversely, a Weierstrass model satisfying either of these conditions is minimal, with singular fibre of the corresponding Kodaira type.

Note that Theorem 5.39 relies on making appropriate coordinate transformations which basically amount to moving the singularities in the resolution always to the affine origin, just as in Sect. 5.8. The advantages of Theorem 5.39 become even more evident when one considers the problem of constructing elliptic surfaces with given additive fibres because then one can choose to work with a suitable Weierstrass form where the above conditions are easy to impose.

We conclude this section with a refinement for fibres of Kodaira type I_n^* that recovers the precise integer n as well:

Proposition 5.40 *An elliptic curve $E/k(t)$ with singular fibre of Kodaira type I_n^*, $n > 0$ at $t = 0$ has a minimal model with*

$$v(a_2) = 1, \quad v(a_i) \geq \tfrac{i}{2} + \left\lfloor \tfrac{i-1}{2} \right\rfloor \tfrac{n}{2}, \quad \begin{cases} v(d) = n+6, \ \ v(b_6) \geq n+3 & \text{if } 2 \mid n, \\ v(d) \geq n+6, \ \ v(b_6) = n+3 & \text{if } 2 \nmid n. \end{cases}$$

Conversely, a Weierstrass equation satisfying these conditions defines an elliptic curve with singular fibre of Kodaira type I_n^* at $t = 0$.

5.10 Minimal Weierstrass Models

Consider a Weierstrass equation that is integral at $t = 0$, say, such that Δ vanishes at $t = 0$ where t is a local parameter on the base curve C. In general, Tate's algorithm tells us how to convert the Weierstrass equation to a minimal form and then determine the singular fibre. This process is called *minimalization*, and the resulting Weierstrass model is called minimal at $t = 0$.

When we work with a general Weierstrass form (5.4), then the concept of minimality may not be very intuitive. Explicitly, apply some *admissible coordinate transformation*

$$(x, y) \mapsto (x + \alpha_2, y + \alpha_1 x + \alpha_3), \quad \alpha_i \in K \tag{5.19}$$

to derive an equivalent Weierstrass form with new coefficients $a_i' \in K = k(C)$. Then non-minimality amounts to the existence of some admissible transformation as above such that the a_i' satisfy the divisibility properties $v(a_i') \geq i$ from Sect. 5.8.2 (or more directly, such that the new coefficients violate Theorem 5.39). If so, then the key reduction step in Tate's algorithm in Sect. 5.8 tells us to apply another admissible transformation, namely $(x, y) \mapsto (t^2x, t^3y)$, and subsequently to check for minimality again.

The concept of minimality becomes much more transparent when we can work with a short Weierstrass form

$$y^2 = x^3 + a_4x + a_6, \tag{5.20}$$

for instance in characteristic $p \neq 2, 3$. Except for scaling as in (3.8) (or in characteristic 2 or 3), this Weierstrass model does not admit any admissible transformations. Hence it is non-minimal if and only if

$$v(a_4) \geq 4 \quad \text{and} \quad v(a_6) \geq 6.$$

We can be more specific if the base curve is $\mathbb{P}^1$. As discussed in Sect. 5.7, we can take the coefficients affinely to be polynomials, $a_i \in k[t]$. Since $k[t]$ is a principal ideal domain, non-minimality at some finite place is equivalent (after some admissible transformation (5.19)) to a polynomial divisibility

$$b^i \mid a_i \quad \text{for some non-constant } b \in k[t].$$

But then the admissible transformation $(x, y) \mapsto (b^2x, b^3y)$ not only reduces the degrees of the coefficients of the Weierstrass form (by $i\deg(b)$ each), but also preserves the polynomial shape. This way, one obtains a polynomial Weierstrass form which is minimal at all finite places simultaneously.

To see how one can cover the place at ∞ as well without leaving the given affine chart of $\mathbb{P}^1$, we define

$$d := \min\{6n;\ n \in \mathbb{N},\ \deg(a_i) \leq ni\ \forall i\}. \tag{5.21}$$

Lemma 5.41 *One obtains the same Weierstrass model from homogenizing the a_i in $k(\mathbb{P}^1)$ and from homogenizing each a_i in $\mathcal{O}_{\mathbb{P}^1}(di)$.*

Proof The polynomials $a_i(t)$ homogenize as rational functions $a_i(t/s) \in k(\mathbb{P}^1)$. In the chart $t = 1$ containing ∞, an integral Weierstrass equation is obtained from the admissible transformation

$$x = x/s^{2d}, \quad y = y/s^{3d}. \tag{5.22}$$

By the definition of d, the resulting polynomial coefficients $s^{id}a_i(1/s)$ can also be obtained from homogenizing in $\mathcal{O}_{\mathbb{P}^1}(di)$ and then dehomogenizing by setting $t = 1$. □

Remark 5.42 The interpretation with each a_i homogeneous of degree di also lends itself to defining a birational model of the elliptic surface S as the degree $6d$ hypersurface in weighted projective space $\mathbb{P}[1, 1, 2d, 3d]$ defined by (5.4). Over $\mathbb{C}$, this comes with the advantage that Hodge theoretic techniques can be adapted to this situation, cf. [Klo07a].

Returning to the problem of minimalizing at ∞, we infer from the proof of Lemma 5.41, especially from (5.22), that this can be achieved by admissible transformations (5.19) with

$$\alpha_i \in k[t], \quad \deg(\alpha_i) \leq di. \tag{5.23}$$

Here the original Weierstrass form was non-minimal at ∞ if the new coefficients a_i' satisfy the condition

$$\deg(a_i') \leq (d-1)i \quad \forall i$$

(so we have lowered the minimum from (5.21)). In summary, this gives a global approach to minimalizing elliptic surfaces over $\mathbb{P}^1$:

Lemma 5.43 *Any elliptic surface over $\mathbb{P}^1$ admits a globally minimal Weierstrass form with polynomial coefficients $a_i(t) \in k[t]$:*

$$y^2 + a_1(t)xy + a_3(t)y = x^3 + a_2(t)x^2 + a_4(t)x + a_6(t). \tag{5.24}$$

Given a globally minimal Weierstrass form of an elliptic surface S over $\mathbb{P}^1$, the degree d from (5.21) defines an important invariant of S. Indeed, the discriminant Δ has degree $12d$, including the contribution at ∞. That is, in the homogeneous interpretation of the a_i from Lemma 5.41, we have $\Delta \in \mathcal{O}_{\mathbb{P}^1}(12d)$. Note also the implication that a given polynomial Weierstrass equation is minimal when Δ does not have a 12th power of a non-constant polynomial as factor (in $\mathcal{O}_{\mathbb{P}^1}(12d)$).

Lemma 5.43 gives a very convenient description for practical issues. Recall that minimality states that all the fibres of the Kodaira–Néron model, both smooth and singular, are obtained from Tate's algorithm (see Sect. 5.8) without any extra rounds of reductions accounting for joint divisibilities of all coefficients of the Weierstrass form, possibly after some admissible transformation.

In contrast, if C has positive genus, then $k[C]$ fails to be a principal ideal domain. Hence there might not be a globally minimal Weierstrass form; instead one has to patch together local minimal models. As an application, one can construct elliptic surfaces without singular fibres which are not isomorphic to a product of an elliptic curve with another curve (see [SSh10, Sect. 4.9]).

As another twist that may feel strange at first, Tate's algorithm itself can very well have the converse effect and make a (globally) minimal Weierstrass form non-minimal (at some place). For instance, consider the multiplicative case where we proposed to perform a translation in x by some polynomial of degree $m-1$ yielding (5.13). The resulting Weierstrass form turns out to be non-minimal at ∞ if the integer m is large compared to the degrees of the coefficients a_i, namely if $m > d$ (defined as in (5.21)).

5.11 Canonical Divisor

We conclude this chapter by collecting some important invariants of elliptic surfaces. In this section we study the canonical divisor.

As before, we let $S \to C$ denote an elliptic surface with section. We want to compute the *canonical divisor* K_S (or rather its class in $\text{Pic}(S)$). The following formula goes back to Kodaira for complex elliptic surfaces (not necessarily with section nor even algebraic surfaces, cf. [Kod63]). A characteristic-free proof was given in [BM76/77]. The statement involves the Euler characteristic

$$\chi(S) = \chi(S, \mathcal{O}_S)$$

which we will investigate further in the next section.

Theorem 5.44 (Canonical bundle formula) *The canonical bundle of an elliptic surface $f : S \to C$ is given by*

$$\omega_S = f^*(\omega_C \otimes \mathcal{L}^{-1}),$$

where $\mathcal{L}$ is a certain line bundle of degree $-\chi(S)$ on C. In particular, we have

$$K_S \approx (2g(C) - 2 + \chi(S))\, F, \quad K_S^2 = 0.$$

We sketch one line of argument. First show that K_S is vertical, i.e. it is supported on fibre components. This stems from the fact that $F.K_S = 0$ by adjunction. Then apply Zariski's lemma ([BHPV04, III. (8.2))] to deduce that K_S is in fact a sum of fibre multiples. Finally, compute the degree with the help of spectral sequences and the Riemann–Roch theorem. For details, the reader may consult [BM76/77] or [Kod63].

Recall that, by convention, all elliptic surfaces considered here have a section. We remark that the canonical bundle formula admits a general formulation for elliptic surfaces without section. Then there are some further terms accounting for possible multiple fibres. In the presence of a section, however, there cannot be any multiple fibres. Moreover, we can work with a Weierstrass model. This allows for an alternative approach which uses differential forms. We will outline this approach for elliptic surfaces over $\mathbb{P}^1$ in Sect. 5.13.

As an application of the canonical bundle formula, we compute the self-intersection of a section (P) by the adjunction formula (Theorem 4.11):

$$2g(P) - 2 = P^2 + (P.K_S) = P^2 + 2g(C) - 2 + \chi(S).$$

Since $g(P) = g(C)$ by construction, we obtain:

Corollary 5.45 *For any $P \in E(K)$, we have $P^2 = -\chi(S)$.*

Remark 5.46 We will see in Corollary 5.50 that $\chi(S) > 0$ always.

5.12 Euler Characteristic and Euler Number

We continue this chapter with a formula for the Euler number (or Euler–Poincaré characteristic) $e(S)$ of an elliptic surface S. Here one can think of the topological Euler number in case we are working over the complex numbers; in general, we work with the alternating sum of the Betti numbers, the dimensions of the ℓ-adic étale cohomology groups.

For a fibre F_v of an elliptic surface with m_v components, we thus obtain

$$e(F_v) = \begin{cases} 0, & \text{if } F_v \text{ is smooth;} \\ m_v, & \text{if } F_v \text{ is multiplicative;} \\ m_v + 1, & \text{if } F_v \text{ is additive.} \end{cases}$$

By (5.14), the Euler number of the fibre agrees exactly with the vanishing order of the discriminant if there is no wild ramification. Otherwise it involves a contribution from the index of wild ramification δ_v at v. Recall from Sect. 5.9 that $\delta_v = 0$ unless $\mathrm{char}(k) = 2, 3$ and F_v is additive.

Theorem 5.47 ([CD89, Prop. 5.16]) *For an elliptic surface S over C, we have*

$$e(S) = \sum_{v \in C} (e(F_v) + \delta_v).$$

Note that the sum in the theorem is always finite, as it runs over the singular fibres of $S \to C$, i.e. $v \in \Sigma$.

In Sect. 5.9 the index of wild ramification was defined in such a way that the summand at $v \in C$ contributing to $e(S)$ equals exactly the vanishing order of Δ at v for a model which is minimal at v. We will indicate this by $\Delta_{\min}$ and use somewhat ambiguously v for points on C and the corresponding valuations. Thus we find

$$e(S) = \sum_{v \in C} v(\Delta_{\min}).$$

For a globally minimal model, this sum coincides with the degree of Δ as a divisor on the curve C. For $C \cong \mathbb{P}^1$, this leads to the following statement using Lemma 5.43 and the discussion succeeding it:

Theorem 5.48 *Elliptic surfaces over $\mathbb{P}^1$ with Euler number $12d$ depend on $10d - 2$ moduli.*

Proof Let $S \to \mathbb{P}^1$ be an elliptic surface and consider a globally minimal Weierstrass equation (5.24). With $d \in \mathbb{N}$ defined by (5.21), we know from Sect. 5.10 that $\deg(\Delta) = 12d > 0$. By the above discussion, this gives the Euler number of S.

The isomorphism classes of the Weierstrass models form a coarse moduli space for fixed d (which can be studied by means of geometric invariant theory (at least in characteristic $\neq 2, 3$), see [Mir81]). A parameter count shows that the corresponding elliptic surfaces have $10d - 2$ moduli, accounting for the $4d + 1$ coefficients of a_4 and $6d + 1$ coefficients of a_6 in the short Weierstrass form (5.20) compared against the admissible transformation (3.8) (with $u \in k$ only!) and Möbius transformations acting with three moduli on the base.

The same parameter count holds in characteristics 2, 3, working with extended or generalized Weierstrass forms, but then also taking admissible transformations (5.19) into account which preserve the shape of the Weierstrass form and the degrees of the coefficients (i.e. satisfying (5.23)). □

Remark 5.49 If $d > 1$ and $k = \mathbb{C}$, there are similar results for the dimensions of the strata of the moduli space where $\rho \geq r$ for a given integer $r > 2$ (see [Klo07a]). However, we will only touch upon this kind of result in a different set-up, namely for $d = 1$ (rational elliptic surfaces) and in any characteristic, see Theorem 8.39.

We shall now come to the Euler characteristic χ. This is the point where our assumption enters that the elliptic surface has a singular fibre (Convention 5.10). In consequence $e(S) > 0$ by Theorem 5.47. As a corollary, we obtain the Euler characteristic $\chi(S)$ through Noether's formula (4.13). Compare Kodaira ([Kod63, Theorem 12.2]).

Corollary 5.50 *For an elliptic surface S, we have*

$$\chi(S) = \frac{1}{12} e(S) > 0.$$

An alternative elementary proof of this formula was given by Oguiso in [Ogui90].

5.13 Differential Forms

We conclude this chapter by explaining how to compute regular 2-forms on an elliptic surface with section. We focus on the case of a rational base curve where we can also work out, starting from a minimal Weierstrass form, where the elliptic surface fits into the Enriques–Kodaira classification.

Given an elliptic surface $S \to \mathbb{P}^1$, there is a generalized Weierstrass form (5.24),

$$S: \quad y^2 + a_1 xy + a_3 y = x^3 + a_2 x^2 + a_4 x + a_6$$

which is globally minimal by Lemma 5.43. That is, $a_i \in k[t]$ for each i and there are at worst rational double point singularities (which do not affect the 2-forms on the surface or the canonical divisor K_S, see Sect. 4.10). Moreover, the same holds at ∞. With d defined by (5.21), we have seen in Sect. 5.12 that

$$e(S) = 12d, \quad \chi(S) = d.$$

We now consider the 2-form

$$\omega = dt \wedge dx/(2y + a_1 x + a_3),$$

which arises naturally from the standard regular 1-form $dx/(2y + a_1 x + a_3)$ on the fibres (much like the regular 2-forms in (4.6)). On the affine patch that we started with, this form has neither zeroes nor poles and extends regularly to the point O of the fibre at ∞. It remains to consider the fibre at ∞. Here the 2-form transforms as

$$\omega = -s^{d-2} ds \wedge dx'/(2y' + a_1' x' + a_3').$$

Thus we find $K_S = (d-2)F_\infty$ in perfect agreement with Theorem 5.44.

In addition, the above computations enable us to determine an affine basis of the space $H^0(S, K_S)$ of regular 2-forms for $n \geq 2$:

$$H^{2,0}(S) = \langle \omega, t\omega, \ldots, t^{d-2}\omega \rangle \cong k^{d-1}.$$

Thus we find the geometric genus $p_g(S) = d - 1$, and the invariant $q(S) = 0$ follows from $\chi(S) = d$ (or alternatively from Lüroth's Theorem). This puts us in the position to determine how these elliptic surfaces over $\mathbb{P}^1$ fit into the surface classification:

Proposition 5.51 *In the above set-up, S is*

1. *a rational surface, if $d = 1$,*
2. *a K3 surface, if $d = 2$,*
3. *an honestly elliptic surface (Kodaira dimension $\kappa(S) = 1$), if $d > 2$.*

Proof We analyse case by case. If $d = 1$, then $p_g(S) = 0$. In fact, all higher plurigenera also vanish because $-K_S > 0$ is effective. In particular, $P_2(S) = q(S) = 0$, so S is rational by Castelnuovo's criterion (Theorem 4.9).

If $d \geq 2$, then K_S is automatically effective and nef (since $K_S^2 = 0$), so S is seen to be minimal. If $d = 2$, then $K_S = 0$ and $q(S) = 0$. Hence S is a K3 surface. If $d > 2$, then $K_S > 0$. By Theorem 4.26 we have an honestly elliptic surface with Kodaira dimension $\kappa(S) = 1$. This completes the proof of Proposition 5.51. □

Rational elliptic surfaces will reappear throughout this book. For instance, we will study their Mordell–Weil groups and lattices in Chaps. 7 and 8. Elliptic K3 surfaces are investigated in detail in Chaps. 11 and 12. Honestly elliptic surfaces are much less understood. They will play a central role when discussing elliptic curves of large rank and an application to the sphere packing problem in Chap. 13.

Note that rational elliptic surfaces are only relatively minimal, but not strictly minimal as by adjunction any section (P) has $P^2 = -1$. In contrast, as we have seen above, the elliptic surfaces for $d > 1$ are always minimal.

The above calculations carry over to elliptic surfaces with section over curves of positive genus, but we omit the details here.

Historical Notes

Let us comment once again on the two faces of elliptic surfaces: once as surfaces with a fibration, the other time as elliptic curves over function fields.

The study of elliptic curves over function fields with finite fields of constants was ignited by E. Artin's search for an analogue of the Riemann hypothesis in this setting ([Art24]). This was to be proved by Hasse in the genus one case and by Weil in general ([Wei48a]). Of course, there were many important subsequent developments and findings which we cannot possibly do justice to in this context. Let us just point out the contributions of Shafarevich, ranging from the famous Shafarevich conjecture and the Tate–Shafarevich group to the Shafarevich–Ogg–Néron formula.

Switching to the surface side, elliptic fibrations, of course, have been known to Enriques and his colleagues (see [Enr49]), but much of the theory of elliptic surfaces goes back to [Kod63] for the analytical part, and Néron [Ner64] and Shafarevich's seminar ([Sha+65]) for the algebraic part. Especially, this applies to the classification of singular fibres where another major contribution is due to Tate and his algorithm. The connection with Dynkin diagrams seems to be due to Shafarevich and his seminar ([Sha+65]). Néron's theory from [Ner64] (cf. also [BLR90] where the exposition is based on scheme theory) applies to abelian varieties of arbitrary dimension, but Néron was careful to check that in the one-dimensional case his results were compatible with Kodaira's. The elliptic curve case can also be considered as an instance of the theory of minimal models of two-dimensional schemes ([Sha66, Lic68]).

The Weierstrass model is the essential step towards the explicit construction of the Kodaira–Néron model; first detailed treatments (over an arbitrary base) can be found in [MS72, Del75]. It is remarkable how elegantly Tate's algorithm can handle the subtleties regarding minimalization and isolated singularities of the Weierstrass model corresponding to configurations of (-2)-curves on the Kodaira–Néron model being blown down (cf. also M. Artin's fundamental work on rational singularities in [Art66] and later also [Art77]). We highlight, and this will also be apparent from later chapters in this book, that Tate's algorithm, both its existence and its streamlined shape, are one of the key features of elliptic surfaces. In fact, this is one of the main reasons why elliptic surfaces are so handy for us these days, especially when compared with higher genus fibrations where dealing with the singular fibres can be quite delicate—already in genus two (work of Ogg, Namikawa–Ueno, Horikawa, and Liu among others). Note, however, that there still persists a theory of Mordell–Weil lattices for higher genus fibrations, cf. [Shio92c, Shio99b], and in a more concrete situation ([Shio15a]).

The canonical bundle formula was discovered by Kodaira over $\mathbb{C}$ ([Kod63]) and later extended to positive characteristic by [BM76/77], introducing in particular the contribution from wild fibres. The formula for the Euler–Poincaré characteristic has a longer history; often it can be derived from the Noether–Zeuthen formula. A proof over $\mathbb{C}$ can be found in [Sha+65], and the positive characteristic case is covered in [Dol72, Del73].

What may be a bit surprising is that the bijective correspondence between sections of the elliptic surface and rational points of the generic fibre entered the picture pretty late. Indeed, this part of the global theory which forms one of, if not the central ingredient of Mordell–Weil lattices, seems to have been recorded for the first time, in the form of the rank formula, in [Shio72], although certainly M. Artin and Tate must have been aware of it (compare [Tat66b]), and the same goes for Shafarevich. Without it, it was almost unthinkable to create independent sections in a systematic manner (although some mathematicians still managed, for instance Néron, as we will revisit in Sect. 13.1.1).

For genus one fibrations without section, the standard reference seems to be [CD89]. In fact, many properties are shared by the fibration and its jacobian, such as the Euler–Poincaré characteristic, the Picard number, and the configuration of singular fibres (except when they are multiple). We will touch upon this only very briefly in Sect. 12.4.

Chapter 6
Mordell–Weil Lattices

In this chapter, we give the definition of Mordell–Weil lattice (in Sect. 6.5). First, we bring together the concepts from Chaps. 4 and 5 in order to gain a better understanding of the Néron–Severi lattice of an elliptic surface. This will lead to the announced notion of Mordell–Weil lattice which will be studied in detail in this chapter, but also throughout the remainder of this book. We continue to work over an algebraically closed field k.

6.1 The Trivial Lattice

Let S be an elliptic surface with section and at least one singular fibre over some algebraic curve C defined over k, an algebraically closed field of characteristic $p \geq 0$. We have seen in greater generality that, up to torsion, the Néron–Severi group $\mathrm{NS}(S)$ becomes an integral lattice with respect to the intersection pairing. In fact, we will see shortly (Theorem 6.4) that $\mathrm{NS}(S)$ is torsion-free for any elliptic surface (with section). Hence it makes sense to regard $\mathrm{NS}(S)$, equipped with the intersection form, as a lattice—which we call the *Néron–Severi lattice* of S. In terms of its rank ρ (the Picard number of S), $\mathrm{NS}(S)$ is an indefinite lattice of signature $(1, \rho - 1)$ by the Hodge Index Theorem 4.14. We start out by studying an important sublattice of $\mathrm{NS}(S)$.

Definition 6.1 The *trivial lattice* $\mathrm{Triv}(S)$ is the sublattice of $\mathrm{NS}(S)$ generated by the zero section and fibre components.

Since any two fibres are algebraically equivalent, the only fibre components we have to consider for the trivial lattice are a general fibre F and fibre components not met by the zero section.

In the notation of Sect. 5.5, this leads to the following decomposition of the trivial lattice $\mathrm{Triv}(S) \subseteq \mathrm{NS}(S)$ as orthogonal sum

M. Schütt and T. Shioda, *Mordell–Weil Lattices*, Ergebnisse der Mathematik und ihrer Grenzgebiete. 3. Folge / A Series of Modern Surveys in Mathematics 70,
https://doi.org/10.1007/978-981-32-9301-4_6

$$\mathrm{Triv}(S) = \langle (O), F\rangle \oplus \bigoplus_{v\in R} T_v^- \tag{6.1}$$

where R denotes the finite subset $\{v \in C(k) \mid T_v \neq 0\}$ of points on the base curve where the singular fibres are located.

Example 6.2 (*Hexagonal example cont'd*) Let E over $k(t)$ denote the elliptic curve from (1.1). Then by Examples 5.26, 5.28, 5.29, the corresponding elliptic surface has trivial lattice

$$\mathrm{Triv}(S) = \begin{cases} \langle (O), F\rangle \oplus E_6^- & \text{if the } \alpha_i \text{ are pairwise distinct,} \\ \langle (O), F\rangle \oplus E_6^- \oplus A_1^- & \text{if there is exactly one duplicate } \alpha_i = \alpha_j, \\ \langle (O), F\rangle \oplus E_6^- \oplus A_2^- & \text{if all } \alpha_i \text{ are the same.} \end{cases}$$

Proposition 6.3 *The divisor classes of* $\{(O),\ F,\ \Theta_{v,i};\ \ v \in R,\ 1 \le i \le m_v - 1\}$ *form a* $\mathbb{Z}$*-basis of* Triv(S). *In particular,*

$$\mathit{rank}(\mathrm{Triv}(S)) = 2 + \sum_{v\in R}(m_v - 1). \tag{6.2}$$

Proof To see that the given divisor classes form a basis, one can compute the intersection matrix M of Triv(S) and verify that $\det(M) \neq 0$. Here it suffices to consider the orthogonal summands from (6.1) separately. Recall from Corollary 5.45 that any section P has self-intersection $(P)^2 = -\chi(S)$ (which is negative by Corollary 5.50). Hence the first summand has intersection form

$$\langle (O), F\rangle \cong \begin{pmatrix} -\chi & 1 \\ 1 & 0 \end{pmatrix}.$$

This matrix has determinant -1 and signature $(1, 1)$. It is indefinite and isometric to either $\langle 1\rangle \oplus \langle -1\rangle$ if χ is odd, or to the hyperbolic plane $\mathbb{Z}^2$ with intersection form $\begin{pmatrix} 0 & 1 \\ 1 & 0 \end{pmatrix}$ if χ is even.

From Sect. 5.5 we already know that all other summands T_v^- are negative-definite even lattices. Hence Triv(S) has signature $(1, 1 + \sum_{v\in R}(m_v - 1))$. □

By definition, the trivial lattice comprises all fibre components. These are often also called *vertical* divisors. On the other hand, we have *horizontal* divisors. These are linear combinations of irreducible curves with integer coefficients where each curve meets any fibre with a given and fixed multiplicity $d > 0$; we also refer to such a curve as a *multisection* of the degree d, and if $d = 1$, then it is a section. Clearly, the Néron–Severi lattice is generated by vertical and horizontal divisors as any divisor decomposes into the corresponding parts. One of the key features of elliptic surfaces with section is that in practice we do really not have to consider multisections (of degree greater than one), but only sections: NS(S) is generated by fibre components

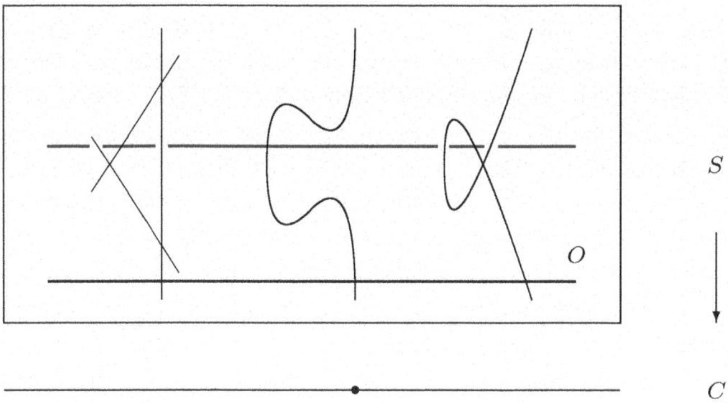

Fig. 6.1 Horizontal and vertical divisors

and sections (see Theorem 6.5 for the precise statement). Figure 6.1 may serve as an illustration.

6.2 The Néron–Severi Lattice

The following result is fundamental for a better understanding of NS(S).

Theorem 6.4 *Let S be an elliptic surface. Then* NS(S) *is finitely generated and torsion-free. That is, algebraic and numerical equivalence coincide.*

As explained before, the finiteness part is valid in greater generality for projective varieties as a special case of the theorem of the base [Lan83, Sect. 5] (see Theorem 4.3). Referring back to this, however, may be not so desirable in the present situation since the proof of the theorem of the base in [Lan83, Sect. 5] is reduced to the Mordell–Weil theorem for the higher-dimensional Jacobian variety Jac(C) of an auxiliary curve C on the variety X in question. This is technically more complicated than is strictly necessary for treating NS(S) where S is an elliptic surface, so we decided to pursue a direct approach following [Shio90a].

Sketch of Proof of Theorem 6.4. As only essential input, we use the existence of the cycle map γ as in (4.7). Since γ preserves the pairing and has kernel all those classes numerically equivalent to zero, we obtain at once that NS(S) modulo numerical equivalence is finitely generated and torsion-free. To prove Theorem 6.4, it thus suffices to show that numerical and algebraic equivalence agree.

Let $D \in \mathrm{Div}(S)$ such that $D \equiv 0$. From Riemann–Roch (4.9), we infer

$$h^0(D) + h^2(D) \geq \chi(D) = \chi(S) > 0$$

using Corollary 5.50. Hence $h^0(D) > 0$ or $h^2(D) > 0$. In the former case, D is effective and at the same time $D \equiv 0$, so we conclude the linear equivalence $D \sim 0$ (which clearly implies the algebraic equivalence $D \approx 0$, cf. Lemma 4.4).

Thus we may assume that $D \not\sim 0$ and $h^2(D) > 0$. Serre duality gives $h^0(K_S - D) > 0$, i.e. there is an effective divisor $D' \geq 0$ such that $D' \sim K_S - D$. We claim that D' is orthogonal to all fibre components of S. Indeed, for D this holds by assumption, while for K_S it is a consequence of the canonical bundle formula (Theorem 5.44) and the moving lemma. But then, since the pairing on the $\mathbb{Z}$-submodule of $\mathrm{NS}(S)$ generated by the fibre components is negative-semi-definite (as the summands T_v^- of the trivial lattice $\mathrm{Triv}(S)$ are negative-definite), $K_S - D$ is algebraically equivalent to some fibre multiple. Again by the canonical bundle formula (Theorem 5.44), the same applies to D. But then the multiplicity has to be zero since $(O.D) = 0$ by assumption. Hence $D \approx 0$ as claimed, and Theorem 6.4 follows. □

One of the obvious advantages of this result is that when dealing with divisor classes in $\mathrm{NS}(S)$ on elliptic surfaces it suffices to compute intersection numbers. Along the same lines, we do not have to use $\mathrm{Num}(S)$ anymore to eliminate the torsion in $\mathrm{NS}(S)$ for an elliptic surface S.

6.3 Horizontal and Vertical Divisors

The trivial lattice includes all the divisors on an elliptic surface S which are obvious:

- the fibre components, i.e. all vertical divisors, and
- the zero section.

While the rest of $\mathrm{NS}(S)$ cannot stem from sections exclusively as it comprises multisections of any degree, we claim that modulo the trivial lattice everything can be understood in terms of sections:

Theorem 6.5 *Let $f : S \to C$ be an elliptic surface with generic fibre E over $K = k(C)$. The map $P \mapsto (P) \mod \mathrm{Triv}(S)$ defines an isomorphism of abelian groups*

$$E(K) \cong \mathrm{NS}(S)/\,\mathrm{Triv}(S).$$

We will discuss the main ideas to prove the theorem below following [Shio90a]. Before going into any details, we note the following consequences:

Theorem 6.6 *Let E be an elliptic curve over the function field $K = k(C)$ of an algebraic curve C. Assume that the Kodaira–Néron model S of E has a singular fibre. Then the abelian group $E(K)$ is finitely generated.*

This result is a special case of the Mordell–Weil theorem, which holds in general for abelian varieties over suitable global fields (cf. [Lan83, Sect. 6], [Ser97, Sect. 4]). Our approach is singled out by its purely geometric nature.

The following formula is sometimes attributed to one of us and Tate; it seems that at least Shafarevich and his school were also aware of it:

Corollary 6.7 *Assumptions as in Theorem 6.6, let $r = rank(E(K))$. Then*

$$\rho(S) = r + 2 + \sum_{v \in R} (m_v - 1).$$

Note also that this agrees with the trivial bounds from (6.2):

$$\rho(S) \geq 2 + \sum_{v \in R} (m_v - 1) \geq 2.$$

Example 6.8 (*Hexagonal example cont'd*) Consider the elliptic curve E over $k(t)$ from (1.1). In Example 5.9, we have interpreted E in terms of a cubic pencil. It follows that the associated elliptic surface S can be identified with $\mathbb{P}^2$ blown up in the nine base points of the cubic pencil (some of them infinitely near). Hence

$$\rho(S) = 10.$$

Using Example 6.2 we infer from Corollary 6.7 that E has rank

$$\text{rank } E(k(t)) = \begin{cases} 2 & \text{if the } \alpha_i \text{ are pairwise distinct,} \\ 1 & \text{if there is exactly one duplicate } \alpha_i = \alpha_j, \\ 0 & \text{if all } \alpha_i \text{ come together.} \end{cases}$$

We emphasize that this result does not depend at all on the base field (once we tacitly extend our notions to (the smooth locus of the generic fibre of) quasi-elliptic surfaces for characteristic 3 (and 2), cf. the discussion around Corollary 7.22).

We shall now elaborate on the proof of Theorem 6.5. It will be achieved by exhibiting the inverse of the map

$$\begin{aligned} E(K) &\to \text{NS}(S)/\text{Triv}(S) \\ P &\mapsto (P) \mod \text{Triv}(S). \end{aligned}$$

For this purpose, we follow the approach employed in the proof of Proposition 5.4 and consider the scheme $\mathcal{S}$ underlying the surface S (compare [Liu06, Sect. 8.3]). Clearly, $\mathcal{S}$ contains the generic fibre E (though E is not a curve on $\mathcal{S}$ in the topological sense as its Zariski closure fills out all of $\mathcal{S}$). We start by defining a homomorphism

$$\text{Div}(\mathcal{S}) \to \text{Div}(E)$$

as follows: Any divisor D on S decomposes into a horizontal part, consisting of sections and multisections, and a vertical divisor consisting of fibre components:

$$D = D' + D'', \quad D' \text{ horizontal}, \quad D'' \text{ vertical}.$$

The vertical divisor D'' remains closed on the scheme $\mathcal{S}$, so in particular it is disjoint from E (as a subscheme). Meanwhile any irreducible component Y of the horizontal divisor D' naturally spreads out to a divisor $\mathcal{Y}$ on $\mathcal{S}$. We claim that $\mathcal{Y}$ and E do not only intersect as sets on the scheme $\mathcal{S}$, but their intersection carries a scheme structure giving a K-rational divisor

$$\mathcal{Y}_\eta = \mathcal{Y} \cap E$$

of degree $Y.F$ on E (where F denotes the general fibre as usual). To see this, it suffices to note that, as in the proof of Proposition 5.4 (or in [Liu06, Proposition 8.4]), $\mathcal{Y}$ is the closure of $\mathcal{Y}_\eta$ in $\mathcal{S}$, so the degrees indeed have to match. (This is also compatible with the specialization property that the intersection of Y and a special fibre F_s is obtained by specializing $\mathcal{Y}_\eta$ at s.) The precise scheme structure on $\mathcal{Y}_\eta$ (with multiplicities) can be derived by regarding $\mathcal{Y}$ (or Y) as a zero-dimensional scheme over K (just as we did with E). Again, this explains why the closure of $\mathcal{Y}_\eta$ returns $\mathcal{Y}$ and Y.

Express the horizontal divisor D' in terms of its irreducible components:

$$D' = \sum_{Y \text{ irred.}} a_Y Y, \quad a_Y \in \mathbb{Z}.$$

By abuse of notation, we define the restriction of D to E to be the following K-rational divisor on E:

$$D|_E := \sum_Y a_Y \mathcal{Y}_\eta \in \operatorname{Div}(E)$$

We shall now distinguish between linear equivalence on E and on S, indicated by the subscripts in $\sim_E$ and $\sim_S$. Thanks to the equality of functions fields

$$k(S) = K(E),$$

one directly verifies that

$$D|_E \sim_E 0 \iff D \sim_S D'' \text{ for some vertical divisor } D''.$$

In particular, the map

$$\begin{aligned} \operatorname{Pic}(S) &\to \operatorname{Pic}(E) \\ D &\mapsto D|_E \end{aligned}$$

is well-defined. By Abel's theorem for E over K (Lemma 3.5), the divisor D thus determines a unique point $P \in E(K)$ by the following linear equivalence of degree zero divisors:

$$D|_E - (D.F)\, O \sim_E P - O.$$

Writing $\psi(D) = P$, we obtain a surjective homomorphism

$$\psi : \mathrm{Div}(S) \to E(K). \tag{6.3}$$

The kernel of ψ is calculated as

$$\ker(\psi) = \langle D \in \mathrm{Div}(S);\ D \approx 0\rangle + \mathbb{Z}(O) + \langle D \in \mathrm{Div}(S);\ D \text{ vertical}\rangle. \tag{6.4}$$

Hence ψ induces the claimed isomorphism

$$\psi :\quad \mathrm{NS}(S)/\,\mathrm{Triv}(S) \cong E(K).$$

This completes the proof of Theorem 6.5. □

The above construction identifies any equivalence class of a divisor D (or of its horizontal component D') with the horizontal divisor given by the section (P). All these considerations will be made more explicit in Sect. 6.5 when we endow the Mordell–Weil group (modulo torsion) with a lattice structure.

Remark 6.9 Alternatively, one could employ Weil's viewpoint of a generic point and fibre from [Wei62] to prove Theorem 6.5 so to say directly on the elliptic surface S (cf. [Shio90a, Theorem 4.1]). Another possible approach consists in base changing to another elliptic surface S' where the horizontal divisor D decomposes into sections (as discussed around (5.18) or, for instance, in [RS15a]), and then argue with each section separately using Proposition 5.4, but then extra care has to be applied to the inseparable case.

6.4 Essential Lattice and Frame

For our further investigations we set up some more notation and terminology:

Definition 6.10 The orthogonal complement $\mathrm{Triv}(S)^\perp$ of $\mathrm{Triv}(S)$ in $\mathrm{NS}(S)$, with sign reversed, is called the *essential sublattice* of $\mathrm{NS}(S)$ and denoted by

$$L = L(S) = (\mathrm{Triv}(S)^\perp)^-.$$

Theorem 6.11 *The essential sublattice of the Néron–Severi group* $\mathrm{NS}(S)$ *is a positive-definite even integral lattice. Its rank is equal to the Mordell–Weil rank of* $E(K)$*:*

$$r = \rho - 2 - \sum_{v \in R} (m_v - 1), \tag{6.5}$$

and its determinant is given by

$$\det L = (-1)^{r-1} \det \mathrm{NS}(S) \cdot [\mathrm{NS}(S) : L^- \oplus \mathrm{Triv}(S)]^2 / \det \mathrm{Triv}(S). \tag{6.6}$$

Proof Since the signature of the lattice $\mathrm{NS}(S)$ is $(1, \rho - 1)$ and that of $\mathrm{Triv}(S)$ is $(1, \mathrm{rk}T - 1)$, L^- is negative-definite. Next, for any element of L^-, take a representative divisor D and write

$$D = \sum_j n_j \Gamma_j \quad (\Gamma_j : \text{irreducible curves}).$$

By the adjunction formula (Theorem 4.11), we have

$$(K.\Gamma_j) + (\Gamma_j^2) = 2p_a(\Gamma_j) - 2$$

for each j, where $p_a(\Gamma_j)$ denotes the arithmetic genus of Γ_j. It follows that

$$(\Gamma_j^2) \equiv (K.\Gamma_j) \equiv \chi \cdot (F.\Gamma_j) \bmod 2$$

by the canonical bundle formula from Theorem 5.44. Therefore we have

$$\begin{aligned}(D^2) &= \sum_j n_j^2(\Gamma_j^2) + 2\sum_{j<k} n_j n_k (\Gamma_j.\Gamma_k) \\ &\equiv \sum_j n_j \chi \cdot (F.\Gamma_j) \bmod 2 \\ &= \chi \cdot (F.D) = 0,\end{aligned}$$

because $D \perp \mathrm{Triv}(S)$ and $F \in \mathrm{Triv}(S)$. This shows that L^- is an even lattice, and so is L.

The rank of $L^- = \mathrm{Triv}(S)^\perp$ is equal to $\mathrm{rk}\,\mathrm{NS}(S) - \mathrm{rk}\ \mathrm{Triv}(S)$, hence (6.5) is a restatement of Corollary 6.7. Similarly, formula (6.6) follows from (2.3). □

Example 6.12 (*Hexagonal example cont'd*) We continue our discussion of the elliptic surface S arising from (1.1). Recall the Mordell–Weil ranks depending on the configuration of the α_i from Example 6.8. Since S was identified with $\mathbb{P}^2$ blown up in nine points in Example 5.9, one can show that $\mathrm{NS}(S)$ is unimodular (and odd, cf. Proposition 7.1). Hence L shares the same determinant, up to sign, with $\mathrm{Triv}(S)'$, the primitive closure of the trivial lattice $\mathrm{Triv}(S)$ of S inside $\mathrm{NS}(S)$, by Lemma 2.12 (we will study $\mathrm{Triv}(S)'$ in detail in Sect. 6.6). It follows that

$$L \cong \begin{cases} A_2 & \text{if the } \alpha_i \text{ are pairwise distinct,} \\ \langle 6 \rangle & \text{if there is exactly one duplicate } \alpha_i = \alpha_j. \end{cases}$$

Indeed, the former case corresponds to the binary even positive-definite quadratic form of minimal determinant (a subject of class group theory, see [Cox13]).

The same proof as for Theorem 6.11 gives the following result, which will be useful in the later application:

Theorem 6.13 *Let W denote the orthogonal complement of the rank 2 sublattice $U = \langle (O), F \rangle$ in* NS(S) *with sign reversed:*

$$W^- = U^\perp = \langle (O), F \rangle^\perp.$$

Then W is a positive-definite even integral lattice of rank $\rho - 2$, with $\det W = |\det \mathrm{NS}(S)|$. *The essential sublattice L of* NS(S) *is isomorphic to the orthogonal complement of $T = \oplus_{v \in R} T_v$ in W, i.e. $L \cong T^\perp \subset W$.*

Definition 6.14 We call W the *frame* of the elliptic surface S.

We remind the reader that the frame W is always an even integral lattice (by Theorem 6.11) even though NS(S) has the same parity as $\chi(S)$, i.e. NS(X) is even if and only of $\chi(S)$ is even.

Remark 6.15 Some papers, e.g. [Elk07], use the term "essential lattice" for what we call the frame.

We will discuss in Sect. 6.6.2 (and also in the framework of elliptic K3 surfaces in Sect. 11.5) how one can recover information about the elliptic surface such as singular fibres and sections from the frame.

6.5 Mordell–Weil Lattices

By Theorem 6.5, we know that the Mordell–Weil group $E(K)$ of an elliptic surface S is canonically isomorphic to the quotient group $\mathrm{NS}(S)/\mathrm{Triv}(S)$. In order to define a good pairing on $E(K)$, the first thing to try would be to split this isomorphism and embed $E(K)$ into NS(S).

We shall show that this can be done with a slight modification, namely by passing from NS(S) to $\mathrm{NS}(S)_{\mathbb{Q}} = \mathrm{NS}(S) \otimes \mathbb{Q}$. Note that there will be no harm in doing so since NS(S) is torsion-free (Theorem 6.4).

Recall the notation for the non-identity components of a reducible fibre at $v \in C(k)$: numbering them $\Theta_{v,1}, \ldots, \Theta_{v,m_v-1}$ ($\Theta_{v,0}$ being the identity component meeting the zero section O), we let A_v denote the Gram matrix of these components.

Lemma 6.16 *For any $P \in E(K)$, there exists a unique element of* $\mathrm{NS}(S)_{\mathbb{Q}}$, *say $\varphi(P)$, satisfying the following conditions:*

$$\text{(i) } \varphi(P) \equiv (P) \bmod \mathrm{Triv}(S)_{\mathbb{Q}}, \quad \textit{and} \quad \text{(ii) } \varphi(P) \perp \mathrm{Triv}(S).$$

Explicitly, $\varphi(P)$ is the class of the divisor D_P with $\mathbb{Q}$-coefficients:

$$D_P = (P) - (O) - ((P.O) + \chi)F \\ + \sum_{v \in R} (\Theta_{v,1}, \cdots, \Theta_{v,m_v-1})(-A_v^{-1}) \begin{pmatrix} (P.\Theta_{v,1}) \\ \vdots \\ (P.\Theta_{v,m_v-1}) \end{pmatrix}. \quad (6.7)$$

Proof Suppose that $\varphi(P)$ has a representative $D \in \mathcal{D}(S) \otimes \mathbb{Q}$. Then condition (i) says that

$$D \approx (P) + d(O) + nF + \sum_{v \in R, i \geq 1} b_{v,i}\Theta_{v,i} \quad (d, n, b_{v,i} \in \mathbb{Q}).$$

By condition (ii), the intersection number of D with F, (O) and $\Theta_{v,j}$ should be 0, which gives $d = -1, n = -((P.O) + \chi)$ and a unique solution of $b_{v,i}$ so that D is numerically equivalent to D_P given by (6.7). Conversely, the class of D_P satisfies the conditions (i) and (ii). □

Lemma 6.17 *The map*

$$\varphi : E(K) \to \mathrm{NS}(S)_{\mathbb{Q}}$$

is a group homomorphism such that

$$\mathrm{Ker}(\varphi) = E(K)_{\mathrm{tors}}. \quad (6.8)$$

Proof By Abel's theorem on E (Theorem 3.9), we see that

$$\varphi(P) + \varphi(Q) \equiv (P + Q) \bmod \mathrm{Triv}(S)_{\mathbb{Q}} \; (P, Q \in E(K)).$$

By the uniqueness in Lemma 6.16, we have then $\varphi(P + Q) = \varphi(P) + \varphi(Q)$. Hence φ is a group homomorphism. For $P \in E(K)$, we have $\varphi(P) = 0$ if and only if some multiple of $(P) - (O)$ is algebraically equivalent to an integral linear combination of F and $\Theta_{v,i}$. By Theorem 6.5, the latter holds if and only if that multiple is linearly equivalent to a vertical divisor. Restricting to E and using Abel's theorem, we see that this is the case if and only if P is a torsion point. □

Lemma 6.18 *Let L be the essential lattice of an elliptic surface S, and let m be the least common multiple of $m_v^{(1)}$ defined just after (5.2):*

$$m = lcm\{m_v^{(1)} | v \in R\}.$$

Then we have

$$\mathrm{Im}(\varphi) \subset \frac{1}{m}L^{-} \quad (6.9)$$

and φ induces an injection:

$$\varphi' : E(K)/E(K)_{\text{tors}} \hookrightarrow \frac{1}{m}L^- \subset L^-_{\mathbb{Q}} = L^- \otimes \mathbb{Q}. \tag{6.10}$$

Proof By (5.3), we have

$$\det(-A_v) = \det T_v = m_v^{(1)}.$$

Hence, by (6.7), mD_P is an integral divisor for any $P \in E(K)$, which implies that $m\varphi(P) \in L^-$. This proves (6.9), and (6.10) follows from Lemma 6.17. □

Remark 6.19 The constant m in Lemma 6.18 can sometimes be improved slightly by analysing the precise group structure on the singular fibres (or the underlying discriminant groups, see Sect. 5.6.1).

Theorem 6.20 *For any* $P, Q \in E(K)$, *let*

$$\langle P, Q \rangle = -(\varphi(P).\varphi(Q)). \tag{6.11}$$

Then this defines a $\mathbb{Q}$*-valued symmetric bilinear pairing on* $E(K)$ *which induces the structure of a positive-definite lattice on* $E(K)/E(K)_{\text{tors}}$.

Proof The pairing (6.11) is obviously symmetric, and it is bilinear because φ is a homomorphism, see Lemma 6.17. Now the essential lattice L, with the intersection pairing $(\cdot)$, is positive-definite (Theorem 6.11), and hence we obtain

$$\langle P, P \rangle = -(\varphi(P).\varphi(P)) \geq 0 \quad \text{for any } P \in E(K).$$

From (6.8), we derive the equivalence

$$\langle P, P \rangle = 0 \iff \varphi(P) = 0 \iff P \subset E(K)_{\text{tors}}. \tag{6.12}$$

This concludes the proof of Theorem 6.20. □

Definition 6.21 The pairing (6.11) on the Mordell–Weil group $E(K)$ will be called the *height pairing*, and the lattice $(E(K)/E(K)_{\text{tors}}, \langle \cdot, \cdot \rangle)$ will be called the *Mordell–Weil lattice* of the elliptic curve E/K, or of the elliptic surface $f : S \to C$ over k.

Notation 6.22 (i) We will often abbreviate the Mordell–Weil lattice as MWL(S) or even MWL if no confusion is likely.
(ii) When dealing with different elliptic curves, we may distinguish the height pairing on each of them by suitable subscripts such as $\langle \cdot, \cdot \rangle_{E(K)}$.

With the concept of Mordell–Weil lattice at our disposal, we shall now work out an explicit formula for the height pairing. For this purpose, we need some extra terms accounting for the way the singular fibres are met by the sections:

Definition 6.23 Let $P, Q \in E(K)$. Then we define the *local contribution* from the singular (reducible) fibre at $v \in R$, denoted by $contr_v(P, Q)$, as follows: suppose that (P) intersects the irreducible component $\Theta_{v,i}$ and (Q) intersects $\Theta_{v,j}$. Then we let

$$contr_v(P, Q) = \begin{cases} (-A_v^{-1})_{i,j} & \text{if } i \geq 1, j \geq 1, \\ 0 & \text{otherwise.} \end{cases} \tag{6.13}$$

where the first alternative refers to the (i, j)-entry of the matrix $(-A_v^{-1})$. Further, we set

$$contr_v(P) = contr_v(P, P).$$

The contribution terms put us in the position to make the height pairing explicit. As before, we denote by χ the Euler characteristic of $\mathcal{O}_S$, while $(P.Q)$ is the intersection number of given sections (P) and (Q), especially for $P = O$ or $Q = O$.

Theorem 6.24 (Explicit formula for the height pairing) *For any $P, Q \in E(K)$, we have*

$$\langle P, Q \rangle = \chi + (P.O) + (Q.O) - (P.Q) - \sum_{v \in R} contr_v(P, Q) \tag{6.14}$$

and, in particular,

$$\langle P, P \rangle = 2\chi + 2(P.O) - \sum_{v \in R} contr_v(P). \tag{6.15}$$

Proof By formula (6.7), we have

$$\langle P, Q \rangle = -(D_P.D_Q) = -(D_P.Q)$$

because $D_Q \equiv (Q) \bmod \mathrm{Triv}(S)_{\mathbb{Q}}$ and $D_P \perp \mathrm{Triv}(S)$. Hence

$$\begin{aligned} \langle P, Q \rangle &= -[(P.Q) - (Q.O) - ((P.O) + \chi)] \\ &\quad - \sum_{v \in R} ((Q.\Theta_{v,1}), \ldots, (Q.\Theta_{v,m_v-1}))(-A_v^{-1}) \begin{pmatrix} (P.\Theta_{v,1}) \\ \vdots \\ (P.\Theta_{v,m_v-1}) \end{pmatrix}. \end{aligned}$$

This is equivalent to (6.14) in view of the definition of $contr_v$. By taking $Q = P$ and using the adjunction formula, (6.14) reduces to (6.15). □

Let us write down the explicit values of the local terms in (6.13). To avoid confusion, we fix the numbering of the irreducible components $\Theta_i = \Theta_{v,i} (i = 0, 1, .., m_v - 1)$ of a reducible fibre F_v as in Sect. 5.4. The identity component meeting the zero section (O) is always denoted by $\Theta_{v,0}$ or Θ_0. For a singular fibre of type I_n $(m_v = m_v^{(1)} = n)$, we number Θ_i cyclically, i.e. Θ_i and Θ_j intersect each other if

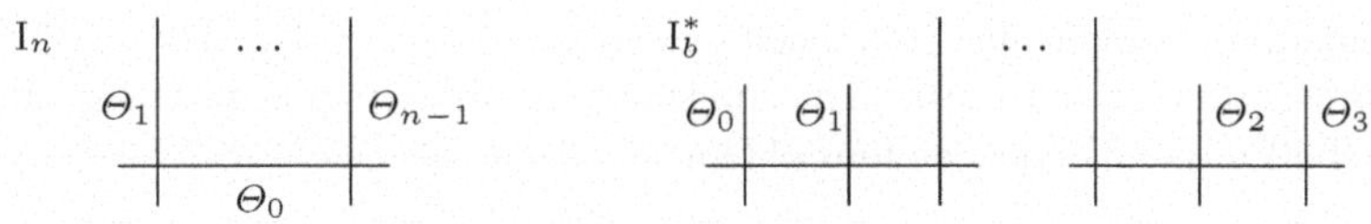

Fig. 6.2 Numbering for fibres of type $\mathrm{I}_n (n > 1)$ and $\mathrm{I}_b^* (b \geq 0)$

Table 6.1 Local contributions from singular fibres to the height pairing

T_v	A_1	E_7	A_2	E_6	A_{n-1}	D_{n+4}
Type of F_v	III	III*	IV	IV*	I_n $(n \geq 2)$	I_n^* $(n \geq 0)$
$contr_v(P)$	$\frac{1}{2}$	$\frac{3}{2}$	$\frac{2}{3}$	$\frac{4}{3}$	$\frac{i(n-i)}{n}$	$\begin{cases} 1 & (i = 1) \\ 1 + \frac{n}{4} & (i > 1) \end{cases}$
$contr_v(P, Q)$	–	–	$\frac{1}{3}$	$\frac{2}{3}$	$\frac{i(n-j)}{n}$	$\begin{cases} \frac{1}{2} & (i = 1) \\ \frac{1}{2} + \frac{n}{4} & (i > 1) \end{cases}$

and only if $i - j \equiv 1 \bmod n$. For a singular fibre of type I_b^* $(m_v = b + 5, m_v^{(1)} = 4)$, we number the four simple components $\Theta_0, \Theta_1, \Theta_2, \Theta_3$ as in Notation 5.13 so that Θ_0 meets the zero section and Θ_1, the near component, meets the same double component as Θ_0 (see Fig. 6.2).

For the other types of reducible fibres, the numbering is irrelevant as long as Θ_0 denotes the identity component.

Now, suppose that (P) intersects $\Theta_{v,i}$ and (Q) intersects $\Theta_{v,j}$ and assume that $i \geq 1, j \geq 1$. Then the contribution terms are given in Table 6.1: the third row gives the value of $contr_v(P, Q)$ for the case $i = j$ as well as $contr_v(P)$, and the fourth row concerns the case $i < j$ (interchange the order of P, Q if necessary).

Remark 6.25 The Mordell–Weil lattice need not be an integral (nor an even) lattice. By inspection of the height pairing and of Table 6.1, it is exactly the local contribution terms which account for this property.

Example 6.26 (*Hexagonal example cont'd*) Recall the elliptic curve E over $k(t)$ from (1.1) with rational points P_i $(i = 1, 2, 3)$. In order to compute their height pairing, we start by computing how they intersect each other and the reducible fibre(s). Obviously the distinct sections are disjoint over $\mathbb{A}^1$, since the x-coordinates do not match. At ∞, we argue as in Example 5.9 with the Weierstrass form

$$y^2 + s^2 y = (x - \alpha_1 s^2)(x - \alpha_2 s^2)(x - \alpha_3 s^2) \tag{6.16}$$

and with the sections $P_i = (\alpha_i s^2, 0)$. Note that the sections are disjoint from O. After two blow-ups, first at the singularity $(x, y, s) = (0, 0, 0)$ and then at $(x', y', s) = (0, 0, 0)$ in the chart $s' = 1$ of the blow-up, we obtain an affine model

$$S' : \quad y''^2 + y'' = s^2(x'' - \alpha_1)(x'' - \alpha_2)(x'' - \alpha_3). \tag{6.17}$$

This contains the simple fibre components given by $(s = y'' = 0)$ and $(s = y'' + 1 = 0)$ and the sections $P_i = (\alpha_i, 0)$, all meeting the same fibre component, but in different points. If all α_i are distinct, we thus obtain

$$\langle P_i, P_i\rangle = 2 - \frac{4}{3} = \frac{2}{3}, \quad \langle P_i, P_j\rangle = 1 - \frac{4}{3} = -\frac{1}{3} \ (i \neq j).$$

Thus either two of them generate the lattice $A_2^\vee$ (to which we already alluded in the introduction), and together the sections P_i and Q_i form the six minimal vectors of $A_2^\vee$ by symmetry.

Next assume that $\alpha_i = \alpha_j \neq \alpha_k$. Then $P_i = P_j$ meets the cusp at $t = 0$ of the Weierstrass model of E, thus it meets the non-identity component of the singular fibre of type I_2. While the height of P_k stays unchanged, we find

$$\langle P_i, P_i\rangle = 2 - \frac{4}{3} - \frac{1}{2} = \frac{1}{6}, \quad \langle P_i, P_k\rangle = 1 - \frac{4}{3} = -\frac{1}{3}.$$

In particular, the height matrix $(\langle P_i, P_j\rangle)_{1\leq i,j\leq 3}$ has rank one only, as predicted in Example 6.8.

Finally, if all α_i are the same, then the fibre at $t = 0$ degenerates to Kodaira type IV as in Example 5.28. As before, it is met by P_i at a non-identity component, so we get

$$\langle P_i, P_i\rangle = 2 - \frac{4}{3} - \frac{2}{3} = 0.$$

In view of Theorem 6.20, this implies that P_i is a torsion section in this case. We study this situation more systematically in the next section (see especially Remark 6.38 which will show that P_i has order 3).

We leave it as an exercise for the reader to compute the height pairing with the inverse sections Q_i from the introduction.

Remark 6.27 We point out that except for the lack of an explicit description of the singular fibres, the concept of Mordell–Weil lattices makes perfect sense for higher genus fibrations as well (cf. [Shio92c]).

6.6 Torsion Sections

Torsion points provide a decisive tool in the study of elliptic curves. In fact, their existence over a given non-closed field is rather restrictive, as we shall illustrate in Example 6.30. By now, there are many results in this direction in all kinds of settings; as a sample we refer to the initial comments in [Cas66, esp. p. 264], and as one of the first results, to [Lev68], plus the monograph [KM85].

In order to study the torsion subgroup of the Mordell–Weil group of an elliptic surface S, we are led to consider the primitive closure (or saturation) $\mathrm{Triv}(S)'$ of the trivial lattice $\mathrm{Triv}(S)$ in $\mathrm{NS}(S)$:

$$\mathrm{Triv}(S)' = (\mathrm{Triv}(S) \otimes \mathbb{Q}) \cap \mathrm{NS}(S).$$

With this overlattice of $\mathrm{Triv}(S)$, we easily deduce the following statements from Theorem 6.5:

Lemma 6.28 *Let $P \in E(K)$ be torsion. Then $D_P \in \mathrm{Triv}(S)'$.*

Corollary 6.29 *In the above notation,*

$$E(K)_{\mathrm{tors}} \cong \mathrm{Triv}(S)'/\,\mathrm{Triv}(S).$$

The above results may give the impression that the primitive closure of the trivial lattice is used to compute the torsion inside $E(K)$. In practice, however, it is often exactly the opposite approach that prevails: we obtain the primitive closure $\mathrm{Triv}(S)'$ of $\mathrm{Triv}(S)$ from the torsion inside $E(K)$. The hexagonal example should illustrate this well:

Example 6.30 (*Hexagonal example cont'd*) Consider the degenerate case for the elliptic curve E over $k(t)$ from (1.1) where all α_i coincide. Then the section P_i has height zero by Example 6.26, hence it is torsion by Theorem 6.20. Of course, this can be checked explicitly without problems, and the order turns out to be 3 (see Remark 6.38). In this direction, we would like to highlight that D_P has coefficients in $\frac{1}{3}\mathbb{Z}$, giving the glue vector for the imprimitive embedding of the trivial lattice into the unimodular Néron–Severi lattice,

$$\mathrm{Triv}(S) = \langle (O), F \rangle \oplus E_6^- \oplus A_2^- \hookrightarrow \mathrm{NS}(S),$$

as implied by the discriminant groups discussed in Sect. 2.3.5.

Conversely, one can often use the trivial lattice to rule out torsion inside the Mordell–Weil group. To this end, we collect some useful results that hold in general for elliptic surfaces.

Proposition 6.31 *(i) Let n be the order of the torsion subgroup $E(K)_{\mathrm{tors}}$. Then*

$$n^2 \mid |\det(\mathrm{Triv}(S))| = \prod_{v \in R} \det T_v = \prod_{v \in R} m_v^{(1)}. \tag{6.18}$$

In particular, if $\det(\mathrm{Triv}(S))$ *is square-free, then the Mordell–Weil group is torsion-free.*

(ii) An element $P \in E(K)$ is torsion if and only if $\langle P, P \rangle = 0$, i.e.,

$$\sum_{v \in R} \mathrm{contr}_v(P) = 2\chi + 2(P.O). \tag{6.19}$$

In particular, $E(K)$ is torsion-free if

$$\sum_{v\in R} \max\{contr_v(P);\ P \in E(K)\} < 2\chi. \tag{6.20}$$

(iii) There is a natural injection

$$E(K)_{\text{tors}} \hookrightarrow \text{Triv}(S)^\vee / \text{Triv}(S) \cong \oplus_{v\in R} T_v^\vee / T_v. \tag{6.21}$$

We briefly indicate how to prove Proposition 6.31. *(i)* follows from (2.3) and Corollary 6.29. *(ii)* is a direct application of the height pairing (6.15) and (6.12). As for *(iii)*, the injectivity of (6.21) is implied by Corollary 6.29 since the primitive closure $\text{Triv}(S)'$ of the trivial lattice $\text{Triv}(S)$ clearly embeds into the dual $\text{Triv}(S)^\vee$.

Example 6.32 (*Hexagonal example cont'd*) If for the elliptic curve E from (1.1) not all α_i come together, then the analysis of the singular fibres in Examples 5.26, 5.28, 5.29 shows by (6.20) that $E(K)$ is torsion-free.

In the above study of torsion sections, we should include three further properties which are well-known in the theory of Kodaira–Néron models ([Ner64], cf. also [Sil94]). Like Lemma 5.23, they do not follow directly from the theory of Mordell–Weil lattices, so we decided to list them separately, but we continue the numbering.

Proposition 6.33 *(iv) At any additive fibre F_v, any torsion subgroup of $E(K)$ with order prime to char(k) injects into the discriminant group of the fibre $T_v^\vee / T_v$.*

(v) For any torsion element P of order not divisible by char(k), *the sections (P) and (O) are disjoint, i.e., $(P.O) = 0$. In particular, if* char$(k) = 0$, *then $P \neq O$ being torsion implies $\sum_{v\in R} \text{contr}_v(P) = 2\chi$.*

(vi) If char$(k) = p$, then the p-torsion subgroup of $E(K)$ may only have order 1 *or p (but the p-torsion group scheme always has length p^2).*

Remark 6.34 We point out that the statements *(iv), (v)* from Proposition 6.33 do generally not hold for torsion sections whose order is a multiple of the characteristic. A counterexample in terms of an elliptic K3 surface in characteristic 2 was given in [OS90, Appendix 2] (see also Sect. 12.4.4.2). However, we shall see (by virtue of a complete classification, see Theorem 8.8) that on rational elliptic surfaces property *(v)* holds regardless of the characteristic (Corollary 8.30).

Remark 6.35 While we highlighted before that almost all properties of elliptic surfaces carry over to quasi-elliptic surfaces, the statements of Proposition 6.33 definitely do not: quasi-elliptic surfaces only admit p-torsion sections (see Corollary 7.22), all fibres are additive, sections tend to intersect, and already in the simplest K3 case, the torsion group may have order 2^4 or 3^4 (see [ElS15, Sen17]).

Example 6.36 (*Full* 3*-torsion*) The Hesse pencil from Example 3.13 has full 3-torsion outside characteristic 3, given by the nine base points of the cubic pencil.

This can be easily verified by the above methods, but, of course, it also follows at once since the base points are points of inflection.

Note that the degeneration in characteristic 3, visible as base points coming together in threes, is forced by Proposition 6.33 *(vi)*.

Example 6.37 (*Full 2-torsion*) Another classical example consists in the Legendre family of cubics

$$y^2 = x(x-1)(x-\lambda). \tag{6.22}$$

Outside characteristic 2, this defines an elliptic curve over $k(\lambda)$ with full 2-torsion given by the sections $(0,0)$, $(1,0)$, $(\lambda,0)$. In characteristic 2, the Kodaira–Néron model of (6.22) defines a quasi-elliptic surface, but rather intriguingly, the same structure (singular fibres, torsion sections) continues to persist.

We point out that the above examples should be considered in the general framework of elliptic modular surfaces as introduced in [Shio72] (analytically over $\mathbb{C}$; see [SSh10] for a quick algebraic treatment).

Remark 6.38 Once we have verified some section P to be torsion by computing its height to be zero, we can easily compute the order of P. For instance, this is achieved by calculating the minimum integer $N \in \mathbb{N}$ such that the $\mathbb{Q}$-divisor $N \cdot D_P$ from (6.7) has in fact integer coefficients (so that it lies in the trivial lattice $\mathrm{Triv}(S)$). This can be read off from the fibre components met by (P), using the group structure on the singular fibres as explained in Sect. 5.6.1. Often, though not always, the order in fact appears as the least common multiple of the denominators of the contribution terms in the height (6.15) of P (such as for the 2-torsion sections in Example 6.37, or in the fully degenerate hexagonal case in Example 6.30).

We conclude this section with the two very useful applications (or instances) of the theory of Mordell–Weil lattices.

6.6.1 Determinant Formula

The first application concerns the precise computation of the Néron–Severi group, and especially of its determinant:

Corollary 6.39 (Determinant formula) *In the above notation, we have for an elliptic surface S of* MW*-rank r:*

$$\det \mathrm{NS}(S) = (-1)^r \det \mathrm{Triv}(S) \cdot \det \mathrm{MWL}(S)/|\,\mathrm{MW}(S)_{\mathrm{tors}}|^2. \tag{6.23}$$

Proof In the terminology of Lemmas 6.16, 6.17 we can regard φ as the orthogonal projection in the $\mathbb{Q}$-vector space $\mathrm{NS}(S) \otimes \mathbb{Q}$ with respect to the subspace $\mathrm{Triv}(S) \otimes \mathbb{Q}$. For the lattice $\mathrm{NS}(S)$ inside this vector space, we thus find

$$\det \mathrm{NS}(S) = \det \mathrm{Triv}(S)' \cdot \det \varphi(\mathrm{NS}(S)), \tag{6.24}$$

since $\mathrm{Triv}(S)'$ embeds primitively into $\mathrm{NS}(S)$ by definition. On the one hand, by Corollary 6.29, we have

$$\det \mathrm{Triv}(S) = \det \mathrm{Triv}(S)' \cdot |\mathrm{MW}(S)_{\mathrm{tors}}|^2.$$

On the other hand, the equality

$$\det \varphi(\mathrm{NS}(S)) = (-1)^r \det \mathrm{MWL}(S)$$

holds by Theorem 6.5 and by definition of (the sign in) the height pairing on $\mathrm{MWL}(S)$. Together, the last two equalities substitute into (6.24) to give (6.23). □

Remark 6.40 The same formula (6.23) applies to any sublattice $N \subset \mathrm{NS}(S)$ generated by fibre components and sections (including general fibre and zero section), with intuitive notation $\mathrm{Triv}(N)$, $\mathrm{MWL}(N)$ etc. obtained by replacing S in formula (6.23) by N.

It is in fact hard to overestimate the practical relevance of Corollary 6.39 (and the extension in Remark 6.40); often it forms a key ingredient for the computation of the Néron–Severi group of an elliptic surface—or even for engineering a surface, say in a family, with prescribed Néron–Severi lattice. Arguably, this technique develops its full power in the framework of K3 surfaces (as we will indicate in Chap. 12, see also [Elk08, EK14, ElS13] for important applications in this direction). Here we content ourselves with showing how (6.23) may be used to compute the full Mordell–Weil group of a given elliptic curve:

Example 6.41 (*Hexagonal example cont'd*) For the elliptic curve $E/k(t)$ from (1.1), we have computed the rank of $E(k(t))$ in Example 6.8 depending on the configuration of the α_i. Since $\mathrm{NS}(S)$ is unimodular as explained before, (6.23) implies that the determinants of the trivial lattice and of the Mordell–Weil lattice are multiplicative inverses to each other up to sign (unless all α_i come together as in Example 6.30). Hence the height pairing computations in Example 6.26 yield the following:

$$\mathrm{MWL}(S) = E(k(t)) = \begin{cases} \langle P_i, P_j\rangle \cong A_2^\vee \ (i \neq j) & \text{if the } \alpha_i \text{ are pairwise distinct,} \\ \mathbb{Z}P_i \cong \langle \frac{1}{6}\rangle & \text{if } \alpha_i = \alpha_j \neq \alpha_k. \end{cases}$$

In particular, this confirms the claim about the Mordell–Weil lattice of E made in the introduction.

6.6.2 Frame Versus Singular Fibres

The second application is directed towards determining the singular fibres abstractly from the frame.

Recall that for an elliptic fibration with a section, the lattice generated by (O) and F is unimodular. Hence we obtain a decomposition

$$\mathrm{NS}(S) = \langle (O), F\rangle \oplus W^{-}. \tag{6.25}$$

Here W is the frame of the elliptic fibration from Sect. 6.4. By Theorem 6.13 it is an even and positive-definite lattice of rank $\rho(S) - 2$. The next proposition explains how one can retrieve the singular fibres and the Mordell–Weil lattice from the frame (as soon as the surface is not rational or otherwise we have $\chi(S) > 1$):

Proposition 6.42 *Let S be an elliptic surface with $\chi(\mathcal{O}_S) > 1$. Consider the corresponding decomposition $\mathrm{NS}(S) = \langle (O), F\rangle \oplus W^{-}$ from (6.25). Then the root lattices of the singular fibres and the sections can be recovered from W as follows:*

(i) $T = \sum_v T_v = \mathcal{R}(W)$, *the sublattice spanned by the roots inside W;*
(ii) $\mathrm{MW}(S)_{\mathrm{tors}} \cong \mathcal{R}(W)'/\mathcal{R}(W)$ *where $\mathcal{R}(W)'$ denotes the primitive closure of the root lattice $\mathcal{R}(W)$ in W;*
(iii) $\mathrm{MWL}(S) \cong W/\mathcal{R}(W)'$.

Proof Let ψ denote the orthogonal projection in $\mathrm{NS}(S)$ with respect to the unimodular lattice $\langle (O), F\rangle$. Then $W^{-} = \operatorname{im}\psi$ clearly contains the given orthogonal sum of root lattices $T^{-} = \sum_v T_v^{-}$, and Theorem 6.5 implies

$$E(K) \cong \mathrm{NS}(S)/\operatorname{Triv}(S) \cong W^{-}/\left(\oplus_v T_v^{-}\right).$$

As a consequence, any vector $w \in W^{-} \setminus (\sum_v T_v^{-})$ corresponds to some section $P \in E(K)$; that is,

$$w = \psi(P) = (P) - (O) - (\chi(\mathcal{O}_S) + (P.O))F.$$

This gives

$$\psi(P)^2 = -2\chi(\mathcal{O}_S) - 2(P.O)$$

(the *naive height* of P without correction terms from singular fibres). In particular

$$w^2 = \psi(P)^2 < -2 \quad \text{if} \quad \chi(\mathcal{O}_S) > 1.$$

This shows *(i)* (after changing signs, in agreement with the fact that W is positive-definite).

For *(ii)* recall that the torsion sections lie inside the primitive closure $\operatorname{Triv}(S)'$ of the trivial lattice $\operatorname{Triv}(S)$, and in fact by Corollary 6.29

$$\mathrm{MW}(S)_{\mathrm{tors}} \cong \mathrm{Triv}(S)'/\,\mathrm{Triv}(S).$$

Presently, $\mathrm{Triv}(S) = \langle O, F\rangle \oplus \mathcal{R}(W)^-$ by *(i)*, and since the first summand is unimodular, we infer the claimed isomorphism

$$\mathrm{MW}(S)_{\mathrm{tors}} \cong \mathrm{Triv}(S)'/\,\mathrm{Triv}(S) \cong \varphi(\mathrm{Triv}(S)')/\varphi(\mathrm{Triv}(S)) \cong \mathcal{R}(W)'/\mathcal{R}(W).$$

Similarly, we can rephrase the relations for the Mordell–Weil lattice based on Theorem 6.5:

$$\mathrm{MWL}(S) \cong \mathrm{NS}(X)/\mathrm{Triv}(S)' \cong W/\mathcal{R}(W)',$$

again using the unimodularity of $\langle O, F\rangle$. This proves *(iii)*. □

Remark 6.43 Compare the case of a rational elliptic surface (so that $\chi = 1$, see the next two chapters): then any section P has $P^2 = -1$, so in the above notation, we have $\psi(P)^2 = -2 - 2(P.O)$. Thus $\psi(P)$ gives a root of the frame W^- (with sign reversed) if and only if (P) is disjoint from (O), i.e. $(P.O) = 0$. (In Sect. 8.7, this will be called an integral section.) This perfectly agrees with the fact that $W \cong E_8$ regardless of the precise configuration of singular fibres.

The decomposition (6.25) is in general not unique, but depends on the choice of an elliptic fibration. In fact, an algebraic surface has a unique elliptic fibration with section up to isomorphism, if any, unless the canonical bundle is trivial; it follows from the canonical bundle formula that this exception occurs exactly for K3 surfaces (and for the constant fibrations on an abelian surface of product type $E \times E'$, but these would not satisfy Convention 5.10). For details, the reader might start from Sect. 11.5.

6.7 Narrow Mordell–Weil Lattice

Now we turn our attention to an important subgroup $E(K)$ which will be called the narrow Mordell–Weil group. It is defined as follows:

$$E(K)^0 = \{P \in E(K) |\ (P) \text{ meets } \Theta_{v,0} \text{ for all } v \in R\}. \tag{6.26}$$

$E(K)^0$ is of finite index in $E(K)$; in fact, $mE(K) \subset E(K)^0$ if m is a common multiple of $m_v^{(1)}$ $(v \in R)$ in analogy of (the proof of) Lemma 6.18. By (6.14) and (6.15), we have:

$$\langle P, Q\rangle = \chi + (P.O) + (Q.O) - (P.Q) \quad \text{if } P \text{ or } Q \in E(K)^0. \tag{6.27}$$

and

$$\langle P, P\rangle = 2\chi + 2(P.O) \quad \text{for any } P \in E(K)^0. \tag{6.28}$$

Theorem 6.44 *$E(K)^0$ is a torsion-free subgroup of finite index in $E(K)$. Viewed as a lattice of rank r with respect to the height pairing, it is a positive-definite even integral lattice with minimal norm $\geq 2\chi$. Further, we have*

$$\det E(K)^0 = (-1)^r \det \mathrm{NS}(S) \cdot [E(K) : E(K)^0]^2 / \det \mathrm{Triv}(S).$$

Proof For any $P \in E(K)$, $P \neq O$, we have $(P.O) \geq 0$ because the sections (P) and (O) are distinct irreducible curves on the surface S. Hence, if $P \in E(K)^0$, $P \neq O$, then we have by Corollary 5.50 and (6.28)

$$\langle P, P \rangle \geq 2\chi > 0.$$

Thus P is not a torsion section by Lemma 6.17. Further, (6.28) obviously shows that the lattice $E(K)^0$ is even and positive-definite, with minimal norm $\geq 2\chi$. The determinant formula for $E(K)^0$ follows from (6.6). □

Definition 6.45 The even integral lattice $(E(K)^0, \langle , \rangle)$ will be called the *narrow Mordell–Weil lattice* of the elliptic curve E/K or of the elliptic surface $f : S \to C$.

Example 6.46 (*Hexagonal example cont'd*) Consider the elliptic curve E over $k(t)$ from (1.1). In Example 6.41, we computed the Mordell–Weil lattice depending on the configuration of the α_i. Here we compute the narrow Mordell–Weil lattice, using the group structure on the smooth locus of the reducible fibres.

If all α_i are distinct, then the only reducible fibre has Kodaira type IV* by Example 5.26. Hence $3P_i \in E(k(t))^0$ for all $i = 1, 2, 3$. By Example 6.26, these sections generate the lattice $A_2^\vee(3)$ of determinant 3. In order to find a basis of $E(k(t))^0$, we consider the sections

$$R_i = P_i - P_k, \quad R_j = P_j - P_k$$

which all meet the identity component of the IV* fibre (since the P_i meet the same (non-identity) component). One easily verifies that they generate the lattice A_2 with the height pairing. Since this has no integral overlattices of finite index, we deduce

$$E(k(t))^0 = \langle R_i, R_j \rangle \cong A_2.$$

If $\alpha_i = \alpha_j \neq \alpha_k$, then we have seen in Example 6.41 that P_i generates $E(k(t)) \cong \langle \frac{1}{6} \rangle$. Recall from Example 6.26 that P_i meets the I_2 fibre at the non-identity component. Hence $6P_i \in E(k(t))^0$, generating a sublattice of finite index and determinant 6. As the determinant is square-free, so there is no integral overlattice of finite index, we infer as before that

$$E(k(t))^0 = \mathbb{Z}(6P_i) \cong \langle 6 \rangle.$$

We continue by drawing the connection between narrow Mordell–Weil lattice and essential lattice.

Theorem 6.47 *Via the map $\varphi : E(K) \to \mathrm{NS}(S)_{\mathbb{Q}}$, the narrow Mordell–Weil lattice $E(K)^0$ is isomorphic to the essential sublattice L of $\mathrm{NS}(S)$. Furthermore, the Mordell–Weil lattice $E(K)/E(K)_{\mathrm{tors}}$ is embedded via φ into the dual lattice of L^-, and we have a commutative diagram of lattices:*

$$\begin{array}{ccc} E(K)/E(K)_{\mathrm{tors}} & \subset & L^\vee \\ \cup & & \cup \\ E(K)^0 & \simeq & L \end{array} \tag{6.29}$$

Proof If $P \in E(K)^0$, then $(P.\Theta_{v,i}) = 0$ for all $v \in R$, all $i \geq 1$. By (6.7), $\varphi(P)$ is the class of the (integral) divisor

$$D_P = (P) - (O) - ((P.O) + \chi)F. \tag{6.30}$$

Hence $\varphi(P) \in \mathrm{Triv}(S)^\perp = L^-$. Conversely, take any element of L^- and represent it by a divisor D. Letting ψ be as in (6.3), we let $P = \psi(D) \in E(K)$. Then, by (6.4), we have $D \equiv (P) \bmod \mathrm{Triv}(S)$, and $D \perp \mathrm{Triv}(S)$. By the uniqueness in Lemma 6.16, we have $D \approx D_P$. This shows that D_P is an integral divisor, which is the case only if $(P.\Theta_{v,i}) = 0$ for all $i \geq 1$, i.e., $P \in E(K)^0$. It is easy to see that the maps $\varphi|_{E(K)^0}$ and $\psi|_{L^-}$ are inverse to each other. This proves the first assertion in view of the definition of the height pairing.

Next observe that (6.27) implies:

$$\langle P, Q\rangle \in \mathbb{Z} \text{ for any } P \in E(K)^0 \text{ and any } Q \in E(K).$$

Hence the Mordell–Weil lattice is contained in the dual lattice of the narrow Mordell–Weil lattice, and we have proven the diagram in the theorem. □

Corollary 6.48 *Assume that $f : S \to C$ is an elliptic surface without any reducible fibres. Then $E(K)$ coincides with the narrow Mordell–Weil lattice $E(K)^0$, and hence it is a positive-definite even integral lattice such that*

$$\det E(K) = |\det \mathrm{NS}(S)|.$$

Proof This is immediate from Theorem 6.44. □

Example 6.49 Recall the elliptic surface S arising from the elliptic curve

$$E: \quad y^2 = x^3 + t^5 + 1.$$

In Example 5.8, we computed that outside characteristics 2, 3, 5, there are no reducible fibres (only type II in Kodaira's notation). Hence $E(k(t)) = E(k(t))^0$ by Corollary 6.48 (and the following chapters will show that $E(k(t))$ is isometric to E_8).

Remark 6.50 The inclusion $E(K)/E(K)_{\text{tors}} \subset L^{\vee}$ in the above theorem cannot be replaced by equality, in general. Some examples are given in [OS90] or [Shio89b, Sect. 4]. On the other hand, the equality in question holds if the Néron–Severi lattice is unimodular. This will be discussed below.

6.8 The Case of a Unimodular Néron–Severi Lattice

The computation of Mordell–Weil lattices is particularly effective if we know the Néron–Severi lattice of the elliptic surface S (that is, its rank and determinant), since then we can apply Corollary 6.39 (as we have done in Example 6.41). In this section, we consider the most fundamental setting where $\mathrm{NS}(S)$ is unimodular. Our results will be instrumental for the study of rational elliptic surfaces in the next two chapters.

Theorem 6.51 *Let S be an elliptic surface such that* $\mathrm{NS}(S)$ *is a unimodular lattice, i.e.,*

$$|\det \mathrm{NS}(S)| = 1.$$

Then the Mordell–Weil lattice $E(K)/E(K)_{\text{tors}}$ is isomorphic to the dual lattice of the narrow Mordell–Weil lattice $E(K)^0$. In other words, with the notation of Theorem 6.47, we have

$$E(K)^0 \simeq L, \quad E(K)/E(K)_{\text{tors}} \simeq L^{\vee}.$$

Proof By Theorem 6.47, it suffices to prove

$$[E(K)/E(K)_{\text{tors}} : E(K)^0] = [L^{\vee} : L]. \tag{6.31}$$

As before let $\mathrm{Triv}(S)'$ denote the primitive closure of the trivial lattice $\mathrm{Triv}(S)$ in $\mathrm{NS}(S)$, i.e., $\mathrm{Triv}(S)' = (\mathrm{Triv}(S) \otimes \mathbb{Q}) \cap \mathrm{NS}(S)$. Then we have

$$\mathrm{Triv}(S)'^{\perp} = \mathrm{Triv}(S)^{\perp} = L^{-}.$$

Applying (2.6) to $\mathrm{Triv}(S)' \oplus L^{-} \subset \mathrm{NS}(S)$, we have by primitivity and unimodularity

$$\det L^{-} = \det \mathrm{Triv}(S)' = [\mathrm{NS}(S) : L^{-} \oplus \mathrm{Triv}(S)']. \tag{6.32}$$

By (2.5), this gives

$$[L^{\vee} : L] = [\mathrm{NS}(S)/\,\mathrm{Triv}(S)' : L^{-}]. \tag{6.33}$$

On the other hand, Theorem 6.5 implies

$$E(K) \simeq \mathrm{NS}(S)/\,\mathrm{Triv}(S) \text{ and } E(K)_{\text{tors}} \simeq \mathrm{Triv}(S)'/\,\mathrm{Triv}(S). \tag{6.34}$$

Therefore we have

$$E(K)/E(K)_{\text{tors}} \simeq \text{NS}(S)/\text{Triv}(S)'. \tag{6.35}$$

Now (6.33) and (6.35) prove the desired equality (6.31) completing the proof of Theorem 6.51. □

It will be convenient to reformulate Theorem 6.51 in the following way, by using Theorem 6.13. Recall the frame W of S from Theorem 6.13. By definition, we have

$$\text{NS}(S) = U \oplus W^-, \quad U = \langle (O), F \rangle, \quad W^- = U^\perp.$$

By Theorem 6.13, W is a positive-definite *even unimodular* lattice. As before, let

$$T = \oplus_{v \in R} T_v \subset W$$

be the root lattice associated with the reducible fibres of $f : S \to C$. By Sect. 5.5, this is a direct sum of simple root lattices of type A, D, E. By (6.1), the trivial lattice $\text{Triv}(S)$ is equal to the direct sum

$$\text{Triv}(S) = U \oplus T^-,$$

where U is a unimodular indefinite lattice of rank two which is odd or even depending on the parity of χ. Letting $\text{Triv}(S)'$ (or T') be the primitive closure of $\text{Triv}(S)$ (or T), we have

$$\text{Triv}(S)' = U \oplus (T')^-.$$

Then (6.34) implies

$$E(K) \simeq W/T, \quad E(K)_{\text{tors}} \simeq T'/T. \tag{6.36}$$

Let $T^\perp$ be the orthogonal complement of T in W. Obviously we have

$$L = (\text{Triv}(S)^\perp \text{ in } \text{NS}(S)) = (T^\perp \text{ in } W)^-,$$

and the isomorphism of the narrow Mordell–Weil lattice, $E(K)^0 \simeq L$, via the map φ.

We can summarize the above as follows:

Theorem 6.52 *Let S be an elliptic surface such that* NS(S) *is a unimodular lattice. Then the frame W of S is a positive-definite even unimodular lattice. Let $T = \oplus_{v \in R} T_v$ be the root lattice associated with the reducible singular fibres of $f : S \to C$.*

Then the narrow Mordell–Weil lattice $E(K)^0$ is isomorphic to the essential lattice lattice $L = T^\perp$, the orthogonal complement of T in W, and the Mordell–Weil lattice $E(K)/E(K)_{\text{tors}}$ is isomorphic to the dual lattice $L^\vee$ of L. Thus we have

$$\begin{array}{ccc} E(K)/E(K)_{\text{tors}} & \simeq & L^{\vee} \\ \cup & & \cup \\ E(K)^0 & \simeq & L \end{array} \tag{6.37}$$

The index of L in $L^{\vee}$ as well as the determinant of L is given by

$$[L^{\vee} : L] = \det L = \det T/|E(K)_{\text{tors}}|^2 \tag{6.38}$$

where

$$\det T = \prod_{v \in R} m_v^{(1)}, \quad |E(K)_{\text{tors}}| = [T' : T].$$

As a group, the Mordell–Weil group $E(K)$ is isomorphic to the quotient group W/T, and furthermore, there is an isomorphism:

$$E(K) \simeq L^{\vee} \oplus (T'/T),$$

which preserves the height pairing $\langle \cdot, \cdot \rangle$.

Proof We have already proven everything in the above, except for the formula (6.38), which follows from (6.32), by using (2.3), (2.5) and (6.36). □

6.9 Functorial Properties

In this section, we prove some basic functorial properties of the height pairing. We continue to work with an algebraically closed base field $k = \bar{k}$ and then consider a one-dimensional function field $K = k(C)$. Hence finite field extensions K'/K correspond to morphisms

$$C' \to C$$

of degree $[K' : K]$ such that $K' = k(C')$.

Theorem 6.53 *Let K' be a finite extension of K. For an elliptic curve E over K, consider the base extension to K', i.e. the elliptic curve $E' = E \otimes_K K'$ over K'. For any $P, Q \in E(K)$, let $P', Q' \in E'(K')$ be induced by P, Q. Then*

$$\langle P', Q' \rangle_{E'(K')} = [K' : K] \cdot \langle P, Q \rangle_{E(K)}. \tag{6.39}$$

Proof It suffices to consider the case $P = Q \in E(K)^0$, because both sides of (6.39) are symmetric and bilinear. Let $f : S \to C$ (or $f' : S' \to C'$) be the elliptic surface associated with E/K (or E'/K'). Then S' is birational to the fibre product $S \times_C C'$, and there is a natural rational map

$$g : S' \dashrightarrow S.$$

Denote by S_1 the open set of S which is the complement of the union of $\Theta_{v,i}(v \in R, i \geq 1)$; in other words, we work with the smooth locus of a Weierstrass model of S. By construction (since $P \in E(K)^0$!), there is a suitable open set S_1' of S' containing the complete curves (P') and (O') such that the restriction of g to S_1' is a morphism

$$g_1 : S_1' \to S_1.$$

Now, by the definition of the height pairing, we have

$$\langle P, P \rangle = -(D_P.D_P),$$

where D_P is the divisor on S (or S_1) defined by (6.30). Observe that

$$g_1^*((P)) = (P'), \quad g_1^*((O)) = (O'), \quad g_1^*(F) = dF' \quad (d = [K' : K]),$$

where we can equivalently work on S' or on S_1' (since the curves are fully contained in S_1'). Letting $D' = g_1^*(D_P)$, we infer

$$D' = (P') - (O') - d((P.O) + \chi)F'. \tag{6.40}$$

By the standard property of intersection numbers, we have

$$(D'.O') = (g_1^*(D_P).g_1^*((O))) = d(D_P.O) = 0. \tag{6.41}$$

Then (6.40) and (6.41) imply

$$(P'.O') + \chi' = d((P.O) + \chi).$$

This proves (6.39), by the formula (6.28) for E and for E'. □

Remark 6.54 In more suggestive notation, identifying P, Q in $E(K)$ and in $E'(K') = E(K')$ and dropping superscripts, we may rewrite (6.39) as

$$\langle P, Q \rangle_{E(K')} = [K' : K] \cdot \langle P, Q \rangle_{E(K)}.$$

Turning to the narrow Mordell–Weil lattice in the proof of Theorem 6.53 is indeed instrumental for avoiding having to deal with reducible fibres in too much detail. In fact, the explicit computations can be rather tedious—especially if there are additive fibres involved where the base change ramifies. We illustrate this with the following example which we will return to in Chap. 12.

Example 6.55 Consider the rational elliptic surface given in affine Weierstrass form

$$S: \quad y^2 = x^3 + t(t-1)^4. \tag{6.42}$$

This fibration is isotrivial with $j = 0$, the singular fibres have type II at $t = 0, \infty$ and IV* at $t = 1$ by Sect. 5.8. For each solution $\beta^3 = 1/4$, there is a pair of sections $P, -P$ given by

$$(x, y) = (\beta(t-1)^2, \pm(t+1)(t-1)^2/2).$$

Each has height $h(P) = h(-P) = 2 - 4/3 = 2/3$. (To see conceptually how the section P was found, confer Example 8.14.)

Now consider a base change $C \to \mathbb{P}^1$ of degree n. On the resulting elliptic surface S' over C, this gives sections $P', -P'$ of height $2n/3$. Note that there may be some minimalizations involved. For example, let $C = \mathbb{P}^1$ with the cyclic base change $t = s^n$. Writing $n = 6m + r$ for some $r = 0, \dots, 5$, we derive the minimal model of the base changed elliptic surface S' over $\mathbb{P}^1_s$ as

$$S': \quad y^2 = x^3 + s^r(s^n - 1)^4.$$

Recall that p denotes the characteristic of the ground field. If $p \nmid n$, this model is also minimal, so we find $e(S') = 8n + 4r$. (For the inseparable case $p \mid n$, see Sect. 12.4.3.) On the above minimal model of S', the section P' takes the shape

$$(x, y) = (\beta(s^n - 1)^2/s^{2m}, \pm(s^n + 1)(s^n - 1)^2/(2s^{3m})).$$

For the height, one thus indeed verifies

$$h(P') = e(S')/6 + 4m - 4n/3 = 2n/3.$$

The next result will become important later (from Chap. 9 on) when we also consider an arithmetical situation where the ground field is not algebraically closed.

Theorem 6.56 *Suppose that k is the algebraic closure of a perfect field k_0. Let C_0 be a geometrically irreducible curve defined over k_0 and let E_0 be an elliptic curve defined over $k_0(C_0)$. Further, let E be the elliptic curve over $K = k(C_0)$ obtained from E_0 by base extension. Then, for any automorphism $\sigma \in Gal(k/k_0)$, we have*

$$\langle P^\sigma, Q^\sigma \rangle = \langle P, Q \rangle \text{ for any } P, Q \in E(K). \tag{6.43}$$

Namely, the height pairing is stable under the action of $Gal(k/k_0)$,

Remark 6.57 Another way to phrase Theorem 6.56 is that $\mathrm{Gal}(k/k_0)$ acts by isometries on $\mathrm{MWL}(S)$ for the elliptic surface S associated with E.

Proof Under the assumption, the elliptic surface $f : S \to C$ associated with E is defined over k_0. Letting φ be as in Lemma 6.16, we claim

$$\varphi(P)^\sigma = \varphi(P^\sigma) \text{ for any } P \in E(K). \tag{6.44}$$

Recall that $\varphi(P)$ is uniquely determined as the divisor class $D \in \mathrm{NS}(S) \otimes \mathbb{Q}$ such that

$$D \equiv (P) \bmod \mathrm{Triv}(S)_{\mathbb{Q}} \text{ and } (D.\Theta) = 0 \text{ for any } \Theta \in \mathrm{Triv}(S).$$

Now let us consider the effect of an automorphism $\sigma \in Gal(k/k_0)$ on these relations. Note that the trivial lattice $\mathrm{Triv}(S)$ is stable under any σ, because σ fixes the classes of (O) and F and interchanges the irreducible components $\Theta_{v,i}$ of fibres. Hence the above relations imply

$$D^\sigma \equiv (P^\sigma) \bmod \mathrm{Triv}(S)_{\mathbb{Q}} \text{ and } (D^\sigma.\Theta) = 0 \text{ for any } \Theta \in \mathrm{Triv}(S)$$

by the invariance of the intersection number under a field automorphism (e.g. [Wei62]). By the uniqueness in Lemma 6.16, this proves (6.44).

Then (6.43) immediately follows from the definition (6.11):

$$\langle P^\sigma, Q^\sigma \rangle = -(\varphi(P^\sigma).\varphi(Q^\sigma)) = -(\varphi(P)^\sigma.\varphi(Q)^\sigma) = \langle P, Q \rangle$$

again by the invariance of the intersection number under σ. □

Example 6.58 (*Hexagonal example in arithmetic setting*) Recall the elliptic curve E over $k(t)$ from (1.1) with Mordell–Weil lattice $A_2^\vee$ (by Example 6.41). Starting with a generalized Weierstrass form

$$y^2 + ty = f(x) \quad \text{for some cubic polynomial } f \in k[x]$$

with roots $\alpha_i \in k$. There are sections $(\alpha_i, 0), (\alpha_i, -t)$ resembling the P_i, Q_i. One infers different Galois actions (identity, quadratic, cyclic cubic or symmetric on the three roots) on the Mordell–Weil group $E(\bar{k}(t))$ depending on the splitting of f over k.

Example 6.59 In Example 6.55, there is an obvious Galois action on the six given sections. One easily checks the compatibilities as recorded in Theorem 6.56. Note that over $k(\omega)$, with $\omega^2 + \omega + 1 = 0$, but $\beta \notin k(\omega)$, the Galois action is exactly accounted for by the order 3 automorphism $(x, y) \mapsto (\omega x, y)$ of the generic fibre.

Historical Notes

The lattice structure underlying the Mordell–Weil group of sections of an elliptic surface has guided a number of mathematicians, at least in some intuitive sense, before it was formally introduced by Elkies and one of us in 1989 [Elk90, Shio89b, Shio90a]. For instance, one may see its influence in Tate's specialization arguments [Tat83] (for the narrow part of the Mordell–Weil group), in Manin's argument concerning the base points of a general cubic pencil [Man64] (which will be the subject of Sect. 7.6) or in the cycle of ideas around the Birch–Swinnerton-Dyer conjecture and the Tate conjecture [BS-D65, Tat66b] (which will appear in Sect. 12.1 and again in Sect. 13.3.1). Also the work of Cox and Zucker [CZ79], even if it did not lead to the definition of Mordell–Weil lattices, clearly points in the direction of a height pairing (over $\mathbb{C}$); not to forget, it was lending itself to some first explicit applications for small rank, see [Cox79].

The same goes for the ingredients, like the decomposition of a divisor into horizontal and vertical components, which, of course, has been used explicitly or implicitly before. A famous instance would be Zariski's lemma, concerning the intersection form on the components of a reducible fibre of a fibration from a surface to a curve (cf. [BHPV04, III.8.2]). Also torsion sections have played an important role in the study of elliptic surfaces, as can be seen in Tate's representations of elliptic curves with a point of given order (see [Kub76]), or in the study of elliptic modular surfaces [Shio72]. Note that all these considerations are guided by the moduli perspective, in connection with level structures, just like the previous work of Deligne [Del71] and Igusa [Igu59, Igu68] (see also [KM85, IL13]). In comparison, the approach to torsion sections through the height pairing is much more elementary, without reference to moduli problems or universal elliptic curves. In practice, it thus proves very useful.

Heights are a fundamental concept and tool in arithmetic, Arakelov and diophantine geometry (cf. [BG06, HS00]). The height of a rational point in projective space over a number field (now often called naive or classical height) was introduced by Northcott [Nor50] and Weil [Wei51]. Néron used this to define a height on an abelian variety in terms of local contributions, which in fact is a quadratic form [Ner65] (cf. also [Sil88]); this is nowadays called the canonical height, or the Néron–Tate height, giving credit to Tate's insight that the canonical height can be obtained as a limit of naive heights (cf. [HS88]). For an elliptic surface, the resulting notion of height agrees (up to a constant factor) with the one derived from the intersection theoretical height paring on the Mordell–Weil group. (For instance, Elkies in [Elk94] worked with the canonical height.)

Personally for us, the turning point in introducing Mordell–Weil lattices was the rational elliptic surface from Examples 5.8, 6.49. Back in 1989 it was only barely possible to compute any or at least one section by hand or on a small Macintosh computer. When it pointed towards (what are now identified with) the 240 minimal sections (see Theorem 8.29), the connection with the E_8 lattice and its 240 roots became clear. This in turn prompted the present notion of Mordell–Weil lattices. To amplify this, compare the last pre-MWL paper [Shio89a] and the first post-MWL paper [Shio89b].

Chapter 7
Rational Elliptic Surfaces

As the first application of the general results developed in the previous chapter, we study the Mordell–Weil lattices of *rational elliptic surfaces* with section. It will be seen that these lattices form a hierarchy dominated by the root lattice E_8, the unique positive-definite even unimodular lattice of rank 8; the details will be discussed in full in Chap. 8. It turns out that the information supplied from the lattice structure has a very strong consequence for the original Diophantine problem of a Mordell–Weil group.

The results in this chapter will have quite rich applications to various fields in mathematics such as algebraic equations, number theory, and algebraic geometry, as will be discussed in Chaps. 9 and 10.

7.1 Preliminaries

By a *rational elliptic surface*, we mean a (smooth projective) rational surface over k, say S, which is given with a relatively minimal elliptic fibration $f : S \to C$. For simplicity we assume that the ground field k is an arbitrary algebraically closed field. Recall that S is called a *rational surface* over k if its function field $k(S)$ is a purely transcendental extension of dimension 2 over k, or equivalently, if S is birationally equivalent to the projective plane $\mathbb{P}^2$.

In the case of a rational elliptic surface, the base curve C is the projective line: $C = \mathbb{P}^1$ and $K = k(\mathbb{P}^1) = k(t)$ is a rational function field. This follows, for instance, from Sect. 5.13, or directly from Lüroth's theorem stating that the only function field $k(C)$ contained in a purely transcendental extension of k is a rational function field.

We always assume that f has a global section $O : C \to S$. On the other hand, we note that the Convention 5.10 that f has singular fibres is automatically satisfied in this case as a product $E \times C$ of an elliptic curve E and another curve C is never rational; similarly, there are non-product surfaces admitting elliptic fibrations with section, but without singular fibres; however, this implies that the base has positive

M. Schütt and T. Shioda, *Mordell–Weil Lattices*, Ergebnisse der Mathematik und ihrer Grenzgebiete. 3. Folge / A Series of Modern Surveys in Mathematics 70,
https://doi.org/10.1007/978-981-32-9301-4_7

genus. The reason for this lies in the minimal Weierstrass form: over $\mathbb{P}^1$, there is a globally minimal Weierstrass form by Lemma 5.43, so a non-product elliptic fibration must have a singular fibre.

Let E/K be the generic fibre of $f : S \to \mathbb{P}^1$. Observe that the function field of the elliptic curve E over $K = k(t)$, $K(E)$, is a purely transcendental extension of the ground field k. For example, if E/K is given by a Weierstrass equation with affine coordinates x, y, the function field $K(E) = K(x, y)$ is equal to the field of fractions of the integral domain $k[t, x, y]/g$ with irreducible polynomial $g \in k[t, x, y]$ obtained from the Weierstrass equation. The assumption that S is a rational elliptic surface means exactly that we should have $K(x, y) = k(\xi, \eta)$ for two suitable elements $\xi, \eta \in k(t, x, y)$.

7.2 Basic Properties

Now let us recall the basic invariants of a rational elliptic surface without actually using Weierstrass forms or Theorem 7.25. Indeed, the approach below does not assume any knowledge about algebraic surface theory beyond what was laid out in Chap. 4.

Proposition 7.1 *Let S be a rational elliptic surface. Then:*

(i) The Euler characteristic χ of S is equal to 1.
(ii) The Néron–Severi lattice NS(S) *is unimodular and of rank $\rho = 10$.*

Proof We have $\chi = p_g - q + 1$ with p_g the geometric genus, q the irregularity, both of which are 0 for a rational surface. Hence $\chi = 1$. Then, by Noether's formula,

$$e(S) + K_S^2 = 12\chi,$$

where $e(S)$ is the Euler number of S. This reduces now to $e(S) = 12$ by Theorem 5.44. On the other hand,

$$e(S) = 2 - 2b_1 + b_2$$

(the alternating sum of Betti numbers) gives $b_2 = 10$, since $b_1 = 0$ as the Picard variety of S is trivial by Sect. 4.2. For a rational surface, $\rho = b_2$ holds since $\rho(\mathbb{P}^2) = b_2(\mathbb{P}^2)$ and the Lefschetz number $\lambda(S) = b_2(S) - \rho(S)$ is a birational invariant. (Over $\mathbb{C}$, one can just use $p_g = 0$ and apply Lefschetz' $(1, 1)$-theorem.) Finally, also $|\det \mathrm{NS}(S)|$ is a birational invariant of smooth algebraic surfaces (blowing up a point introduces an exceptional curve Γ with $(\Gamma^2) = -1$, which is perpendicular to the pull-back of NS(S), see Sect. 4.5). Presently, NS$(\mathbb{P}^2)$ is unimodular. Hence $|\det \mathrm{NS}(S)| = 1$. □

The reader is invited to check that this result is consistent with our previous considerations, such as the Weierstrass form of a rational elliptic surface in Sect. 5.13

with $\chi = 1$. Compare also the standard model as a cubic pencil, which will be discussed in Sect. 7.5. Indeed, this leads to an interpretion of S as $\mathbb{P}^2$ blown up in the nine base points, and this has exactly the invariants of Proposition 7.1.

Lemma 7.2 *The frame W of S is the root lattice E_8.*

Proof As we remarked in Theorem 6.52, W is a positive-definite even unimodular lattice, and in the present case, of rank $\rho - 2 = 8$. Hence we have $W \simeq E_8$ by the uniqueness of such a lattice mentioned in Remark 2.33. □

The Mordell–Weil rank formula from Corollary 6.7 reduces in this case to

$$r = 8 - \sum_{v \in R} (m_v - 1), \tag{7.1}$$

which implies that $r \leq 8$. In this chapter and the next one, we will study all possible cases in detail.

7.3 The Case of High Rank

Let us first consider the most basic case for rational elliptic surfaces, that is the case of high rank $r = 8, 7, 6$. The above formula (7.1) easily shows:

Lemma 7.3 *(i)* $r = 8 \iff R = \emptyset$ *(i.e. f has no reducible fibres).*

(ii) $r = 7 \iff R = \{v\}, m_v = 2$ *(i.e. there is only one reducible fibre $f^{-1}(v)$ and it has two components: thus a singular fibre of type I_2 or III).*

(iii) $r = 6 \iff$

- *(a)* $R = \{v\}, m_v = 3$ *(i.e. there is only one reducible fibre $f^{-1}(v)$, and it has three components: thus of type I_3 or IV), or*
- *(b)* $R = \{v, v'\}, m_v = m_{v'} = 2$ *(i.e. there are exactly two reducible fibres, each with two components, thus of type I_2 or III.)*

Now we can state the structure theorem of the Mordell–Weil lattice (as well as of the underlying group) $E(K)$ of rank $r = 8, 7$ or 6 arising from rational elliptic surfaces. As in Chap. 2, E_8, E_7, E_6 and D_6 denote the root lattices of designated type (the subscript being the rank) and superscripts $^\vee$ indicate their dual lattices.

Theorem 7.4 *Suppose $f : S \to \mathbb{P}^1$ is a rational elliptic surface with generic fibre E/K. In case the Mordell–Weil group $E(K)$ has rank $r \geq 6$, then it is torsion-free and the structure of the Mordell–Weil lattice is as follows.*

(i) If f has no reducible fibres ($R = \emptyset$), then $r = 8$ and

$$E(K) = E(K)^0 \simeq E_8.$$

(ii) If there is exactly one reducible fibre $f^{-1}(v)$ and $m_v = 2$, then $r = 7$ and

$$\begin{array}{ccc} E(K) & \simeq & E_7^\vee \\ \cup & & \cup \;\; \textit{index } 2 \\ E(K)^0 & \simeq & E_7. \end{array}$$

(iii) If there is exactly one reducible fibre $f^{-1}(v)$ and $m_v = 3$, then $r = 6$ and

$$\begin{array}{ccc} E(K) & \simeq & E_6^\vee \\ \cup & & \cup \;\; \textit{index } 3 \\ E(K)^0 & \simeq & E_6. \end{array}$$

(iv) If there are exactly two reducible fibres $R = \{v, v'\}$ and $m_v = m_{v'} = 2$, then $r = 6$ and

$$\begin{array}{ccc} E(K) & \simeq & D_6^\vee \\ \cup & & \cup \;\; \textit{index } 4 = 2^2 \\ E(K)^0 & \simeq & D_6. \end{array}$$

Proof It suffices to prove that the narrow Mordell–Weil lattice $E(K)^0$ is isomorphic to the root lattice E_8, E_7, E_6 or D_6 according to the four cases $(i), \ldots, (iv)$ and that the Mordell–Weil group $E(K)$ is torsion-free. In fact, we can then apply Theorem 6.52 since the Néron–Severi lattice is unimodular by Proposition 7.1.

With the notation of Theorem 7.4, the part of the trivial sublattice generated by the components of reducible fibres is equal, up to sign, to

$$T = \{0\}, A_1, A_2 \text{ or } A_1^2$$

according to the cases $(i), \ldots, (iv)$. Then, by Proposition 2.36 and a little bit of linear algebra (or the classification of even definite lattices of small determinant), the essential lattice L, which is the orthogonal complement of T in $W = E_8$ is equal, respectively, to

$$L = T^\perp \simeq E_8, E_7, E_6, \text{ or } D_6.$$

This proves the claim on the narrow Mordell–Weil lattice.

Next, by the explicit formula of the height pairing (6.15), we have for $P \in E(K)$

$$\langle P, P\rangle = 2 + 2(P.O) - \begin{cases} 0 & \text{for } P \in E(K)^0, \\ 1/2 & \text{for } P \in E(K) - E(K)^0, \text{case } (ii), \\ 2/3 & \text{for } P \in E(K) - E(K)^0, \text{case } (iii), \\ 1 \text{ or } 1/2 & \text{for } P \in E(K) - E(K)^0, \text{case } (iv). \end{cases} \tag{7.2}$$

For any $P \neq O$, we thus have $(P.O) \geq 0$, and hence

$$\langle P, P \rangle \geq \begin{cases} 2 & \text{for case } (i), \\ 3/2 & \text{for case } (ii), \\ 4/3 & \text{for case } (iii), \\ 1 & \text{for case } (iv). \end{cases} \tag{7.3}$$

In particular, we have $\langle P, P \rangle > 0$ in each case and P cannot be a torsion element. Hence $E(K)$ is torsion-free. Since $\mathrm{NS}(S)$ is moreover unimodular by Proposition 7.1, we deduce with the essential lattice L that $E(K)^0 \cong L$ and $E(K) \cong L^\vee$. This proves the structure theorem. □

We will see in the next chapter that the same kind of argument will go through for many other configurations of singular fibres of rational elliptic surfaces.

Example 7.5 Recall from Examples 5.8, 6.49 that the elliptic curve

$$y^2 = x^3 + t^5 + 1$$

gives rise to a rational elliptic surface without reducible fibres outside characteristics 2, 3, 5, i.e. with $R = \emptyset$. Hence $E(\bar{k}(t)) \cong E_8$ by Theorem 7.4.

Let us recall the relevant invariants of the lattices in question (see Chap. 2 and the references given there). The entries in brackets (for $D_6^\vee$) give the next-to-minimal norm and the number of vectors with this norm (since this will be used later, e.g. in Remark 7.9).

Definition 7.6 We call $P \in E(K)$ a *minimal rational point* or a *minimal section* if $\langle P, P \rangle$ has the smallest positive value in the Mordell–Weil lattice. Also we sometimes use the word *minimal vector* to emphasize that such a rational point of an elliptic curve (or a section of an elliptic surface) is regarded as a vector in the lattice in question.

By comparing the formula (7.2) or (7.3) with the above table, we obtain some remarkable consequences.

For example, consider the case (i) where S is a rational elliptic surface without reducible fibres. Then $E(K) \simeq E_8$ has rank $r = 8$, and it has 240 minimal vectors of minimal norm 2. On the other hand, (7.2) shows that $\langle P, P \rangle = 2$ holds if and only if $(P.O) = 0$, i.e. the two sections (P) and (O) do not intersect at all (as curves on the surface S). Therefore there exist exactly 240 rational points $P \in E(K)$ such that (P) is disjoint from the zero section (O).

For the case (ii), S has a single reducible fibre with two irreducible components: $f^{-1}(v) = \Theta_{v,0} + \Theta_{v,1}$. In this case, $E(K) \simeq E_7^\vee$ has rank $r = 7$, and it has 56 minimal vectors of minimal norm 3/2. On the other hand, (7.2) shows that $\langle P, P \rangle = 3/2$ holds if and only if $(P.O) = 0$ and $contr_v(P) = 1/2$. The latter is equivalent to the condition that (P) is disjoint from (O) and intersects the non-identity component $\Theta_{v,1}$ of the fibre $f^{-1}(v)$.

In the same way, we have the following result, which will be used later to obtain more definitive results formulated in terms of the Weierstrass form (see Sect. 7.4).

Lemma 7.7 *For each of the cases (i), ..., (iv), if $P \in E(K)$ is a minimal section, then we have $(P.O) = 0$, i.e., (P) is disjoint from (O). The converse is also true in case (i). The additional condition for the other cases is that (P) should pass through a non-identity component $\Theta_{v,i}$ $(i \geq 1)$ of every reducible fibre $f^{-1}(v)$.*

Theorem 7.8 *The number of the minimal sections (with minimal norm μ) in the Mordell–Weil lattice $E(K)$ is:*

(i) 240 $(\mu = 2)$,
(ii) 56 $(\mu = 3/2)$,
(iii) 54 $(\mu = 4/3)$,
(iv) 12 $(\mu = 1)$,

according to the cases (i), (ii), (iii), and (iv) above. They contain a set of free generators of the Mordell–Weil group $E(K)$ in case (i), (ii) or (iii). In case (iv), the minimal sections generate a subgroup of index 2 in $E(K)$, and together with any section of norm 3/2 (there are 64 such vectors for $D_6^\vee$) they generate the full Mordell–Weil group $E(K)$.

Proof The fact corresponding to the second assertion for the lattices E_8, $E_7^\vee$, $E_6^\vee$ or $D_6^\vee$ is well-known and was easily verified in Chap. 2 (see Table 7.1). □

Remark 7.9 The purely lattice-theoretic facts contained in Theorem 7.8 admit a nice geometric explanation from the viewpoint of Mordell–Weil lattices. For example, in case *(iv)* above, the elliptic surface has two reducible fibres, say $f^{-1}(v) = \Theta_{v,0} + \Theta_{v,1}$ and $f^{-1}(v') = \Theta_{v',0} + \Theta_{v',1}$. Take any $P \in E(K)$ such that $(P.O) = 0$. Such a P is a minimal section ($\langle P, P \rangle = 1$) if and only if (P) passes through both $\Theta_{v,1}$ and $\Theta_{v',1}$. It follows for any two minimal sections P, Q that $\langle P, Q \rangle \in \mathbb{Z}$ by inspection of the correction terms. Thus minimal sections generate an integral (though not even) sublattice of $E(K)$. Since the lattice $D_6^\vee$ is not integral (it has determinant $1/4 \notin \mathbb{Z}$), this sublattice cannot be $E(K)$ itself.

In comparison, we have a section of next-to-minimal norm $\langle P, P \rangle = 3/2$ if (P) passes through either $\Theta_{v,1}$ or $\Theta_{v',1}$, but not through both. Such sections do actually occur; together with the minimal sections they generate $E(K)$.

The hexagonal elliptic curve, to be revisited below, is located at the opposite end of the scale as the Mordell–Weil rank is rather low, thanks to the fibre of Kodaira type IV* (using Corollary 6.7).

Table 7.1 High Mordell–Weil rank data

	E_8	E_7	$E_7^\vee$	E_6	$E_6^\vee$	D_6	$D_6^\vee$
Determinant	1	2	1/2	3	1/3	4	1/4
Minimal norm	2	2	3/2	2	4/3	2	1 (3/2)
Kissing number	240	126	56	72	54	60	12 (64)

Example 7.10 (*Hexagonal example cont'd*) Recall the minimal sections of the elliptic curve (1.1) from Examples 6.26, 6.41:

$$\{\text{minimal sections}\} = \begin{cases} \{P_i, Q_i;\ i = 1, 2, 3\} & \text{if all } \alpha_i \text{ are distinct,} \\ \{P_i, Q_i\} & \text{if } \alpha_i = \alpha_j \neq \alpha_k. \end{cases}$$

For another (rather) explicit example, see Example 7.13. The other lattices will be analysed within the framework of excellent families of elliptic curves in Sect. 10.1.

7.4 Weierstrass Form

We can make the preceding results more explicit by writing down the equation of the elliptic curve E over $K = k(t)$ in the Weierstrass form. Recall from Lemma 5.43 that there is a globally minimal Weierstrass form

$$E: \quad y^2 + a_1(t)xy + a_3(t)y = x^3 + a_2(t)x^2 + a_4(t)x + a_6(t) \tag{7.4}$$

where $a_i(t) \in k[t]$ for each i. By Proposition 5.51 the associated elliptic surface $f: S \to \mathbb{P}^1$ is a rational surface if and only if

$$\deg a_i(t) \leq i \quad \forall i. \tag{7.5}$$

Minimality is ensured as soon as the discriminant Δ is not a twelfth power (including the constant case). Conversely, one can show that this is a sufficient condition outside characteristics 2 and 3. The exceptions in those characteristics are due to the existence of wild ramification (see Sect. 5.9).

Lemma 7.11 *Let E be an elliptic curve over some field K, given in terms of a minimal Weierstrass form, such that the associated elliptic surface is rational. Let $P = (\mathrm{x}, \mathrm{y}) \in E(K)$, $P \neq O$. Then the section (P) is disjoint from the zero section (O) if and only if* x *and* y *are polynomials in t of degree ≤ 2 resp. ≤ 3, i.e., of the form:*

$$\mathrm{x} = gt^2 + at + b, \quad \mathrm{y} = ht^3 + ct^2 + dt + e, \tag{7.6}$$

with $a, b, \ldots, g, h \in k$.

Proof If (P) and (O) intersect at some point lying over $v \in \mathbb{P}^1$, $v \neq \infty$, then the x-coordinate of P must have a pole at v. Hence if (P) and (O) are disjoint over $\mathbb{P}^1 - \{\infty\}$, then x is a polynomial in t, and so is y. At $v = \infty$, rewrite (7.4) in terms of $s = 1/t$, $\mathrm{x}' = \mathrm{x}/t^2$, $\mathrm{y}' = \mathrm{y}/t^3$ (the coordinates of the other chart from (5.22) for $d = 1$), and apply the same argument to see that x is of degree ≤ 2 in t, and y is of degree ≤ 3 in t. □

Now we formulate the Weierstrass version of Theorem 7.8. Here it is often convenient to apply coordinate transformations so that the reducible singular fibre(s) sit(s) at ∞ (or 0) and the singularity of the Weierstrass model is at $(0, 0)$. Note that such simultaneous normalizations (which form the key simplification in Tate's algorithm, see Sect. 5.8) are generally only possible at a given number of singular fibres due to the degree constraints in (5.23) for the admissible transformations in (5.19) (which are necessary to preserve the minimality of the Weierstrass form at ∞, compare the discussion at the end of Sect. 5.10).

Theorem 7.12 *Let $E/K = k(t)$ be an elliptic curve given by a minimal Weierstrass form such that the associated elliptic surface $f : S \to \mathbb{P}^1$ is a rational elliptic surface. Then:*

(i) If f has no reducible fibres, then there are exactly 240 rational points $P = (\mathrm{x}, \mathrm{y}) \in E(K)$ of the form (7.6),

$$\mathrm{x} = gt^2 + at + b, \quad \mathrm{y} = ht^3 + ct^2 + dt + e$$

and they contain a set of eight free generators of the Mordell–Weil group $E(K)$.

(ii) If there is only one reducible fibre $f^{-1}(v)$, we may take $v = \infty$. Suppose that it is of type I_2 or III ($m_\infty = 2$) and the corresponding Weierstrass cubic has a singular point (a node or a cusp) at $(0, 0)$. Then there are exactly 56 rational points (x, y) of the form:

$$\mathrm{x} = at + b, \quad \mathrm{y} = ct^2 + dt + e, \tag{7.7}$$

and they contain a set of seven free generators of $E(K)$.

(iii) Similarly, suppose that $f^{-1}(\infty)$ is the only reducible fibre, of type I_3 or IV ($m_\infty = 3$) and the corresponding Weierstrass cubic has a singular point at $(0, 0)$. Then there are exactly 54 points of the form (7.7), and they contain a set of six free generators of $E(K)$.

(iv) If there are only two reducible fibres, we may assume they lie over $v = 0$ and ∞. Suppose that $m_0 = m_\infty = 2$, i.e. both fibres have Kodaira type I_2 or III, and $(0, 0)$ is the singular point for the Weierstrass cubic at both $v = 0$ and ∞. Then there are 12 points of the form

$$\mathrm{x} = at, \quad \mathrm{y} = ct^2 + dt, \tag{7.8}$$

and 32 points of the form

$$\mathrm{x} = at + b, \quad \mathrm{y} = ct^2 + dt + e, \quad (b, e) \neq (0, 0). \tag{7.9}$$

This set of 44 points contains a set of six free generators of $E(K)$.

Proof By Lemma 7.7 the minimal sections are disjoint from the zero section (O) in either case, so Lemma 7.11 applies to show that the sections take the shape (7.6). This already proves *(i)* using Theorem 7.8.

For *(ii)* and *(iii)*, we infer from Theorem 7.4 that the minimal sections meet the reducible fibre at the non-identity component (different from Θ_0 which is met by (O)). In terms of the (singular minimal) Weierstrass model given by (7.4), this translates as (P) meeting the node (or cusp) of the corresponding singular fibre. With the given normalizations, this exactly gives (7.7).

For *(iv)*, the same argument gives the 32 sections of the form (7.9), again building on Theorems 7.4, 7.8, while the 12 minimal sections meet both reducible fibres non-trivially, at once yielding the shape of (7.8). □

Example 7.13 As an explicit illustration, consider the case where the elliptic surface has two singular fibres of type I_2 and Mordell–Weil lattice $D_6^\vee$ as in Theorem 7.12 (iv). We proceed as above to assume that the singular fibres are located at $t = 0, \infty$. Moving the node in each fibre to the affine origin, we derive the extended Weierstrass form (outside char$(k) = 2$)

$$y^2 = x^3 + Ax^2 + tBx + t^2C, \quad A, B, C \in k[t] \text{ of degree 2.} \tag{7.10}$$

Then a section of minimal height takes exactly the shape of (7.8). Substituting for x into (7.10), we solve for the RHS to be a square. That is, we compute the discriminant of the quadratic factor $a^3t + Aa^2 + Ba + C$ of the RHS (in t); this gives a polynomial of degree 6 in a. Since every root corresponds uniquely to the x-coordinate of the section, it gives rise to two choices for the y-coordinate, so we obtain the 12 minimal sections as claimed.

One can proceed similarly for the sections of next-to-minimal height, and also modify the above approach to work in characteristic 2.

7.5 Cubic Pencils

We investigate one of the standard examples of rational elliptic surfaces: cubic pencils (which we have already encountered before in this book, for instance in Examples 5.7, 5.9). Let $F, G \in k[X, Y, Z]$ be homogeneous cubic polynomials without common factor. Consider the cubic pencil

$$S: \quad sF + tG = 0, \quad [s, t] \in \mathbb{P}^1. \tag{7.11}$$

Indeed, S is a rational surface, since the ratio s/t is expressed by F and G, so the function field of S in the (often to be taken affine) coordinates $x = X/Z,\ y = Y/Z$ is simply $k(S) = k(x, y)$. If the pencil contains at least one smooth cubic curve, then the cubic pencil defines a genus one fibration over the $\mathbb{P}^1$-line with homogeneous coordinates $[s, t]$ (possibly after resolving singularities of S as a projective surface in $\mathbb{P}^2 \times \mathbb{P}^1$). Since k is algebraically closed, S gives in fact an elliptic surface with sections given by the base points of the cubic pencil. Here we have to pay special attention when there are infinitely near (i.e. multiple) base points: then there are singularities involved, and each multiple base points represents the components of the exceptional divisor of the resolution ((-2)-curves contained in some fibre) plus a section meeting one of these components (a (-1)-curve not contained in any fibre). We discuss a few examples, some of which we have encountered before:

Example 7.14 The Hesse pencil from Example 3.13 gives a smooth projective elliptic surface in $\mathbb{P}^2 \times \mathbb{P}^1$ (see Example 5.7). One can easily verify that the nine base points of the pencil are all points of inflection on the generic fibre. Choosing one of them as zero element for the group structure, the others thus correspond to points of order 3 on the generic fibre, or equivalently, they give 3-torsion sections on the elliptic surface.

Example 7.15 One can construct the quotient of the above example by the subgroup of the automorphism group generated by translation by a point of order 3 (i.e. by a torsion section). Ultimately, this results in the following elliptic curve over $k(t)$:

$$S: \quad X^2Y + Y^2Z + Z^2X + tXYZ = 0. \tag{7.12}$$

For general formulas to check this, the reader is referred to [Top93]. Presently, there is also a direct way to compute the isogeny. To this end, we exhibit the rational map of degree 3 from the Hesse pencil to the rational elliptic surface S. Writing the Hesse pencil as

$$U^3 + V^3 + W^3 - tUVW = 0,$$

the rational map can be given by

$$X = U^2V, \;\; Y = W^2V, \;\; Z = V^2W.$$

Again (7.12) defines a cubic pencil, but this time with three triple base points $[1, 0, 0], [0, 1, 0], [0, 0, 1]$—which are exactly the image points of the nine base points of the Hesse pencil. In the singular model (7.12), each base point corresponds to an A_2 surface singularity in the singular fibre $XYZ = 0$ at $t = \infty$ where two of the three lines intersect. The resolution thus results in a cycle of nine $\mathbb{P}^1$'s, i.e. in a fibre of Kodaira type I_9, and there remain three sections induced by the base points. As these are again points of inflection on the generic fibre, they continue to have order 3. In fact, quotienting by an automorphism given by a translation by such a point of order 3 induces the dual isogeny, leading back to the Hesse pencil.

In the above examples, there are always three collinear base points. This is far from a coincidence as the following result shows:

Lemma 7.16 *If the cubic pencil S contains a reducible member, then three of its base points are collinear.*

Proof Suppose that the cubic pencil S contains some reducible member C. It necessarily contains a component L which is a line. The base points of the pencil can be computed by intersecting C with any other member C' of the pencil. By Bezout's

Theorem, L intersects the cubic C' in exactly three points (counted with multiplicities). Thus we obtain three collinear base points. □

We can also extend these considerations to the fibres of the induced genus one fibration $S \to \mathbb{P}^1$:

Lemma 7.17 *If the genus one fibration $S \to \mathbb{P}^1$ admits a reducible fibre, then there are three collinear base points or a multiple base point.*

Proof If there are no three collinear base points, then the cubic pencil S contains no reducible members by Lemma 7.16. Hence the genus one fibration $S \to \mathbb{P}^1$ can only acquire reducible fibres by means of resolution of surface singularities. After performing a Möbius transformation, we may assume that the singular member of the cubic pencil in question lies at $t = 0$ with singular point P. But then for P to be a surface singularity requires

$$0 = \left(\frac{\partial}{\partial t}(F + tG)\right)(P) = G(P).$$

Since F vanishes quadratically at P by assumption, we infer that P is a base point with multiplicity at least two. □

Example 7.18 Consider the cubic pencil S over $\mathbb{Q}$ given by the polynomials

$$F = 30909x^2y - 12983x^2z - 45834xy^2 + 32923xyz - 5911xz^2 + 896yz^2,$$
$$G = 36055x^2y - 19461x^2z - 53726xy^2 + 45233xyz - 8997xz^2 + 896y^2z.$$

Using standard formulas (for instance converting the pencil to Weierstrass form or applying the formulas for the jacobian of a plane cubic, cf. [AK+01, ARVT05]), one verifies that the discriminant has only simple zeroes over $\bar{\mathbb{Q}}$, so the corresponding elliptic surface has no reducible fibres. Alternatively, we can read this off the base points

$$[0,0,1], [0,1,0], [1,0,0], [1,1,1], [1,2,3], [1,5,7],$$
$$[1,11,13], [4,3,1], [-91637444, 7974341, 223433111],$$

using Lemma 7.17 (and the same holds true over any field of characteristic $p > 7639$). After a choice of zero section we can compute the height pairing of the sections given by the base points to obtain a sublattice of $\mathrm{MWL}(S)$ of rank 8 and determinant 9. We will show in Theorem 7.26 in a more general framework that the base points generate a subgroup of $\mathrm{MW}(S)$ of index 3.

In Lemma 7.17 we deliberately did not refer to an elliptic fibration because in general the fibres of $S \to \mathbb{P}^1$ need not be smooth. This is obvious when F and G are at most linear in one of the variables (in which case S is not even normal), but there are also other special cases as the following example shows.

Example 7.19 Let $F = y^2z - x^3$, $G = z^3$ with nine-fold base point $[0, 1, 0]$. Blowing up uniformly gives a smooth surface $\tilde{S}$ admitting a genus one fibration $\tilde{S} \to \mathbb{P}^1$ with a fibre of type II* at $t = \infty$. The fibration is elliptic outside characteristics 2, 3, but the general fibre is a cuspidal cubic in characteristics 2 and 3.

It turns out to be a general fact that exceptions as in the above example only occur in characteristics 2, 3:

Theorem 7.20 *If a smooth projective surface S admits a morphism f to a smooth projective curve C whose general fibre F is an irreducible curve of arithmetic genus one, then F is either elliptic or the characteristic is 2 or 3 and F is a cuspidal cubic.*

In characteristic zero, Theorem 7.20 can be regarded as a consequence of Sard's theorem; in positive characteristic it follows from Tate's theorem on genus change in inseparable extensions of functions fields [Tat52]. For later reference, we include a sketch of the proof.

Proof Assume that the general fibre F is singular. We obtain a horizontal curve $B \subset S$ as the closure of the singular locus on the irreducible fibres. B thus defines a multisection of the fibration of degree $n > 1$. If F were nodal, then a local argument would show that the smoothness of S prevents the node from extending to singularities of the neighbouring fibres (as encoded in B). Similarly, when F is cuspidal, then the intersection multiplicity $n = (F.B)$ can be calculated to equal 2 or 3. On the other hand, n is a power the characteristic of the base field, since the morphism

$$f|_B : B \to C \tag{7.13}$$

of degree n is purely inseparable by construction. □

Definition 7.21 A fibration from a smooth projective surface S to a smooth projective curve such that the general fibre is a cuspidal curve is called *quasi-elliptic*.

Along the same lines, we let the terminology *genus one fibration* (as in Lemma 7.17) encompass both elliptic and quasi-elliptic fibrations (including those without section). This will be of relevance in the study of K3 surfaces, see Sects. 11.4 and 12.4.4.1.

Note that the sections of a quasi-elliptic fibration continue to form a finitely generated abelian group by (3.4.1)—which we again refer to as Mordell–Weil group. The group structure, however, is very restrictive:

Corollary 7.22 *The Mordell–Weil group of a quasi-elliptic surface in characteristic p consists of p-torsion only.*

Proof By Theorem 7.20 the general fibre F is a cuspidal cubic. Hence the smooth locus $F^\#$ carries the group structure $\mathbb{G}_a$ of the additive group, and any section is annihilated by p. □

With these preparations, we can return to cubic pencils. Consider S as a projective variety in $\mathbb{P}^2 \times \mathbb{P}^1$ defined by (7.11). The projection onto the first factor exhibits S as the blow-up of $\mathbb{P}^2$ in the nine base points of the cubic pencil. This property will be useful in proving that the projection onto the second factor defines not only a genus one fibration, but an elliptic fibration.

Proposition 7.23 *Let S be a cubic pencil without multiple base points such that no three base points are collinear. Then S carries the structure of an elliptic fibration.*

Proof By Lemma 7.16, the general fibre of the genus one fibration $S \to \mathbb{P}^1$ is irreducible. Since the nine base points are distinct, the surface S, as a blow-up of $\mathbb{P}^2$ in the base points, is smooth with

$$b_2(S) = \rho(S) = 10. \tag{7.14}$$

By Theorem 7.20, the fibration is either elliptic or quasi-elliptic. In the latter case (where the characteristic can only be 2 or 3), $MW(S)$ is only torsion by Corollary 7.22. Hence the rank formula from Corollary 6.7 implies that $\mathrm{Triv}(S)$ has rank 10. But by Lemma 7.17, all fibres are irreducible, contradiction. □

Remark 7.24 The $b_2 = \rho$ part of (7.14) holds true for any quasi-elliptic surface $S \to \mathbb{P}^1$. The reason for this lies in the purely inseparable morphism (7.13) which presently implies $B \cong \mathbb{P}^1$. Hence base change results in a rational surface S' dominating S (as explained in Sect. 5.9.2). That is, S is unirational, and in particular, supersingular as stated.

The interpretation as blow-up of $\mathbb{P}^2$ in nine points turns out to be valid not only for cubic pencils, but for all rational elliptic surfaces. As in (7.14), this gives a way to compute their invariants (which we did in Proposition 7.1).

Theorem 7.25 *Over an algebraically closed field every rational elliptic surface (with section) admits a model as a cubic pencil.*

For a proof, the reader may consult [CD89, Theorem 5.6.1].

7.6 General Cubic Pencil

To conclude this chapter, we go back to a coordinate-free situation and give a simple proof for the following result of Manin and Shafarevich (cf. [Man64, Theorem 6], [Man86, Chap. IV, 26.4]).

Theorem 7.26 *Consider a pencil of cubic curves in $\mathbb{P}^2$ such that*

(i) every member of the pencil is irreducible and
(ii) the nine base points are distinct.

If we denote by S the blow-up of $\mathbb{P}^2$ at the nine base points, the pencil defines an elliptic fibration $f : S \to \mathbb{P}^1$ such that the nine exceptional curves arising from the blow-up give nine sections $P_0, \ldots, P_8$ of f. Choose P_0 as the zero section O, and let E be the generic fibre of f. Then $P_1, \ldots, P_8$ generate a subgroup of index 3 in $E(K)$, and there is a unique point $Q \in E(K)$ such that $\sum_{i=1}^{8} P_i = 3Q$. Together with $P_1, \ldots, P_7$, the point Q generates the full Mordell–Weil group $E(K)$.

Proof Since $f : S \to \mathbb{P}^1$ has no reducible fibres by *(i)*, the fibration cannot be quasi-elliptic (by the same argument as in the proof of Proposition 7.23). Hence the fibration is elliptic as claimed, and we have $r = 8$ and $E(K) \simeq E_8$ by Theorem 7.4. Since the nine exceptional curves (P_i) are disjoint from each other, (6.14) shows for any $i, j \geq 1$:

$$\langle P_i, P_j \rangle = \chi + (P_i.P_0) + (P_j.P_0) - (P_i.P_j) = \begin{cases} 1 & i \neq j, \\ 2 & i = j. \end{cases}$$

Thus one calculates that the Gram matrix of the height pairing has determinant

$$\det(\langle P_i, P_j \rangle) = 9 = 3^2.$$

Hence we see first that $P_1, \ldots, P_8$ are linearly independent and second that they generate a subgroup, say H, of index 3 in $E(K)$ since the latter lattice is isometric to E_8 and thus unimodular.

Take $Q \in E(K)$ which is not in H. Since $3Q \in H$, we can write

$$3Q = \sum_{i=1} n_i P_i \quad (n_i \in \mathbb{Z}).$$

We may assume (by adjusting Q by some element of H) that

$$-1 \leq n_i \leq 1 \quad \forall i = 1, \ldots, 8.$$

Since $E(K)$ is an integral lattice, $\langle Q, P_i \rangle$ is an integer. Now

$$\begin{aligned} 3\langle Q, P_i \rangle &= \langle 3Q, P_i \rangle \\ &= 2n_i + \sum_{j \neq i} n_j \\ &= n_i + N, \quad N = \sum_j n_j. \end{aligned}$$

Reducing modulo 3, we find that $n_i \equiv -N \pmod 3$ for all i, and hence $n_i \equiv n_j$ mod 3 and even $n_i = n_j$ for all i, j since $|n_i| \leq 1$ by our choice. Hence we see that $3Q$ is either O or $\pm \sum_i P_i$. Since $E(K)$ is torsion-free, the latter must hold. Then it is obvious that $Q, P_1, \ldots, P_7$ generate the full Mordell–Weil group $E(K)$. □

The above result is quite striking since it concerns a general rational elliptic surface (with section): all cubic pencils with Mordell–Weil rank less than 8 lie on a hypersurface inside the moduli space of cubic pencils (the discriminant divisor). It might therefore come as a little surprise that the general case seems to be exceptional within rational elliptic surfaces. Indeed, papers by [Fus06, Salg09, Pas12] show that for any cubic pencil of Mordell–Weil rank 4 to 7, the base points generate the full Mordell–Weil group. At the other end of the scale, [Bea82, Nar87] solved the case of Mordell–Weil rank zero where also there is a model as a cubic pencil whose base points generate the full (finite) Mordell–Weil group.

Historical Notes

When introducing the Mordell–Weil lattices, one of the key cases consisted in unimodular Néron–Severi lattices (see Sect. 6.8). Naturally this led to considering those surfaces where this automatically holds true: rational elliptic surfaces, the subject of this (and the next) chapter. Among elliptic surfaces, they form the most basic and classical class (besides the trivial case of products or fibrations without singular fibres which we ruled out by Convention 5.10). The main incarnation of rational elliptic surfaces consists in cubic pencils which can be studied by almost elementary means. Indeed, cubic pencils show up early, and naturally, in the study of elliptic curves (without original reference to surface theory) as they form a standard instrument in the geometric proof that the group law on an plane cubic curve (with distinguished rational point) is associative (cf. [Cas91, Sect. 7], [Hus14, Sect. 3.1]).

The basic properties of rational elliptic surfaces are all standard, and it appears that many researchers basically knew all the ingredients to investigate the structure of Mordell–Weil lattices. That is, with the exception, maybe, that the frame, the lattice E_8 was still missing from the picture, which is governing the theory so instructively. With its remarkable properties within all lattices, the frame naturally guides the investigation, and eventually the classification, of rational elliptic surfaces. From this viewpoint, the cases of high Mordell–Weil rank are conceptually easiest, without or with only few reducible fibres preventing the Mordell–Weil lattice to fill out the whole frame, i.e. MWL being as close to the frame E_8 as possible. In practice, however, one may argue that the cases of low Mordell–Weil rank at the opposite end of the scale are easier to handle; for instance, this holds true if it comes to determining generators or all (integral) sections in the Mordell–Weil group. (This statement will be justified by the study of (semi-stable) extremal rational elliptic surfaces, i.e. with Mordell–Weil rank zero, in Sect. 8.8.)

Weierstrass models of rational elliptic surfaces have been studied systematically to some extent before, for instance in [Cox79]. They are indispensable for translating properties of the Mordell–Weil lattice (considered inside the frame E_8) into polynomial equations encoding the sections. This seemingly innocent property will be essential soon for the definition and study of excellent families.

Quasi-elliptic surfaces have been included partly for completeness, but partly also for later reference as they will show up in our study of supersingular elliptic K3 surfaces (so including them is required in order to allow for a uniform treatment in terms of lattice theory). They were first introduced by Bombieri and Mumford as a special feature (or pathology) of the Enriques–Kodaira classification in characteristics 2 and 3 [BM76/77]. Their very special properties have first been recorded (and used) to large extent by [RS78] (see also [Ito94, ElS15]).

A first illustration of the powerful machinery of Mordell–Weil lattices is given by the alternative proof of Manin's result on the base points of general cubic pencils in Sect. 7.6. Indeed, it is quite remarkable that with a little bit of theory borrowed from Mordell–Weil lattices, the proof ends up being this streamlined and relatively easy.

Chapter 8
Rational Elliptic Surfaces and E_8-Hierarchy

So far we have focused our attention on the first remarkable cases of higher Mordell–Weil rank for rational elliptic surfaces. In this chapter, we follow [OS90] to work out the complete classification of all other possible cases over an algebraically closed field k of arbitrary characteristic (Theorem 8.8). Whenever possible, we also prove the existence (Theorem 8.9).

8.1 Singular Fibres

In Lemma 7.2 we have computed the frame of a rational elliptic surface S as $W = E_8$. Thus there is an embedding of the orthogonal sum of the root lattices associated with the singular fibres into this lattice:

$$T = \bigoplus_{v \in C(k)} T_v \hookrightarrow E_8. \tag{8.1}$$

Note that this embedding need not be primitive. In fact, we know that the torsion in the Mordell–Weil group is encoded in the primitive closure $\mathrm{Triv}(S)'$ of the trivial lattice $\mathrm{Triv}(S)$ inside $\mathrm{NS}(S)$. Alternatively, we can compute the primitive closure T' of the above direct sum (8.1) inside E_8. That is to say, the embedding (8.1) is primitive if and only if $T = T'$ if and only if $\mathrm{MW}(S)$ is torsion-free.

Naturally, (8.1) bounds the sum of the ranks of the root lattices T_v by eight. Thus the singular fibres of a rational elliptic surface may have types II, III, IV, IV*, III*, II*, but the other types are limited to I_n $(n \leq 9)$ and I_m^* $(m \leq 4)$. The fibre types with a maximum number of fibre components correspond to the irreducible root lattices of rank 8:

$$\mathrm{I}_9 \leftrightarrow A_8, \quad \mathrm{I}_4^* \leftrightarrow D_8, \quad \mathrm{II}^* \leftrightarrow E_8.$$

We will briefly discuss these maximal cases at the end of this chapter in Sect. 8.10.

M. Schütt and T. Shioda, *Mordell–Weil Lattices*, Ergebnisse der Mathematik und ihrer Grenzgebiete. 3. Folge / A Series of Modern Surveys in Mathematics 70,
https://doi.org/10.1007/978-981-32-9301-4_8

Table 8.1 Root lattices embedding into E_8

$s = 8$	$A_8,\ D_8,\ A_7 \oplus A_1,\ A_5 \oplus A_2 \oplus A_1,\ A_4^2,\ A_2^4,\ E_6 \oplus A_2,\ E_7 \oplus A_1$
	$D_6 \oplus A_1^2,\ D_5 \oplus A_3,\ D_4^2,\ D_4 \oplus A_1^4,\ A_3^2 \oplus A_1^2,\ A_1^8$
$s = 7$	$A_6 \oplus A_1,\ A_4 \oplus A_2 \oplus A_1,\ A_5 \oplus A_2,\ A_2^3 \oplus A_1,\ E_6 \oplus A_1,\ E_7,$
	$D_7,\ D_5 \oplus A_1^2,\ D_4 \oplus A_1^3,\ A_3^2 \oplus A_1,\ A_1^7,\ D_6 \oplus A_1,\ D_5 \oplus A_2,$
	$A_3 \oplus A_2 \oplus A_1^2,\ D_4 \oplus A_3,\ A_3 \oplus A_1^4,\ A_4 A_3,\ A_5 \oplus A_1^2,\ A_7$
$s = 6$	$A_2^3,\ E_6,\ D_6,\ D_4 \oplus A_1^2,\ A_3^2,\ D_5 \oplus A_1,\ A_3 \oplus A_1^3,\ D_4 \oplus A_2,\ A_1^6,$
	$A_2 \oplus A_1^4,\ A_4 \oplus A_1^2,\ A_6,\ A_3 \oplus A_2 \oplus A_1,\ A_5 \oplus A_1,\ A_4 \oplus A_2,\ A_2^2 \oplus A_1^2$
$s = 5$	$D_5,\ A_3 \oplus A_1^2,\ A_3 \oplus A_2,\ A_5,\ A_1^5,\ A_4 \oplus A_1,$
	$D_4 \oplus A_1,\ A_2 \oplus A_1^3,\ A_2^2 \oplus A_1$
$s = 4$	$D_4,\ A_1^4,\ A_2 \oplus A_1^2,\ A_2^2,\ A_3 \oplus A_1,\ A_4$
$s = 3$	$A_3,\ A_2 \oplus A_1,\ A_1^3$
$s = 2$	$A_2,\ A_1^2$
$s = 1$	A_1

In fact, one can determine all root lattices embedding into E_8 (or equivalently into E_8^-) explicitly. This was first carried out by Dynkin [Dyn57]:

Theorem 8.1 *Let T be a root lattice of rank s which is embedded as a sublattice of E_8, other than $\{0\}$ and E_8. Then $s \leq 8$, and T is isomorphic to one of the 70 root lattices specified in Table 8.1.*

For our purposes, we have to take into account two problems:

1. whether the embedding (8.1) is unique up to isometries;
2. whether there indeed is a rational elliptic surface corresponding to each embedding.

The first problem has been solved by Dynkin as follows:

Theorem 8.2 *Each sublattice of E_8 as above admits a unique embedding into E_8 modulo the action of the Weyl group $W(E_8)$, except for the following five types:*

$$A_7,\quad A_3^2,\quad A_5 \oplus A_1,\quad A_3 \oplus A_1^2,\quad A_1^4. \tag{8.2}$$

Moreover, in each of the exceptional five cases, there exist exactly two inequivalent embeddings into E_8 modulo $W(E_8)$.

It will be interesting to note that in all ambiguous cases, there is one primitive and one imprimitive embedding into E_8. Namely, each lattice from (8.2) has a unique even integral proper overlattice of finite index, given in the order of (8.2) by

$$E_7,\quad D_6,\quad E_6,\quad D_5,\quad D_4.$$

(This is easily checked by an explicit gluing argument based on the discriminant forms, see Sect. 2.3.5.) It follows that the rational elliptic surfaces supporting a given root lattice from (8.2) on the singular fibres (up to sign) will be distinguished by the torsion in the Mordell–Weil groups.

As for the second problem, we give a partial answer by ruling out all those root lattices which are a priori impossible on a rational elliptic surface for *all* characteristics:

Lemma 8.3 *There are exactly three cases in Table 8.1 that cannot be associated with the singular fibres of a rational elliptic surface:*

$$T = D_4 \oplus A_1^4, \quad A_1^8 \quad or \quad A_1^7. \tag{8.3}$$

Proof By Theorem 5.4.7 the sum of the local Euler number of the reducible singular fibres cannot exceed the Euler number (= 12) of a rational elliptic surface. Presently, the local Euler numbers of the reducible fibres would already sum up to 14 or 16. □

On the other hand, Theorem 8.1 rules out many root lattices right away that could a priori have been possible. For instance, this applies to the root lattice $A_4 \oplus A_1^3$ that does not appear in Table 8.1. In other approaches to configurations of singular fibres on rational elliptic surfaces, alternative arguments have been developed using monodromy representations, for instance, to rule out such configurations (see the papers by Persson [Per90] and Miranda [Mir90], and also Miranda's book [Mir89] or [SSh10]).

Remark 8.4 We point out that two of the root lattices ruled out in Lemma 8.3, namely $D_4 \oplus A_1^4$ and A_1^8, are supported on singular fibres of certain rational quasi-elliptic surfaces in characteristic 2, compare ([Ito94, Proposition 5.1]).

Summary 8.5 Including $\{0\}$ and E_8, there are 69 root lattices that could a priori be associated with the singular fibres of a rational elliptic surface. Among them, exactly the five root lattices from (8.2) do not admit a unique embedding into E_8 up to the action of $W(E_8)$. In total, there are therefore 74 embeddings of root lattices into E_8 to consider.

8.2 Structure of Mordell–Weil Lattice

Recall that $\mathrm{NS}(S)$ is unimodular for a rational elliptic surface S by Proposition 7.1. Thus Theorem 6.5.2 gives a complete description for the structure of the Mordell–Weil lattice of a rational elliptic surface in terms of the reducible singular fibres. For easier reference within this chapter, we collect the data in the following theorem:

Theorem 8.6 *Given a rational elliptic surface $f : S \to \mathbb{P}^1$ with a section O, let R be the subset of $C(k)$ such that $f^{-1}(v)$ is reducible if and only if $v \in R$. For*

$v \in R$, let m_v be the number of irreducible components $\Theta_{v,i}$ of $f^{-1}(v)$, numbered $\Theta_{v,0}, \ldots, \Theta_{v,m_v-1}$ such that $\Theta_{v,0}$ meets (O). Let $T_v^- = \langle \Theta_{v,i} | 1 \leq i \leq m_v - 1 \rangle$ and $T = \oplus_{v \in R} T_v$ with embedding (8.1). Let $L = T^\perp$ be the orthogonal complement of T in E_8; L is a positive-definite even lattice. Then the Mordell–Weil group $M := E(K)$ has rank

$$r = \mathrm{rk} E(K) = 8 - \sum_{v \in R} (m_v - 1)$$

$$E(K)_{\mathrm{tors}} \simeq T'/T, \quad n = |E(K)_{\mathrm{tors}}| = [T' : T]$$

$$E(K)^0 \simeq L, \quad \det L = \prod_{v \in R} m_v^{(1)} / n^2$$

$$E(K)/E(K)_{\mathrm{tors}} \simeq L^\vee, \quad \det L^\vee = 1/\det L$$

$$M = E(K) \simeq L^\vee \oplus (T'/T).$$

In the above, T' is the primitive closure of T in E_8, and $m_v^{(1)}$ is the number of simple components in $f^{-1}(v)$.

Corollary 8.7 *$E(K)$ is torsion-free if* $\det T = \prod_{v \in R} m_v^{(1)}$ *is square-free.*

As a consequence, the determination of the Mordell–Weil group (or lattice) of a rational elliptic surface reduces to the study of the sublattices T in E_8 which are direct sums of root lattices. We have already examined the case $\mathrm{rk}\, T \leq 2$ which corresponds to the case of higher Mordell–Weil rank $r = 8, 7$ or 6 in the previous chapter (Theorem 7.4).

For the remaining cases, we can make use of Dynkin's result on the classification of root lattices $T \hookrightarrow E_8$ which we have seen in Sect. 8.1. Thus it suffices to apply Theorem 8.6 to all possible embeddings (8.1). This programme has been carried out in joint work of one of us with Keiji Oguiso [OS90] which we follow closely throughout this chapter.

8.3 Classification

We start by stating the classification of Mordell–Weil lattices of rational elliptic surfaces from [OS90]. (N.B. The misprints about No. 32 and No. 70 in the table of [OS90] are corrected, following [Shio92b, p. 251].)

Theorem 8.8 (Classification theorem) *The structure of the Mordell–Weil lattice of a rational elliptic surface is classified into 74 types in Table 8.2, in terms of the triple $\{T, L, M\}$ where T is the fibre contribution to the trivial lattice (with a fixed embedding into E_8), $L = E(K)^0$ the narrow Mordell–Weil lattice and $M = E(K)$ the Mordell–Weil group with height pairing.*

Table 8.2 Types of Mordell–Weil lattices of rational elliptic surfaces

No.	r	rk$T$	T	L	M
1	8	0	$\{0\}$	E_8	E_8
2	7	1	A_1	E_7	$E_7^\vee$
3	6	2	A_2	E_6	$E_6^\vee$
4			A_1^2	D_6	$D_6^\vee$
5	5	3	A_3	D_5	$D_5^\vee$
6			$A_2 \oplus A_1$	A_5	$A_5^\vee$
7			A_1^3	$D_4 \oplus A_1$	$D_4^\vee \oplus A_1^\vee$
8	4	4	A_4	A_4	$A_4^\vee$
9			D_4	D_4	$D_4^\vee$
10			$A_3 \oplus A_1$	$A_3 \oplus A_1$	$A_3^\vee \oplus A_1^\vee$
11			A_2^2	A_2^2	$A_2^{\vee\oplus 2}$
12			$A_2 \oplus A_1^2$	$(A_5 : A_1)$	$(A_5 : A_1)^\vee$
13			A_1^4	D_4	$D_4^\vee \oplus \mathbf{Z}/2\mathbf{Z}$
14			A_1^4	A_1^4	$A_1^{\vee 4}$
15	3	5	A_5	$A_2 \oplus A_1$	$A_2^\vee \oplus A_1^\vee$
16			D_5	A_3	$A_3^\vee$
17			$A_4 \oplus A_1$	$(A_4 : A_1)$	$(A_4 : A_1)^\vee$
18			$D_4 \oplus A_1$	A_1^3	$(A_1^\vee)^3$
19			$A_3 \oplus A_2$	$(D_5 : A_2)$	$(D_5 : A_2)^\vee$
20			$A_2^2 \oplus A_1$	$A_2 \oplus \langle 6\rangle$	$A_2^\vee \oplus \langle 1/6\rangle$
21			$A_3 \oplus A_1^2$	A_3	$A_3^\vee \oplus \mathbf{Z}/2\mathbf{Z}$
22			$A_3 \oplus A_1^2$	$A_1^2 \oplus \langle 4\rangle$	$(A_1^\vee)^2 \oplus \langle 1/4\rangle$
23			$A_2 \oplus A_1^3$	$A_1 \oplus A_2(2)$	$A_1^\vee \oplus A_2^\vee(1/2)$
24			A_1^5	A_1^3	$(A_1^\vee)^3 \oplus \mathbf{Z}/2\mathbf{Z}$
25	2	6	A_6	$\begin{pmatrix} 4 & 1 \\ 1 & 2 \end{pmatrix}$	$\begin{pmatrix} 2/7 & 1/7 \\ 1/7 & 4/7 \end{pmatrix}$
26			D_6	A_1^2	$(A_1^\vee)^2$
27			E_6	A_2	$A_2^\vee$
28			$A_5 \oplus A_1$	A_2	$A_2^\vee \oplus \mathbf{Z}/2\mathbf{Z}$
29			$A_5 \oplus A_1$	$A_1 \oplus \langle 6\rangle$	$A_1^\vee \oplus \langle 1/6\rangle$
30			$D_5 \oplus A_1$	$A_1 \oplus \langle 4\rangle$	$A_1^\vee \oplus \langle 1/4\rangle$
31			$A_4 \oplus A_2$	$\begin{pmatrix} 8 & -1 \\ -1 & 2 \end{pmatrix}$	$\begin{pmatrix} 2/15 & 1/15 \\ 1/15 & 8/15 \end{pmatrix}$
32			$D_4 \oplus A_2$	$A_2(2)$	$A_2^\vee(1/2)$
33			$A_4 \oplus A_1^2$	$\begin{pmatrix} 6 & -2 \\ -2 & 4 \end{pmatrix}$	$\begin{pmatrix} 1/5 & 1/10 \\ 1/10 & 3/10 \end{pmatrix}$

(continued)

Table 8.2 (continued)

No.	r	rk$T$	T	L	M
34	2	6	$D_4 \oplus A_1^2$	A_1^2	$(A_1^\vee)^2 \oplus \mathbf{Z}/2\mathbf{Z}$
35			A_3^2	A_1^2	$(A_1^\vee)^2 \oplus \mathbf{Z}/2\mathbf{Z}$
36			A_3^2	$\langle 4\rangle^2$	$\langle 1/4\rangle^2$
37			$A_3 \oplus A_2 \oplus A_1$	$A_1 \oplus \langle 12\rangle$	$A_1^\vee \oplus \langle 1/12\rangle$
38			$A_3 \oplus A_1^3$	$A_1 \oplus \langle 4\rangle$	$A_1^\vee \oplus \langle 1/4\rangle \oplus \mathbf{Z}/2\mathbf{Z}$
39			A_2^3	A_2	$A_2^\vee \oplus \mathbf{Z}/3\mathbf{Z}$
40			$A_2^2 \oplus A_1^2$	$\langle 6\rangle^2$	$\langle 1/6\rangle^2$
41			$A_2 \oplus A_1^4$	$A_2(2)$	$A_2^\vee(1/2) \oplus \mathbf{Z}/2\mathbf{Z}$
42			A_1^6	A_1^2	$(A_1^\vee)^2 \oplus (\mathbf{Z}/2\mathbf{Z})^2$
43	1	7	E_7	A_1	$A_1^\vee$
44			A_7	A_1	$A_1^\vee \oplus \mathbf{Z}/2\mathbf{Z}$
45			A_7	$\langle 8\rangle$	$\langle 1/8\rangle$
46			D_7	$\langle 4\rangle$	$\langle 1/4\rangle$
47			$A_6 \oplus A_1$	$\langle 14\rangle$	$\langle 1/14\rangle$
48			$D_6 \oplus A_1$	A_1	$A_1^\vee \oplus \mathbf{Z}/2\mathbf{Z}$
49			$E_6 \oplus A_1$	$\langle 6\rangle$	$\langle 1/6\rangle$
50			$D_5 \oplus A_2$	$\langle 12\rangle$	$\langle 1/12\rangle$
51			$A_5 \oplus A_2$	A_1	$A_1^\vee \oplus \mathbf{Z}/3\mathbf{Z}$
52			$D_5 \oplus A_1^2$	$\langle 4\rangle$	$\langle 1/4\rangle \oplus \mathbf{Z}/2\mathbf{Z}$
53			$A_5 \oplus A_1^2$	$\langle 6\rangle$	$\langle 1/6\rangle \oplus \mathbf{Z}/2\mathbf{Z}$
54			$D_4 \oplus A_3$	$\langle 4\rangle$	$\langle 1/4\rangle \oplus \mathbf{Z}/2\mathbf{Z}$
55			$A_4 \oplus A_3$	$\langle 20\rangle$	$\langle 1/20\rangle$
56			$A_4 \oplus A_2 \oplus A_1$	$\langle 30\rangle$	$\langle 1/30\rangle$
57			$D_4 \oplus A_1^3$	A_1	$A_1^\vee \oplus (\mathbf{Z}/2\mathbf{Z})^2$
58			$A_3^2 \oplus A_1$	A_1	$A_1^\vee \oplus \mathbf{Z}/4\mathbf{Z}$
59			$A_3 \oplus A_2 \oplus A_1^2$	$\langle 12\rangle$	$\langle 1/12\rangle \oplus \mathbf{Z}/2\mathbf{Z}$
60			$A_3 \oplus A_1^4$	$\langle 4\rangle$	$\langle 1/4\rangle \oplus (\mathbf{Z}/2\mathbf{Z})^2$
61			$A_2^3 \oplus A_1$	$\langle 6\rangle$	$\langle 1/6\rangle \oplus \mathbf{Z}/3\mathbf{Z}$
62	0	8	E_8	0	0
63			A_8	0	$\mathbf{Z}/3\mathbf{Z}$
64			D_8	0	$\mathbf{Z}/2\mathbf{Z}$
65			$E_7 \oplus A_1$	0	$\mathbf{Z}/2\mathbf{Z}$
66			$A_5 \oplus A_2 \oplus A_1$	0	$\mathbf{Z}/6\mathbf{Z}$
67			A_4^2	0	$\mathbf{Z}/5\mathbf{Z}$
68			A_2^4	0	$(\mathbf{Z}/3\mathbf{Z})^2$
69			$E_6 \oplus A_2$	0	$\mathbf{Z}/3\mathbf{Z}$
70			$A_7 \oplus A_1$	0	$\mathbf{Z}/4\mathbf{Z}$
71			$D_6 \oplus A_1^2$	0	$(\mathbf{Z}/2\mathbf{Z})^2$
72			$D_5 \oplus A_3$	0	$\mathbf{Z}/4\mathbf{Z}$
73			D_4^2	0	$(\mathbf{Z}/2\mathbf{Z})^2$
74			$A_3^2 \oplus A_1^2$	0	$\mathbf{Z}/4\mathbf{Z} \oplus \mathbf{Z}/2\mathbf{Z}$

We will prove separately that the classification of Theorem 8.8 is independent of the characteristic with just a few exceptions in characteristics 2 and 3. Here we just state the result; details can be found in Sects. 8.8 and 8.9).

Theorem 8.9 *1. Each type from Theorem 8.8 occurs over any algebraically closed field of characteristic* $\neq 2, 3$.
2. *In characteristic* 2, *all types occur except for Nos. 42, 57, 60, 61, 71, 73, 74.*
3. *In characteristic* 3, *all types occur except for No. 68.*

Remark 8.10 There are numerous models over $\mathbb{Z}$ (see e.g. Table 8.3). Exceptions occur when the type degenerates upon reducing modulo 2 or 3, as predicted by Proposition 6.33(vi). Some of the exceptional types can be realized on rational quasi-elliptic surfaces (like the extra types from Remark 8.4).

Table 8.3 Extremal rational elliptic surfaces

No.	Weierstrass eqn.	Sing. fibres	char(k)	MW
62	$y^2 + xy = x^3 + t$	$\mathrm{II}^*, 2 \times \mathrm{I}_1$	$p \neq 2, 3$	0
		$\mathrm{II}^*, \mathrm{I}_1$	$p = 2, 3$	
	$y^2 + y = x^3 + t$	$\mathrm{II}^*, \mathrm{II}$	$p \neq 2, 3$	0
		II^*	$p = 2$	
	$y^2 = x^3 + x + t$	II^*	$p = 3$	
63	$y^2 + (tx + 1)y = x^3$	$\mathrm{I}_9, 3 \times \mathrm{I}_1$	$p \neq 3$	$\mathbb{Z}/3\mathbb{Z}$
		$\mathrm{I}_9, \mathrm{II}$	$p = 3$	
64	$y^2 = x(x^2 + tx + t^4)$	$\mathrm{I}_4^*, 2 \times \mathrm{I}_1$	$p \neq 2$	$\mathbb{Z}/2\mathbb{Z}$
	$y^2 + xy = x^3 + tx^2 + \lambda x \quad (\lambda \in k^\times)$	I_4^*	$p = 2$	
65	$y^2 + xy = x(x^2 + t)$	$\mathrm{III}^*, \mathrm{I}_2, \mathrm{I}_1$	$p \neq 2$	$\mathbb{Z}/2\mathbb{Z}$
		$\mathrm{III}, \mathrm{I}_2$	$p = 2$	
	$y^2 = x^3 + tx$	$\mathrm{III}^*, \mathrm{III}$	$p \neq 2$	$\mathbb{Z}/2\mathbb{Z}$
66	$y^2 + txy = x(x^2 + (t-2)x + 1 - t)$	$\mathrm{I}_6, \mathrm{I}_3, \mathrm{I}_2, \mathrm{I}_1$	$p \neq 2, 3$	$\mathbb{Z}/6\mathbb{Z}$
		$\mathrm{I}_6, \mathrm{I}_3, \mathrm{III}$	$p = 3$	
		$\mathrm{I}_6, \mathrm{IV}, \mathrm{I}_2$	$p = 2$	
67	$y^2 + ((1-t)x - t)y = x^3 - tx^2$	$2 \times \mathrm{I}_5, 2 \times \mathrm{I}_1$	$p \neq 5$	$\mathbb{Z}/5\mathbb{Z}$
		$2 \times \mathrm{I}_5, \mathrm{II}$	$p = 5$	
68	$y^2 + (3tx + t^3 - 1)y = x^3$	$4 \times \mathrm{I}_3$	$p \neq 3$	$(\mathbb{Z}/3\mathbb{Z})^2$
69	$y^2 + (x + t)y = x^3$	$\mathrm{IV}^*, \mathrm{I}_3, \mathrm{I}_1$	$p \neq 3$	$\mathbb{Z}/3\mathbb{Z}$
		$\mathrm{IV}^*, \mathrm{I}_3$	$p = 3$	
	$y^2 + ty = x^3$	$\mathrm{IV}^*, \mathrm{IV}$	$p \neq 3$	$\mathbb{Z}/3\mathbb{Z}$
70	$y^2 + txy = x(x^2 + 2x + 1)$	$\mathrm{I}_8, \mathrm{I}_2, 2 \times \mathrm{I}_1$	$p \neq 2$	$\mathbb{Z}/4\mathbb{Z}$
		$\mathrm{I}_8, \mathrm{III}$	$p = 2$	
71	$y^2 = x(x-1)(x-t)$	$\mathrm{I}_2^*, 2 \times \mathrm{I}_2$	$p \neq 2$	$(\mathbb{Z}/2\mathbb{Z})^2$
72	$y^2 + xy = x(x+t)^2$	$\mathrm{I}_1^*, \mathrm{I}_4, \mathrm{I}_1$	$p \neq 2$	$\mathbb{Z}/4\mathbb{Z}$
		$\mathrm{I}_1^*, \mathrm{I}_4$	$p = 2$	
73	$y^2 = x(x^2 + tx + \lambda t^2) \quad (\lambda \in k^\times)$	$2 \times \mathrm{I}_0^*$	$p \neq 2$	$(\mathbb{Z}/2\mathbb{Z})^2$
74	$y^2 = x(x-1)(x-t^2)$	$2 \times \mathrm{I}_4, 2 \times \mathrm{I}_2$	$p \neq 2$	$\mathbb{Z}/4\mathbb{Z} \times \mathbb{Z}/2\mathbb{Z}$

Before we state the classification, we introduce some terminology and recall the usual notation for the reader's convenience.

Notation 8.11 • $r = \operatorname{rk} E(K)$: the Mordell–Weil rank.
- $T = \oplus_{v \in R} T_v \hookrightarrow E_8$: the sublattice of the trivial lattice inside the frame (with sign reversed) generated by fibre components disjoint from O.
- $L = E(K)^0 \simeq T^\perp$: the narrow Mordell–Weil lattice.
- $M = E(K) \simeq L^\vee \oplus E(K)_{\text{tors}}$: the Mordell–Weil group with the height pairing (i.e. the Mordell–Weil lattice unless there is torsion inside $E(K)$).
- $\langle m \rangle$: rank 1 lattice $\mathbb{Z}\xi$ such that $\langle \xi, \xi \rangle = m \in \mathbb{Q}_{>0}$.
- $L(m)$: a lattice L with intersection form scaled by a positive rational number $m \in \mathbb{Q}_{>0}$. For instance, $A_2^\vee \cong A_2(1/3)$.
- $(A : B)$: the orthogonal complement of a sublattice $B \subset A$. (This symbol is used only in three cases where the orthogonal complement itself cannot be written as a direct sum of a root lattice and a lattice of rank one; in each case, the embedding $B \hookrightarrow A$ is unique up to isometries.)
- I: a lattice with the Gram matrix I.

For Nos. 12, 17, 19, we will need the following lattices in the above notation (given together with determinants and dual lattices):

$$L = (A_5 : A_1) \simeq \begin{pmatrix} 4 & -1 & 0 & 1 \\ -1 & 2 & -1 & 0 \\ 0 & 1 & 2 & -1 \\ 1 & 0 & -1 & 2 \end{pmatrix}, \quad L^\vee \simeq \frac{1}{6} \begin{pmatrix} 2 & 1 & 0 & -1 \\ 1 & 5 & 3 & 1 \\ 0 & 3 & 6 & 3 \\ -1 & 1 & 3 & 5 \end{pmatrix}, \quad \det(L) = 12,$$

$$L = (A_4 : A_1) \simeq \begin{pmatrix} 4 & -1 & 1 \\ -1 & 2 & -1 \\ 1 & -1 & 2 \end{pmatrix}, \quad L^\vee \simeq \frac{1}{10} \begin{pmatrix} 3 & 1 & -1 \\ 1 & 7 & 3 \\ -1 & 3 & 7 \end{pmatrix}, \quad \det(L) = 10,$$

$$L = (D_5 : A_2) \simeq \begin{pmatrix} 2 & 0 & -1 \\ 0 & 2 & -1 \\ -1 & -1 & 4 \end{pmatrix}, \quad L^\vee \simeq \frac{1}{12} \begin{pmatrix} 7 & 1 & 2 \\ 1 & 7 & 2 \\ 2 & 2 & 4 \end{pmatrix}, \quad \det(L) = 12.$$

Remark 8.12 Sometimes it can be beneficial to employ a different interpretation of the above lattices. For instance, $(D_5 : A_2)$ looks, up to sign, like two copies of A_1 with a section of height 3 attached.

Example 8.13 (*Hexagonal example cont'd*) Recall the elliptic curve from (1.1). In Example 6.41, building on Examples 5.26, 5.28, 5.29 etc., we have seen that this elliptic curve realizes the following cases from Theorem 8.8:

No. 27 if the α_i are pairwise distinct,
No. 49 if $\alpha_i = \alpha_j \neq \alpha_k$,
No. 69 if all α_i come together.

(In characteristic 3, the fully degenerate case only gives a quasi-elliptic fibration, though.)

Example 8.14 Consider No. 27 from Theorem 8.8. As it happens, Example 6.55 provides such an elliptic surface (albeit quite a special one which will be used later again, with constant j-invariant and outside characteristics 2 and 3). Take sections P, Q for two different roots of $\beta^3 - 2$, normalized with the same sign in the y-coordinate such that both sections meet the IV* fibre in the same non-identity component. Thus we find the height pairing

$$\langle P, Q\rangle = 1 - 0 - 2/3 = 1/3.$$

Note that this agrees with the announced shape of the Mordell–Weil lattice which in particular is generated by P, Q.

We conclude with a brief word as to how these sections were originally found. By Theorem 8.8, there are sections of height 2/3. By inspection of the height formula, this requires that these sections are disjoint from the zero section and meet the IV* fibre in a non-identity component. Then the Weierstrass form in (6.42) leads to the ansatz

$$(x, y) = (\beta(t-1)^2, (\gamma t + \delta)(t-1)^2).$$

Solving directly leads to the given sections: $\beta^3 = 1/4, \gamma = \delta = \pm 1/2$.

8.4 Torsion Sections on Rational Elliptic Surfaces

We have already proved the first four entries of the classification in Sect. 7.3. In order to prove the classification completely, it suffices to compute the possibilities for torsion in the Mordell–Weil group and the narrow Mordell–Weil lattice—both depending on the embedding (8.1).

Here we will achieve this through an investigation of contraction terms. For other approaches see the references at the end of Sect. 8.1.

In the following we will often use the criterion of (6.20). For this purpose we recall the maximum possible value of the term $\text{contr}_v(P)$ when T_v is a root lattice of a given type. The given values follow immediately from Table 6.1 in Sect. 6.5.

$$\max \text{contr}_v(P) = \begin{cases} (n+1)/4 & A_n \quad (n : \text{odd}), \\ n(n+2)/4(n+1) & A_n \quad (n : \text{even}), \\ n/4 & D_n \quad (n \geq 4), \\ 4/3 & E_6, \\ 3/2 & E_7. \end{cases} \tag{8.4}$$

Recall that the existence of torsion in $E(K)$ is equivalent to the property that the trivial lattice Triv (or its orthogonal summand T^-) has its primitive closure Triv′ (or T') as a proper overlattice. If n denotes the index of Triv $\subset$ Triv′ (or equivalently, of $T \subset T'$), then $\det(\text{Triv}') = \det(\text{Triv})/n^2$ as predicted by Proposition 6.31(i). Sometimes general lattice results then suffice to rule out torsion. This particularly holds for the unimodular case due to Theorem 2.9 and Corollary 2.10, see also the discussion around Remark 2.33 in Sect. 2.3.4.

Our lattice theoretic approach has the advantage that it is independent of the characteristic. If one restricts to the situation where the torsion order is relatively prime to the characteristic, then there are other methods to investigate the possible torsion given the singular fibres. Namely, translation by a torsion section defines an automorphism of the elliptic surface whose quotient surface has a resolution of the same kind (again rational for instance). Here the singular fibres of the quotient can be determined a priori from the fibre components met by the torsion section, and thus they would also have to appear in the list. On the level of generic fibres, this is reflected by an isogeny, as we have seen in Example 7.15.

We shall now illustrate the proof of Theorem 8.8 by discussing several instructive cases. They should enable the attentive reader to fill in the remaining steps without problems.

8.5 Proof of Classification Theorem 8.8

The columns in Table 8.2 are arranged in the following order: they descend with respect to

1. the rank $r = \text{rk } E(K)$,
2. the determinant $\det T$,
3. the size of the torsion subgroup $n = |E(K)_{\text{tors}}|$.

Thus the enumeration follows in principle from Theorem 8.1. To verify the table, it remains to compute (the possibilities for) the index $n = [T' : T]$ (plus the structure of $E(K)_{\text{tors}}$ in case $n = 4$, 8 or 9) and the structure of the orthogonal complement $L = T^\perp$ of T in E_8, since $E(K)$ is then completely determined by Theorem 8.6.

In the following subsections, we divide the proof into five cases:

$$r \geq 4, \quad r = 3, \quad r = 2, \quad r = 1, \quad r = 0.$$

Throughout the proof, we will freely use facts about embeddings of lattices. In particular, this concerns uniqueness of the embeddings up to isometries and orthogonal complements. For instance, A_k always admits a unique embedding into A_n for $k < n$ up to the action of the Weyl group $W(A_n)$. A good reference to look up these properties is [Nis96], but essentially all computations can also easily be carried out by hand with some elementary linear algebra.

8.5.1 $r \geq 4$ (Nos. 1–14)

Note that we already covered the cases $r \geq 6$ in Theorem 7.4. Generally, all cases with $r \geq 4$ are fairly easy to analyse since there can only be torsion for one configuration T of root lattices:

Lemma 8.15 *If $r \geq 4$, then $n = 1$ (i.e. $E(K)$ is torsion-free), except possibly for No.* 13 *or* 14.

Proof Here is one possible way to prove the lemma:

(a) If $\det T$ is square-free, then the claim follows from Proposition 6.31(i).
(b) Next suppose $\det T = 4$ or 9. If $n \neq 1$, then $n = 2$ or 3 by (6.18). Thus the primitive closure T' of T in E_8 would have $\det T' = 1$. Since $\operatorname{rk} T = 8 - r$, this contradicts Corollary 2.10. Hence $n = 1$.
(c) If $\det T = 8$ or 12, we can easily check the condition (6.20) in Proposition 6.31(ii). Namely, by (8.4), we have

$$\sum_{v \in R} \operatorname{contr}_v(P) \leq \begin{cases} 3/2 & \text{(No. 7 or 10)}, \\ 5/3 & \text{(No. 12)}. \end{cases}$$

(d) Finally, if $\det T = 16$ (No. 13 or 14), then $n = 1$ or 2. Here the possibility $n = 4$ is excluded by the same argument as in (b), or alternatively by (6.21). □

It might be instructive to take a closer look at the two inequivalent embeddings of A_1^4 into E_8 corresponding to case Nos. 13 and 14. One is obvious: pick any four disjoint roots from the Dynkin diagram associated with E_8 (see Fig. 8.1). By definition, this embedding is primitive, so it gives rise to case No. 14 with $n = 1$. On the other hand, note that D_4 also admits an obvious primitive embedding into E_8 (and also into E_6 and E_7), as depicted in Fig. 8.1.

In comparison, A_1^4 can be viewed as index 2 sublattice of D_4, or in fact, any A_1 embedded into D_4 has A_1^3 as orthogonal complement. Thus we obtain an imprimitive embedding of A_1^4 into E_8. This corresponds to the $n = 2$ case of No. 13.

Along the same lines, one can work out primitive embeddings $T \hookrightarrow E_8$ for many root lattices T, such as for $T = A_k, D_k (k < 8)$ or E_6, E_7. These primitive embeddings will be used without further specification in the sequel.

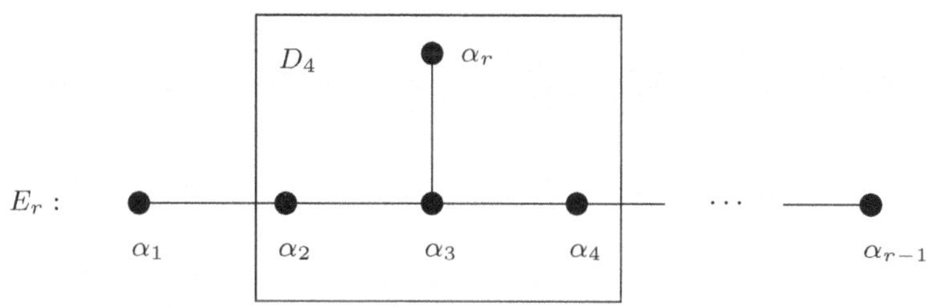

Fig. 8.1 Dynkin diagram of type E_r ($r = 6, 7, 8$) (with D_4 embedding)

The following concept of a companion root lattice will simplify our analysis significantly (and can be extended directly to all other lattices).

Definition 8.16 Given a root sublattice T of E_8, we say that a root lattice M is a *companion* of T if $T \oplus M$ is a root lattice of rank 8 appearing in the case $s = 8$ of Theorem 8.1.

We will now discuss a direct application of this concept (where we will take $V = T$):

Lemma 8.17 *Suppose that a lattice V has a companion M such that*

$$\det V / \det M = \nu^2 \text{ for some } \nu = 1, 2, \ldots \tag{8.5}$$

Assume that M (or V) is primitively embedded in E_8. Then we have

$$V^\perp = M \quad \text{and} \quad [V' : V] = \nu.$$

(The orthogonal complement $V^\perp$ and the primitivity of V are to be taken with respect to the embedding $V = V \oplus \{0\} \hookrightarrow V \oplus M \hookrightarrow E_8$ here.)

Proof Since $V \oplus M$ is a direct sum contained in E_8, it is clear that $M \subset V^\perp$. If M is primitive, then $M = V^\perp = V'^\perp$. By (2.3) and Lemma 2.12,

$$\det V/\nu^2 \overset{8.5}{=} \det M = \det V' = \det V/[V' : V]^2,$$

which implies $[V' : V] = \nu$. On the other hand, if V is primitive, then we have $V' = V$ and

$$\det V = \det V' = \det M'$$

for M' the primitive closure of M inside E_8. Letting $m = [M' : M]$, we have $\det M' = \det M/m^2$. By (8.5), $m = 1/\nu \in \mathbb{N}$. Hence $m = \nu = 1$, which implies that $M = M' = V^\perp$. □

Remark 8.18 As already foreshadowed before Definition 8.16, the validity of Lemma 8.17 does not at all depend on the lattices involved being root lattices (in contrast to the definition of the companion).

In what follows, we shall apply Lemma 8.17 to the lattice T corresponding to the reducible fibres of a rational elliptic surface (with sign reversed), with primitive closure T' inside E_8 and orthogonal complement $T^\perp = L = E(K)^0$ the narrow Mordell–Weil lattice. Throughout we set

$$\nu = [T' : T]$$

as in Lemma 8.17.

With the given concepts, we are now ready to prove the cases No. 1 through No. 14 of Theorem 8.8. In the first 11 cases, we have $\nu = 1$, $T' = T$ by Lemma 8.15. In order to apply Lemma 8.17, observe that each T has a companion M with $\det T = \det M$ ($\nu = 1$). The companion is listed in the column for $L = E(K)^0$ in Table 8.2. Hence $T^\perp = M$ by Lemma 8.17.

For the remaining root lattices T, there is either no companion available or the companion depends on the embedding $T \hookrightarrow E_8$. Instead, we start by embedding a sublattice of T (a root lattice which is an orthogonal summand of T) with companion into E_8; here we can use the results from the previous cases. Then we continue with the missing orthogonal summand from T.

In detail, consider No. 12. Here $T = A_2 \oplus A_1^2$ has no companion. Thus the narrow Mordell–Weil lattice is not a root lattice. Instead of the full lattice T, we choose to work with the sublattice $A_2 \oplus A_1$. For No. 6, we have used its companion A_5. It follows that $T^\perp$ is isomorphic to the orthogonal complement of A_1 in A_5. Since that embedding is unique up to isometries, this settles the case No. 12 as claimed.

Finally, we come to Nos. 13, 14 where $T = A_1^4$. As above we find by No. 7 that $T^\perp$ is isomorphic to $A_1^\perp$ in $L = D_4 \oplus A_1$. There are two cases to consider (in agreement with our calculations before Definition 8.16). If $A_1 \hookrightarrow L$ is embedded as the second factor, it is obvious that $A_1^\perp = D_4$; then we have $\nu = 2$ by Lemma 8.17, hence No. 13. If $A_1 \hookrightarrow D_4 \subset L$, then $A_1^\perp$ in L is equal to A_1^4, since $(D_4 : A_1) \cong A_1^3$ as discussed above. Hence $T^\perp = A_1^4$ and $\nu = 1$, proving No. 14. This completes the proof of Classification Theorem 8.8 for Mordell–Weil rank $r \geq 4$. □

8.5.2 $r = 3$ (Nos. 15–24)

We turn to the case of Mordell–Weil rank $r = 3$ in Classification Theorem 8.8. In the same way as in Lemma 8.15 for rank $r = 4$, one finds that ν is either 1 or 2 and that $\nu = 2$ is possible only for Nos. 21, 22 or 24.

We first consider the cases where T has a companion and thus $\nu = 1$. Presently, this applies to Nos. 15, 16 or 18 where $T = A_5$, D_5 or $D_4 \oplus A_1$ have companions $M = A_2 \oplus A_1$, A_3 or A_1^3. The primitivity of the (unique) embeddings is easily verified, as neither T nor M has an even integral overlattice of finite index, thus Lemma 8.17 proves the claim.

For all other cases, $T^\perp$ can be computed by making use of the proven cases as follows:

No. 17: $T^\perp = (A_4 \oplus A_1)^\perp = (A_1 \hookrightarrow A_4)^\perp$ (by No. 8).
No. 19: $T^\perp = (A_3 \oplus A_2)^\perp = (A_2 \hookrightarrow D_5)^\perp$ (by No. 5).
No. 20: $T^\perp = (A_2^2 \oplus A_1)^\perp = (A_1 \hookrightarrow A_2^2)^\perp = A_2 \oplus \langle 6 \rangle$ (by No. 11).
No. 23: $T^\perp = (A_2 \oplus A_1^3)^\perp = (A_2 \hookrightarrow D_4 \oplus A_1)^\perp = (A_2 \hookrightarrow D_4)^\perp \oplus A_1$ (by No. 7).
No. 24: $T^\perp = (A_1^5)^\perp = (A_1 \hookrightarrow A_1^4)^\perp = A_1^3$ (by No. 14).

Remark 8.19 By Theorem 8.2, the embedding $T \hookrightarrow E_8$ is unique in each of the above cases. Hence the above argument is independent of the choice of auxiliary sublattice. For instance, consider case No. 24 with $T = A_1^5$. If we use No. 13 instead of No. 14 above, then we obtain $T^\perp = (A_1 \hookrightarrow D_4)^\perp$. However, in the end, this results again in $T^\perp = A_1^3$.

In retrospect, the five copies of A_1 involve exactly one non-trivial 2-divisible class, given by a chosen set of four A_1's. Whether we pick exactly these four A_1's or a different configuration for the auxiliary embedding $A_1^4 \hookrightarrow E_8$ determines precisely whether the orthogonal complement inside E_8 is A_1^4 (as in No. 14) or D_4 (as in No. 13).

It remains to consider the case $T = A_3 \oplus A_1^2$ (No. 21, 22). By No. 10, we have $T^\perp = (A_1 \hookrightarrow A_3 \oplus A_1)^\perp$. According to whether A_1 is embedded into the first or second factor of the auxiliary root lattice, we have

$$T^\perp = \begin{cases} (A_3 : A_1) \oplus A_1 \cong A_1^2 \oplus \langle 4 \rangle \\ A_3, \end{cases}$$

which corresponds to No. 21 or 22. This concludes the proof of Classification Theorem 8.8 for rank $r = 3$. □

8.5.3 $r = 2$ (Nos. 25–42)

In the case of Mordell–Weil rank $r = 2$, most cases can be treated by exactly the same arguments as above. We will go into details only for two instructive cases: Nos. 25 and 42.

One useful remark concerns the narrow Mordell–Weil lattice L. As it has rank two in the present situation, we can draw on the classification of positive definite integral even binary quadratic forms. Through group theory, there is a close connection to imaginary quadratic fields (see, for instance, [Cox13]) that will turn up incidentally below.

For No. 25, $T = A_6$ is primitive in E_8 since $\det T = 7$ is square-free by Proposition 6.31(i), so $\nu = 1$. Hence $T^\perp$ is of rank 2 and $\det T^\perp = 7$. Since the field $\mathbb{Q}(\sqrt{-7})$ has class number one, it follows that the quadratic form $\begin{pmatrix} 2 & 1 \\ 1 & 4 \end{pmatrix}$ gives the unique such lattice up to isomorphism.

For No. 42, we have $T^\perp = A_1^2$ and $\nu = 4$ by Lemma 8.17. It remains to determine whether $E(K)_{\text{tors}}$ is cyclic or not. For this, we can employ (6.21) in Proposition 6.31: since

$$T^\vee/T \cong (A_1^\vee/A_1)^6 \cong (\mathbb{Z}/2\mathbb{Z})^6,$$

we can only have $E(K)_{\text{tors}} \simeq (\mathbb{Z}/2\mathbb{Z})^2$ as claimed. □

8.5.4 $r = 1$ (Nos. 43–61)

Coming to Mordell–Weil rank $r = 1$, we only touch upon a few relevant cases and leave the rest to the interested reader.

Nos. 44 and 45 with $T = A_7$: Note that T does not admit a unique embedding into E_8 by Theorem 8.2. On the one hand, we view T with companion A_1. Then Lemma 8.17 directly leads to No. 44 with 2-torsion ($\nu = 2$). On the other hand, we consider T primitively embedded into E_8. By [Nis96], the orthogonal complement is $\langle 8 \rangle$ as stated in No. 45.

Note that all other cases for $r = 1$ come with a unique embedding of the rank 7 root lattice T into E_8 by Theorem 8.2.

Nos. 55 and 56: In view of No. 8, we only have to compute the orthogonal complement of A_3 and $A_2 \oplus A_1$ in A_4. This can easily be obtained from [Nis96]. In detail, we fix the standard basis $\{\alpha_i | i = 1, \cdots, 4\}$ of A_4. Then A_3 can be spanned by α_1, α_2 and α_3, which has $A_3^\perp = \mathbb{Z}x$ with $x = \sum_{i=1}^4 i\alpha_i$. Here $x^2 = 20$. Similarly, $A_2 \oplus A_1$ can be spanned by α_1, α_2 and α_4. The orthogonal complement $\mathbb{Z}y$ is generated by $y = 2\alpha_1 + 4\alpha_4 + 6\alpha_3 + 3\alpha_4$ with norm 30.

Nos. 57 and 58: In these cases, T has companion A_1. Lemma 8.17 gives $\nu = 4$. We have to show that

$$E(K)_{\text{tors}} \simeq \begin{cases} (\mathbb{Z}/2\mathbb{Z})^2 & \text{(No. 57)}, \\ \mathbb{Z}/4\mathbb{Z} & \text{(No. 58)}. \end{cases}$$

For No. 57, this follows from (6.21) as $T^\vee/T \cong (\mathbb{Z}/2\mathbb{Z})^5$. For No. 58, we examine the condition (6.19):

$$\sum_{v \in R} \text{contr}_v(P) = 2 + 2(P.O). \tag{8.6}$$

For a reducible fibre with $T_v \simeq A_3$ (resp. A_1), we have by Table 6.1

$$\text{contr}_v(P) = \begin{cases} 0, \ \ 3/4 \ \text{ or } \ 1 & (A_3), \\ (0 \text{ or } 1/2) & (A_1). \end{cases}$$

Thus the solution of the relation (8.6) can be either

$$1 + 1 + 0 = 2 \quad \text{or} \quad 3/4 + 3/4 + 1/2 = 2,$$

with $(P.O) = 0$ in both cases. Comparing the fibre components met by P to (6.21), we find that the cases correspond to 2-torsion and 4-torsion. But the former occurs at most for a single 2-torsion point, because again by (6.21) the sum of two such sections will be in $E(K)^0$ which is torsion-free. As we need $\nu = 4$ in the end, the second alternative must also hold for some section P. Thus P is 4-torsion and $2P$ has order 2, covering the first alternative.

Finally, we come to *No. 60*. The claimed shape of L follows from No. 38. Thus we again have $\nu = 4$. We claim that

$$E(K)_{\text{tors}} \simeq (\mathbb{Z}/2\mathbb{Z})^2.$$

Assume to the contrary that there is a section P of order 4. By (6.21), the section P has to meet the component Θ_1 or Θ_3 of the A_3 fibre. Equivalently, the local correction term is $3/4$. But then, since all other reducible fibres have type A_1, all other correction terms are in $\frac{1}{2}\mathbb{Z}$. In particular, this implies that $\sum_{\upsilon \in R} \text{contr}_\upsilon(P) \notin \mathbb{Z}$. Therefore P cannot have height zero; in particular, it cannot be of order 4. The claim follows. □

8.5.5 $r = 0$ *(Nos. 62–74)*

In this case, T is of finite index in $T' = E_8$ and $\det T = \nu^2$ with $\nu = |E(K)_{\text{tors}}|$. Moreover, the exponent of the torsion group is bounded by $T^\vee/T$ in view of (6.21). This settles all the cases other than Nos. 70, 72 and 74. In the former two cases, we can again use (6.19) to figure out that $E(K)$ is cyclic. In the latter case, we only need that the torsion of an elliptic curve has p-length at most two. □

Conclusion 8.20 *This completes the proof of Classification Theorem 8.8.* □

8.6 Corollaries

Let us note two corollaries concerning the structure of the Mordell–Weil group of a rational elliptic surface. They follow immediately from Classification Theorem 8.8.

Corollary 8.21 *As an abstract group, the Mordell–Weil group of a rational elliptic surface is one of the following 26 groups:*

$$\begin{array}{llll}
\mathbb{Z}^r \ (1 \le r \le 8), & \mathbb{Z}^r \oplus \mathbb{Z}/2\mathbb{Z} \ (1 \le r \le 4), & \mathbb{Z}^r \oplus \mathbb{Z}/3\mathbb{Z} \ (1 \le r \le 2), & \\
\mathbb{Z}^r \oplus (\mathbb{Z}/2\mathbb{Z})^2 \ (1 \le r \le 2), & \mathbb{Z} \oplus \mathbb{Z}/4\mathbb{Z}, & & \\
\mathbb{Z}/4\mathbb{Z} \oplus \mathbb{Z}/2\mathbb{Z}, & (\mathbb{Z}/3\mathbb{Z})^2, & (\mathbb{Z}/2\mathbb{Z})^2, & \\
\mathbb{Z}/6\mathbb{Z}, \quad \mathbb{Z}/5\mathbb{Z}, & \mathbb{Z}/4\mathbb{Z}, \quad \mathbb{Z}/3\mathbb{Z}, & \mathbb{Z}/2\mathbb{Z}, & \{0\}.
\end{array}$$

In particular, the Mordell–Weil group is torsion-free if the rank r is greater than 4.

Remark 8.22 By Theorem 8.9, all cases indeed occur unless the ground field has characteristic $p > 0$ and the Mordell–Weil group has full p-torsion. The result of Corollary 8.21 has been proven in [Cox82] in the case $k = \mathbb{C}$ under some additional assumption.

Remark 8.23 In the realm of complex K3 surfaces, a classification corresponding to Corollary 8.21 has been given by Shimada [Shima00]. We will discuss this result among other things in Chap. 11, see in particular Theorem 11.36 in Sect. 11.5.

Corollary 8.24 *The narrow Mordell–Weil lattice $E(K)^0$ of a rational elliptic surface is a direct sum of root lattices if the rank r is greater than 4, but not necessarily so if $r \leq 4$.*

Later we will call the types $\{T, L, M\}$ where the narrow Mordell–Weil lattice L is a root lattice, *admissible* (Definition 10.5).

8.7 Integral Sections

A famous result of Siegel [Sie29] affirms that an elliptic curve (or in fact any plane curve of positive genus, and even genus zero curves with a certain behaviour at ∞) with a fixed affine model in $\mathbb{A}^2_K$ for some number field K, admits only finitely many integral points, i.e. points with coordinates in the ring of integers of K. In this section, we discuss the natural analogue for elliptic curves over one-dimensional function fields, or elliptic surfaces.

Definition 8.25 Let S be an elliptic surface. A section $P \in \mathrm{MW}(S)$ is called *integral* if it is disjoint from the zero section, i.e. $(P.O) = 0$.

We now state the easy analogue of Siegel's theorem:

Proposition 8.26 *Any elliptic surface admits only finitely many integral sections.*

Proof Let S be an elliptic surface and $P \in \mathrm{MW}(S)$ an integral section. Then

$$h(P) = \langle P, P \rangle \leq 2\chi(S)$$

by (6.15). In particular, the class of P in the Mordell–Weil lattice $MWL(S)$ lies in the bounded set

$$\{v \in MWL(S);\ v^2 \leq 2\chi\}.$$

Since $MWL(S)$ is positive-definite, this set is finite for any S. In $\mathrm{MW}(S)$, each vector from $MWL(S)$ has exactly $|\,\mathrm{MW}(S)_{\mathrm{tors}}|$ representatives, so the claim follows. □

Remark 8.27 Of course, the same argument applies to any other bounded set in $\mathrm{MW}(S)$, for instance to sections with "bounded poles", i.e. with $(P.O) \leq m$ for any given m (so that $h(P) \leq 2\chi + 2m$).

For a rational elliptic surface, we can be much more precise about the number of integral sections:

Lemma 8.28 *For any rational elliptic surface, there are at most 240 integral sections. Every such section P has norm $\langle P, P\rangle \leq 2$.*

Proof Since a rational elliptic surface S has $\chi = 1$, we have already seen the second claim above.

Given an integral section $P \in \mathrm{MW}(S)$, consider the divisor

$$D'_P := (P) - (O) - F \tag{8.7}$$

on the elliptic surface. Here we have applied an orthogonal projection with respect to the unimodular sublattice $U = \langle (O), F\rangle$ in $\mathrm{NS}(S)$; note that this works completely on the level of lattices, mapping (P) into the frame W of S, so there is no need to tensor with $\mathbb{Q}$. Indeed, we can easily compute the intersection numbers as follows:

$$(D'_P.F) = 0, \quad (D'_P.O) = 0, \quad (D'_P)^2 = -2.$$

Hence if we denote by $\xi_P \in \mathrm{NS}(S)$ the divisor class of D'_P, then ξ_P defines a "root vector" of the frame $W^- \simeq E_8^-$ (with sign reversed). Moreover, the correspondence

$$P \longleftrightarrow \xi_P$$

is one-to-one by Theorem 6.5 and Lemma 8.17. Hence the number of integral sections P is bounded by the number of root vectors of E_8, $\tau(E_8) = 240$. □

Theorem 8.29 *The Mordell–Weil group of a rational elliptic surface is generated by the integral sections (which are at most 240 in number), hence by those sections P with norm $\langle P, P\rangle \leq 2$.*

Proof As above, let $U = \langle (O), F\rangle_{\mathbb{Z}}$ be the unimodular rank two sublattice of $\mathrm{NS}(S)$ generated by (the classes of) the zero section and the fibre, and let $W^- = U^\perp$ denote the orthogonal complement of U in N, i.e. W is the frame. By Lemma 7.2, W is isometric to the negative root lattice E_8; of course, the latter is spanned by the set of roots

$$\mathcal{D} = \mathcal{R}(W) = \{D \in W;\ D^2 = 2\}.$$

We will use that the canonical surjection

$$\pi : \mathrm{NS}(S) \to \mathrm{NS}(S)/\,\mathrm{Triv}(S) \simeq \mathrm{MW}(S)$$

from Theorem 6.5 factors through W (since U is a sublattice of the trivial lattice $\mathrm{Triv}(S)$; explicitly, one can directly extend the $\mathbb{Z}$-divisor D'_P perpendicular to U from (8.7) to the $\mathbb{Q}$-divisor D_P perpendicular to $\mathrm{Triv}(S)$ from (6.7)). This induces the map

$$\pi' : \quad W \to W/(\oplus_{v\in C} T_v) \cong \mathrm{MW}(S),$$

which would be an orthogonal projection once tensored with $\mathbb{Q}$. It maps the subset $\mathcal{D}$ of W into $\mathcal{P} \cup \{O\}$, where $\mathcal{P}$ denotes the set of integral sections (cf. [Shio10]).

Since $W \simeq E_8$ is generated by $\mathcal{D}$, the image $\mathrm{MW}(S)$ of W is generated by the image of $\mathcal{D}$ under π', thus by $\mathcal{P}$. □

We note the following immediate corollary whose prime-to-characteristic part could have been deduced from Proposition 6.33(v).

Corollary 8.30 *On a rational elliptic surface S, torsion sections are always disjoint, i.e. $(P.Q) = 0$ for any $P \neq Q \in \mathrm{MW}(S)_{\mathrm{tors}}$.*

Remark 8.31 Of course, for non-rational elliptic surfaces, one can neither expect that the Mordell–Weil group is generated by integral sections, nor that all sections with height bounded by 2χ will be integral, nor that torsion sections are disjoint. The easiest counterexample to all three claims (though very special) is provided by peculiar torsion sections, compare Sect. 12.4.4.2.

For later use, we also note a consequence for the precise fibre types of rational elliptic surfaces with torsion sections. Namely, Theorem 8.8 and Corollary 8.30 allow us sometimes to distinguish between the multiplicative and additive fibres with the same associated Dynkin diagram (including irreducible singular fibres).

Corollary 8.32 *Let S be a rational elliptic surface over an algebraically closed field of characteristic p, equipped with a torsion section of order exactly p. Then S admits an additive fibre.*

Proof By Corollary 8.30, the p-torsion section P is disjoint from O. It thus follows from Lemma 5.23 that P meets every multiplicative fibre in a non-identity component. In particular, there are no I_1 fibres. If S were *semi-stable*, i.e. if S had only multiplicative fibres $\mathrm{I}_{n_1}, \dots, \mathrm{I}_{n_l}$ with $n_i > 1$ for each i, then we would have

$$T = \bigoplus_{i=1}^{l} A_{n_i-1}. \tag{8.8}$$

By Theorem 5.4.7, we would have

$$n_1 + \ldots + n_l = 12,$$

so $l \leq 6$. Since $n_1 + \ldots + n_l - l = \mathrm{rk}(T) = 8 - r$, with $r = \mathrm{rk}(\mathrm{MW}(S)$, we read off $r = l - 4 \leq 2$, and we easily find the following types in Theorem 8.8 which potentially meet all the given requirements:

$$\text{Nos.} \quad 42, 60, 68, 74.$$

In each case, however, there is full 2- or 3-torsion, so existence in the respective characteristic is ruled out by Proposition 6.33(vi). □

By Lemma 8.28 and Theorem 8.29, we can restate the above in terms of minimal Weierstrass equations as follows:

Theorem 8.33 (Generator theorem) *For any elliptic curve E over $K = k(t)$ defined by a minimal Weierstrass equation of the form (7.4) satisfying the condition (7.5), i.e. associated with a rational elliptic surface, there are at most 240 K-rational points $P = (\mathrm{x}, \mathrm{y})$ of the form*

$$\mathrm{x} = gt^2 + at + b, \quad \mathrm{y} = ht^3 + ct^2 + dt + e, \qquad a, \dots, h \in k$$

and they generate the Mordell–Weil group $E(K)$.

Remark 8.34 The bound 240 in the above Lemma 8.28 or Theorem 8.29 is sharp. For example, consider the elliptic curve

$$E: \quad y^2 = x^3 + t^5 + 1$$

over $k(t)$ as in Example 3.51, where we let $k = \bar{\mathbb{Q}}, \mathbb{C}$ or $\bar{\mathbb{F}}_p$ for $p > 5$. Then there are exactly 240 integral sections $P = (\mathrm{x}, \mathrm{y})$ with $\deg_t \mathrm{x} \leq 2$ and $\deg_t \mathrm{y} \leq 3$ (which can be computed explicitly, for instance over $\mathbb{F}_{31}$).

Just like with Siegel's theorem which immediately extends to rational numbers with bounded denominators, we can also weaken the integrality notion from Definition 8.25 by allowing for sections meeting O at a given finite set of places of the base curve C (cf. Remark 8.27). Most instructively, consider the rational case $C \cong \mathbb{P}^1$ and allow for ∞-*integral sections*, that is sections disjoint from O outside ∞. Starting from a minimal Weierstrass form over $k[t]$, this translates as the condition that the section $P = (\mathrm{x}, \mathrm{y})$ must have polynomial coordinates $\mathrm{x}, \mathrm{y} \in k[t]$. Surprisingly, one can sometimes still make very precise statements about such *polynomial sections*, including finiteness, but this tends to be much more subtle than what we have seen above. We will come back to this in the context of excellent families for the elliptic curve from Remark 8.34 in Example 9.37.

Section 8.10, though seemingly not connected to this one, will provide yet another view on integral and polynomial sections (see Sect. 8.10.5). Before getting there, we turn to the existence problem, i.e. Theorem 8.9.

8.8 Extremal Rational Elliptic Surfaces

For Classification Theorem 8.8, we first discussed the high rank cases in detail in Chap. 7. For the corresponding existence problem (Theorem 8.9), it is the opposite end of the scale which is most critical, namely the rational elliptic surfaces with Mordell–Weil rank zero. Following standard terminology, they are often called *extremal*. Here existence (and uniqueness/moduli) is covered by work of Miranda–Persson over $\mathbb{C}$ ([MP86]) and Lang in positive characteristic ([Lan91, Lan94]):

Theorem 8.35 *Let k be an algebraically closed field of characteristic $p \geq 0$. For any rank zero type from Theorem 8.8, there exists an extremal rational elliptic surface over k except for Nos. 68 if $p = 3$, and Nos. 71, 73, 74 if $p = 2$.*

Moreover, each type is realized by one or two unique rational elliptic surfaces except for No. 73 if $p \neq 2$, and No. 64 if $p = 2$, each of which has one-dimensional moduli.

Remark 8.36 As in the proof of Corollary 8.32, it follows from Proposition 6.33(vi) that the four excluded types cannot exist in the respective characteristic.

For later reference, Table 8.3 records generalized Weierstrass forms for the extremal rational elliptic surfaces which work in most characteristics not excluded a priori (these have been used extensively in [Schü17, Schü19]). Whenever MW $\neq 0$, we choose a section to be $(0, 0)$. A few of the entries will be discussed in more detail in Sect. 8.10.

Remark 8.37 Extremal rational elliptic surfaces have interesting applications, for instance for the classification of Enriques surfaces with finite automorphism group ([Kon86, Mar19]).

Remark 8.38 Over $\mathbb{C}$, one can extend the notion of being extremal to all elliptic surfaces by requiring maximal Picard number, $\rho = h^{1,1}$, and finite MW, see e.g. [SZ01]. In positive characteristic, however, the analogue with $\rho = b_2$ and finite MW is quite restrictive, as shown in [Ito02].

8.9 Existence of Types of Rational Elliptic Surfaces

After the preparations for extremal rational elliptic surfaces, we now turn to the proof of Theorem 8.9. In characteristic zero, there have been several approaches worked out to give solutions.

Over $\mathbb{C}$, Persson gave a complete classification of all fibre configurations of rational elliptic surfaces in [Per90] by geometric arguments, using plane curves with prescribed singularities. Soon after, Miranda gave an alternative proof using the combinatorial data of the j-invariant in [Mir90]; considered as a self-map on $\mathbb{P}^1$ of degree at most 12, this has prescribed ramification structures which can be used for existence and non-existence results.

The notion of *excellent families*, which we will discuss in detail in the next two chapters, gives rise to examples over $\mathbb{Q}$ in a very natural way, together with information about the Mordell–Weil groups (which is a priori not provided in the above two approaches, but only a posteriori by way of Theorem 8.8). In fact, excellent families even give rise to $\mathbb{Q}$-split examples, i.e. with a basis of NS defined by sections and fibre components over $\mathbb{Q}$. Combined with the concept of vanishing roots (see Sect. 10.3.2), this allows one to find explicit families of rational elliptic surfaces over $\mathbb{Q}$ for each type in Theorem 8.8. We will explain an exemplary case in Sect. 10.5.6, see especially Remark 10.62.

Here we will pursue an intermediate approach combining algebraic and geometric aspects, which lends itself directly to cover positive characteristic as well. It also gives control over the moduli dimensions. All three methods are essentially independent from each other; hence the readers may follow their own preferences and go, for instance, directly to the approach through excellent families and vanishing roots explained in Sect. 10.5.6.

To ease the arguments and their presentation, we concentrate on the *semi-stable* case, i.e. on rational elliptic surfaces without additive fibres. This has two big advantages in determining the types of singular fibres I_{n_i} uniquely by

- the vanishing orders n_i of the discriminant Δ, or equivalently, by
- the root lattice T from (8.1), the direct sum of the Dynkin diagrams associated with the reducible fibres, taking the shape (8.8).

A semi-stable rational elliptic surface S with root lattice T as in (8.8) has reducible fibres of Kodaira types $\mathrm{I}_{n_1}, \dots, \mathrm{I}_{n_l}$, and by Theorem 5.4.7, there are $l' = 12 - n_1 - \dots - n_l$ fibres of type I_1. The configuration of singular fibres is thus encoded in the following partition of 12:

$$\mathcal{P} = [n_1, \dots, n_l, 1^{l'}].$$

With this notation, we can introduce the strata $\mathcal{M}_{\mathcal{P}}$ in the moduli space $\mathcal{M}$ of rational elliptic surfaces consisting of all those semi-stable rational elliptic surfaces with fibre configuration given by $\mathcal{P}$.

Theorem 8.39 *Let k be an algebraically closed field of characteristic p. There exists a semi-stable rational elliptic surface of given type $\{T, L, M\}$ from Theorem 8.8 if and only if T contains only A_m summands as in* (8.8) *and M contains no p-torsion.*

Moreover, if $\mathcal{P}$ denotes the corresponding partition, then $\mathcal{M}_{\mathcal{P}}$ has a component of dimension $r = \mathrm{rk}(L) = \mathrm{rk}(M)$.

Proof The "only if" part follows directly from Corollary 8.32, so we only have to prove existence and dimension.

For $r = 8$, we have $\mathcal{P} = [1^{12}]$ and $\overline{\mathcal{M}_{\mathcal{P}}} = \mathcal{M}$, so the claims follow from Theorem 5.48. Similarly, the case $r = 0$ is covered by Theorem 8.35 (but the semi-stable case considered here goes back to Beauville originally [Bea82]).

Over $\mathbb{C}$, one can proceed as Miranda in [Mir89] by deforming the j-map. This amounts to proving combinatorially that, if a semi-stable rational elliptic surface with partition $\mathcal{P}$ exists, and if the partition $\mathcal{P}'$ is finer, i.e. $\mathcal{P}' < \mathcal{P}$, then there is a semi-stable rational elliptic surface with partition $\mathcal{P}'$. This way, one can derive the existence of all semi-stable fibre configurations from Theorem 8.8 over $\mathbb{C}$ (without regarding torsion and moduli dimensions) as deformations of the extremal ones.

Here we will pursue an algebraic analogue of the above approach based on the discriminant and specialization, i.e. working in the opposite direction. Assume that $\mathcal{M}_{\mathcal{P}'} \neq \emptyset$ and that $\mathcal{P}' < \mathcal{P}$. Then $\mathcal{M}_{\mathcal{P}} \subset \overline{\mathcal{M}_{\mathcal{P}'}} \subset \mathcal{M}$ can be obtained by merging certain singular fibres such that the corresponding vanishing orders of Δ add up correctly.

We explain this for the case where $\#\mathcal{P}' = \#\mathcal{P} + 1$ and $\mathcal{M}_{\mathcal{P}'} \neq \emptyset$. Here two fibres have to be merged to obtain the partition $\mathcal{P}$; for any component Z of $\mathcal{M}_{\mathcal{P}'}$ (which we mostly suppress to ease notation), this merging is encoded in a single equation on $\bar{Z}$. Either this defines the empty set or a hypersurface with components the strata $V_1, \ldots, V_m \subset \bar{Z}$. To find a component of $\mathcal{M}_{\mathcal{P}}$, it remains to check, for each Z:

(1) whether the equation has a solution on $\bar{Z}$, i.e. $V_1 \neq \emptyset$, and if so,
(2) whether the strata $V_1, \ldots, V_m$ qualify to parametrize semi-stable elliptic fibrations (as opposed to acquiring additive fibres or degenerating), and if so,
(3) whether the resulting partition on some qualifying stratum V_i is $\mathcal{P}$ and not coarser (so that $V_i \subset \mathcal{M}_{\mathcal{P}} \neq \emptyset$), and
(4) what the torsion in MW is on V_i.

The first two questions can be approached through suitable terminal objects in the spirit of [Mir89]. Namely, if $\mathcal{P} < \mathcal{P}_0$ and $\mathcal{M}_{\mathcal{P}_0} \neq \emptyset$, then also the equation arising from the specialization $\mathcal{P} > \mathcal{P}'$ has some semi-stable elliptic solution, for one can successively specialize further to $\mathcal{P}_0$ from some qualifying stratum V_i, merging more and more fibres. Presently, it thus suffices to verify that any partition $\mathcal{P}$ satisfies $\mathcal{P} < \mathcal{P}_0$ for the partition $\mathcal{P}_0$ of some extremal semi-stable rational elliptic surface, and this is easily achieved.

The same approach can also be used to answer (3). If the resulting partition $\mathcal{P}_i$ were coarser than $\mathcal{P}$ for every qualifying stratum V_i, i.e. $\mathcal{P}_i > \mathcal{P}$, then some component of $\mathcal{M}_{\mathcal{P}_i}$ would have codimension one in $\bar{Z}$ (inside $\overline{\mathcal{M}_{\mathcal{P}'}}$) by construction, i.e. the dimension of $\mathcal{M}_{\mathcal{P}_i}$ would be greater than expected. But then the same would carry over to any terminal partition

$$\mathcal{P}_0 > \mathcal{P},$$

since the specialization process from $\mathcal{M}_{\mathcal{P}'}$ would necessarily factor through one of the qualifying strata V_i. In consequence, $\mathcal{M}_{\mathcal{P}_0}$ would have dimension greater than zero, contradicting Theorem 8.35. At the same time, this proves that $\mathcal{M}_{\mathcal{P}}$ is non-empty and has a component of the expected dimension (which is easily seen to be r).

To complete the proof of Theorem 8.39, it remains to cover the five partitions with two different types. By inspection, one of the types is always distinguished by the property that M contains 2-torsion, so different types for the same partition can only occur in characteristic $p \neq 2$ by Corollary 8.32. We will now use the fact that torsion sections are preserved upon specialization. Presently, this is obvious from the extended Weierstrass form (5.10) which we can work with (since $p \neq 2$): 2-torsion sections are in correspondence with linear factors of the RHS of (5.10). In particular, there is a 2-torsion section if and only if the RHS is reducible, and of course, this property is preserved by specialization as stated.

To prove the claims of Theorem 8.39 for the five types without torsion (Nos. 14, 22, 29, 36, 45), it thus suffices to specialize to a (terminal) partition without torsion, and again, this is easily achieved. For the five types with 2-torsion (Nos. 13, 21, 28, 35, 44), we conclude by working with an extended minimal Weierstrass form as above. Moreover, by Corollary 8.30, we can apply an admissible transformation so that the 2-torsion section is $(0, 0)$. But then one directly finds the following Weierstrass forms, with polynomial coefficients of degree given by the index, and moduli dimensions calculated as in the proof of Theorem 5.48:

No.	Weierstrass eqn.	Moduli
13	$y^2 = x(x^2 + a_2x + a_4)$	4
21	$y^2 = x(x^2 + a_2x + b_2)$	3
28	$y^2 = x(x^2 + a_2x + a_1)$	2
35	$y^2 = x(x^2 + a_2x + a_0t^2)$	2
44	$y^2 = x(x^2 + a_2x + a_0)$	1

This completes the proof of Theorem 8.39. □

Remark 8.40 The strata V_i in proof of Theorem 8.39 show that all components of $\mathcal{M}_\mathcal{P}$ have dimension at least r.

Remark 8.41 The moduli part of Theorem 8.39 resembles results on Noether–Lefschetz loci of complex elliptic surfaces over $\mathbb{P}^1$ with $p_g > 0$ in [Klo07a], or Artin's stratification of the moduli space of K3 surfaces in positive characteristic by height and Artin invariant in [Art74a], see Sect. 12.4.2. In fact, our strategy of proof turns out quite close to Artin's line of argument.

To complete the proof of Theorem 8.9, it remains to study those types from Classification Theorem 8.8 which are supported on rational elliptic surfaces with additive singular fibres—either by inspection of the root lattice T (containing summands D_k or E_l) or by virtue of Corollary 8.32.

In the former case, there is a non-reduced fibre, so one can pursue a similar approach to the one above, starting from the generic Weierstrass forms for a rational elliptic surface with a fibre of type I_0^* (moving in a four-dimensional family) resp. of type IV* (a two-dimensional family). The Weierstrass forms can be read off directly from Tate's algorithm 5.8 which is of great help computing the strata (so one can avoid using the discriminant which, in characteristics 2 and 3, would not be sufficient anyway due to wild ramification).

In the latter case, but only with reduced fibres, there is nothing left to do in the extremal case (by Theorem 8.35). For the remaining types, one can proceed as in the proof of Theorem 8.39 by using the torsion section to write out the respective strata. For instance, for a non-isotrivial rational elliptic surface with a 3-torsion section (disjoint from O by Corollary 8.30), one may follow Tate to use the normal form

$$y^2 + a_1xy + a_3y = x^3$$

(see [Kub76, Table 8.3]). Readily, this gives types Nos. 39, 51, 61 in any characteristic (including $p = 3$). A similar approach works for the exceptional cases in characteristic 2. We omit the details for the sake of brevity.

8.10 Maximal Singular Fibres

The maximal irreducible root lattices embedding into E_8 are A_8, D_8 and obviously E_8 itself (see Sect. 8.1). In this section, we discuss existence and uniqueness of rational elliptic surfaces with such a trivial lattice. This should be seen as an illustration of how one can work with (extremal) rational elliptic surfaces explicitly. At the same time, these cases indicate some of the peculiarities in characteristics 2 and 3.

8.10.1 $T = A_8$

The embedding $A_8 \hookrightarrow E_8$ has index 3. In consequence, any rational elliptic surface with a singular fibre of the corresponding type I_9 will have a 3-torsion section, and in fact essentially no other section: $MW = \mathbb{Z}/3\mathbb{Z}$.

We have seen the corresponding rational elliptic surface as a cubic pencil related to the Hesse pencil in Example 7.15. Brought to normal form by standard formulas (e.g. [Cas91]), we derive the equation

$$y^2 + txy + y = x^3 \tag{8.9}$$

with point $(0, 0)$ of order three, which is unique up to admissible transformations in any characteristic (cf. Table 8.3).

8.10.2 $T = D_8$

Similarly, the embedding $D_8 \hookrightarrow E_8$ has index 2. Hence a rational elliptic surface with a singular fibre of type I_4^* will have $MW = \mathbb{Z}/2\mathbb{Z}$. In characteristic $\neq 2$, the fibre has no wild ramification, see Schütt and Schweizer ([SSc13, Sect. 5]). It is thus not surprising that an explicit argument shows uniqueness and existence; translating the section to $(0, 0)$ and the I_4^* fibre to $t = 0$ (as in the last step of the proof of Theorem 8.39), we can assume that the rational elliptic surface is given by an extended Weierstrass form

$$y^2 = x^3 + tax^2 + t^2bx \quad \deg(a) \leq 1, \deg(b) \leq 2. \tag{8.10}$$

It is immediate that the fibre type I_4^* is equivalent to $b = \beta t^2$. Here $\beta \neq 0$, since otherwise all fibres are singular, so we can rescale to achieve $\beta = 1$. There are two further singular fibres, both of type I_1, so we apply a Möbius transformation to move them to $t = 1$ and ∞. This implies that $a(\infty) = a(1) = \pm 2$ and we find, up to sign, $a = 2t - 4$. This settles the existence of a rational elliptic surface with an I_4^*

fibre outside characteristic 2. The sign choices can be identified under $(x, y, t) \mapsto (-x, \sqrt{-1}y, t)$, so uniqueness follows as well.

In characteristic 2, however, fibres of type I_4^* necessarily come with wild ramification by Schütt and Schweizer ([SSc13, Sect. 5]). Moreover, we cannot work with an extended Weierstrass form. In fact, by [Ito02] rational elliptic surfaces with this fibre occur in the following one-dimensional family:

$$S_\alpha : \quad y^2 + xy = x^3 + tx^2 + \alpha, \quad \alpha \neq 0,$$

with 2-torsion section $(0, \sqrt{\alpha t^3})$. Note that the elliptic surface S_α has no singular fibres other than the I_4^* fibre at $t = 0$ by inspection of the discriminant $\Delta = t^{12}$. In fact, it is a peculiarity of this family of rational elliptic surfaces that each member remains constant under purely inseparable base change.

8.10.3 $T = E_8$

It remains to discuss the trivial case of $T = E_8$. We distinguish two cases, depending on the fibration being isotrivial or not. Both come with a unique rational elliptic surface with a fibre of type II* regardless of the characteristic (although the fibre type implies wild ramification in characteristics 2 and 3 by [SSc13]).

If the elliptic fibration is isotrivial and $\text{char}(k) \neq 3$, then its Weierstrass form is uniquely determined up to admissible transformations as

$$y^2 + y = x^3 + t.$$

By Theorem 8.8 there are no sections other than O. Note that in characteristic 2, there are no singular fibres other than the special fibre of Kodaira type II* at $t = \infty$, while otherwise there is a second singular fibre (Kodaira type II) at $t = -1/4$. By Jarvis ([JLR12, Lemma 1.1]), there cannot be a model over $\mathbb{Q}$ with good reduction at 3, but the model

$$y^2 = x^3 + x + t$$

works in characteristic 3, uniquely by [Ito02], again without singular fibres outside $t = \infty$.

If the fibration is not isotrivial, there is a characteristic-free unique model

$$y^2 + xy = x^3 + t.$$

In addition to the II* fibre at $t = \infty$, this elliptic surface has one ($\text{char}(k) = 2, 3$) or two ($\text{char}(k) \neq 2, 3$) further singular fibres (Kodaira type I_1) and again no non-zero sections.

8.10.4 Comment on K3 Surfaces and Beyond

Already for the next class of elliptic surfaces, namely elliptic K3 surfaces, the corresponding classifications become quite non-trivial. In detail, the maximal singular fibres (i.e. those with maximum number of components, as before) have been worked out in [Shio03] over $\mathbb{C}$ and in [Schü06] for any characteristic, while uniqueness was proved in [SSc13]. Notably, the bounds on the Picard number imposed by the Lefschetz inequality (4.15) limit the singular fibres to maximally types I_{19}, I_{14}^* over $\mathbb{C}$. As it happens, the above types also describe the maxima in odd positive characteristic, while in characteristic 2, the maximal possible singular fibres have one component less (Kodaira types I_{18}, I_{13}^*).

We conclude this section by pointing out that beyond K3 surfaces, i.e. for (honestly) elliptic surfaces of Kodaira dimension $\kappa = 1$, it need no longer be true that the maximal singular fibres over $\mathbb{C}$ and in positive characteristic coincide. In essence this is due to the influence of inseparable base change as analysed in detail in [SSc08].

8.10.5 Connection with Polynomial Sections

In this concluding section, we open up a different perspective onto maximal singular fibres by drawing a connection to polynomial equations, and ultimately to very special polynomial sections. To this end, we assume that the characteristic is different from 2 and 3, so that we can work with a short Weierstrass form (3.3), which we rescale here according to our needs:

$$S: \quad y^2 = x^3 - 3Ax + 2B. \tag{8.11}$$

Then the discriminant (3.2) evaluates as

$$\Delta = 12^3(A^3 - B^2). \tag{8.12}$$

Coming back to the maximal singular fibres of elliptic surfaces over $\mathbb{P}^1$, we can work over $k[t]$ and locate the special fibre (to be maximal) at $t = \infty$. Assume that the Euler–Poincaré characteristic is given as

$$e(S) = \deg \Delta = 12N,$$

where, of course, we interpret Δ homogeneously, i.e. taking the contribution at $t = \infty$ into account. For the second Betti number and the central Hodge number (over $\mathbb{C}$), this implies

$$b_2(S) = 12N - 2, \quad h^{1,1}(S) = 10N.$$

For the coefficients of (8.11), this translates as

$$\deg A = 4N, \quad \deg B = 6N.$$

Over $\mathbb{C}$, Lefschetz' inequality (4.15) for the Picard number, together with Corollary 6.7, shows that the special fibre at $t = \infty$ (or in fact any fibre) can have an associated Dynkin diagram T_∞ of rank maximally $10N - 2$, i.e. not exceeding

$$A_{10N-2} \leftrightarrow \mathrm{I}_{10N-1}, \quad D_{10N-2} \leftrightarrow \mathrm{I}^*_{10N-6} \quad (\text{and } E_8 \text{ if } N = 1) \qquad (8.13)$$

confirming Sects. 8.10.1–8.10.3. Since the contribution of the singular fibre to the Euler–Poincaré characteristic exactly equals the vanishing order of the discriminant (presently at ∞), the maximal singular fibre types translate into affine degrees, i.e. in $k[t]$, as

$$\deg \Delta = 2N + 1 \text{ (for } \mathrm{I}_{10N-1}) \quad \text{resp.} \quad 2N \text{ (for } \mathrm{I}^*_{10N-6})$$

Such triples (A, B, Δ) are rather special as they attain a degree bound (for Δ) originally due to Davenport [Dav65]. (Much later these triples were coined "Davenport–Stothers triples" by one of us in [Shio05b]).

Let us quickly draw the connection to polynomial sections. For this purpose, consider the elliptic surface over $\mathbb{P}^1$ given in affine Weierstrass form,

$$S' : \quad Y^2 = X^3 - \Delta/12^3.$$

Then the triple (A, B, Δ) gives a polynomial section $P = (A, B)$ of S'.

Example 8.42 Starting from the extended Weierstrass form (8.9), we obtain a polynomial section

$$P = (t(t^3 - 24), t^6 - 36t^3 + 216)$$

on the rational elliptic surface given by

$$y^2 = x^3 - 12^3(t - 3)(t^2 + 3t + 9).$$

Similarly, the I^*_4 example derived from (8.10) yields an integral section

$$P = (16t^2 - 16t + 1, (2t - 1)(32t^2 - 32t - 1))$$

on the rational elliptic surface given in Weierstrass form,

$$y^2 = x^3 - 108t(t - 1).$$

We conclude this chapter by pointing out how differently the situation presents itself in positive characteristic (>3, say). Since Lefschetz' inequality (4.15) fails to hold in general, there is no bound such as (8.13) anymore, and indeed the above considerations for maximal fibres are known to lose validity (outside the initial cases of $N = 1$ and 2). Essentially this is due to inseparable base change (intertwining with the presence of additive fibres), as explored in [SSc08].

Historical Notes

The starting point of our classification of rational elliptic surfaces in terms of their Mordell–Weil lattices was Dynkin's classification of semi-simple subalgebras of semi-simple Lie algebras ([Dyn57]), restated in terms of root lattices, which shows in particular the classification of the root lattices embedding into E_8 (Theorems 8.1 and 8.2). Essentially, this fundamental work only had to be combined with the restrictions imposed by the Euler–Poincaré characteristic of the surface.

The hierarchy among rational elliptic surfaces arising in this way may be thought to resemble the hierarchy of del Pezzo surfaces, grouped by degree (cf. [Man86, CvGL09]), but arguably the picture provided by rational elliptic surfaces is more harmonic since everything takes place and is described in terms of the fixed reference lattice E_8 given by the frame (while for the del Pezzo surfaces, the Picard number will vary). We will come back to del Pezzo surfaces briefly in Sect. 10.5.3.

The existence part of the classification of rational elliptic surfaces (Theorem 8.9) could have been deduced (over $\mathbb{C}$) from work by Persson [Per90], and independently, Miranda [Mir90] (compare also [Mir89]). Here we pursued a more algebraic approach, inspired by work on Noether–Lefschetz loci (see [Klo07a]), but adapted to the positive characteristic setting.

Extremal elliptic surfaces have been studied starting from Beauville [Bea82], and from different viewpoints such as classification in terms of equations or lattices. Results range from rational elliptic surfaces ([MP86, Nar87]), through elliptic K3 surfaces ([SZ01, BM08, Schü07a]), both over $\mathbb{C}$, to the positive characteristic setting, which takes a very different flavour ([Ito02]).

Closely related to extremal elliptic surfaces, there is the study of torsion sections, one of the most classical aspects of elliptic curves in general. We have commented on the curve case over $\mathbb{Q}$ in Chap. 3. For the general case, the standard reference seems to be the monograph by Katz and Mazur [KM85]. Within the realm of elliptic surfaces, there are two instances where torsion sections appear most naturally: on the one hand, obviously, on the universal elliptic curve with a point of given order (with nice explicit models due to Tate, see [Kub76]); on the other hand, from a more complex angle, on elliptic modular surfaces [Shio72]. Note that all of the resulting complex elliptic surfaces will be extremal! It should be emphasized that these very special surfaces have some relevance to higher-dimensional geometry. For instance, Deligne considers fibre products of universal elliptic curves with several copies of themselves to construct Galois representations attached to modular forms ([Del71]), and following Schoen [Scho88] one can setup different fibre products to construct rigid Calabi–Yau threefolds (which are quite hard to approach and have interesting arithmetic properties, see [Schü04], for instance).

Siegel's theorem on integral points on a smooth curve of positive genus ([Sie29]) is often regarded as the first result in diophantine geometry which is completely independent of the chosen equations. For curves of genus $g > 1$, this was much later superseded by Faltings' finiteness theorem for all *rational* points ([Fal83]). Siegel's theorem per se is not effective, but it can be made effective using Baker's method on linear forms in the logarithms of algebraic numbers ([Bak66/67]; which also has important applications to the class number one problem, see Sect. 13.1.3.2). In comparison, the framework of Mordell–Weil lattices facilitates a much easier study of integral points. We have already seen this in Sect. 8.7, and we will come back to it in the next two chapters (see Sects. 9.5.3, 10.3.2 and 10.4.4). In fact, on a rational elliptic surface, the integral sections can be inscribed in a zero-dimensional scheme of length 240 which, by way of the primary decomposition of the underlying ideal, even encodes multiplicities in the presence of reducible fibres (see Remark 10.6.3 or [Shio10]).

Originally, we were led by the idea that there should exist an approach to produce more explicit examples for all rational elliptic surfaces with given Mordell–Weil lattice through Weierstrass forms—potentially even split, i.e. with all sections defined over a given field (with a view towards arithmetic applications). Naturally, this led to the notion of excellent families as a standard way to produce rational elliptic surfaces with fixed MWL—one of the main subjects of the next two chapters.

Chapter 9
Galois Representations and Algebraic Equations

In this and the next chapter, we discuss Galois representations and algebraic equations which arise naturally from Mordell–Weil lattices. This subject has been one of the central themes in developing the Mordell–Weil lattice theory.

We organize this chapter as follows. First we review the main ideas from the papers [Shio91d, Part II], [Shio91c], and then we formulate the notion of excellent families (cf. [SU92]), first in the additive setting, to describe a common deep connection between Mordell–Weil lattices of rational elliptic surfaces, algebraic equations and invariant theory of Weyl groups.

In the next chapter we will proceed to discuss more recent developments about the multiplicative excellent families, especially of type E_6, E_7 or E_8 ([Shio12, KumS13]). Both chapters conclude with various applications.

9.1 Galois Representations Arising From Mordell–Weil Lattices

So far we have mostly considered the geometric setting in the sense that the base field is algebraically closed (for an exceptional case, see Theorem 6.56).

From now on, we consider the following setting. The base field k_0 is an arbitrary perfect field (for example, take $\mathbb{Q}$; see also the appendix Sect. 9.6 at the end of this chapter concerning the perfectness assumption) and we let k be its algebraic closure and $G = \mathrm{Gal}(k/k_0)$ the Galois group of the extension k/k_0. Take an absolutely irreducible smooth projective curve C defined over k_0 and denote by $K_0 = k_0(C)$ the function field of C over k_0, and similarly, denote by $K = k(C)$ the function field of $C \otimes k$ over k. For example, if $C = \mathbb{P}^1$ is a projective line with affine coordinate t, then we have $K_0 = k_0(t)$ and $K = k(t)$.

Now let E denote an elliptic curve defined over K_0 and consider the base extension $\bar{E}$ over K, i.e. the geometric setting

M. Schütt and T. Shioda, *Mordell–Weil Lattices*, Ergebnisse der Mathematik und ihrer Grenzgebiete. 3. Folge / A Series of Modern Surveys in Mathematics 70,
https://doi.org/10.1007/978-981-32-9301-4_9

$$\bar{E} = E \otimes_{K_0} K.$$

By construction, the elliptic surface associated with $\bar{E}$ comes with a model

$$f : S \longrightarrow C$$

over k_0 supplied by E.

Obviously, the Galois group G acts on the Mordell–Weil group $E(K)$, and $E(K_0)$ coincides with $E(K)^G$, the subgroup of G-invariant rational points in $E(K)$. First we recall from Theorem 6.56 that the height pairing on $E(K)$ is stable under G:

$$\langle P^\sigma, Q^\sigma \rangle = \langle P, Q \rangle \text{ for any } P, Q \in E(K),\ \sigma \in G. \tag{9.1}$$

Therefore we obtain naturally a Galois representation on the Mordell–Weil lattice $M = (E(K)/E(K)_{\text{tors}}, \langle , \rangle)$:

$$\varrho : G = \text{Gal}(k/k_0) \longrightarrow \text{Aut}(M).$$

Note that the target space is a finite group, since it consists of isometries of a positive definite (not necessarily integral) lattice (cf. Theorem 2.14). Of course we also have a very natural variant by restricting the Galois representations to $M^0 = E(K)^0$, the narrow Mordell–Weil lattice of E/K (or any other G-stable sublattice). We point out that either of these two choices for the positive definite lattice M gives an integral Galois representation with finite image, e.g.

$$\varrho' : G \longrightarrow \text{Aut}(M) \subset \text{GL}_r(\mathbb{Z}) \quad (r = \text{rk}(E(K))).$$

Remark 9.1 Note how this contrasts with the now classical setting of the Galois representations on the ℓ^n-torsion points of an elliptic curve E over a number field: upon passing to the Tate module $T_\ell(E)$, these representations become infinite, and in fact, outside the CM case, they tend to have full image in $\text{GL}_2(\mathbb{Z}_\ell)$ by [Ser72].

In comparison, a good analogy might be the Galois representation of the Néron–Severi group of a del Pezzo surface as studied by Manin in [Man86]. We will see an instance of this in Sect. 10.5.3.

Remark 9.2 It is possible to extend the above approach to the full Mordell–Weil group $E(K)$, that is, including the torsion (defined over k_0 to be on the safe side). Then $E(K)$ is not necessarily a lattice in the strict sense, but the Galois representation

$$\varrho : G = \text{Gal}(k/k_0) \longrightarrow \text{Aut}(E(K))$$

will still respect the height pairing and have finite image. Since throughout this chapter and the next, there will be no torsion sections involved, we decided to restrict to the proper Mordell–Weil lattices for ease of presentation.

Let $\mathcal{K}/k_0$ be the finite extension corresponding to $\mathrm{Ker}(\varrho)$, i.e. the fixed field of $\mathrm{Ker}(\varrho)$; equivalently, $\mathcal{K}$ is the smallest extension of k_0 such that $E(\mathcal{K}(C)) = E(K)$. By definition, $\mathcal{K}/k_0$ is a finite Galois extension such that

$$\mathrm{Gal}(\mathcal{K}/k_0) = \mathrm{Im}(\varrho).$$

Definition 9.3 We call $\mathcal{K}$ the *splitting field* of the Mordell–Weil group of E/K_0, or in short, of E/K_0.

Remark 9.4 For non-perfect base fields k_0, the splitting field of an elliptic curve over K_0 need not be separable over k_0, compare Appendix Sect. 9.6.

Example 9.5 (*Hexagonal example cont'd*) Consider the elliptic curve E from (1.1) with hexagonal Mordell–Weil lattice, and with k replaced by a possibly non-algebraically closed field k_0. Then $E(K_0)$ is generated by (any two of) the sections P_i over k_0, so the splitting field is $\mathcal{K} = k_0$.

Switching to the general Weierstrass form of the given shape,

$$y^2 + ty = f(x) \quad \text{for some cubic polynomial } f \in k_0[x]$$

as in Example 6.58, we read off that the splitting field $\mathcal{K} = k_0(f)$ is obtained from k_0 by adjoining the roots of f.

The basic problem on the Galois representations arising from the Mordell–Weil groups is as follows:

Problem 9.6 Determine the image of ϱ. In particular, we ask:

- (big Galois) How big can $\mathrm{Im}(\varrho)$ be?
- (small Galois) How small can $\mathrm{Im}(\varrho)$ be?

The interest of these questions will become clear from the examples in the subsequent sections. However, a few words of explanation may be in order. First, we should like to highlight that Problem 9.6 should usually be seen relative to a given family of elliptic surfaces. For instance, one may think of rational elliptic surfaces and then fix the type from Theorem 8.8 and the ground field k_0, say $k_0 = \mathbb{Q}$. The Mordell–Weil lattice of any rational elliptic surface is governed by the root lattice E_8, so it is not only natural to ask when $\mathrm{Im}(\varrho)$ may be trivial, but also whether there is a uniform bound for $\mathrm{Im}(\varrho)$ given in terms of $\mathrm{Aut}(\mathrm{MWL})$, thus ultimately in terms of the Weyl group $W(E_8)$. This will be achieved later by considering excellent families.

Of course, we can also extend Problem 9.6 to the Galois representation on the whole Mordell–Weil group as in Remark 9.2. Usually, this does not alter the outcome substantially, but in special cases, such as with full N-torsion for $N > 2$, it may very well affect the issue of small Galois due to the Weil pairing (as hinted at in Sect. 10.5.6).

Lastly, it should be remarked that we can avoid the use of the infinite Galois group $\mathrm{Gal}(k/k_0)$ by replacing it by the finite group $\mathrm{Gal}(\mathcal{K}/k_0)$ throughout this chapter and the next, but then, of course, this finite group will usually vary with the elliptic curves in consideration (both as an abstract group, but even more so in terms of the explicit Galois action).

9.2 Algebraic Equations Arising From Mordell–Weil Lattices

In this section, we introduce a canonical way to create interesting (and arithmetically or geometrically meaningful) algebraic equations from an elliptic curve E/K_0. To this end, take a G-equivariant map

$$s: \ E(K) \to k$$

and a G-stable finite subset I of $E(K)$. Then we define

$$\Phi(X) = \Phi(E/K_0, s, I; X) = \prod_{P \in I} (X - s(P)) \in k[X]. \tag{9.2}$$

Here we will soon take s to be the specialization map sp'_v at a k_0-rational singular fibre F_v, but the reader might well think of any k_0-rational expression in the x- and y-coordinates of a section P specialized to F_v. As a typical example of the set I, we have

$$I_h = \{P \in E(K) | \langle P, P \rangle = h\} \quad (h \geq 0). \tag{9.3}$$

This is G-stable by (9.1) and finite for any fixed height h (compare Remark 8.27).

Proposition 9.7 *In the above set-up, the polynomial $\Phi(X)$ has coefficients in k_0, i.e.*

$$\Phi(X) \in k_0[X].$$

The splitting field of the algebraic equation $\Phi(X) = 0$ over k_0, say $\mathcal{K}'$, is a subfield of the splitting field $\mathcal{K}$ of E/K_0 defined above, and we have $\mathcal{K}' = \mathcal{K}$ if

(1) s is injective and
(2) the set I contains a set of generators of $E(K)$.

Proof Immediate by Galois theory. □

To see an application of this result, the reader may want to jump ahead to Example 9.11 (see also Sect. 9.4.1), but for a proper treatment we first have to introduce our preferred candidate for s—the specialization map.

We have already pointed out in Remark 9.1 that the Galois representations and algebraic equations arising from Mordell–Weil lattices are quite different from those arising from the torsion points of an elliptic curve or an abelian variety (e.g. the Tate modules). Admittedly, the algebraic equations look similar to the classical "division equations" or "modular equations" arising from torsion points on an elliptic curve or from "transformations" (i.e. isogenies) of elliptic curves. But we are concerned here with points of *infinite order* on an elliptic curve E, and in general the algebraic equations will lead to highly *non-abelian* extensions, as the examples in the subsequent sections will show (see also the consequence for the zeta functions in Sect. 10.5.1.1).

9.3 The Specialization Maps

Let us define the specialization maps sp_v and sp'_v in order to obtain potential candidates for G-equivariant maps to insert for s into $\Phi(X)$ in (9.2) and Proposition 9.7.

Fix $v \in C(k_0)$ (e.g. $v = \infty$ in $C = \mathbb{P}^1$) and suppose that $F_v := f^{-1}(v)$ is a singular fibre with m_v (geometrically) irreducible components and $m = m_v^{(1)}$ simple ones (i.e. with coefficient 1 in the expression below):

$$F_v = f^{-1}(v) = \Theta_{v,0} + \ldots + \Theta_{v,m-1} + \sum_{k_j>1} k_j \Theta_{v,j}.$$

(cf. (5.2)). Then the smooth part of F_v is

$$F_v^{\#} = f^{-1}(v)^{\#} = \Theta_{v,0}^{\#} \cup \ldots \cup \Theta_{v,m-1}^{\#}$$

where $\Theta_{v,i}^{\#} := \Theta_{v,i} \cap F_v^{\#}$ denotes the open part of $\Theta_{v,i}$ obtained by removing the intersection points with other components in case $m_v > 1$ resp. the singularity if $m_v = 1$. As in Sect. 5.6.1, $F_v^{\#}$ has the structure of an algebraic group over k whose *identity component* $\Theta_0^{\#} := \Theta_{v,0}^{\#}$ is isomorphic (over k) to

- either the additive group $\mathbb{G}_a$
- or the multiplicative group $\mathbb{G}_m$,

according to the type of the singular fibre F_v. The specialization map sp_v is defined by

$$\begin{aligned} sp_v :\ M := E(K) &\to F_v^{\#}(k) \\ P &\mapsto sp_v(P). \end{aligned} \tag{9.4}$$

By definition, this is a $\mathrm{Gal}(k/k_0)$-equivariant group homomorphism. We emphasize that this whole set-up does not require that all components of F_v are defined over k_0. The restriction to the narrow Mordell–Weil lattice gives another G-equivariant group homomorphism

$$sp_v^0 : M^0 := E(K)^0 \longrightarrow \Theta_0^{\#}(k) \simeq k \text{ or } k^{\times}. \tag{9.5}$$

To gain a better understanding of sp_v, we first consider the restriction to the *v-narrow part* of M, i.e. the finite-index subgroup $M' \subseteq M$ consisting of those sections P passing exclusively through $\Theta_{v,0}$:

$$M' := \{P \in M;\ sp_v(P) \in \Theta_0^\#\}.$$

Similarly, we call a section $P \in M$ *v-integral* if $P \in M'$, i.e. if (P) meets Θ_0. Clearly, this leads to the following much simplified specialization homomorphism:

$$sp'_v := sp_v|_{M'}:\ M' \longrightarrow \Theta_0^\#(k) \simeq k \text{ or } k^\times. \tag{9.6}$$

In particular, since the narrow Mordell–Weil lattice M^0 is contained in M', we infer the natural compatibility

$$sp'_v|_{M^0} = sp_v^0.$$

At this point, let us recall the explicit isomorphisms of the smooth locus of a nodal or cuspidal cubic with the multiplicative or additive group. These will be of particular relevance to our issues since for the v-integral sections in M', it suffices to work with a Weierstrass model locally at v to determine the specialization map sp'_v. For simplicity, we work with the affine models even though the isomorphisms should be understood for the projective closures (to which they extend directly):

Lemma 9.8 *(i) Suppose Θ_0 is identified with the cuspidal curve $y^2 = x^3$ with the cusp* $(0, 0)$*, then the point $(x, y) \in \Theta_0^\#$ corresponds to $u = x/y \in \mathbb{G}_a$, so that*

$$x = \frac{1}{u^2}, \quad y = \frac{1}{u^3}.$$

This correspondence preserves the (additive) group law: i.e. three points (x_i, y_i) are collinear if and only if $u_1 + u_2 + u_3 = 0$.

(ii) Suppose Θ_0 is identified with the nodal curve $y^2 + xy = x^3$ with the node $(0, 0)$*, then the point $(x, y) \in \Theta_0^\#$ corresponds to $s = (x + y)/y \in \mathbb{G}_m$, so that*

$$x = \frac{s}{(s-1)^2}, \quad y = \frac{s}{(s-1)^3}.$$

The correspondence preserves the (multiplicative) group law: i.e. three points (x_i, y_i) are collinear if and only if $s_1 s_2 s_3 = 1$.

For the proof, see Sect. 3.4. The above cases *(i)* and *(ii)* correspond to the cases (a) and (b) there up to a coordinate change. Note that in the nodal case Sect. 3.4 (b), the two tangent lines at the node are not necessarily rational over the base field k_0, but may require a quadratic extension. If the two tangent lines are indeed rational over k_0, then the curve Θ_0 can be transformed over k_0 to the normal form $y^2 + xy = x^3$ of *(ii)* above.

Remark 9.9 In the sequel, we will usually fix an isomorphism

$$\Theta_0^\# \cong \mathbb{G}_a \text{ or } \mathbb{G}_m$$

as in Lemma 9.8.

In particular, if $m = 1$, that is, if the singular fibre F_v is of type I_1 or II or II*, then the specialization homomorphism $sp'_v = sp_v : M \to k$ or $k^\times$ is computable by Lemma 9.8, provided that we fix an isomorphism of $\Theta_0^\#(k)$ with the additive group $\mathbb{G}_a(k) = k$ or with the multiplicative group $\mathbb{G}_m(k) = k^\times$, and this can be done in a G-equivariant way, as indicated in Lemma 9.8.

Going back to the general setting, we ask if there exists a $\mathrm{Gal}(k/k_0)$-equivariant homomorphism on the entire M,

$$sp'_v : M \longrightarrow k \text{ or } k^\times,$$

which extends the specialization maps $sp_v|_{M'}$ and sp_v^0 given by (9.6) or (9.5).

Lemma 9.10 *Let the singular fibre F_v be additive with $m = m_v^{(1)}$ simple components. Assume that $char(k) = p \nmid m$. Given a choice of isomorphism $\Theta_0^\#(k) \cong k$, the required specialization map*

$$sp'_v : M \to k$$

exists, and it is unique.

Proof Recall that $m = m_v^{(1)}$ is at most 4 for an additive fibre (see Corollary 5.17). By assumption, m is invertible in k. Since mP belongs to M' for any $P \in M$ (compare Sect. 5.6.1), we obtain a map sp'_v with the required property by defining

$$sp'_v(P) = \frac{1}{m} sp_v^0(mP) \quad (P \in M).$$

If there are two such maps s', s'', then consider the difference $s' - s''$. It will induce a homomorphism $M/M' \to k$, which must be trivial since the order of the quotient group M/M' is invertible in k by assumption. □

We shall illustrate the use of Lemma 9.10 with the following example which at the same time shows that the desired specialization map sp' may not exist in the exceptional characteristic $p \in \{2, 3\}$ such that $p \mid m$ (with the notation of Lemma 9.10).

Example 9.11 (*Hexagonal example cont'd*) The elliptic curve E from (1.1) comes with a singular fibre of Kodaira type IV* at $v = \infty$, where we want to work out the specialization morphism. To this end, we change the affine chart to $s = 1/t$, where the Weierstrass from becomes (6.16),

$$y^2 + s^2 y = (x - \alpha_1 s^2)(x - \alpha_2 s^2)(x - \alpha_3 s^2).$$

This makes the cuspidal cubic at $s = 0$ visible, with standard equation as in Lemma 9.8 *(i)* and thus with specialization map sp_v^0, but also equipped with a singular surface point at $(0, 0)$. After blowing up twice, the Weierstrass form becomes (6.17),

$$E: \quad y'^2 + y' = s^2(x' - \alpha_1)(x' - \alpha_2)(x' - \alpha_3),$$

with sections $P_i = (\alpha_i, 0)$ all meeting the same non-identity component

$$\Theta_1 = \{y' = 0\}$$

of the special fibre. Note that the components intersect outside the given chart, so indeed, the set of smooth k_0-rational points on $\Theta_1^\#$ is

$$\Theta_1^\#(k_0) = \{(x', 0); x' \in k_0\} \cong k_0.$$

We shall normalize the specialization map sp such that restricted to the v-narrow Mordell–Weil group (which generically equals the narrow Mordell–Weil group), it agrees with sp^0. To this end, we apply the approach from the proof of Lemma 9.10 to the sections $3P_i$. An easy computation reveals

$$sp^0(P_i - P_j) = \alpha_j - \alpha_i, \quad sp^0(3P_i) = \alpha_j + \alpha_k - 2\alpha_i \quad (i \neq j \neq k \neq i)$$

Outside characteristic 3 we thus obtain

$$sp'(P_i) = \frac{1}{3} sp^0(3P_i) = \frac{1}{3}(\alpha_j + \alpha_k - 2\alpha_i). \tag{9.7}$$

Note that generally $sp^0(3P_i) \neq 0$. Hence, in characteristic 3, one cannot extend the specialization map sp' from the ∞-narrow part M' to all of the Mordell–Weil lattice M. In particular, this also shows that $F_v^\#$ does *not* decompose as a direct product of $F_{v,0}^\#$ and the group of simple fibre components.

In the following sections, we will see more examples in both additive and multiplicative cases. Since it is unclear whether the specialization map sp_v' exists in general (outside characteristics 2, 3), we shall proceed as follows. If the chosen singular fibre F_v is multiplicative of type I_m with $m > 1$, then we shall verify directly that the required map $sp_v' : M \to k^\times$ exists in the examples under consideration. Otherwise we will appeal to Lemma 9.10.

9.4 Galois Representation and Algebraic Equation of Type E_6, E_7 or E_8

To illustrate the nature of the Galois representations and the algebraic equations arising from Mordell–Weil lattices, we consider some interesting cases in detail, ultimately leading to the notion of excellent families.

The main result of this section is Theorem 9.21 that determines the Galois representations and algebraic equations for the families of elliptic surfaces defined by the Weierstrass equations (9.9), (9.10), or (9.11), with the Mordell–Weil lattices of type E_r for $r = 8, 7, 6$.

To fix the idea, we will soon choose the ground field to be the rational number field $\mathbb{Q}$, but it can be replaced by almost any perfect field (of char $\neq 2, 3$ at least) after some suitable modification. We start by outlining the essence of the theory for the hexagonal case which has been continuously reappearing throughout this book, before going into the details for the provisional (additive) excellent families of types E_6, E_7 and E_8.

In the next chapter, extracting the common properties in Theorem 9.21, we will introduce the notion of *excellent families* properly. In this terminology, for example, the elliptic surfaces (9.9) with the specialization map sp_∞ will define an *additive excellent family with Galois group* $W(E_8)$, or more simply, an *excellent family of additive type* E_8.

9.4.1 *Galois Representations on Hexagonal Elliptic Curves*

In this section, we discuss the hexagonal elliptic curve from (1.1) in the framework of Galois representations arising from Mordell–Weil lattices. This will at the same time serve as an illustration and as a motivation for most of what is to come for more complicated settings in this chapter and the next.

Over a non-algebraically closed (perfect) field k_0, it is more natural to write the elliptic curves with hexagonal Mordell–Weil lattice from (1.1) in terms of a (monic) cubic polynomial $f \in k_0[x]$:

$$E: \quad y^2 + ty = f(x) \tag{9.8}$$

(cf. Example 6.58). From the general set-up, we know that the Galois representation on the Mordell–Weil lattice $M = E(k(t)) \cong A_2^\vee \cong A_2(1/3)$ has image contained in $O(A_2)$, the group of isometries of A_2, due to the stated proportionality. Essentially, this thus leads to the Weyl group of A_2:

$$\varrho: \text{Gal}(k/k_0) \longrightarrow \text{Aut}(M) \cong \text{Aut}(A_2^\vee) \cong \text{Aut}(A_2) = O(A_2) = W(A_2) \times \{\pm 1\}.$$

Indeed, in the present setting we can be a little more precise:

Proposition 9.12 *The Galois representation ϱ has image inside the Weyl group $W(A_2) \cong S_3$. Moreover, E has splitting field $\mathcal{K} = k_0(f)$, where* $\mathrm{Gal}(\mathcal{K}/k_0) \cong$ *$Im(\varrho) \subset S_3$.*

Proof The claim about the $\mathcal{K} = k_0(f)$ follows from our previous considerations, culminating in Example 6.41 (see also Example 9.11). Indeed, if f factors over $\mathcal{K}$ as

$$f(x) = (x - \alpha_1)(x - \alpha_2)(x - \alpha_3), \quad \alpha_i \in \mathcal{K},$$

then the sections $P_i = (\alpha_i, 0)$ generate M. Here $\mathrm{Gal}(k/k_0)$ acts through permutations on the P_i, i.e. without involving their inverses Q_i, so indeed $\mathrm{Im}(\varrho) \subset W(A_2)$ as stated. □

Remark 9.13 By twisting (outside characteristic 2) the Weierstrass form

$$y^2 = f(x) + t^2/4$$

over $k_0(\sqrt{d})$ for some $d \in k_0$ with $\sqrt{d} \notin k_0(f)$, i.e. by considering the $k_0(\sqrt{d})$-isomorphic Weierstrass form

$$dy^2 = f(x) + t^2/4,$$

we can easily modify the set-up such that, for f with Galois group S_3 over k_0, the Galois representation ϱ will map subjectively onto $W(A_2) \times \{\pm 1\}$. This would happen, though, at the price of making the IV* fibre at ∞ non-split over k_0, and even over $k_0(f)$. Accordingly, the sections P_i would not be rational over $k_0(f)$ anymore as $\mathcal{K} = k_0(f)(\sqrt{d})$.

We conclude this brief section by drawing the connection with algebraic equations arising from Mordell–Weil lattices, starting from the hexagonal elliptic curve E given by (9.8). To this end, we define the Galois-stable set I of sections as

$$I = \{P_1, P_2, P_3\}$$

and work with the additive fibre of Kodaira type IV* at $t = \infty$ with specialization map sp' derived in Example 9.11.

Proposition 9.14 *The algebraic equation $\Phi(X) = \Phi(E/k_0(t), \infty, I; X) = 0$ arising from the Mordell–Weil lattice $E(k(t)) \cong A_2^\vee$ is obtained from the cubic equation $f(x) = 0$ by a linear transformation over k. In particular, both polynomials have the same splitting field over k_0:*

$$\mathcal{K} = k_0(f) = k_0(\Phi).$$

Proof We simply spell out $\Phi(X)$ using the explicit shape of sp' deduced in (9.7) and the classical identities in the symmetric functions. For normalization reasons, we will need the coefficient $\beta = -(\alpha_1 + \alpha_2 + \alpha_3)$ of x^2 in f.

$$\begin{aligned}\Phi(X) &= \prod_{P\in I}(X - sp'(P)) = (X - sp'(P_1))(X - sp'(P_2))(X - sp'(P_3)) \\ &= \left(X - \tfrac{1}{3}(\alpha_2 + \alpha_3 - 2\alpha_1)\right)\left(X - \tfrac{1}{3}(\alpha_1 + \alpha_3 - 2\alpha_2)\right)\left(X - \tfrac{1}{3}(\alpha_1 + \alpha_2 - 2\alpha_3)\right) \\ &= -f(-X + \beta/3).\end{aligned}$$

This proves Proposition 9.14. □

To wrap up the preceding discussion, we collect the following well-known results in Galois theory which point-by-point highlight what we will derive in Theorem 9.21 for elliptic curves with Mordell–Weil lattice E_r, $r = 6, 7, 8$:

Theorem 9.15 (Hexagonal example) *Let $f = x^3 - \epsilon_1 x^2 + \epsilon_2 x - \epsilon_3$ be generic over $\mathbb{Q}$ (i.e. the coefficients ϵ_i are algebraically independent over $\mathbb{Q}$) and consider the elliptic curve E given by* (9.8) *over $k_0(t)$, $k_0 = \mathbb{Q}(\epsilon_1, \epsilon_2, \epsilon_3)$. Then:*

(i) the image of the Galois representation ϱ is the full Weyl group $W(A_2)$:

$$\mathrm{Im}(\varrho_\lambda) = W(A_2).$$

(ii) Let $\mathcal{K}/k_0$ be the splitting field of $E_\lambda/k_0(t)$, and let $I = \{P_1, P_2, P_3\}$ be a $W(A_2)$-orbit of minimal vectors in $E(k(t)) \simeq A_2^\vee$. Further, let $u_i = sp'_\infty(P_i) \in \mathcal{K} \subset k$. Then the only algebraic relation between $u_1, \dots, u_3$ over $\mathbb{Q}$ is

$$u_1 + u_2 + u_3 = 0$$

and $\mathcal{K}$ is a purely transcendental extension of $\mathbb{Q}$ generated by any two of u_1, u_2, u_3:

$$\mathcal{K} = k_0(u_1, u_2, u_3) = \mathbb{Q}(u_1, u_2, u_3) = \mathbb{Q}(u_2, u_3).$$

Thus we have a Galois extension between two purely transcendental extensions of $\mathbb{Q}$ with Galois group $W(A_2) \cong S_3$:

$$\mathrm{Gal}(\mathbb{Q}(u_2, u_3)/\mathbb{Q}(\epsilon_2, \epsilon_3)) = W(A_2).$$

(iii) $W(A_2)$ acts on the vector space $\sum \mathbb{Q}u_i$ and hence on the polynomial ring $\mathbb{Q}[u_2, u_3]$, and the subring of invariants is the polynomial ring:

$$\mathbb{Q}[u_2, u_3]^{W(A_2)} = \mathbb{Q}[\epsilon_2, \epsilon_3]$$

generated by $\epsilon_2 = u_1u_2 + u_1u_3 + u_2u_3$, $\epsilon_3 = u_1u_2u_3$.

(iv) *The polynomial $\Phi(E/k_0(t), \infty, I; X)$ is a monic irreducible polynomial in X with coefficients in the polynomial ring $\mathbb{Z}[\epsilon_2, \epsilon_3]$, and the splitting field of $\Phi(E/k_0(t), \infty, I; X)$ over $\mathbb{Q}(\epsilon_2, \epsilon_3)$ is equal to the splitting field $\mathcal{K}$ in (ii).*

Remark 9.16 Compared to Theorem 9.15 *(iii)*, our next main result, Theorem 9.21, will draw an even nicer picture with all fundamental invariants of the Weyl group $W(E_r)$ appearing. Between these two settings, the main difference lies in the fact that presently the linear symmetric polynomial satisfies

$$\epsilon_1(\mathbf{u}) = u_1 + u_2 + u_3 = 0$$

upon substituting $sp(P_i) = u_i$ by Example 9.11, hence the absence of ϵ_1 in the formulae (compare Sect. 9.5.1).

9.4.2 Set-Up for the Additive Type E_r $(r = 6, 7, 8)$

In what follows, we aim to develop an analogue of Theorem 9.15 for elliptic curves with Mordell–Weil lattice $E_r, r = 6, 7, 8$. Consider the following Weierstrass equations with the parameters λ over the rational function field $k_0(t)$ where $k_0 = \mathbb{Q}(\lambda) = \mathbb{Q}(p_i, q_j)$:

$$(add\,E_8)\,y^2 = x^3 + x\left(\sum_{i=0}^{3} p_i t^i\right) + \left(\sum_{i=0}^{3} q_i t^i + t^5\right), \tag{9.9}$$

$$\lambda = (p_0, p_1, p_2, p_3, q_0, q_1, q_2, q_3) \in \mathbb{A}^8.$$

$$(add\,E_7)\,y^2 = x^3 + x(p_0 + p_1 t + t^3) + \left(\sum_{i=0}^{4} q_i t^i\right), \tag{9.10}$$

$$\lambda = (p_0, p_1, q_0, q_1, q_2, q_3, q_4) \in \mathbb{A}^7.$$

$$(add\,E_6)\,y^2 + t^2 y = x^3 + x\left(\sum_{i=0}^{2} p_i t^i\right) + \left(\sum_{i=0}^{2} q_i t^i\right), \tag{9.11}$$

$$\lambda = (p_0, p_1, p_2, q_0, q_1, q_2) \in \mathbb{A}^6.$$

Fix $r = 8, 7, 6$ and let $E = E_\lambda$ be the elliptic curve over $k_0(t)$ (and over $k(t)$ where $k = \bar{k}_0$ denotes the algebraic closure of k_0 as before), defined by one of the above Weierstrass equations (9.9), (9.10) or (9.11). Let

$$f : S_\lambda \longrightarrow \mathbb{P}^1$$

be the associated elliptic surface; it is rational by Proposition 5.51. The fibre over $t = \infty$ is an additive singular fibre of Kodaira type II, III, IV, with $9 - r$ irreducible components in total ($r = 8, 7, 6$), all of which are rational over k_0 by construction. The identity component Θ_0 can be identified with the projective closure of the cuspidal cubic

$$\Gamma : y^2 = x^3$$

of the Weierstrass model as in Lemma 9.8 *(i)*. Literally, this only holds if $r = 8$, while for $r = 7$ or 6 we ought to consider its strict transform in the resolution of the singularity (node) at $(x, y, t) = (0, 0, \infty)$. In all three cases, we have

$$\Theta_0^\# \simeq \mathbb{G}_a$$

and $f^{-1}(\infty)^\#$ is an extension of $\mathbb{G}_a$ by $\mathbb{Z}/(9-r)$. By Lemma 9.10 (valid outside characteristic 2 for $r = 7$, or characteristic 3 for $r = 6$), we have a well-defined specialization homomorphism

$$sp'_\infty : M_\lambda := E_\lambda(k(t)) \to \mathbb{G}_a(k) = k \tag{9.12}$$

which is $\mathrm{Gal}(k/k_0)$-equivariant.

Note that for generic, or general, λ, the elliptic surface S_λ has no reducible fibres outside $t = \infty$. To see this, either spell out the discriminant and factorize or appeal to suitable special members where this holds, e.g. the surfaces with all $p_i = q_j = 0$ except for $q_0 = 1$ (one of which appeared in Example 5.8). Thus it is very natural to assume that λ satisfies the following generality condition:

$$\text{every fibre of } f \text{ over } t \neq \infty \text{ is irreducible.} \tag{9.13}$$

Then the Mordell–Weil lattice $E(k(t))$ is isomorphic to the dual lattice $E_r^\vee$ of the root lattice E_r by Theorem 7.4, and we get the Galois representation

$$\varrho = \varrho_\lambda : \mathrm{Gal}(k/k_0) \longrightarrow \mathrm{Aut}(M) \simeq \mathrm{Aut}(E_r^\vee) = \mathrm{Aut}(E_r). \tag{9.14}$$

Remark 9.17 A general rational elliptic surface with Mordell–Weil lattice isomorphic to E_r ($r = 6, 7, 8$) has only multiplicative singular fibres (especially at $t = \infty$), as follows directly from the discriminant. However, this complicates the analysis substantially, as we shall explore in Sect. 10.2.

Remark 9.18 It will be very useful to view (9.9)–(9.11) as weighted homogeneous equations. For later reference, we record the resulting weights, including the total degree d of the equation, in Table 9.1. It turns out that the given parameters, including their weights, relate extremely well to the invariants of the Weyl group $W(E_r)$; indeed the weights in Table 9.1 match exactly those in Theorem 2.44. We will make this more precise in Theorem 9.21.

Table 9.1 Weights of homogeneous equations (9.9)–(9.11)

R	d	x	y	t	p_0	p_1	p_2	p_3	q_0	q_1	q_2	q_3	q_4
E_8	30	10	15	6	20	14	8	2	30	24	18	12	
E_7	18	6	9	4	12	8			18	14	10	6	2
E_6	12	4	6	3	8	5	2		12	9	6		

Lemma 9.19 *For each of the above families with* $r = 6, 7, 8$, *we have*

$$Im(\varrho_\lambda) \subset W(E_r).$$

Proof Recall from Sect. 2.3.6 that

$$\text{Aut}(E_r) = W(E_r) \quad \text{for } r = 8 \text{ or } 7,$$

and

$$\text{Aut}(E_6) = W(E_6) \cdot \{\pm 1\}$$

(just like the hexagonal case with MWL $\cong A_2$ in Sect. 9.4.1). Clearly, we thus have

$$\text{Im}(\varrho_\lambda) \subset W(E_r)$$

for $r = 7, 8$, and it remains to prove the claim of Lemma 9.19 for $r = 6$. To this end, recall from Theorem 7.12 that E_λ admits 54 minimal sections (of height 4/3), and that they take the polynomial shape $P = (\text{x}, \text{y})$ with

$$\text{x} = at + b, \quad \text{y} = ct^2 + dt + e, \quad a, b, c, d, e \in k.$$

To get a clear picture of the situation (just as in Example 9.11), we work in the affine chart with parameter $s = 1/t$ at ∞, where E_λ reads (letting $\bar{x} = x/t^2$, $\bar{y} = y/t^3$)

$$E_\lambda: \quad \bar{y}^2 + s\bar{y} = \bar{x}^3 + (p_2 s^2 + s^3(\ldots))\bar{x} + q_2 s^4 + s^5(\ldots).$$

Similarly, the coordinates of the sections become

$$\bar{\text{x}} = bs + as^2, \quad \bar{\text{y}} = es + ds^2 + cs^3.$$

Note that the sections meet the cusp of the special fibre. Following Tate's algorithm Sect. 5.8, we blow up the cusp to obtain the following strict transform of E_λ (or rather of S_λ) in the affine chart $\bar{x} = sx'$, $\bar{y} = sy'$:

$$E_\lambda: \quad y'^2 + y' = sx'^3 + (p_2 s + \ldots)x' + q_2 s^2 + \ldots$$

It follows that the two non-identity fibre components are given as

$$\Theta_1 = \{s = y' = 0\}, \quad \Theta_2 = \{s = y' + 1 = 0\}.$$

As the sections' shape transforms to

$$x' = b + as, \quad y' = e + ds + cs^2,$$

we infer that either $e = 0$ or $e = -1$, corresponding to the fibre component met by the respective section. As a consequence, all Galois conjugates of a given section meet the same fibre component while their inverses meet the opposite component. The claim of Lemma 9.19 thus follows. □

In practice, one could set out to compute the Galois representation ϱ_λ directly, say by explicit computations with the minimal sections. Essentially, this is what we will do in the next section, but we will obtain a much cleaner and more streamlined picture if we phrase everything through the specialization homomorphism (compare Example 9.47, where we will (have to) do without).

9.4.3 Generic Galois Representation and Algebraic Equation of Additive Type E_r ($r = 6, 7, 8$)

Now we assume that λ is *generic*, i.e. that the r parameters p_i, q_j are algebraically independent over $\mathbb{Q}$ (though most of the arguments carry over to fields of characteristic $\neq 2, 3$, but even then there are certain subtleties to consider, cf. Remark 9.22). Then the condition (9.13) holds as one easily checks, and the elliptic curves from (9.9)–(9.11) have Mordell–Weil lattices $M \cong E_r^\vee$ ($r = 8, 7, 6$). We can thus study the Galois representation on M. In order to relate this to algebraic equations, we specify how to define the polynomial $\Phi(X)$ from (9.2) in the given setting.

Definition 9.20 Consider the following set I as in (9.3):

- for $r = 8$ or 7, we take the set of minimal vectors in $M = E_r^\vee$; thus $I = I_\mu$ for $\mu = 2$ or $3/2$, and $N = \#I = 240$ or 56 accordingly.
- for $r = 6$, we take $I = I_\mu^\pm$, either one of the two orbits of $W(E_6)$ in I_μ ($\mu = 4/3$, $N = \#I = 27$).

The *universal polynomial of additive type E_r* is defined as a special case of (9.2), for the specialization map $s = sp'_\infty$ defined in (9.12) and the set I defined above:

$$\Phi_{E_r}(X, \lambda) = \Phi(E_r/\mathbb{Q}(\lambda)(t), sp'_\infty, I; X) = \prod_{P \in I}(X - sp'_\infty(P)) \in \mathbb{Q}(\lambda)[X]. \tag{9.15}$$

It has degree $N = 27, 56$ or 240 for $r = 6, 7$ or 8.

By the choice of I, it will be natural to refer to $\Phi_{E_r}(X, \lambda)$ also as the *algebraic equation of the minimal sections* of E_r (up to ± 1, i.e. in a $W(E_r)$-orbit, if $r = 6$).

This puts us in the position to state our main theorem. In each single point, it parallels Theorem 9.15 for the hexagonal case, although the statements themselves become substantially more complicated, and more interesting.

Theorem 9.21 *Fix $r = 6, 7$ or 8, and let λ be generic over $\mathbb{Q}$. Then:*

(i) The image of the Galois representation ϱ_λ is the full Weyl group $W(E_r)$:

$$\mathrm{Im}(\varrho_\lambda) = W(E_r).$$

(ii) Let $\mathcal{K}_\lambda/k_0$ be the splitting field of $E_\lambda/k_0(t)$, and let $\{P_1, \ldots, P_r\}$ be a basis of $E(k(t)) \simeq E_r^\vee$ consisting of minimal vectors. Further, let $u_i = sp'_\infty(P_i) \in \mathcal{K}_\lambda \subset k$. Then $u_1, \ldots, u_r$ are algebraically independent over $\mathbb{Q}$, and $\mathcal{K}_\lambda$ is a purely transcendental extension of $\mathbb{Q}$ generated by $u_1, \ldots, u_r$:

$$\mathcal{K}_\lambda = k_0(u_1, \ldots, u_r) = \mathbb{Q}(u_1, \ldots, u_r). \tag{9.16}$$

Thus we have a Galois extension between two purely transcendental extensions of $\mathbb{Q}$ with Galois group $W(E_r)$:

$$\mathrm{Gal}(\mathbb{Q}(u_1, \ldots, u_r)/\mathbb{Q}(p_i, q_j)) = W(E_r). \tag{9.17}$$

(iii) $W(E_r)$ acts on the vector space $\sum \mathbb{Q}u_i$ and hence on the polynomial ring $\mathbb{Q}[u_1, \ldots, u_r]$. The subring of invariants is the polynomial ring

$$\mathbb{Q}[u_1, \ldots, u_r]^{W(E_r)} = \mathbb{Q}[p_i, q_j] \tag{9.18}$$

generated by $\{p_i, q_j\}$. In particular, p_i and q_j form a set of the fundamental polynomial invariants of the Weyl group $W(E_r)$ and we have

$$p_i \text{ or } q_j = J_d(u_1, \ldots, u_r), \tag{9.19}$$

matched according to their weights or degrees. Here p_i, q_j have weight recorded in Table 9.1 and each J_d denotes a $W(E_r)$-invariant polynomial of degree d, with d running over the degrees from Theorem 2.44:

$$\begin{cases} 2, 5, 6, 8, 9, 12 & (r = 6), \\ 2, 6, 8, 10, 12, 14, 18 & (r = 7), \\ 2, 8, 12, 14, 18, 20, 24, 30 & (r = 8). \end{cases}$$

(iv) The universal polynomial of additive type E_r, $\Phi_{E_r}(X, \lambda)$, is a monic irreducible polynomial in X with coefficients in the polynomial ring $\mathbb{Z}[\lambda] = \mathbb{Z}[p_i, q_j]$, and the splitting field of $\Phi_{E_r}(X, \lambda)$ over $\mathbb{Q}(\lambda)$ is isomorphic to the splitting field $\mathcal{K}_\lambda$ in (ii).

Proof of Theorem 9.21. Let us briefly sketch the idea of proof of the above theorem (for details, we refer to Shioda [Shio91c, Sects. 8–10]. By the results obtained in Sects. 7.4, 7.5 (Theorem 7.4, ...), we know the structure of the Mordell–Weil lattice $M_\lambda := E_\lambda(k(t))$. For $r = 6, 7, 8$, M_λ is isometric to the lattice $E_6^\vee$, $E_7^\vee$ or E_8, respectively, and the number of minimal sections N (and the minimal height μ) are given as follows:

$$N = \begin{cases} 54 \\ 56 \\ 240 \end{cases} \quad \mu = \begin{cases} 4/3 & (r = 6), \\ 3/2 & (r = 7), \\ 2 & (r = 8). \end{cases}$$

The central idea is to consider, for each $r = 6, 7, 8$, the algebraic equation of the minimal sections of M_λ (up to ± 1, i.e. in a $W(E_r)$-orbit, if $r = 6$). Recall from Theorem 7.12 that the minimal sections take a prescribed shape. Hence the idea of the algebraic equation of the minimal sections, as captured in $\Phi_{E_r}(X, \lambda)$, can be made precise systematically as follows.

Step 1. Minimal Sections and Specialization Map

For $\boldsymbol{r} = \mathbf{8}$, there exist 240 minimal sections $P = (\mathrm{x}, \mathrm{y})$ of height 2, which are of the form

$$P:\ \mathrm{x} = \frac{1}{u^2}t^2 + at + b, \quad \mathrm{y} = \frac{1}{u^3}t^3 + ct^2 + dt + e \quad (u \neq 0). \tag{9.20}$$

Then Lemmas 9.8, 9.10 *(i)* give

$$sp_\infty(P) = u.$$

For $\boldsymbol{r} = \mathbf{7}$, there exist $N = 56$ minimal sections $P = (\mathrm{x}, \mathrm{y})$ of height 3/2, which are of the form

$$P:\ \mathrm{x} = at + b, \quad \mathrm{y} = ct^2 + dt + e. \tag{9.21}$$

Note that $2P$ and the difference $P_1 - P_2$ of two such sections belong to the narrow Mordell–Weil lattice $M^0 := E_\lambda(k(t))^0$, and, a direct computation (using the addition theorem) shows that

$$sp_\infty(2P) = -2c, \qquad sp_\infty(P_1 - P_2) = -(c_1 - c_2),$$

where c, c_1, c_2 denote the t^2-coefficient of the y-coordinate of P, P_1, P_2 respectively. It follows that

$$sp'_\infty(P) = -c.$$

For $\boldsymbol{r} = \mathbf{6}$, there exist $N = 54$ minimal sections $P = (\mathrm{x}, \mathrm{y})$ of height 4/3, and half of them are linear by the proof of Lemma 9.19:

$$P: \ \mathrm{x} = at + b, \quad \mathrm{y} = dt + e \tag{9.22}$$

(or more precisely, they are polynomial of degree at most one). Recall that by Lemma 9.19 the *linear sections* exactly form one $W(E_6)$-orbit of the minimal sections of the Mordell–Weil lattice $M = E_6^\vee$ as required for Definition 9.20. Hence we shall define I as the set of linear sections in what follows.

This time, $3P$ and the difference $P_1 - P_2$ of two such sections belong to the narrow Mordell–Weil lattice $M^0 := E_\lambda(k(t))^0$ (since all linear sections meet the same fibre component Θ_1 at ∞ as entered instrumentally in the proof of Lemma 9.19). A direct computation reveals:

$$sp_\infty(3P) = -3a, \qquad sp_\infty(P_1 - P_2) = a_2 - a_1. \tag{9.23}$$

Hence, by defining the specialization maps $sp'_\infty : M \to k$ by

$$sp'_\infty(P) = \begin{cases} u & (r = 8), \\ -c & (r = 7), \\ -a & (r = 6), \end{cases}$$

we arrive at the required form of the universal polynomial for additive type E_r for each r—which essentially only depends on the coefficients u resp. c resp. a of the minimal sections!

Step 2. Universal Polynomial

Recall that the universal polynomial encodes essentially the coefficients u_i resp. c_i resp. a_i of the minimal sections P_i. Hence we can develop a second expression of the universal polynomial for each $r = 6, 7, 8$ by an *elimination* argument solving explicitly for the specified coefficients. We lay out the argument in full detail for $r = 6$ in what follows.

In case $\boldsymbol{r} = \mathbf{6}$, substitute (9.22) into the Weierstrass equation (9.11). We have formally a polynomial of t of degree 3, whose coefficients should be identically zero for P to be a section of (9.11). Thus we obtain four polynomial relations among the coefficients a, b, d, e:

$$\begin{cases} d &= a^3 + ap_2 \\ e &= 3a^2b - d^2 + ap_1 + bp_2 + q_2 \\ 2de &= 3ab^2 + ap_0 + bp_1 + q_1 \\ e^2 &= b^3 + bp_0 + q_0 \end{cases} \tag{9.24}$$

Since d, e are precisely encoded in the first two relations, we can substitute them into the remaining two relations. We obtain two polynomial relations of b of degree 2 and 3:

$$\varphi_1(b) := 3ab^2 + \ldots = 0, \quad \varphi_2(b) := b^3 + \ldots = 0 \tag{9.25}$$

with (unwritten) coefficients in $\mathbb{Z}[\lambda][a] = \mathbb{Z}[p_i, q_j][a]$.

If $\boldsymbol{a = 0}$, then (9.24) gives a relation between the p_i, q_j (since $d = 0$ and b, e can easily be eliminated), contradicting the genericity assumption in Theorem 9.21.

Hence we can assume that $\boldsymbol{a \neq 0}$. Then we can eliminate b, e.g. by taking the resultant of $\varphi_1(b), \varphi_2(b)$ with respect to b, to obtain a monic polynomial of degree 27 in a with coefficients in $\mathbb{Z}[\lambda]$:

$$\Phi(a) = a^{27} + \ldots \tag{9.26}$$

More precisely, let us write down a few terms which are essential for the following argument:

$$\begin{aligned} &\Phi(X) = \\ &X^{27} + 12p_2X^{25} + 60p_2^2X^{23} - 12p_1X^{22} + (24q_2 + 168p_2^3)X^{21} - 84p_1p_2X^{20} \\ &\quad + (30p_0 + 294p_2^4 + 132p_2q_2)X^{19} - (84q_1 + 252p_1p_2^2)X^{18} + \ldots \\ &\quad + (270q_0 + 38p_1^2p_2 + 298p_0p_2^2 + 252p_2^6 + 300p_2^3q_2 - 78q_2^2)X^{15} + \ldots \end{aligned} \tag{9.27}$$

Similarly, in case $\boldsymbol{r = 7}$, substitute (9.21) into (9.10), and argue as above to derive relations among a, b, c, d, e. Then eliminating a, b, e, d we obtain a monic relation of c with coefficients in $\mathbb{Z}[\lambda]$ of degree 56.

In case $\boldsymbol{r = 8}$, substitute (9.22) into the Weierstrass equation (9.9), and we obtain polynomial relations among the coefficients $a, \ldots, u$. Then the successive elimination leads to a monic relation of u with coefficients in $\mathbb{Z}[\lambda]$ of degree 240.

Step 3. Invariants

Now we prove the claim *(iii)* of Theorem 9.21.

Look at the polynomial $\Phi(X)$, (9.27) in case $r = 6$. Observe that each coefficient of X^m, denoted below by $(-1)^w\epsilon_w$, such that

$$m + w = 27, \quad w \in \{2, 5, 6, 8, 9, 12\},$$

contains a term which is either p_i or q_j times a non-zero constant. Namely we have:

$$\begin{cases} \epsilon_2 &= 12p_2 \\ -\epsilon_5 &= -12p_1 \\ \epsilon_6 &= 24q_2 + 168p_2^3 \\ \epsilon_8 &= 30p_0 + 294p_2^4 + 132p_2q_2 \\ -\epsilon_9 &= -84q_1 - 252p_1p_2^2 \\ \epsilon_{12} &= 270q_0 + 38p_1^2p_2 + 298p_0p_2^2 + 252p_2^6 + 300p_2^3q_2 - 78q_2^2 \end{cases} \tag{9.28}$$

Therefore each of p_i and q_j can be expressed in terms of $\{\epsilon_2, \ldots, \epsilon_{12}\}$ as follows:

$$\begin{cases} p_2 = \epsilon_2/12 \\ p_1 = \epsilon_5/12 \\ q_2 = (\epsilon_6 - 168p_2^3)/24 \\ p_0 = (\epsilon_8 - 294p_2^4 - 132p_2q_2)/30 \\ q_1 = (\epsilon_9 - 252p_1p_2^2)/84 \\ q_0 = (\epsilon_{12} - (38p_1^2p_2 + 298p_0p_2^2 + 252p_2^6 + 300p_2^3q_2 - 78q_2^2))/270 \end{cases} \tag{9.29}$$

By definition, ϵ_ν is the ν-th elementary symmetric polynomial of the 27 roots a_i $(1 \le i \le 27)$ of $\Phi(X)$. Here each root a_j can be expressed as a polynomial in $\{a_i\ (1 \le i \le 6)\}$ by a suitable choice of ordering relative to the standard generators of $E_6^\vee$ from Theorem 2.34. (Essentially, this holds since the specialization map sp'_∞ is a homomorphism.) Usually, we would take $a_i \leftrightarrow \alpha_i \leftrightarrow u_i$ (cf. Sect. 10.2.3). In any case, this gives the explicit formulas (9.19) for $r = 6$.

In the same way, looking at the two expressions of the universal polynomial $\Phi_{E_r}(X)$ for $\boldsymbol{r} = \mathbf{7, 8}$ and comparing the coefficients of X^d, where d runs over the set of degrees in (9.20), we deduce the explicit formulas of fundamental invariants (9.19). (For space reasons, we omit the explicit formulas for J_d here, and refer to [Shio91c, Formulas (2.21), (2.25)].) It follows that the coefficients $\{p_i, q_j\}$ in the Weierstrass form, in each case of r, form a set of homogeneous generators of the graded ring $\mathbb{Q}[u_1, \ldots, u_r]^{W(E_r)}$ of $W(E_r)$-invariants. This proves (9.18) and hence the claim *(iii)*.

Step 4. Galois Extension

Finally, taking the field of fractions of both sides of (9.18), we see that the subfield of invariants $\mathbb{Q}(u_1, \ldots, u_r)^{W(E_r)}$ in $\mathbb{Q}(u_1, \ldots, u_r)$ coincides with $k_0 = \mathbb{Q}(p_i, q_j)$. By Galois theory, we conclude that $\mathbb{Q}(u_1, \ldots, u_r)$ is a Galois extension of k_0 with Galois group $W(E_r)$. This proves (9.17), and hence Theorem 9.21 *(i)*. The remaining assertions in the theorem follow easily from the results mentioned above. This concludes the outline of the proof of Theorem 9.21. □

Remark 9.22 Actually, the proof shows that in the above theorem, the ground field $\mathbb{Q}$ can be replaced by any field κ, provided that its characteristic p is different from a few primes which enter in the denominators of the expression of p_i, q_j as fundamental invariants of the Weyl groups, e.g. $p > 7$ suffices in case $r = 6$. In fact, in Step 3 above, the coefficients $\{12, -12, 24, 30, -84, 270\}$ in (9.28) have the prime factors $\{2, 3, 5, 7\}$ only. In particular, all the theorems in this section for $r = 6$ (and the applications to follow) hold true even when $\mathbb{Q}$ is replaced by any field of characteristic zero (in particular by number fields) or of characteristic $p > 7$. Similar remarks apply to the case $r = 7$ or 8 (cf. [Shio91c]).

9.5 Application to Number Theory

Theorem 9.21 has applications to various subjects: number theory, algebraic geometry, singularity theory and others (cf. [Shio91d]). Before discussing some of them, it might be useful to note a remarkable analogy.

9.5.1 *Exceptional Type* E_r $(r = 6, 7, 8)$ *Versus Classical Type* A_{n-1}

Let us compare the results of the previous section with the classical theory of the *generic algebraic equation* of degree n. Let

$$F(X, \epsilon) = X^n + \epsilon_2 X^{n-2} + \ldots + (-1)^n \epsilon_n \quad (\epsilon_i : \text{alg.indep.}/\mathbb{Q})$$

be such an equation, normalized so that the sum ϵ_1 of the roots $x_1, \ldots, x_n$ is 0. Let $\mathcal{K}$ be the minimal splitting field of $F(X, \epsilon)$ over $k_0 = \mathbb{Q}(\epsilon_2, \ldots, \epsilon_n)$. Then, as everyone learns in the first course of Galois Theory, we have:

(i) $\mathcal{K} = k_0(x_1, \ldots, x_n) = \mathbb{Q}(x_2, \ldots, x_n)$.
(ii) $\mathrm{Gal}(\mathbb{Q}(x_2, \ldots, x_n)/\mathbb{Q}(\epsilon_2, \ldots, \epsilon_n)) = \mathcal{S}_n$.
(iii) $\mathbb{Q}[x_2, \ldots, x_n]^{\mathcal{S}_n} = \mathbb{Q}[\epsilon_2, \ldots, \epsilon_n]$.
(iv) ϵ_i = the fundamental invariants of $\mathcal{S}_n$.

If we recall the elementary fact that the symmetric group on n letters $\mathcal{S}_n$ is isomorphic to the Weyl group $W(A_{n-1})$, then Theorem 9.21 (based on the theory of Mordell–Weil lattices) gives a complete analogy for the exceptional type E_r $(r = 6, 7, 8)$ of what the theory of the generic algebraic equation of degree n does to the classical type A_{n-1} (compare Theorem 9.15 where we saw this for A_2 and the hexagonal elliptic curves).

The latter theory has a standard application to number theory related to Problem 9.6:

(a) *big Galois*: construction of Galois extensions of $\mathbb{Q}$ with Galois group the full symmetric group $\mathcal{S}_n$, via specialization *downstairs*, i.e. via $(\epsilon_i) \to (\epsilon_i^0) \in \mathbb{Q}^{n-1}$ and Hilbert's irreducibility theorem (cf. [Ser70, Lan83]);
(b) *small Galois*: construction of an algebraic equation with the prescribed roots, via specialization *upstairs* $(x_i) \to (x_i^0) \in \mathbb{Q}^n$, which is indeed trivial.

In what follows, we describe a similar application of our theory in order to find rather satisfactory answers to Problem 9.6. First it will be shown immediately below in Theorem 9.25 that the analogue of (b), *small Galois*, for our theory is *not* trivial at all; indeed this will lead to a systematic construction of elliptic curves (over a rational function field) of rank as high as 8, together with a full set of generators of the Mordell–Weil group (which will feature as a key ingredient in the explicit construction of elliptic curves of even higher rank over $\mathbb{Q}(t)$, see Sect. 13.1.1, for instance).

Next we discuss the analogue of (a), *big Galois*, which systematically produces every Galois extension of $\mathbb{Q}$ with Galois group $W(E_r)$ $(r = 6, 7, 8)$, the Weyl groups, of respective orders 51840 $(r = 6)$, 2903040 $(r = 7)$, 696729600 $(r = 8)$, from our Mordell–Weil lattice setting just by specializing the parameters downstairs, i.e. via $\lambda \to \lambda^0 \in \mathbb{Q}^r$ and applying Hilbert's irreducibility theorem (which we will also make explicit). Indeed, this method also allows one to produce some concrete numerical examples.

9.5.2 Systematic Construction of Elliptic Curves over $\mathbb{Q}(t)$ of Rank $r = 6, 7, 8$, with Prescribed Generators

Considering the elliptic curves (9.9)–(9.11) over the splitting field $\mathcal{K}_\lambda(t) = \mathbb{Q}(u_1, \ldots, u_r)(t)$, we can reformulate Theorem 9.21 in the following way.

Let

$$\psi : \mathbb{A}^r \to \mathbb{A}^r, \quad \mathbf{u} = (u_1, \ldots, u_r) \mapsto \lambda = (p_i, q_j) := \psi(\mathbf{u}) \tag{9.30}$$

be the map defined by (9.19) in Theorem 9.21 *(iii)*, namely by the explicit formulas (9.29) in case $r = 6$ and by [Shio91c, Formulas (2.21), (2.25)] in case $r = 7, 8$. It is clearly a morphism of the affine space $\mathbb{A}^r_u$ with coordinates $\mathbf{u} = (u_i)$ to the affine space $\mathbb{A}^r_\lambda$ with coordinates $\lambda = (p_i, q_j)$.

Theorem 9.23 *Consider the family of rational elliptic surfaces $S_\mathbf{u}$ obtained by applying the base change ψ to the elliptic curves E_λ from (9.9)–(9.11). Then this is a smooth family of Kodaira–Néron models (i.e. smooth projective elliptic surfaces) parametrized by $\mathbf{u} \in A^r$.*

Remark 9.24 The theorem gives an analogue of Brieskorn's simultaneous resolution of the ADE-singularity in the Weierstrass model. Here the Galois representation (9.14) plays the role of the monodromy representation in the deformation theory of isolated singularities due to Brieskorn and others. In fact, if we set $\lambda = 0$ (i.e. all $p_i = 0, q_j = 0$), then the Weierstrass equations (9.9)–(9.11), reduce to

$$y^2 = x^3 + t^5, \quad y^2 = x^3 + xt^3, \quad \text{and} \quad y^2 + t^2 y = x^3$$

which define exactly the E_8-, E_7- or E_6-singularity at the origin $x = y = t = 0$ (in characteristic zero or ≥ 7, see Sect. 4.10). Each Weierstrass equation (9.9)–(9.11), in case $k = \mathbb{C}$, locally is known as the semi-universal deformation of the E_r-singularity.

All statements of the proposition should be clear. For Remark 9.24, compare [Bri68, Bri71, Slo80, S-B19]; see also [Shio91b] for the first connection of Mordell–Weil lattices to this subject. We shall now describe the sections of these families.

Theorem 9.25 *Fix $r = 8, 7, 6$ and let*

$$\mathbf{u} = (u_1, \ldots, u_r), \quad \mathcal{K} = \mathbb{Q}(\mathbf{u}) = \mathbb{Q}(u_1, \ldots, u_r),$$

where $u_1, \ldots, u_r$ are algebraically independent elements over $\mathbb{Q}$. Define

$$\lambda = (p_i, q_j) := \psi(\mathbf{u})$$

and the elliptic curve $E_{\mathbf{u}} = E_\lambda$ over $\mathbb{Q}(\mathbf{u})(t)$ by (9.9), (9.10) *or* (9.11), *with associated elliptic surface $S_{\mathbf{u}}$ defined over $\mathbb{Q}(\mathbf{u})$. Then the Mordell–Weil group (lattice) of $(\mathbb{Q}(\mathbf{u})(t))$-rational points of $E_{\mathbf{u}}$ (or $(\mathbb{Q}(\mathbf{u}))$-rational sections of $S_{\mathbf{u}}$),*

$$E_{\mathbf{u}}(\mathbb{Q}(u_1, \ldots, u_r)(t)),$$

is of rank $r = 8, 7$ or 6, and it has a basis

$$\{P_1, \ldots, P_r\} \text{ such that } sp'_\infty(P_i) = u_i \tag{9.31}$$

which consists of minimal sections of $S_{\mathbf{u}}$ of height 2 (resp. 3/2, or 4/3) for $r = 8$ (resp. 7 or 6). More explicitly, we have $P_i = (\mathrm{x}, \mathrm{y})$ where x, y are of the form (9.20), *i.e. $\mathrm{x} = gt^2 + at + b$ and $\mathrm{y} = ht^3 + ct^2 + dt + e$ with the following properties:*

1. *If $r = 8$, then we have*

$$u_i = sp_\infty(P_i) = g/h, \ g = u_i^{-2}, \quad h = u_i^{-3}$$

and $a, \ldots, e$ are rational functions of u_i with coefficients in $\mathbb{Q}(p_0, \ldots, q_3)$ which belong at the same time to the subring $\mathbb{Q}[u_1, \ldots, u_8][u_i^{-1}]$.

2. *If $r = 7$, then we have $P_i : \ \mathrm{x} = at + b$ and $\mathrm{y} = ct^2 + dt + e$, and*

$$u_i = sp'_\infty(P_i) = -c$$

and

$$a, b, d, e \in \mathbb{Q}(p_0, \ldots, q_4)[u_i] \cap \mathbb{Q}[u_1, \ldots, u_7].$$

3. *If* $r = 6$, *then we have* $P_i : \ \mathrm{x} = at + b$ *and* $\mathrm{y} = dt + e$, *and*

$$u_i = sp'_\infty(P_i) = -a$$

and

$$b, d, e \in \mathbb{Q}[u_1, \ldots, u_6] \cap \mathbb{Q}(p_0, \ldots, q_2)(u_i).$$

Proof This follows almost immediately from the proof of Theorem 9.21. For example, for $r = 6$, we only have to verify, in addition, that

$$b, d, e \in \mathbb{Q}[u_1, \ldots, u_6] \cap \mathbb{Q}(p_0, \ldots, q_2)(a), \ a = u_i.$$

This can be shown as follows by a closer look at the elimination process in the proof of Theorem 9.21. Note first that b is equal to a rational function of $a = u_i$ with coefficients in $\mathbb{Q}(\lambda)(a) = \mathbb{Q}(p_0, \ldots, q_2)(a)$ for $\mathbf{u}$ generic, since b satisfies a linear equation derived from the two equations of (9.25), the remainder of φ_2 with respect to b when divided by φ_1 (this works unless $a = 0$ or both coefficients of the remainder vanish, but either condition gives an equation in the u_i, thus contradicting their algebraic independence).

Next we note that $a = -u_i$ is integral over $\mathbb{Q}[\lambda] = \mathbb{Q}[p_0, \ldots, q_2]$, since it satisfies the monic equation (9.26) over $\mathbb{Q}[\lambda]$. Then b is integral over $\mathbb{Q}[\lambda][a]$, which follows from the cubic relation of b in (9.25). Thus b is an element of

$$\mathbb{Q}(\lambda)(a) \subset \mathbb{Q}(u_1, \ldots, u_6),$$

which is at the same time integral over $\mathbb{Q}[u_1, \ldots, u_6]$. Hence $b \in \mathbb{Q}[u_1, \ldots, u_6]$ by normality. On the other hand, we have $d \in \mathbb{Q}[\lambda][a]$ and $e \in \mathbb{Q}[\lambda][a, b]$ by (9.24), and hence the claim about d, e follows. The case $r = 7$ or 8 can be treated similarly (cf. [Shio91c]). □

As a direct consequence, we obtain a systematic construction of elliptic curves over $\mathbb{Q}(t)$ of rank $r = 6, 7, 8$, together with a set of r generators, by specializing the parameters $u_1, \ldots, u_r$ to rational numbers $u_1^0, \ldots, u_r^0$ in such a way that the assumption (9.13) is satisfied. Of course, this assumption can in principle be checked explicitly using the discriminant Δ from (3.6) of the given Weierstrass form, but presently we can give a much preferable conceptual characterization thanks to the key fact that the parameters $u_1, \ldots, u_r$ of $\mathbb{Q}(\mathbf{u})$ in Theorem 9.25 exactly feature as the images of the sections $P_1, \ldots, P_r$ under the specialization map sp'_∞ by (9.31).

Definition 9.26 With the notation of Theorem 2.3.4 for positive roots of E_r, we set, with i, j, k running over $\{1, 2, \ldots, r\}$ for each $r = 6, 7, 8$,

$$\Pi(E_6) = \{\alpha_{ij}\ (i < j),\ \beta_{ijk}\ (i < j < k),\ -v_0\},$$
$$\Pi(E_7) = \{\alpha_{ij}\ (i < j),\ \beta_{ijk}\ (i < j < k),\ u_i'\},$$
$$\Pi(E_8) = \{\alpha_{ij}\ (i < j),\ \beta_{ijk}\ (i < j < k),\ -\gamma_{ij}\ (i < j),\ -u_i\},$$

where each element α_{ij}, β_{ijk}, etc. is given in terms of the "standard generators" $\{u_1, \ldots, u_r, v_0\}$ of $E_r^\vee$. We define $\Pi(E_r)|_{\mathbf{u}^0}$ as the set $\Pi(E_r)$ evaluated with $\mathbf{u}$ specialized to $\mathbf{u}^0$.

The decisive properties of specialization are captured by the following theorem. For the background, the reader is referred to Sect. 9.5.3, while the proofs will only be carried out in the multiplicative case in Sect. 10.3.1.

Theorem 9.27 *In the setting of Theorem 9.25, specialize the parameters $u_1, \ldots, u_r$ to rational numbers $u_1^0, \ldots, u_r^0$ to obtain the resulting elliptic curve $E_{\mathbf{u}^0}$ over $\mathbb{Q}(t)$. Then the Mordell–Weil group $E_{\mathbf{u}^0}(\mathbb{Q}(t))$ has rank r if and only if the following non-degeneracy condition holds:*

$$\Pi(E_r)|_{\mathbf{u}^0} \text{ does not contain } 0. \tag{9.32}$$

Further, in this case, the $\mathbb{Q}(t)$-rational points $\{P_i(\mathbf{u}^0);\ i = 1, \ldots, r\}$ form a minimal set of generators of $E_{\mathbf{u}^0}(\mathbb{Q}(t)) \simeq E_r^\vee$.

Remark 9.28 The set of standard generators of $E_r^\vee$ consists of $r + 1$ elements, of which the r elements u_i are minimal vectors, while v_0 is not. One can replace it by a set of "minimal generators" of $E_r^\vee$ which consists of r minimal vectors. For example, for each $r = 6, 7, 8$, consider the set of r elements:

$$\{u_1, \ldots, u_{r-1}, \gamma_{12}\} \quad (\gamma_{12} = v_0 - u_1 - u_2), \tag{9.33}$$

which are minimal vectors for each r and by which the standard generators are expressible as follows:

$$v_0 = u_1 + u_2 + \gamma_{12}, \quad u_r = -(u_1 + \cdots + u_{r-1}) + 3v_0.$$

Thus we can interchangeably use the minimal generators (as in the statement of Theorem 9.21) or the standard generators (as in the description of the (positive) roots as above or for the proof (Sect. 10.2.4) and various applications (e.g. Sects. 10.2.6 or 10.2.7, 10.5.6).

9.5.3 Non-degeneracy Condition

We postpone the detailed discussion of the non-degeneracy conditions of the Mordell–Weil lattice, such as (9.32), till Sect. 10.3.1, where we will treat it in the case of multiplicative excellent families, but the argument is formally the same as the present case of additive excellent families. Here we state the relevant results for additive excellent families without proof, see Theorem 10.21 for details.

Theorem 9.29 *Fix $r = 6, 7$ or 8. Consider the morphism ψ defined by (9.30). Then it is a finite Galois covering with Galois group $W(E_r)$, which is unramified outside the discriminant locus $\{D^2 = 0\}$, where*

$$D = \prod_{\alpha>0} \alpha$$

is the basic anti-invariant (see (2.28) and (2.29)). The morphism ψ induces an isomorphism of the quotient variety $\mathbb{A}^r_u / W(E_r)$ onto $\mathbb{A}^r_\lambda$.

Associated with α, there is a root Q_α in the generic (narrow) Mordell–Weil lattice. This relates to the specialization at some $\mathbf{u}$ as follows:

Lemma 9.30 *Let us denote by $\alpha|_\mathbf{u}$ the linear form corresponding to a root α evaluated at $\mathbf{u}$. Then we have, for any $\mathbf{u}$,*

$$\alpha|_\mathbf{u} = sp_\infty(Q_\alpha) \tag{9.34}$$

(up to a fixed non-zero constant factor).

Corollary 9.31 *For any specialized $\mathbf{u}^0 = (u_i) \in \mathbb{A}^r$, $S_{\mathbf{u}^0}$ is non-degenerate (i.e. it has the Mordell–Weil lattice $\simeq E_r^\vee$) if and only if $\alpha|_{\mathbf{u}^0} \neq 0$ for all roots α. Equivalently, this is the case if and only if the discriminant $\Delta = D^2$ does not vanish.*

For example, to define the discriminant $\Delta = D^2$ of E_r in the present context, fix $r = 6, 7, 8$ and recall the following formulas for the positive roots of E_r from Theorem 2.3.4:

$$\begin{cases} \alpha_{ij} &= u_i - u_j \quad (0 < i < j \le r), \\ \beta_{ijk} &= v_0 - (u_i + u_j + u_k) \quad (0 < i < j < k \le r), \\ \gamma_{ij} &= v_0 - u_i - u_j \quad (0 < i < j \le r), \\ u_i' &= u_i - v_0 \quad (0 < i \le r), \end{cases} \tag{9.35}$$

where $v_0 = (\sum_{i=1}^r u_i)/3$.

Definition 9.32 (*Discriminant of E_r*) With the above notation, we define the discriminant of E_r by

$$\Delta_r(\mathbf{u}) := D_r(\mathbf{u})^2, \tag{9.36}$$

where

$$D_r(\mathbf{u}) = \prod_{i<j} \alpha_{ij} \prod_{i<j<k} \beta_{ijk} \cdot \begin{cases} (-v_0) & (r = 6), \\ \prod_{i=1}^{r} u_i' & (r = 7), \\ \prod_{i<j}(-\gamma_{ij}) \prod_{i=1}^{r}(-u_i) & (r = 8). \end{cases}$$

Remark 9.33 For the other root lattices, we can define a discriminant exactly in the same manner, building on Propositions 2.30 or 2.31. For instance, the root lattices A_n directly lead to the discriminant

$$\Delta_{A_n}(\mathbf{u}) = \prod_{0<i<j\leq n+1} (u_i - u_j)^2.$$

Actually, we find it more useful to consider the set of positive roots itself, denoted by $\Pi_{\mathbf{u}}$, rather than the discriminant $\Delta(\mathbf{u})$. The non-degeneracy criterion can now be stated as

$$S_{\mathbf{u}} \text{ is non-degenerate} \iff \Pi_{\mathbf{u}} \text{ does not contain } 0.$$

We conclude this section by collecting a few corollaries and comments concerning the specialization upstairs. The first application is of rather classical, but implicit nature. It relates to the standard height on $\mathbb{P}^1(\mathbb{Q})$ and to the canonical height $\langle \cdot, \cdot \rangle_{\text{can}}$ on the Mordell–Weil group of an elliptic curve over $\mathbb{Q}$ (cf. [Tat83]).

Corollary 9.34 *Given $E/\mathbb{Q}(t)$ as in Corollary 9.31, we have an infinite family $\{E^{(\tau)} | \tau \in \mathbb{Q} - \Sigma\}$ ($\sharp\Sigma < \infty$) of elliptic curves over $\mathbb{Q}$ of rank $\geq r$, with r independent points $P_i^{(\tau)}$, by specializing $t \longrightarrow \tau \in \mathbb{Q}$. Furthermore*

$$\det(\langle P_i^{(\tau)}, P_j^{(\tau)} \rangle_{\text{can}} / h(\tau)) \longrightarrow 1/2^r d \quad \text{as}\, h(\tau) \longrightarrow \infty,$$

where $d = 1, 2$ or 3 for $r = 8, 7$ or 6.

This follows from a specialization result of Néron, Silverman ([Sil83]), Tate ([Tat83]) which will be discussed in more detail in Chap. 13 (see, for instance, Sect. 13.1.1).

We would like to highlight a subtle point which was rather surprising for us at first. Namely, for the singular fibres not to degenerate (the generality condition (9.13) which was the starting point of Sect. 9.5.3), we do not have to assume that the specialization map sp^0_∞ is injective on $\mathcal{R}(M^0_{\mathbf{u}})$, i.e. that $sp^0_\infty(P) \neq sp^0_\infty(Q)$ for all $P, Q \in \mathcal{R}(M^0_{\mathbf{u}})$. Indeed, there is a numerical example in [Shio91c, Sect. 3] comprising an elliptic curve $E/\mathbb{Q}(t)$ of MW rank $r = 8$ which is constructed from the data $u^0_i = 1$ for all $i = 1, \ldots, 8$, together with 8 $\mathbb{Q}(t)$-rational points $P_i = (\mathrm{x}, \mathrm{y})$ of the form

$$\mathrm{x} = t^2 + \ldots, \quad \mathrm{y} = t^3 + \ldots$$

for all i (i.e. all of them intersecting at the point $(\bar{x}, \bar{y}) = (1, 0)$ over $t = \infty$), which generate the full Mordell–Weil group $E(\mathbb{Q}(t)) \cong E_8$.

In order to wrap up this discussion of the specialization upstairs, somewhere in between the non-vanishing condition for Δ_r from Corollary 9.31 and the genericity assumption from Theorem 9.25, we shall now draw a nice connection to the notion of polynomial (or ∞-integral) sections.

Lemma 9.35 *Let $r = 6, 7, 8$ and K any field of characteristic zero, viewed as an extension of $\mathbb{Q}$. Let $u^0_1, \ldots, u^0_r \in K$ be $\mathbb{Q}$-linearly independent. Then $E_{\mathbf{u}^0}(K(t)) \cong E^\vee_r$ does not contain any polynomial sections other than the integral ones.*

Proof We will prove a substantially stronger statement, namely that no section P intersects the zero section O non-trivially at ∞. Assuming the contrary, we have

$$sp'_\infty(P) = sp'_\infty(O) = 0.$$

Since the $P_i(\mathbf{u}^0)$ generate $E_{\mathbf{u}^0}(K(t))$ (which is isometric to $E^\vee_r$ by the linear independence assumption by Corollary 9.31) we know that there are integers $n_i \in \mathbb{Z}$ such that

$$P = \sum_{i=1}^{r} n_i P_i(\mathbf{u}^0).$$

Hence we obtain a contradiction to the linear independence from spelling out the specialization map:

$$sp'_\infty(P) = \sum_{i=1}^{r} n_i \, sp'_\infty(P_i(\mathbf{u}^0)) = \sum_{i=1}^{r} n_i u^0_i.$$

This contradiction completes the proof of the statement. Applied to polynomial sections, Lemma 9.35 follows. □

Remark 9.36 The same argument is valid in characteristic p, i.e. with $\mathbb{F}_p$-linearly independent $u^0_1, \ldots, u^0_r$ in some extension K of $\mathbb{F}_p$, as soon as all rational expressions involved in $\psi(\mathbf{u}^0)$ und the coordinates of the sections $P_i(\mathbf{u}^0)$ are well-defined.

Example 9.37 Consider $E : y^2 = x^3 + t^5 + 1$ over $k(t)$, where $k = \bar{\mathbb{Q}}$ is the algebraic closure of $\mathbb{Q}$. The Mordell–Weil group $M = E(k(t))$ is E_8, since there are no reducible fibres. We claim that there are no polynomial sections other than the 240 integral sections of height 2 (i.e. the roots of M).

To see this, take and fix any section of height 2, $P = (\mathrm{x}, \mathrm{y}) \in M$, and set

$$u := sp_\infty(P) = \lim_{t\to\infty} \frac{\mathrm{x}t}{\mathrm{y}} \neq 0.$$

Now let σ be the cyclic automorphism of order 30 acting on the elliptic surface (the Kodaira–Néron model of $E/k(t)$) by

$$(x, y, t) \mapsto (\zeta_3 x, -y, \zeta_5 t),$$

where ζ_n denotes a primitive n-th root of unity. Then we have

$$P^\sigma = (\zeta_3 \mathrm{x}(\zeta_5 t), -\mathrm{y}(\zeta_5 t))$$

so that

$$sp_\infty(P^\sigma) = \zeta_{30} \cdot u,$$

where $\zeta_{30} = -\zeta_3\zeta_5^{-1}$. Applying the same argument to $P' = P^{\sigma^n}$ for $n = 1, 2, 3, \ldots$, we see that the image of the specialization homomorphism $sp_\infty : M \to k$ contains $\mathbb{Z}[\zeta_{30}] \cdot u$ for some $u \neq 0$. Since the latter $\mathbb{Z}$-module has rank $\varphi(30) = 8$, we conclude that $sp_\infty : M \to k$ is injective (remember that $rkM = rkE_8 = 8$). Hence the claim follows from Lemma 9.35.

Example 9.38 (*Hexagonal case cont'd*) Arguing analogously for the hexagonal case, we find that the elliptic curve (9.8) over a field k_0 of characteristic different from 3 admits no polynomial sections over $\bar{k}_0$ unless f factors over the base field k_0 or one of the minimal sections P_i has zero image under the specialization map sp'_∞. (Equivalently, the last condition can be phrased as follows: after the translation in x such that the x^2-coefficient of f vanishes, f has x as a factor).

9.5.4 Galois Extensions over $\mathbb{Q}$ with Galois Group $W(E_r)$ $(r = 6, 7, 8)$

Next we consider the specialization "downstairs":

$$\lambda \longrightarrow \lambda^0 \in \mathbb{Q}^r$$

(or in any other intermediate field $\mathbb{Q}(\lambda) \supseteq K \supseteq \mathbb{Q}$). We keep the same notation as before, except that we will allow ourselves to write λ for λ^0 (to ease the presentation). Note that there is little chance for confusion since our results will basically not rely on any specialization arguments. With this convention, the splitting field $\mathcal{K}_\lambda$ gives now a Galois extension $\mathcal{K}_\lambda/\mathbb{Q}$.

Theorem 9.39 *Fix $r = 6$, 7 or 8, and let*

$$\Lambda = \Lambda_r = \{\lambda = (p_i, q_j) \in \mathbb{Q}^r \mid \mathrm{Gal}(\mathcal{K}_\lambda/\mathbb{Q}) = W(E_r)\}.$$

Then Λ is a Zariski dense subset of $\mathbb{Q}^r$ (more precisely, it is the complement of a "thin set" in $\mathbb{Q}^r$ in the sense of [Ser70]). In particular, there exist infinitely many $\{\lambda\} \subset \Lambda$ such that the fields $\{\mathcal{K}_\lambda\}$ are linearly disjoint over $\mathbb{Q}$.

Proof Immediate from Theorem 9.21 (esp. (9.17)) by applying Hilbert's irreducibility theorem [Ser70, Chap. 10]. □

We note that, for each $\lambda \in \Lambda$, we have an elliptic curve $E_\lambda/\mathbb{Q}(t)$ with the Mordell–Weil lattice $E_\lambda(\bar{\mathbb{Q}}(t)) = E_\lambda(\mathcal{K}_\lambda(t)) \simeq E_r^\vee$ and $E_\lambda(\mathbb{Q}(t)) = \{O\}$, and a rational elliptic surface S_λ defined over $\mathbb{Q}$. The surface S_λ can be shown to have zeta function with factor the non-abelian Artin L-function $L(s, \varrho_\lambda, \mathcal{K}_\lambda/\mathbb{Q})$ (compare Sect. 10.5.2, especially Remark 10.52).

Theorem 9.40 *Every Galois extension of $\mathbb{Q}$ with Galois group $W(E_r)$ is obtained as $\mathcal{K}_\lambda$ for some $\lambda \in \Lambda$. In other words, every such extension arises from the Mordell–Weil lattice of the elliptic curve $E_\lambda/\mathbb{Q}(t)$ for some $\lambda \in \mathbb{Q}^r$.*

Proof Let $K/\mathbb{Q}$ be any extension with Galois group $W(E_r)$. Viewed as a $W(E_r)$-module, K is equivalent to the regular representation, and hence it contains a subspace, say U, isomorphic to $E_r \otimes \mathbb{Q}$ (a natural irreducible representation of $W(E_r)$). Fixing such an isomorphism, choose a $\mathbb{Z}$-basis $\{u_i\}$ of $E_r^\vee \subset U$ and define $\lambda = (p_i, q_j)$ using the fundamental invariants (9.19). Then we have $\lambda \in \mathbb{Q}^r$, since p_i, q_j are $W(E_r)$-invariant. The splitting field $\mathcal{K}_\lambda$ of $\Phi_{E_r}(X, \lambda) \in \mathbb{Q}[X]$ is contained in K and contains U. We claim: $\mathcal{K}_\lambda = K$. Indeed, any $\sigma \in \mathrm{Gal}(K/\mathcal{K}_\lambda)$ acts trivially on $U \subset \mathcal{K}_\lambda$. But U is a faithful representation of $W(E_r)$. Hence $\sigma = 1$, which proves the claim. □

Remark 9.41 Our original proof of the claim used the structure of $W(E_r)$. We owe the above simplification to T. Ekedahl (see [Eke90]).

Our method allows us to find explicit numerical examples for Problem 9.6, *big Galois*, which will give, at the same time, some explicit integral polynomials with Galois group isomorphic to the Weyl group $W(E_r)$. Similar examples (geared towards del Pezzo surfaces, compare Sect. 10.5.3) have appeared in [Eke90, Ern94, VAZ09].

Example 9.42 ($r = 6$) Take $\lambda = (0, 0, 1; 0, 1, 0)$. The elliptic curve is:

$$E = E_\lambda : y^2 + t^2 y = x^3 + t^2 x + t.$$

Since the corresponding rational elliptic surface has no reducible singular fibres over $\bar{\mathbb{Q}}$ outside $t = \infty$, the Mordell–Weil lattice is $E(\bar{\mathbb{Q}}(t)) \simeq E_6^\vee$. We claim that $\mathrm{Gal}(\mathcal{K}_\lambda/\mathbb{Q}) = W(E_6)$, i.e. $\lambda \in \Lambda_6$. Further, $\mathcal{K}_\lambda$ is the splitting field of $F(X) = \Phi(X, \lambda)$, where

$$\begin{aligned} F(X) = {} & X^{27} + 12X^{25} + 60X^{23} + 168X^{21} + 294X^{19} - 84X^{18} + 336X^{17} \\ & -348X^{16} + 252X^{15} - 552X^{14} + 120X^{13} - 408X^{12} + 33X^{11} \\ & -132X^{10} - 155X^9 - 12X^8 - 150X^7 - 48X^5 - 10X^3 - X - 1. \end{aligned}$$

Proof This can be verified as follows. By Lemma 9.19, the splitting field $\mathcal{K}_\lambda$ has Galois group a subgroup of $W(E_6)$. Now look at the decomposition of $F(X)$, as obtained from (9.27), mod p into irreducible factors. For $p = 37$, it has three irreducible factors of degree 9 each. Hence the conjugacy class of Frob_p in $\mathrm{Gal}(\mathcal{K}_\lambda/\mathbb{Q})$ consists of three 9-cycles, denoted symbolically by $(9)^3$. Similarly, for $p = 2$, the conjugacy class of Frob_p turns out as $(2)(5)^3(10)$. Then the claim will follow from the following lemma. □

Lemma 9.43 (Serre) *Let*[1] *C_1 or C_2 be the conjugacy classes in $W(E_6)$ of the elements of order* 9 *or* 10 *with cycle type* $(9)^3$ *or* $(2)(5)^3(10)$ *in $\mathcal{S}_{27}$ (cf. [Swi67, p. 57]). Suppose a subgroup H of $W(E_6)$ has the property that $H \cap C_i \neq \emptyset$ for $i = 1, 2$. Then $H = W(E_6)$.*

Proof We use the following information given in the Atlas [Con85] of the maximal subgroups of $W(E_6)$. The Atlas, on p. 26, has $W(E_6)$ as "$G.2$", where $G = W^0(E_6)$ is a simple group of order 25920. The maximal subgroups of $W(E_6)$ are listed, and they are G itself (of index 2) and five others (up to conjugation), of orders 960, 720, 648, 648, 576, respectively.

Suppose that H is a subgroup of $W(E_6)$ satisfying the condition $H \cap C_i \neq \emptyset$ for $i = 1, 2$. Assume that $H \neq W(E_6)$, and take a maximal subgroup, say $\tilde{H}$, containing H. Then its order must be divisible by both 9 and 10. Thus $\tilde{H}$ must be either G itself or it corresponds to one of the maximal subgroups of order 960, 720. But the former does not contain any element of order 10, and the latter do not contain any element of order 9. Hence we found a contradiction. □

[1] Private communication dated March 8, 1990.

Example 9.44 ($r = 7$) Let $\lambda = (1, 1; 1, 1, 1, 0, 1)$ and

$$E = E_\lambda : y^2 = x^3 + x(1 + t + t^3) + (1 + t + t^2 + t^4).$$

Then the Mordell–Weil lattice $E(\bar{\mathbb{Q}}(t)) \simeq E_7^\vee$, and $\mathrm{Gal}(\mathcal{K}_\lambda/\mathbb{Q}) = W(E_7)$, i.e. $\lambda \in \Lambda_7$. The universal polynomial $F(X) = \Phi(X, \lambda)$ is obtained from $f(Y)$ below upon substituting $Y = X^2$, and the splitting field of $f(Y)$ over $\mathbb{Q}$ has the Galois group isomorphic to $W(E_7)/\{\pm 1\} \simeq S_6(2)$.

$$\begin{aligned} f(Y) = Y^{28} &- 36Y^{27} + 594Y^{26} - 6084Y^{25} + 43935Y^{24} - 240192Y^{23} \\ &+ 1039392Y^{22} - 3661764Y^{21} + 10681839Y^{20} - 26088660Y^{19} \\ &+ 53894394Y^{18} - 95282532Y^{17} + 145821463Y^{16} - 194265660Y^{15} \\ &+ 223728462Y^{14} - 216948108Y^{13} + 168475770Y^{12} - 98229852Y^{11} \\ &+ 42796234Y^{10} - 19590492Y^{9} + 14262444Y^{8} - 9949084Y^{7} \\ &+ 4609696Y^{6} - 1118808Y^{5} + 53521Y^{4} + 160Y^{3} + 4288Y^{2} \\ &- 5312Y + 1024. \end{aligned}$$

Proof The proof is similar to the above case of $W(E_6)$. The decomposition type of $F(X)$ mod p is $(7)^8$ for $p = 47$, and $(3)^2(5)^4(15)^2$ for $p = 131$. This suffices to conclude $\mathrm{Gal}(\mathcal{K}_\lambda/\mathbb{Q}) = W(E_7)$ in view of the Atlas [Con85, p. 46], in the same way as above. □

Example 9.45 ($r = 8$) Let $\lambda = (1, 1, 1, 1; 1, 1, 1, 1)$ and

$$E = E_\lambda : y^2 = x^3 + x(1 + t + t^2 + t^3) + (1 + t + t^2 + t^3).$$

Then the Mordell–Weil lattice $E(\bar{\mathbb{Q}}(t)) \simeq E_8$, and $\mathrm{Gal}(\mathcal{K}_\lambda/\mathbb{Q}) = W(E_8)$, i.e. $\lambda \in \Lambda_8$.

See [Shio09] for the detailed proof; here we only give the corresponding integral polynomial $F(X)$ of degree 240 with Galois group $W(E_8)$ as $F(X) = G(X^2)$ for the following polynomial $G \in \mathbb{Q}[Y]$:

$$\begin{aligned} G(Y) = \; & 1 + 60\,Y + 1764\,Y^2 + 33880\,Y^3 + 478890\,Y^4 + 5327856\,Y^5 + 48793140\,Y^6 + 380483064\,Y^7 \\ & + 2598324795\,Y^8 + 15932785020\,Y^9 + 89749362936\,Y^{10} + 473980028160\,Y^{11} + 2387129524492\,Y^{12} \\ & + 11610734817520\,Y^{13} + 54946822132728\,Y^{14} + 253570184893640\,Y^{15} + 1139170471812505\,Y^{16} \\ & + 4966863067888332\,Y^{17} + 20975997259257420\,Y^{18} + 85751930578096488\,Y^{19} \\ & + 338777493097323270\,Y^{20} + 1286110326634556720\,Y^{21} + 4645243511039448812\,Y^{22} \\ & + 15781295779679038440\,Y^{23} + 49982542358210104135\,Y^{24} + 147131229010289262732\,Y^{25} \\ & + 404490375319591401040\,Y^{26} + 1051866512316875968008\,Y^{27} + 2627876995725411369560\,Y^{28} \end{aligned}$$

$$+6344305570523224297840\,Y^{29}+14520529565511555036172\,Y^{30}+29688209277164999351080\,Y^{31}$$

$$+46109411843449203201495\,Y^{32}+11874604308325191705300\,Y^{33}-30452071214048924433244\,Y^{34}$$

$$-171050760749422630542760 0\,Y^{35}-656437328621726202297262 6\,Y^{36}-206407330688695142031789 28\,Y^{37}$$

$$-554853760728702657856535 12\,Y^{38}-128793713071489729206023952\,Y^{39}-257743742783949813779493007\,Y^{40}$$

$$-442469975366494543245531996\,Y^{41}-655052776921613524140717120\,Y^{42}$$

$$-884472107222475356521568192\,Y^{43}-1280575595777430358465307604\,Y^{44}$$

$$-2232747093968418475468736712\,Y^{45}-3709663451251650088528216368\,Y^{46}$$

$$-2787642171110435554924445112\,Y^{47}+10836668436242566282618846201\,Y^{48}$$

$$+58617803145640098757603141084\,Y^{49}+169901174645519701610315084748\,Y^{50}$$

$$+364619363395851165251556074640\,Y^{51}+621760817442592188743112149958\,Y^{52}$$

$$+853996995954752839123279092256\,Y^{53}+936403079514859290353492992584\,Y^{54}$$

$$+813242989399427654276532027720\,Y^{55}+561901601257250767568960750119\,Y^{56}$$

$$+135336745534853793043766886324\,Y^{57}-1359056654022871830736377392248\,Y^{58}$$

$$-6463196369092920757436226504288\,Y^{59}-19274666150603339082780988303322\,Y^{60}$$

$$-43000266962344916776074492071304\,Y^{61}-73878522660917308900155611253488\,Y^{62}$$

$$-93075326288055274100372536594400\,Y^{63}64536984699047964526813581514542\,Y^{64}$$

$$+4930092929380099646206922807646 4\,Y^{65}+25286015965624245209267582936764 0\,Y^{66}$$

$$+489725119826300703593418835619104\,Y^{67}+660488031167958421155995175713950\,Y^{68}$$

$$+685712147170409615430086676252264\,Y^{69}+551683526957510476762145457969296\,Y^{70}$$

$$+288512672181776185291212516963736\,Y^{71}-74267572398719223747581053086990\,Y^{72}$$

$$-500903915968439620592031943919824\,Y^{73}-910549984484808540102258882764384\,Y^{74}$$

$$-1206968873617213245195049682159960\,Y^{75}-1388629308181446796708722329894566\,Y^{76}$$

$$-1569354844716924282753640688567888\,Y^{77}-1823707475660170741752627186852960\,Y^{78}$$

$$-2048864511428828705784772599553992\,Y^{79}-2035166278567233185515268622860248\,Y^{80}$$

$$-1687433627384737395302666916386904\,Y^{81}-1115482759086740014512839085137184\,Y^{82}$$

$$-526761122020931966446367902077448\,Y^{83}-585778648799393756054504186 79390\,Y^{84}$$

$$+250069449965626638272171839412808\,Y^{85}+387136100056401973048117693037508\,Y^{86}$$

$$+357325360560127756205870421947760\,Y^{87}+211364951183975868623079248436736\,Y^{88}$$

$$+51418627724822694637234112345296\,Y^{89}-46773865393411874487242211411132\,Y^{90}$$

$$-71774553474500093078405620618056\,Y^{91}-53829465805447662564570533480300\,Y^{92}$$

$$-23062139597305102540202721147928\,Y^{93}+1816100885334286383938979391048\,Y^{94}$$

$$+11380969645808962145214086704520\,Y^{95}+8855368362001862935225686348201\,Y^{96}$$

$$+3508823119483958223796553026092\,Y^{97}+698584814104495914238234637808\,Y^{98}$$

$$-706078710984494230280625007 28\,Y^{99}-981800556683347475495524341 82\,Y^{100}$$

$$+1390690879311479198589019809 6\,Y^{101}+6129263667769098005360552039 2\,Y^{102}$$

$$+4019612065247427905463964114 4\,Y^{103}+1785518191703922564995192808 3\,Y^{104}$$

$$+633449306664507019256741478 0\,Y^{105}+227878741256845877459023889 6\,Y^{106}$$

$$+73755179620218901165457688 8\,Y^{107}+4329311866110078260715445 06\,Y^{108}$$

$$+429301341790049890169381 68\,Y^{109}+538259733452669451411924\,Y^{110}-849592431713619927676092 0\,Y^{111}$$

$$-10183346161070303085041 27\,Y^{112}+300905188142683299657 00\,Y^{113}+67986368606208563098464\,Y^{114}$$

$$+5895943156273015604992\,Y^{115}-4480198117983524981 76\,Y^{116}-280668086084640358400\,Y^{117}$$

$$-33657833261042688 00\,Y^{118}+1750559212171657216\,Y^{119}+313989595009449984\,Y^{120}.$$

While the idea of the proof is the same as the case of $r = 6, 7$ in Examples 9.42 and 9.44, which were first proven in the early nineties in [Shio91d], the necessary information about the maximal subgroups of $W(E_8)$ was not available at that time. More recently, Jouve et al. [JKZ08] have constructed integral polynomials with Galois group $W(E_8)$ by a different method, based on the following group-theoretic result:

Lemma 9.46 *(Jouve–Kowalski–Zywina [JKZ08]) Consider the conjugacy classes of cycle type* $(15)^{16}$ *and* $(4)^2(8)^{29}$ *in* $W(E_8)$*. Any two elements, one from each conjugacy class, suffice to generate* $W(E_8)$*.*

One can verify Example 9.45 easily using Lemma 9.46. The cycle types appear, for instance, at $p = 11$ and $p = 17$.

In the above discussion (and mostly throughout this section), we have taken $\mathbb{Q}$ as the ground field, but as mentioned before, the same ideas work in more general settings. Here is one such example, and it is a rather special one.

Example 9.47 ($r = 6$). Let $k_0 = \mathbb{F}_3(s)$ be the rational function field over the finite field $\mathbb{F}_3$. Then the following simple equation

$$F(X) = X^{27} + X^{12} - X^7 - X^2 - s^3 \in \mathbb{F}_3[s][X]$$

has the Galois group $W(E_6)$. It arises from the elliptic curve $E/k_0(t)$

$$y^2 + t^2 y = x^3 + xt + st$$

having the Mordell–Weil lattice $E(\bar{k}_0(t)) = E_6^\vee$.

Proof Formally, there is no specialization homomorphism sp'_∞ since we work in the exceptional characteristic 3 (compare Example 9.11), but we can compute the Galois action directly with the usual elimination algorithm for half of the minimal sections. This leads to the above polynomial $F(X) \in \mathbb{F}_3[s][X]$.

Note that regardless of the characteristic, the Galois group sits inside $W(E_6)$ by Lemma 9.19. Hence we can proceed as before and prove Lemma 9.47 by factoring $F(X)$ mod $\mathfrak{p}$ for the prime ideals $\mathfrak{p} = (s - 1)$ and $(s + 1)$ of $\mathbb{F}_3[s]$. We find that $\mathrm{Frob}_\mathfrak{p}$ has the conjugacy type C_1 resp. C_2 from Lemma 9.43 which thus applies as before. □

9.6 Appendix: Non-perfect Fields

At the beginning of this chapter concerning Galois representations, we made the assumption that k, while not necessarily being algebraically closed, is a perfect field. In this appendix, we will see why this assumption is relevant (beyond the fact that there is no Galois action accounting for an inseparable field extension). To this end, let $k_0 = k(u)$ be a rational function field over a field k of characteristic p. For characteristics $p \geq 5$, we give a somewhat uniform, albeit not very direct argument in Example 9.49. Meanwhile, for characteristics $p = 2, 3$, there is an immediate argument as follows:

Example 9.48 Consider the elliptic curve $E/k_0(t)$

$$E: \quad y^2 + txy = x^3 + u.$$

We consider two cases separately. First, if $p = 2$, then we let $s^2 = u$ and have

$$P = (0, s) \in E(k_0(s)(t)),$$

where $k_0(s)/k_0$ is a purely inseparable extension. Secondly, if $p = 3$, then we let $s^3 = u$. As before, we find

$$P = (-s, 0) \in E(k_0(s)(t)),$$

with a purely inseparable extension $k_0(s)/k_0$.

The second part of the next example will see the return of the hexagonal Mordell–Weil lattice, arising from a special member of the family of elliptic curves given in (1.1).

Example 9.49 Let $p \geq 5$ and consider the following elliptic curve $E/k_0(t)$,

$$E: \quad y^2 = x^3 + t(t^n - u^2)^2, \tag{9.37}$$

with associated elliptic surface X. For $n \equiv 1 \mod 6$, this arises from the elliptic curve

$$E': \quad y'^2 = x'^3 + t(t - u^2)^2.$$

via the base change $\pi: t \mapsto t^n$. Note that E' corresponds to an isotrivial rational elliptic surface X' by Proposition 5.51. X' has singular fibers of type II at $t = 0$, IV at $t = u^2$ and I_0^* at $t = \infty$. Its Mordell–Weil group is generated by the section

$$P' = (-(t - u^2), u(t - u^2)).$$

of height $1/3$ and any of the sections $\omega P', \omega^2 P'$ resulting from applying the order 3 automorphism $(x, y) \mapsto (\omega x, y)$ of the generic fiber $(\omega^2 + \omega + 1 = 0)$. Indeed, by Theorem 8.8, $\mathrm{MWL}(X')$ has rank 2 and determinant $1/12$, and so has the lattice generated by P' and $\omega P'$, as one easily verifies. By Theorem 6.53, the base change induces a sublattice

$$\pi^* \operatorname{MWL}(X') \subset \operatorname{MWL}(X)$$

of determinant $n^2/12$.

Now we make the special assumption that $n = p$. Then X attains an additional section over the purely inseparable extension $k(s)/k(u)$ with $u = s^p$ as follows. Considered over $K = k(s)(t)$, E is not minimal anymore. In fact, minimalizing by

$$x = (t - s^2)^{2(p-1)/3} x', \quad y = (t - s^2)^{p-1} y' \tag{9.38}$$

leads back to the elliptic curve E' (with u replaced by s in (9.37)). Hence, in fact,

$$\pi^* \operatorname{MWL}(X) \subset \operatorname{MWL}(X') \cong \operatorname{MWL}(X).$$

The section $P' \in \operatorname{MW}(X')$ of height $1/3$ (with parameter s instead of u) translates to E via (9.38) as

$$\operatorname{MW}(X) \ni P' = (-(t - s^2)^{(2p+1)/3}, s(t - s^2)^p).$$

Visibly, P' is not contained in $E(k(u)(t))$, so the splitting field indeed involves the purely inseparable extension $K/k(u)(t)$, which is exactly the kind of phenomenon that we wanted to demonstrate.

A similar reasoning applies to the case $n \equiv -1 \mod 6$. Here, we can view E as connected via the base change $\pi : t \mapsto t^n$ to the elliptic curve

$$E'' : \quad y^2 = x^3 + t^5(t - u^2)^2$$

whose Kodaira–Néron model X'' is a K3 surface by Proposition 5.51. X'' has singular fibers of type II* at $t = 0, \infty$, and IV at $t = u^2$. We will study X'' in more detail in Sect. 12.4.3.1. In particular, we will see quite explicitly in case $p \equiv -1 \mod 3$ that X'' has Mordell–Weil lattice

$$\operatorname{MWL}(X'') = A_2^\vee(p)$$

(over the separable extension $k(\sqrt[3]{u}, \sqrt[3]{2}, \omega)/k(u)$, compare below), giving rise to a sublattice

$$A_2^\vee(np) \subset \operatorname{MWL}(X)$$

by way of pull-back by Theorem 6.53.

Now assume as before that $n = p$. We claim that X attains an additional section over (a separable extension of) the purely inseparable extension $k(s)/k(u)$ with $u = s^p$. To see this, we note that over $k(s)(t)$, E is not minimal anymore. Minimalizing leads to the elliptic curve

$$E''' : \quad y^2 = x^3 + (t - s^2)^4 t.$$

By Proposition 5.51, the associated elliptic surface X''' is again rational. The section $(\beta^2(t-s^2)^2/w^2, \beta^3(t+s^2)(t-s^2)^2/s)$ over the separable extension $k(w,\beta)/k(s)$ with $s = w^3, \beta^3 = 1/2$ can be found using Example 6.55. In particular, this gives generators of the hexagonal lattice

$$E(k(w,\beta,\omega)) \cong A_2^\vee$$

in contrast to the sublattice $A_2^\vee(p^2)$ defined over $k(\beta,\omega)(\sqrt[3]{u})$ which we derived before.

Remark 9.50 Example 9.49 also illustrates why Tate's algorithm in Sect. 5.8 requires the input of a perfect field as this was necessary for minimalizing.

Historical Notes

Galois representations are a central topic in arithmetic geometry, with close connections to algebraic geometry and number theory, but also to other areas such as modular and automorphic forms. The classical example is provided by the Tate module of an abelian variety defined over some number field, but following the invention of étale cohomology by Artin and Grothendieck, the cohomology groups of any smooth projective variety over some number field admit structures of Galois representations (see [Ser68]). The most famous Galois representations are probably still those attached to elliptic curves over $\mathbb{Q}$—due to the modularity which we mentioned at the end of Chap. 3. Another key theorem, predating essentially all explicit results, is Faltings' finiteness result, stating that any Galois representation over a number field (continuous, finite-dimensional, with a finite number of ramified primes) is determined by a finite number of traces at certain Frobenius elements up to semi-simplification. Effective versions of this result, however, are very hard (for the two-dimensional case, see [Liv87, Ser85]).

Galois representations on spaces of algebraic cycles have rather special behaviour since the action can be described through some finite group, the Galois group of the minimal field of definition of the cycles in question; in our case of Mordell–Weil lattices, the field was the splitting field from Definition 9.3, and this exactly led to the Weyl group as targeted Galois group. The motivation and guiding principle for our considerations, however, came from a different area, namely from Brieskorn's simultaneous resolution of E_r-singularities ($r = 6, 7, 8$) (see Remark 9.24). Even though we did not use this explicitly, but held it as a sort of resource in the back of our mind, the basic idea was that the role of Galois representations for Mordell–Weil lattices, in particular in the framework of excellent families, should replace what the monodromy does for Brieskorn's simultaneous resolution; here the topological covering transformation group would be replaced by the Galois group, which turned out to be the Weyl group. Then the setting of Mordell–Weil lattices came with the great benefit that it does not only cover topological settings, but also has arithmetic applications.

The concentration on the additive case (of excellent families) happened for a simple reason: this was exactly what we could carry out at that time. More precisely, the ring of invariants of a finite reflection group is again a polynomial ring, with generators in known degrees. Basic elimination theory then gave a direct relation with the coefficients of the Weierstrass equation. At first, this was just an abstract and theoretical relation which remained to be carried out explicitly.

At this point, we benefited greatly from the progress in personal computers which enabled naive mathematicians like us, with barely any knowledge of computers at all, to use them for our purposes, namely for carrying out the elimination process and for constructing numerical examples. Indeed, for the root lattice E_6, the given relations were not too complicated so that in June 1989 we could solve them on a small Macintosh computer with 4 MB memory. In contrast, solving the corresponding problem for E_7 and E_8 relied on the help of several colleagues to whom we are greatly indebted (and a more powerful computer, with 8 or 16 MB memory!).

The precise relations between the Weierstrass coefficients and the explicit invariants of the Weyl group (which we regard as the key formulas of the whole result) is of particular importance because once we have it, we can reversely engineer various numerical examples of elliptic curves with certain rational points. Moreover, this verification provides a valuable sanity check for the key formulas and the underlying theoretical reasoning (also against errors possibly introduced by the hardware or software involved).

The reverse engineering aspect of the excellent families of rational elliptic surfaces relates directly to inverse Galois theory. The guiding open problem here is whether every finite group can be realized as the Galois group of a Galois extension of $\mathbb{Q}$ (see [MM18]). Classically solved cases include all cyclic, abelian, symmetric and alternating groups (some of them due to Hilbert). In fact, many proofs proceed geometrically over $\mathbb{Q}(t)$ by constructing Galois covers of the projective line and then specializing to $\mathbb{Q}$ using Hilbert's irreducibility theorem.

Specialization is also a very natural and useful ingredient for many other problems in algebraic geometry and number theory. For instance, we will later use Silverman's specialization theorem [Sil83] in Chap. 13. This concerns the behaviour of the rank of a family of elliptic curves (over $\mathbb{Q}(t)$ or $\mathbb{Q}(C)$ for a base curve C of genus one and positive rank) under specialization to $\mathbb{Q}$. Note, however, that in the given setting, we always specialize to smooth elliptic curves, while in Sect. 9.3 we deliberately aimed for specialization to a singular fibre.

Specialization morphisms are built very naturally into scheme theory—just think of the comparison between generic fibre and special fibre (which exactly underlies the above problems). Generally, one is led to ask what properties carry over between generic fibre, general fibre and special fibre—in one direction or the other. For instance, one may consider basic properties such as smoothness or more advanced invariants such as fundamental groups—or the (rank of the) Picard group, as we will exploit in Sect. 12.2.7 and in Sect. 13.1.4 following [vLui07, EJ11].

Along these lines, one can also view the Kodaira–Néron model as a topic on specialization—this time again to singular fibres within a family of abelian varieties. Indeed, this viewpoint was already pursued systematically before Néron [Ner64], for instance by Shimura in [Shimu55].

Chapter 10
Applications to Classical Topics

This chapter forms a unit with the previous one in the sense that it continues to discuss more recent developments arising from Mordell–Weil lattices from the point of view of Galois representations and algebraic equations. We start by elaborating on excellent families properly and detailing on the multiplicative case, especially of type E_6, E_7 or E_8 (paralleling Theorem 9.21). This will be followed by various applications of Mordell–Weil lattices. In particular, we will shed new light on such a classical topic as the 27 lines on a cubic surface, and on the algebraic equation of degree 27 with Galois group $W(E_6)$, both seen from the Mordell–Weil lattice point of view. We also discuss applications to other del Pezzo surfaces, and to the 28 bitangents of a plane quartic curve.

10.1 Multiplicative Excellent Families

10.1.1 Excellent Families

First we introduce a provisional notion of an *excellent family* of varieties with Galois group G (cf. [Shio92b, SU92]), and then make a more precise definition of *additive or multiplicative excellent families* in the case of rational elliptic surfaces. Throughout the reader may take the additive examples from Theorem 9.21, and also the hexagonal case from Theorem 9.15 as a guiding motivation. For simplicity and ease of presentation, we will continue to work over $\mathbb{Q}$, although most arguments carry over literally to many other perfect fields.

Suppose

$$\mathcal{X} \to \mathbb{A}^n$$

M. Schütt and T. Shioda, *Mordell–Weil Lattices*, Ergebnisse der Mathematik und ihrer Grenzgebiete. 3. Folge / A Series of Modern Surveys in Mathematics 70,
https://doi.org/10.1007/978-981-32-9301-4_10

is a family of algebraic varieties, parametrized by the affine n-space $\mathbb{A}^n$ over $\mathbb{Q}^n$ (or over $\mathbb{C}^n$ as we will do later), i.e. in terms of n independent parameters $\lambda_1, \dots, \lambda_n$; set

$$\lambda = (\lambda_1, \dots, \lambda_n).$$

The generic member X_λ of the family $\mathcal{X}$ is a variety over the rational function field $k_0 = \mathbb{Q}(\lambda)$. Also, suppose that $\mathcal{C}(X_\lambda)$ is a group of algebraic cycles on X_λ, with an isomorphism

$$\phi_\lambda : \mathcal{C}(X_\lambda) \otimes \mathbb{Q} \cong V$$

for a fixed $\mathbb{Q}$-vector space V, and that $\mathcal{C}(X_\lambda)$ is preserved by the Galois group $\mathrm{Gal}(k/k_0)$, where $k = \overline{k_0}$ is the algebraic closure. Then we have a Galois representation

$$\rho_\lambda : \mathrm{Gal}(k/k_0) \to \mathrm{Aut}(\mathcal{C}(X_\lambda)) \to \mathrm{Aut}(V).$$

We let $\mathcal{K}_\lambda$ be the smallest extension of k_0 over which the cycles of $\mathcal{C}(X_\lambda)$ are defined (i.e. where a set of generators of $\mathcal{C}(X_\lambda)$ or a basis of V is defined). In parallel with Definition 9.3, we call $\mathcal{K}_\lambda$ the *splitting field* of $\mathcal{C}(X_\lambda)$. Obviously, it is a Galois extension of k_0 with

$$\mathrm{Gal}(\mathcal{K}_\lambda/k_0) \cong \mathrm{Im}(\rho_\lambda).$$

(In case the base field $\mathbb{Q}$ is replaced by any perfect field, the following definitions can be adapted easily. For simplicity, we continue to work over $\mathbb{Q}$.)

In what follows, we let G be a finite reflection group acting on the vector space V.

Definition 10.1 We say $\{X_\lambda\}$ is an *excellent family* with Galois group G if the following conditions hold:

1. The image of ρ_λ is equal to G.
2. There is a $\mathrm{Gal}(k/k_0)$-equivariant evaluation map $s : \mathcal{C}(X_\lambda) \to k$.
3. There exists a $\mathbb{Q}$-basis $\{Z_1, \dots, Z_n\}$ of $V = \mathcal{C}(X_\lambda) \otimes \mathbb{Q}$ such that if we set $u_i = s(Z_i)$, then $u_1, \dots, u_n$ are algebraically independent over $\mathbb{Q}$.
4. $\mathbb{Q}(u_1, \dots, u_n)^G = \mathbb{Q}(\lambda_1, \dots, \lambda_n)$.

Remark 10.2 In our setting of (rational) elliptic surfaces, we will take the evaluation map s to be the specialization map sp', see Definition 10.6. In particular, it will be a group homomorphism, with image in $k = \mathbb{G}_a(k)$ or $k^\times = \mathbb{G}_m(k)$.

Example 10.3 ($G = W(E_8)$) For $G = W(E_8)$, consider the family of elliptic surfaces S_λ defined by the Eq. (9.9) where $\lambda = (p_0, p_1, p_2, p_3, q_0, q_1, q_2, q_3)$. Take the group of algebraic cycles $\mathcal{C}(S_\lambda)$ to be the Mordell–Weil lattice $E_\lambda(k(t))$ and the evaluation map sp to be the specialization map sp_∞. Then Theorem 9.21 shows in particular that $\{S_\lambda\}$ is an excellent family of elliptic surfaces with Galois group $G = W(E_8)$.

Example 10.4 ($G = W(E_6), W(E_7)$) Similarly, we obtain an excellent family for $G = W(E_7)$ or $W(E_6)$, by the same Theorem 9.21, using the Weierstrass Eq. (9.10) or (9.11) and the specialization map sp'_∞.

These examples for the Weyl groups $G = W(E_6), W(E_7), W(E_8)$ are the motivating examples for introducing the idea of "excellent" families. The applications to number theory, as already alluded to in Sect. 9.5, will be further exploited in Sect. 10.4.

At any rate, these examples suggest a refined definition in the case where we take the group of algebraic cycles $\mathcal{C}(S_\lambda)$ to be the Mordell–Weil lattice (or group) of a rational elliptic surface. Thus we are led to consider a family of rational elliptic surfaces

$$\{S_\lambda \,|\lambda = (\lambda_1, \ldots, \lambda_n)\}$$

such that the generic member S_λ has type $\{T, L, M\}$ in the notation of Theorem 8.8, i.e. T is the fibre contribution to the trivial lattice, L the narrow Mordell–Weil lattice and M the Mordell–Weil group with the height pairing.

Definition 10.5 The type $\{T, L, M\}$ is called *admissible* if L is (isometric to) a root lattice of positive rank.

Assume that the type $\{T, L, M\}$ is admissible. Let $W = W(L)$ be the Weyl group of L. Take

$$\mathcal{C}(S_\lambda) = M := \mathrm{MW}(S_\lambda) = E_\lambda(k(t))$$

and a k_0-rational point $v \in \mathbb{P}^1$ such that the fibre over v is a singular fibre (of either additive or multiplicative type). Assume that there is a specialization map

$$s = sp'_v : M \longrightarrow k \text{ or } k^\times$$

extending $sp^0_v : L \to k$ or $k^\times$. (Recall that in the additive case we have checked this for many characteristics in Lemma 9.10.)

Definition 10.6 In the above setting with a family of rational elliptic surfaces $\{S_\lambda\}$ of admissible type $\{T, L, M\}$ and $\mathrm{rank}(L) = n$, suppose furthermore that:

1. the map $s = sp'_v$ is a $\mathrm{Gal}(k/k_0)$-equivariant group homomorphism;
2. there exists a basis $\{P_1, \ldots, P_n\}$ of $L^\vee \cong M/M_{\mathrm{tors}}$ such that $\{s(P_1), \ldots, s(P_n)\}$ are algebraically independent over $\mathbb{Q}$.

Then we distinguish the following two cases:

(i) We call $\{S_\lambda\}$ an *additive excellent* family of type L (or with Galois group $W = W(L)$) if

$$sp'_v : M \to \mathbb{G}_a(k) \cong k$$

is a $\mathrm{Gal}(k/k_0)$-equivariant homomorphism of M into the additive group k such that, if we set $u_i = sp'_v(P_i)$, the subring of W-invariants in the graded polynomial ring $\mathbb{Q}[u_1, \ldots, u_n]$ is a polynomial ring in the parameters $\lambda = (\lambda_1, \ldots, \lambda_n)$:

$$\mathbb{Q}[u_1, \ldots, u_n]^W = \mathbb{Q}[\lambda_1, \ldots, \lambda_n].$$

(ii) We call $\{S_\lambda\}$ a *multiplicative excellent* family of type L (or with Galois group $W = W(L)$) if

$$sp'_v : M \to \mathbb{G}_m(k) \cong k^\times$$

is a $\mathrm{Gal}(k/k_0)$-equivariant homomorphism of M into the multiplicative group $k^\times$ such that, if we set $s_i = sp'_v(P_i)$, then the Weyl group acts on the ring of Laurent polynomials $\mathbb{Q}[s_1, s_1^{-1}, \ldots, s_n, s_n^{-1}]$ and the subring of W-invariants is again a polynomial ring in the parameters λ:

$$\mathbb{Q}[s_1^{\pm 1}, \ldots, s_n^{\pm 1}]^W = \mathbb{Q}[\lambda_1, \ldots, \lambda_n].$$

In the notation of the above definition, the splitting field $\mathcal{K}_\lambda$ of an additive or multiplicative excellent family $\{S_\lambda\}$ is a Galois extension of $k_0 = \mathbb{Q}(\lambda)$ with Galois group $W = W(L)$, and the Galois representation ρ_λ induces an isomorphism of $\mathrm{Gal}(\mathcal{K}_\lambda/k_0)$ with W. Further, $\mathcal{K}_\lambda$ is a purely transcendental extension of the base field $\mathbb{Q}$, namely it is equal to either $\mathcal{K}_\lambda = \mathbb{Q}(u_1, \ldots, u_n)$ in the case *(i)* or $\mathcal{K}_\lambda = \mathbb{Q}(s_1, \ldots, s_n)$ in the case *(ii)*.

We have already seen some examples of additive excellent families with Galois groups $W(E_r)$ for $r = 6, 7, 8$ in Examples 10.3, 10.4. In the next section, we treat multiplicative excellent families with the same Weyl groups $W(E_r)$, whose study in the recent work [Shio12, KumS13] played indeed the role of the motivating examples for the formulation of an *multiplicative excellent family* as given in Definition 10.6 *(ii)*.

More generally, by Theorem 8.8, the candidate lattices L for an excellent family belong, besides $L = E_8, E_7, E_6$, to the following list of root lattices:

$$L = D_6, D_5, A_5, D_4 \oplus A_1, A_4, D_4, A_3 \oplus A_1, A_2^2, A_1^4, A_3, A_2 \oplus A_1, A_1^3, A_2, A_1^2, A_1.$$

For examples of additive excellent families with some of these root lattices, see [SU92, Shio92b, Usu93]. Later in Sect. 10.4.3, we will give some examples in the multiplicative case by applying the idea of degeneration via "vanishing roots" (see also Example 10.37). Note that these families give canonical ways to prove part of Theorem 8.9 (see Remark 10.62).

Before going into the details for the lattices $L = E_8, E_7, E_6$, we shed more light on our prototype case of the hexagonal Mordell–Weil lattice A_2, this time both for the additive and the multiplicative case.

10.1.2 Excellent Families for the Hexagonal Case

As a guiding example, we return to the case of the hexagonal Mordell–Weil lattice, but this time we will include both the additive and the multiplicative case (corresponding to two of the entries with Mordell–Weil lattice $A_2^\vee$ in Theorem 8.8).

We start with the *additive* hexagonal case which has made its first appearance already in the introduction in (1.1). In order to study Galois representations, however, we will rather use the Weierstrass equation in its general form given by (9.8) in (9.4.1). Previously, we decided to retain part of the symmetry by working with the three minimal sections P_1, P_2, P_3 of height $2/3$ such that $P_1 + P_2 + P_3 = O$. In view of Definition 10.6 we shall now pick two of these sections, say P_2, P_3. In view of the relation between their images $u_i = sp'_\infty(P_i)$ under the specialization homomorphism sp'_∞ deduced in Example 9.11,

$$u_1 + u_2 + u_3 = 0,$$

we shall also normalize f by a translation in x such that the coefficient of x^2 vanishes, i.e. $\epsilon_1 = 0$. Then it follows readily from Theorem 9.15 that the family of elliptic curves (or rational elliptic surfaces)

$$E_{(\epsilon_2,\epsilon_3)}: \quad y^2 + ty = x^3 + \epsilon_2 x - \epsilon_3$$

forms an additive excellent family of type A_2.

For the *multiplicative* case, the following cubic pencil was brought up in discussions with Spencer Bloch (in a different context, see [BKV17]):

$$E: \quad (\mu_1 x_1 + \mu_2 x_2 + \mu_3 x_3)(x_1x_2 + x_1x_3 + x_2x_3) - t x_1x_2x_3 = 0. \qquad (10.1)$$

Indeed, this defines an elliptic curve over the field $k_0(t)$, where $k_0 = \mathbb{Q}(\mu_1, \mu_2, \mu_3)$. We claim that for generic μ_1, μ_2, μ_3, the Mordell–Weil group will be

$$E(k_0(t)) \cong A_2^\vee \oplus \mathbb{Z}/2\mathbb{Z}. \qquad (10.2)$$

To see this, one may note that among the nine base points of the cubic pencil (10.1), there are generically three with multiplicity two, $[1,0,0], [0,1,0], [0,0,1]$, and three simple ones, $[0, \mu_3, -\mu_2]$ etc. Blowing up in the three double base points, we derive a fibre of Kodaira type I_6 at $t = \infty$, while the apparent fibre of type I_2 is located at $t = 0$. Picking, say, $[1,0,0]$ as zero section O, the other double base points yield minimal sections (with height matrix corresponding to $A_2^\vee$), and the section $[0, \mu_3, -\mu_2]$ gives a 2-torsion section (by direct calculation, or by inspection of the height pairing (6.15), e.g. arguing as in Remark 6.38). Hence the claimed shape of the Mordell–Weil group (10.2) follows.

With a view towards the excellence of our family, we shall now exploit the symmetry of the given cubic pencil. Indeed, converting to Weierstrass form (e.g. using [Cas91]), we arrive at

$$E: \quad y^2 + (\epsilon_1 - t)xy - 2\epsilon_3 y = x^3 - \epsilon_2 x^2 + \epsilon_1\epsilon_3 x - \epsilon_3^2,$$

where ϵ_i denotes the elementary symmetric polynomial of μ_1, μ_2, μ_3 of degree i. Here the two-torsion section transforms to $P_0 = (0, \epsilon_3)$, and the two remaining double base points map to P_2 and $-P_3$, where

$$P_1 = (\mu_2\mu_3, 0), \quad P_2 = (\mu_1\mu_3, 0), \quad P_3 = (\mu_1\mu_2, 0).$$

Indeed, one easily checks that the sections P_i all meet the same component of the I_6 fibre at $t = \infty$ (say Θ_2), and only the identity component of the I_2 fibre at $t = 0$. It follows that the six roots of the narrow Mordell–Weil lattice $E(k_0(t))^0 \cong A_2$ are uniquely represented by the integral sections $P_i - P_j$ for $i \neq j$. In detail, we find that $P_i - P_j$ has coordinates

$$\begin{aligned} \mathrm{x} &= \mu_i\mu_j(t-\mu_k)^2/(\mu_i-\mu_j)^2, \\ \mathrm{y} &= -\mu_i\mu_j^2 t^3/(\mu_i-\mu_j)^3 + \dots, \end{aligned}$$

where $\{i, j, k\} = \{1, 2, 3\}$. With the specialization morphism sp_∞^0 from Lemma 9.8 (with the sign of y adjusted), we obtain

$$sp_\infty^0(P_i - P_j) = \frac{\mu_i}{\mu_j}, \quad sp_\infty^0(3P_i) = \frac{\mu_i^2}{\mu_j\mu_k} = \mu_i^3/\epsilon_3 \quad (\{i, j, k\} = \{1, 2, 3\}). \tag{10.3}$$

After normalizing the equation such that $\epsilon_3 = 1$ (for instance by scaling t), we find the unique Galois invariant specialization homomorphism sp'_∞ extending sp_∞^0 to the full Mordell–Weil group such that

$$sp'_\infty(P_i) = \mu_i \ \ \forall i = 1, 2, 3.$$

Taking $\{P_1, P_2\}$, say, as basis of the Mordell–Weil lattice MWL, we exactly arrive at a multiplicative excellent family of type A_2.

10.2 Multiplicative Excellent Families of Type E_6, E_7 or E_8

In comparison with the additive excellent families of type E_r discussed in Examples 10.3, 10.4 (based on Sect. 9.4.2), let us consider the following families which will turn out to be multiplicative excellent families of type E_r.

10.2.1 Set-Up for the Multiplicative Type E_r ($r = 6, 7, 8$)

Consider the following Weierstrass equations with the parameters λ over the rational function field $k_0(t)$, where $k_0 = \mathbb{Q}(\lambda) = \mathbb{Q}(p_i, q_j)$.

$$(mult\,E_6) \quad y^2 + txy = x^3 + x\left(\sum_{i=0}^{2} p_i t^i\right) + \left(\sum_{i=0}^{2} q_i t^i + t^3\right), \tag{10.4}$$

$$\lambda = (p_0, p_1, p_2, q_0, q_1, q_2) \in \mathbb{A}^6.$$

$$(mult\,E_7) \quad y^2 + txy = x^3 + x\left(p_0 + p_1 t + p_2 t^2\right) + \left(\sum_{i=0}^{3} q_i t^i - t^4\right), \tag{10.5}$$

$$\lambda = (p_0, p_1, p_2, q_0, q_1, q_2, q_3) \in \mathbb{A}^7.$$

$$(mult\,E_8) \quad y^2 = x^3 + t^2x^2 + x\left(\sum_{i=0}^{2} p_i t^i\right) + \left(\sum_{i=0}^{4} q_i t^i + t^5\right), \tag{10.6}$$

$$\lambda = (p_0, p_1, p_2, q_0, q_1, q_2, q_3, q_4) \in \mathbb{A}^8.$$

Fix $r = 6, 7, 8$ and let $E = E_\lambda$ be the elliptic curve over $k_0(t)$, defined by one of the above Weierstrass Eqs. (10.4), (10.5) or (10.6), and let

$$f : S_\lambda \longrightarrow \mathbb{P}^1$$

be the associated elliptic surface. It is easy to check that the fibre $f^{-1}(\infty)$ at $t = \infty$ is a multiplicative singular fibre of type I_d, where $d = 9 - r$, i.e. $d = 3, 2$ or 1 for $r = 6, 7$ or 8:

$$f^{-1}(\infty) = \begin{cases} \Theta_0 & (r = 8), \\ \Theta_0 + \Theta_1 & (r = 7), \\ \Theta_0 + \Theta_1 + \Theta_2 & (r = 6). \end{cases}$$

By definition, the irreducible components Θ_i $(0 \le i \le d - 1)$ are rational over k_0, i.e. the fibre is split multiplicative. The identity component Θ_0 can be identified with the projective closure of the nodal cubic

$$\Gamma : y^2 + xy = x^3$$

of the Weierstrass model as in Lemma 9.8 *(ii)*. (Formally, this only holds for the smooth locus if $r = 8$, while for $r = 7$ or 6 we ought to consider its strict transform in the resolution of the singularity of Γ, as in the proof of Lemma 9.19.) In all three cases,

$$\Theta_0^\# \simeq \mathbb{G}_m$$

and $f^{-1}(\infty)^{\#}$ is an extension of $\mathbb{G}_m$ by $\mathbb{Z}/d\mathbb{Z}$. For $r = 6, 7$, we can find a specialization homomorphism

$$sp'_{\infty} : M_{\lambda} := E_{\lambda}(k(t)) \to \mathbb{G}_m(k) = k^{\times} \tag{10.7}$$

which is $\mathrm{Gal}(k/k_0)$-equivariant and which extends $sp^0 : M^0 \to k^{\times}$ (see Lemma 10.10 for $r = 6$; for $r = 8$, there is no problem as $M = M^0$).

If a special value $\lambda^0 \in \mathbb{A}^r$ satisfies the generality condition (9.13) that every fibre of f over $t \neq \infty$ is irreducible, then the Mordell–Weil lattice M_{λ} is isomorphic to the dual lattice $E_r^{\vee}$ of the root lattice E_r by Theorem 8.8, and we obtain a Galois representation:

$$\varrho = \varrho_{\lambda} : \mathrm{Gal}(k/k_0) \longrightarrow \mathrm{Aut}(M) \simeq \mathrm{Aut}(E_r^{\vee}) = \mathrm{Aut}(E_r).$$

Lemma 10.7 *The image of ϱ_{λ} is contained in the Weyl group $W(E_r) \subset \mathrm{Aut}(E_r)$ for $r = 6, 7, 8$:*

$$G := \mathrm{im}(\rho_{\lambda}) \subseteq W(E_r) \ (r = 6, 7, 8) \tag{10.8}$$

Proof This is clear for $r = 7, 8$ since then $\mathrm{Aut}(E_r) = W(E_r)$. For $r = 6$, where $\mathrm{Aut}(E_6) = W(E_6) \times \{\pm 1\}$, the same reasoning as in the proof of Lemma 9.19 applies to show that $-1 \notin \mathrm{im}(\rho_{\lambda})$. Indeed, the minimal sections from Theorem 7.12 appear for the Weierstrass form (10.4) again as one half of linear sections

$$P = (\mathrm{x}, \mathrm{y}), \quad \mathrm{x} = at + b, \quad \mathrm{y} = dt + e; \tag{10.9}$$

the other half is given by the inverses, i.e. they have the same x-coordinate, but

$$\mathrm{y}(-P) = -at^2 + (d - a)t + e.$$

Clearly, either half of the minimal sections is preserved under any Galois automorphism. Note that, as before, the difference of any two such sections, say $P - P'$, generically belongs to the narrow Mordell–Weil lattice $M^0 \cong E_6$ (or, without generality assumption (9.13), to the ∞-narrow part of the Mordell–Weil group). Moreover, Tate's algorithm (5.8) reveals that the linear sections meet the fibre component $\Theta_1 = \{s = y' = 0\}$ in an affine chart of the blow-up of (10.4) at the node of the fibre at $t = 1/s = \infty$. □

10.2.2 *Generic Galois Representation and Algebraic Equation of Multiplicative Type E_r ($r = 6, 7, 8$)*

Now we assume that λ is *generic*, i.e. that the r parameters p_i, q_j are algebraically independent over $\mathbb{Q}$. Then the condition (9.13) again holds. The main result here is the *multiplicative* analogue of Theorem 9.21, and it shows that each of the Weierstrass Eqs. (10.4)–(10.6) defines a multiplicative excellent family with Galois group $W(E_r)$ for $r = 6, 7$ and 8.

Theorem 10.8 *Fix $r = 6, 7$ or 8, and let λ be generic over $\mathbb{Q}$. Then:*

(i) The Galois representation ϱ_λ induces an isomorphism:

$$\mathrm{Im}(\varrho_\lambda) \simeq W(E_r).$$

(ii) Let $\mathcal{K}_\lambda/k_0$ be the splitting field of $E_\lambda/k_0(t)$, and let $\{P_1, \ldots, P_r\}$ be a basis of $E(k(t)) \simeq E_r^\vee$ consisting of minimal vectors. Furthermore let $s_i = sp'_\infty(P_i) \in \mathcal{K}_\lambda \subset k$. Then $s_1, \ldots, s_r$ are algebraically independent over $\mathbb{Q}$, and $\mathcal{K}_\lambda$ is a purely transcendental extension of $\mathbb{Q}$ generated by $s_1, \ldots, s_r$:

$$\mathcal{K}_\lambda = \mathbb{Q}(s_1, \ldots, s_r)$$

and we have

$$\mathrm{Gal}(\mathbb{Q}(s_1, \ldots, s_r)/\mathbb{Q}(p_i, q_j)) = W(E_r).$$

(iii) $W(E_r)$ acts on the ring of Laurent polynomials $\mathbb{Q}[s_1, s_1^{-1}, \ldots, s_r, s_r^{-1}]$, and the subring of invariants is isomorphic to the polynomial ring generated by p_i and q_j over $\mathbb{Q}$:

$$\mathbb{Q}[s_1, s_1^{-1}, \ldots, s_r, s_r^{-1}]^{W(E_r)} = \mathbb{Q}[p_i, q_j].$$

In particular, we have explicit formulas [such as (10.30) for $r = 6$] expressing p_i and q_j as $W(E_r)$-invariant Laurent polynomials, say $\lambda_n = J_n$:

$$\lambda_n := p_i \text{ or } q_j = J_n(s_1, \ldots, s_r) \ (1 \le n \le r). \tag{10.10}$$

(iv) The universal polynomial of multiplicative type E_r, $\Psi_{E_r}(X, \lambda)$ defined as in Definition 9.20 (see below), is a monic irreducible polynomial in X with coefficients in the polynomial ring $\mathbb{Z}[\lambda] = \mathbb{Z}[p_i, q_j]$ and with constant term equal to ± 1, and the splitting field of $\Psi_{E_r}(X, \lambda)$ over $\mathbb{Q}(\lambda)$ is equal to $\mathcal{K}_\lambda$.

The *universal polynomial of multiplicative type E_r* is defined formally in the same way as in the additive case in Definition 9.20:

$$\Psi_{E_r}(X, \lambda) = \Psi(E_\lambda/\mathbb{Q}(\lambda)(t), \infty, I; X) = \prod_{P \in I}(X - sp'_\infty(P)) \in \mathbb{Q}(\lambda)[X].$$

In particular, it is based on the same sets of minimal sections as in Definition 9.20. The only essential difference is that here $sp'_\infty : M_\lambda \to k^\times$ is a $k^\times$-valued multiplicative homomorphism defined as in (10.7), instead of the k-valued additive homomorphism from (9.12).

The proof in the case $r = 6$ ([Shio12]) will be reviewed in the next two sections. We refer to [KumS13] for the proofs of the more involved cases $r = 7, 8$ which require substantial machine assistance provided by standard computer algebra packages. (We remark, though, that there is no direct symbolic approach feasible, but similar formulas have been derived by Eguchi and Sakai [EgS03] using calculations from string theory and mirror symmetry.) Since this also makes it much harder to present the proofs, we decided to omit this part for space reasons. Anyway, the interested reader is recommended to take a look at the paper [KumS13], since it gives, in the cases $r = 7, 8$, not only the proof of Theorem 10.8, but also some explicit numerical examples for various problems such as big (or small) Galois and the degeneration of Mordell–Weil lattices which we will touch upon, mostly for E_6 ($r = 6$), later in this chapter.

10.2.3 Preparation: Standard Generators of $E_6^\vee$

Recall from Chap. 2 (esp. Sect. 2.3.4; see also Remark 9.28) that the dual lattice $L = E_6^\vee$ of E_6 has a set of generators $\{u_1, \ldots, u_6, v_0\}$ with a single relation $u_1 + \ldots + u_6 = 3v_0$ such that

$$\langle u_i, u_j \rangle = \delta_{ij} + \frac{1}{3}, \quad \langle v_0, v_0 \rangle = 2.$$

Suppose that S is a rational elliptic surface with only one reducible fibre, of Kodaira type I_3 or IV; without loss of generality we locate it at a fixed point of the base curve $\mathbb{P}^1$, say at $t = \infty$. Then its Mordell–Weil lattice M is isometric to $E_6^\vee$ by Theorem 7.4 (as included in Theorem 8.8). By the above, it has a set of standard generators

$$\{P_1, \ldots, P_6, R\} \quad \text{with relation} \quad P_1 + \ldots + P_6 = 3R, \tag{10.11}$$

such that the height pairing among them is given as follows:

$$\langle P_i, P_j \rangle = \delta_{ij} + \frac{1}{3}, \quad \langle R, R \rangle = 2. \tag{10.12}$$

In particular, R is a section of height 2 (a root in the narrow Mordell–Weil lattice $M^0 \cong E_6$).

On the other hand, the height formula (6.14) says:

$$\langle P_i, P_j \rangle = 1 + (P_i.O) + (P_j.O) - (P_i.P_j) - \gamma, \quad \gamma = contr_\infty(P_i, P_j),$$

where the value of γ is either 2/3 or 1/3 or 0 for type I_3 or IV (by Table 6.1). Presently, (10.12) gives $-\gamma \equiv 1/3$ modulo $\mathbb{Z}$, which implies $\gamma = 2/3$; this holds if and only if the two sections (P_i) and (P_j) hit one and the same non-identity component, say Θ_1, of the I_3 (or IV) fibre (exactly as it happened for the linear sections in the proof of Lemma 9.19). It thus follows from (10.12) that

$$(P_i.O) = 0 \quad \text{and} \quad (P_i.P_j) = -\delta_{ij},$$

that is, each section (P_i) is disjoint from the zero section (O), and (P_i) is disjoint from (P_j) if $i \neq j$.

Next we define further 21 sections which by inspection meet the same fibre component Θ_1:

$$\begin{cases} P_i' := P_i - R & (1 \le i \le 6), \\ P_{i,j}'' := R - P_i - P_j & (1 \le i < j \le 6). \end{cases} \tag{10.13}$$

Then the set of 27 elements

$$\Omega_M := \{P_1, P_2, \ldots, P_{27}\} = \{P_1, \ldots, P_6, \ P_1', \ldots, P_6', \ P_{1,2}'', \ldots, P_{5,6}''\} \tag{10.14}$$

is invariant under the action of the Weyl group $W(E_6)$; it thus gives one half of the minimal sections of height $4/3$ (compare Proposition 2.36, Lemma 9.19 and the discussion just before 10.2.2).

Similarly, the following sections correspond to the 36 positive roots of E_6 given in the table of Theorem 2.34:

$$\begin{cases} -R, \\ A_{i,j} := P_i - P_j \ \ (i < j \le 6), \\ B_{i,j,k} := R - P_i - P_j - P_k \ \ (i < j < k \le 6). \end{cases} \tag{10.15}$$

It is easy to check that $\{P_i' \ (1 \le i \le 6)\}$ has the same intersection behaviour as $\{P_i \ (1 \le i \le 6)\}$, and that

$$(P_i.P_j') = 1 - \delta_{ij} \quad (1 \le i, j \le 6),$$

i.e. the two sections (P_i) and (P_j') are disjoint if and only if $i = j$, and they intersect transversally in a single point otherwise. Thus we get two sets of six disjoint curves $\{P_i\}$, $\{P_j'\}$ on S such that each curve of one set intersects exactly five curves in the other set. This important configuration is called a *double six* in the sense of Schläfli; the first and most famous example of such a configuration is a double six of lines on a smooth cubic surface. Eventually, our approach via Mordell–Weil lattices will naturally lead us to this classical setting (see Sect. 10.5, especially Example 10.54).

Remark 10.9 In the sequel the set $\{u_1, \ldots, u_6, v_0\}$ will be called the *standard generators* of $E_6^\vee$. Of course, we could choose a set of free generators of L consisting of minimal vectors (see Remark 9.28), but we find it more useful to use the standard ones, as they have more symmetry and capture, for instance, the structure of the *double six* as explained above.

10.2.4 Proof of Theorem 10.8 in the Case $r = 6$

In this section we consider in detail the family of rational elliptic surfaces

$$S \to \mathbb{A}^6$$

with fibres S_λ defined by the Weierstrass Eq. (10.4), i.e.

$$S_\lambda : \quad y^2 + txy = x^3 + (p_0 + p_1 t + p_2 t^2)\, y + q_0 + q_1 t + q_2 t^2 + t^3 \tag{10.16}$$

with parameter $\lambda = (p_0, p_1, p_2, q_0, q_1, q_2)$. The surface S_λ has generically only one reducible fibre, located at $t = \infty$, of Kodaira type I_3. Therefore, the Mordell–Weil lattice M_λ of S_λ is isomorphic to $E_6^\vee$. There are 54 minimal sections of height $4/3$; recall from the proof of Lemma 10.7 how one can extract from Theorem 7.12 that exactly half of the minimal sections have the property that both the x- and y-coordinate are linear in t. If we thus assume that the minimal section P takes the linear shape from (10.9), then substituting P in the Weierstrass equation, we get a system of four equations in a, b, d, e with coefficients in $\mathbb{Z}[\lambda] = \mathbb{Z}[p_i, q_j]$:

$$\begin{cases} ad = a^3 + ap_2 + 1 \\ ae = (3a^2 - d + p_2)b + (ap_1 + q_2 - d^2) \\ 0 = 3ab^2 - be - 2de + ap_0 + bp_1 + q_1 \\ 0 = b^3 - e^2 + bp_0 + q_0 \end{cases} \tag{10.17}$$

The first equation implies that $a \neq 0$; hence the first two equations determine $d, e \in \mathbb{Z}[\lambda][a, b, a^{-1}]$. Then the remaining two equations become polynomials in b with coefficients in $\mathbb{Z}[\lambda][a, a^{-1}]$, say

$$\phi_2(b) = \phi_3(b) = 0, \tag{10.18}$$

of respective degree 2 and 3. Using the resultant (or basic elimination), it follows that b also is a rational function of a for generic λ, and the field of definition of P is

$$k_0(P) := k_0(a, b, d, e) \;=\; k_0(a) \quad (k_0 = \mathbb{Q}(\lambda)).$$

As a consequence, the splitting field $\mathcal{K}_\lambda$ of the Mordell–Weil lattice $E_\lambda(k(t))$ is equal to

$$\mathcal{K}_\lambda = k_0(P_n \mid 1 \le n \le 27) = k_0(a_n \mid 1 \le n \le 27). \tag{10.19}$$

Then, eliminating b from (10.18), the resultant of $\phi_2(b)$, $\phi_3(b)$ with respect to b gives a monic equation of degree 27 in a, say

$$\Phi_\lambda(a) = 0.$$

The polynomial $\Phi(X) = \Phi_\lambda(X)$ has coefficients in $\mathbb{Z}[\lambda]$, a handful of which are given as follows:

$$\Phi(X) = \begin{cases} X^{27} + \left(p_2{}^2 - q_2\right) X^{26} + (-2p_1p_2 + 6p_2 + q_1)X^{25} \\ + \left(8p_2{}^3 + 2p_0p_2 + p_1{}^2 - 6p_1 - q_0 + 9\right) X^{24} \\ + \left(-14p_1p_2{}^2 + 20p_2{}^2 - 6q_1p_2 + 7p_0 - 2p_0p_1 + 2p_1q_2 - 8q_2\right) X^{23} \\ + \dots \\ + \left(p_0q_2 - 14p_2{}^4 - p_0p_2{}^2 + 16q_2p_2{}^2 + 4p_1{}^2p_2 \right. \\ \quad \left. -26p_1p_2 + 48p_2 - 2q_2{}^2 - 2p_1q_1 + 7q_1\right) X^4 \\ + \left(8p_2{}^3 + 2p_0p_2 + p_1{}^2 - 6p_1 - q_0 + 9\right) X^3 \\ + \left(13p_2{}^2 + p_0 - q_2\right) X^2 + 6p_2X + 1. \end{cases} \tag{10.20}$$

On the other hand, the specialization map sp_∞ takes all the linear sections P of height 4/3 to one and the same fibre component Θ_1 at $t = \infty$. The restriction sp^0_∞ of sp_∞ to the narrow Mordell–Weil lattice is a group homomorphism

$$sp^0_\infty : M^0_\lambda \longrightarrow \Theta^\#_0(k) \simeq \mathbb{G}_m(k) = k^\times.$$

Note that there is an ambiguity in the choice of the specialization map as $s \mapsto 1/s$ is an automorphism of $\mathbb{G}_m(k) = k^\times$; to fix this, we use the map $(x, y) \mapsto s$ from Lemma 9.8 *(ii)* for the isomorphism $\Theta^\#_0(k) \simeq k^\times$ above.

Lemma 10.10 *There is a unique* $\mathrm{Gal}(k/k_0)$*-equivariant homomorphism*

$$sp'_\infty : M_\lambda \to k^\times$$

extending the map sp^0_∞ *which satisfies*

$$sp'_\infty(P) = -\frac{1}{a} \tag{10.21}$$

for any linear section $P = (\mathrm{x}, \mathrm{y})$, $\mathrm{x} = at + b$, $\mathrm{y} = dt + e$.

Proof For ease of presentation, we will drop the suffix from $sp_\infty, \dots$ below. Since the index of M^0_λ in M_λ is equal to a prime number, $[E_6^\vee : E_6] = 3$ (so that M_λ/M^0_λ is cyclic), the required homomorphism sp' extending sp^0 is uniquely determined by the value $sp'(P)$ for a single section P, say P_1, not belonging to the narrow part M^0. Presently, we have $3P \in M^0$ and $P_i - P_j \in M^0$ for all $1 \le i, j \le 27$ by definition. A direct computation shows

$$sp^0(3P) = -1/a^3, \quad sp^0(P_i - P_j) = a_j/a_i \tag{10.22}$$

for a linear section P given by (10.9), and for any linear sections P_i, P_j with the corresponding coefficient a_i, a_j. Since the P_i generate M_λ, by defining $sp'(P)$ by (10.21), we obtain a well-defined homomorphism $sp' : M_\lambda \to k^\times$ with all required properties. □

Remark 10.11 Lemma 10.10 holds in any characteristic (contrary to Lemma 9.10, Example 9.11). A similar result can be derived for the multiplicative excellent family of type E_7.

Returning to the proof of Theorem 10.8 for E_6, we continue to take standard generators $P_1, \ldots, P_6, R_0$ of $M \simeq E_6^\vee$ as in (10.11) and define $\Omega_M = \{P_1, \ldots, P_{27}\}$ as in (10.14). Let

$$sp'(P_n) = s_n \ \ (1 \le n \le 27), \quad sp'(R) = r. \tag{10.23}$$

By (10.11), we have

$$s_1 \cdots s_6 = r^3. \tag{10.24}$$

Further, letting $sp'(P_i') = s_i'$, $sp'(P_{ij}'') = s_{ij}''$, we read off from (10.13) that we have, for $n > 6$,

$$s_n = s_i' = \frac{s_i}{r} \ \ (n = i + 6, 1 \le i \le 6), \quad s_n = s_{ij}'' = \frac{r}{s_i s_j} \ \ (13 \le n \le 27, 1 \le i < j \le 6), \tag{10.25}$$

where the n bijectively correspond to the pairs $\{i, j\}$ in the second relation.

Collecting all the data derived so far, we can spell out the universal polynomial $\Phi_\lambda(X)$:

$$\begin{aligned}\Phi_\lambda(X) &= \prod_{n=1}^{27}(X - a_n) \\ &= \prod_{n=1}^{27}(X + 1/s_n) \\ &= X^{27} + \epsilon_{-1}X^{26} + \epsilon_{-2}X^{25} + \cdots + \epsilon_4 X^4 + \epsilon_3 X^3 + \epsilon_2 X^2 + \epsilon_1 X + 1.\end{aligned} \tag{10.26}$$

Here ϵ_i is the i'th elementary symmetric polynomial of the s_n and ϵ_{-i} that of the $1/s_n$. By (10.20), the coefficients of $\Phi_\lambda(X)$ are polynomials in the coordinates of λ, and we have:

$$\begin{aligned}\epsilon_1 &= 6p_2 \\ \epsilon_{-1} &= p_2^2 - q_2 \\ \epsilon_2 &= 13p_2^2 + p_0 - q_2 \\ \epsilon_{-2} &= -2p_1p_2 + 6p_2 + q_1\end{aligned}$$

$$\epsilon_3 = \epsilon_{-3} = 8p_2^3 + 2p_0p_2 + p_1^2 - 6p_1 - q_0 + 9$$
$$\epsilon_4 = p_0q_2 - 14p_2^4 - p_0p_2^2 + 16q_2p_2^2 + 4p_1^2p_2 - 26p_1p_2 + 48p_2 - 2q_2^2 - 2p_1q_1 + 7q_1$$

Then we can solve the above equations for $p_0, \ldots, q_2$. However, the resulting solution has ϵ_{-2} in the denominator, e.g.

$$p_1 = \frac{-\epsilon_4 + \epsilon_1 - \epsilon_{-1}\epsilon_2 - \epsilon_{-1}^2 + 7\epsilon_{-2}}{2\epsilon_{-2}}.$$

At any rate, this implies that

$$\mathbb{Q}(p_i, q_j) = \mathbb{Q}(\epsilon_1, \epsilon_2, \epsilon_3, \epsilon_{-1}, \epsilon_{-2}, \epsilon_4) \subseteq \mathbb{Q}(s_1, \ldots, s_{27})^{W(E_6)}. \tag{10.27}$$

Furthermore, we have from (10.19), (10.21), (10.27),

$$\mathcal{K}_\lambda = \mathbb{Q}(p_i, q_j)(s_1, \ldots s_{27}) = \mathbb{Q}(s_1, \ldots s_{27}) = \mathbb{Q}(s_1, \ldots s_6, r) = \mathbb{Q}(s_1, \ldots s_5, s_7).$$

By Galois theory, we have $G = Gal(\mathcal{K}_\lambda/k_0) = W(E_6)$, in view of (10.8) and (10.27). We have thus proven the claims *(i)*, *(ii)* and *(iv)* of Theorem 10.8 ($r = 6$).

To prove *(iii)*, we need a new $W(E_6)$-invariant. Observe that, by (10.15) and (10.23), the image set of the 36 positive roots of $M^0 \cong E_6$ under the specialization homomorphism sp' is equal to

$$\Pi := \left\{ \frac{1}{r}, \ \frac{s_i}{s_j} \ (i < j), \ \frac{r}{s_i s_j s_k} \ (i < j < k) \right\}. \tag{10.28}$$

Let δ_1 denote the sum of all the elements in Π and their inverses, i.e.

$$\delta_1 = r + \frac{1}{r} + \sum_{i \neq j} \frac{s_i}{s_j} + \sum_{i<j<k} \left(\frac{r}{s_i s_j s_k} + \frac{s_i s_j s_k}{r} \right).$$

Then δ_1 belongs to the $W(E_6)$-invariants of

$$\mathbb{Q}[s_1, \ldots, s_6, r, s_1^{-1}, \ldots, s_6^{-1}, r^{-1}] \cong \mathbb{Q}[E_6^\vee],$$

and we can easily show that $p_1 = -\delta_1/2$ by checking the following identity in the Laurent polynomial ring:

$$\delta_1\epsilon_{-2} = -\epsilon_4 + \epsilon_1 - \epsilon_{-1}\epsilon_2 - \epsilon_{-1}^2 + 7\epsilon_{-2}.$$

Therefore we have

$$\mathbb{Q}[s_1^{\pm 1}, \ldots, s_6^{\pm 1}, r^{\pm 1}]^{W(E_6)} \supseteq \mathbb{Q}[\delta_1, \epsilon_1, \epsilon_2, \epsilon_3, \epsilon_{-1}, \epsilon_{-2}] = \mathbb{Q}[p_0, p_1, p_2, q_0, q_1, q_2], \tag{10.29}$$

where the relations showing the second equality are given by

$$\begin{cases} \delta_1 = -2p_1 \\ \epsilon_1 = 6p_2 \\ \epsilon_{-1} = p_2^2 - q_2 \\ \epsilon_2 = 13p_2^2 + p_0 - q_2 \\ \epsilon_{-2} = -2p_1p_2 + 6p_2 + q_1 \\ \epsilon_3 = 8p_2^3 + 2p_0p_2 + p_1^2 - 6p_1 - q_0 + 9 \end{cases}$$

and

$$\begin{cases} p_2 = \frac{1}{6}\,\epsilon_1 \\ p_1 = -\frac{1}{2}\,\delta_1 \\ p_0 = \epsilon_2 - \frac{1}{3}\,\epsilon_1^2 - \epsilon_{-1} \\ q_2 = -\epsilon_{-1} + \frac{1}{36}\,\epsilon_1^2 \\ q_1 = -\epsilon_1 + \epsilon_{-2} - \frac{1}{6}\delta_1\epsilon_1 \\ q_0 = 9 + 3\delta_1 + \frac{1}{4}\delta_1^2 - \frac{1}{3}\epsilon_{-1}\epsilon_1 - \frac{2}{27}\,\epsilon_1^3 + \frac{1}{3}\epsilon_1\epsilon_2 - \epsilon_3. \end{cases} \tag{10.30}$$

On the other hand, the ring $A := \mathbb{Q}[s_1^{\pm 1}, \ldots, s_{27}^{\pm 1}]$ is integral over $\mathbb{Q}[\lambda] = \mathbb{Q}[p_i, q_j]$, as each $-1/s_n$ satisfies $\Phi = 0$ (which is a monic equation with constant term ± 1 with coefficients in $\mathbb{Q}[\lambda]$). Hence $A^{W(E_6)}$ is also integral over $\mathbb{Q}[\lambda]$. But the fraction field of $A^{W(E_6)}$ is equal to that of $\mathbb{Q}[\lambda]$ by (10.27). This implies that $A^{W(E_6)} = \mathbb{Q}[\lambda]$, since $\mathbb{Q}[\lambda]$ is integrally closed. Together with (10.29), this proves

$$\mathbb{Q}[s_1^{\pm 1}, \ldots, s_6^{\pm 1}, r^{\pm 1}]^{W(E_6)} = \mathbb{Q}[\delta_1, \epsilon_1, \epsilon_2, \epsilon_3, \epsilon_{-1}, \epsilon_{-2}] = \mathbb{Q}[p_0, p_1, p_2, q_0, q_1, q_2]. \tag{10.31}$$

This gives the claim *(iii)* of Theorem 10.8 ($r = 6$) in the present precise form. Thus we have indeed verified that the family $\{S_\lambda\}$ is a multiplicative excellent family with Galois group $W(E_6)$ (or of type E_6). □

Remark 10.12 As for the first equality of (10.31), we record the following observation (cf. [KumS13]) for a similar statement for the root lattices E_7, E_8). The six fundamental representations of the Lie algebra $\mathbb{E}_6$ correspond to the fundamental weights in the following diagram which displays the standard labelling of these representations ([Bou68]):

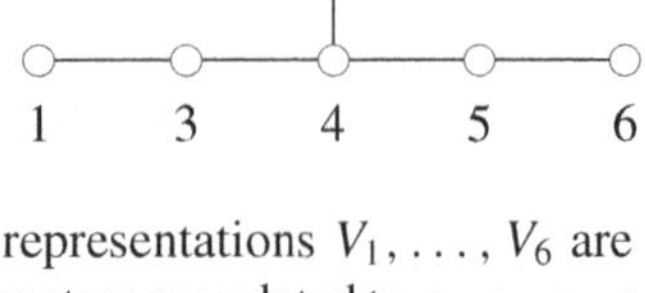

The dimensions of these representations $V_1, \ldots, V_6$ are $27, 78, 351, 2925, 351, 27$ respectively, and their characters are related to $\epsilon_1, \epsilon_2, \epsilon_3, \epsilon_{-1}, \epsilon_{-2}, \delta_1$ by the following nice formulas:

$$\chi_1 = \epsilon_1, \qquad \chi_2 = \delta_1 + 6, \qquad \chi_3 = \epsilon_2,$$
$$\chi_4 = \epsilon_3, \qquad \chi_5 = \epsilon_{-2}, \qquad \chi_6 = \epsilon_{-1}. \tag{10.32}$$

Therefore the first equality in (10.31) is nothing but the statement of Theorem 2.45 in the case of the present root lattice $L = E_6$.

10.2.5 Reverse Engineering of Linear Sections

In this section, we will show how one can reverse the construction given in the previous section to explicitly find rational elliptic surfaces S with Mordell–Weil lattice $E_r^\vee$ (here mostly for $r = 6$) and with a full set of generators of MWL(S) defined over the ground field. This is one of the reasons why the families in the preceding sections were coined *excellent.*

We keep the notation of the previous section, where we have studied the linear sections of the elliptic surface S_λ for generic λ. Fixing a set of standard generators $\{P_1, \ldots, P_6, R\}$ of the Mordell–Weil lattice $M \cong E_6^\vee$ as in (10.11), and taking their images under the specialization homomorphism sp'_∞ from (10.23), we obtain a 7-tuple

$$\xi = (s_1 \ldots, s_6; r)$$

satisfying the relation (10.24). I.e., the extended parameter ξ lies on the toric hypersurface

$$\mathcal{T} = \{s_1 \cdots s_6 = r^3\} \subset \mathbb{G}_m^7. \tag{10.33}$$

Conversely, the data $\xi = (s_1 \ldots, s_6; r)$ determines the elliptic surface S_λ by (10.30) after computing $s_7, \ldots, s_{27}$ by (10.25). We obtain the linear sections $P_n (n = 1, \ldots, 27)$ by reversing the elimination process combined with the key property (10.21) of the specialization homomorphism sp'_∞.

Following the definition of the linear sections P_n $(1 \le n \le 27)$ comprising the set Ω_M in (10.14), we distinguish three cases:

(i) $n \le 6$,
(ii) $7 \le n \le 12$ and
(iii) $n > 12$.

In the following, i, j, k, l will be distinct elements of the index set $I := \{1, 2, 3, 4, 5, 6\}$.

Theorem 10.13 *The coefficients of the sections $P_n = (a_n t + b_n, d_n t + e_n)$ $(1 \le n \le 27)$ of the elliptic surface S_λ are given in each case in a relatively simple, symmetric form:*

(i) For $n = i \le 6$, $P_n = P_i$ has

$$a_n = -\frac{1}{s_i}, \quad b_n = s_i + \frac{1}{s_i^2} + \sum_{j \ne i} (s'_j + s''_{ij}) - \frac{1}{3}\epsilon_1.$$

(ii) For $n = 6 + i, i \leq 6$, we have $s_n = s'_i$ and $P_n = P'_i$ has

$$a_n = -\frac{1}{s'_i}, \quad b_n = s'_i + \frac{1}{s'^2_i} + \sum_{j \neq i}(s_j + s''_{ij}) - \frac{1}{3}\epsilon_1.$$

(iii) For $n > 12$, we have $s_n = s''_{ij}$ for some $i \neq j$, and $P_n = P''_{ij}$ has

$$a_n = -\frac{1}{s''_{ij}}, \quad b_n = s''_{ij} + \frac{1}{s''^2_{ij}} + (s_i + s'_i + s_j + s'_j) + \sum_{\{k,l\} \cap \{i,j\} = \emptyset} s''_{kl} - \frac{1}{3}\epsilon_1.$$

In each case, the coefficients d_n, e_n of P_n are determined by a_n, b_n by the relations:

$$\begin{cases} d_n = a_n^{-1} + a_n^2 + p_2, \\ e_n = ((3a_n^2 - d_n + p_2)b_n - (a_n p_1 + q_2 - d_n^2))/a_n. \end{cases}$$

Proof The coefficients a_n are uniquely determined by s_n as $a_n = -1/s_n$ for all $n = 1, \ldots, 27$. Then the other coefficients b_n, d_n, e_n are obtained from the generic reverse elimination process upon substitution. Indeed, it is straightforward to verify that each $P_n = (a_n t + b_n, d_n t + e_n)$ thus defined satisfies the Weierstrass Eq. (10.4). □

Remark 10.14 With the above notation, let $\xi = (s_1, \ldots, s_6; r)$ and define the set Π_ξ by (10.28). Then the elliptic surface $S = S_\lambda$ is non-degenerate (in the sense that it has the Mordell–Weil lattice $M \simeq E_6^\vee$) if and only if the set Π_ξ does not contain 1, equivalently, if and only if the following condition holds:

$$r \neq 1, \; s_i \neq s_j \; (i < j), \; s_i s_j s_k \neq r \; (i < j < k). \tag{10.34}$$

For the proof, see Theorem 10.21 which treats more generally the non-degeneracy condition, but also points towards a better understanding of degenerations.

10.2.6 Formulation of an Algorithm (M)

For the sake of later applications, let us formulate the above results in the form of an *algorithm*:

Algorithm 10.15 (*Algorithm (M)*) The *input* is a 7-tuple of non-zero elements in any given field K of characteristic $\neq 2, 3$:

$$\xi = (s_1, \ldots, s_6; r) \quad \text{such that} \quad s_1 \cdots s_6 = r^3.$$

The *output* is:

(i) the parameter $\lambda = (p_i, q_j)$ defined by the formula (10.30) determining the elliptic surface S_λ from (10.4), and

(ii) the 27 linear solutions P_n : $\mathrm{x} = at + b, \mathrm{y} = dt + e$ of the Weierstrass equation of S_λ given by Theorem 10.13.

Remark 10.16 For any other Weierstrass equation which defines an (additive or multiplicative) excellent family of rational elliptic surfaces (also for other A, D, E-type), one can formulate a similar algorithm. They will be called in general *MWL-algorithms*, following [Shio15b].

Below we will develop a variant of Algorithm (M) giving a slight refinement.

10.2.7 Variant: A Refined Algorithm (M')

In this section we will refine Algorithm (M) slightly so that it even applies to include fields of characteristic 3. To this end, consider the following Weierstrass equation:

$$(M') \quad y^2 + txy = x^3 + m_0x^2 + (p_0 + p_1t)x + q_0 + q_1t + q_2t^2 + t^3 \quad (10.35)$$

with a new parameter:

$$\lambda' = (m_0, p_0, p_1, q_0, q_1, q_2).$$

It will very soon turn out that this time the base field K can be any field of characteristic $\neq 2$.

Consider the condition for a linear section

$$P: \ \mathrm{x} = at + b, \ \ \mathrm{y} = dt + e$$

to satisfy the new Weierstrass Eq. (10.35). Eliminating d and e and then taking the resultant of the two relations with respect to b as in 10.2.4, we obtain, in exactly the same way as before, an algebraic equation

$$\Phi(a) = 0$$

of degree 27 with coefficients in $\mathbb{Z}[m_0, p_0, p_1, q_0, q_1, q_2]$; explicitly, we have

$$\Phi(X) := X^{27} - q_2X^{26} + (q_1 - m_0)X^{25} + (9 - 6p_1 + p_1^2 - q_0)X^{24} + \dots \\ + (9 - 6p_1 + p_1^2 - q_0)X^3 + (p_0 - q_2)X^2 - m_0X + 1. \quad (10.36)$$

By comparing the coefficients of X^d in the two expressions (10.36) and (10.26), we find the first output for the algorithm (M'):

$$\begin{cases} m_0 = -\epsilon_1 \\ q_2 = -\epsilon_{-1} \\ p_0 = \epsilon_2 - \epsilon_{-1} \\ q_1 = \epsilon_{-2} - \epsilon_1 \\ q_0 = -\epsilon_3 + \frac{1}{4}(\delta_1 + 6)^2 \\ p_1 = -\frac{1}{2}\,\delta_1 \end{cases} \tag{10.37}$$

where ϵ_n, ϵ_{-n} or δ_1 are the same as before.

The second output is given by the following:

(i) For $n = i \leq 6$, $P_n = P_i$ has

$$a = -\frac{1}{s_i}, \quad b = s_i + \frac{1}{s_i^2} + \sum_{j \neq i} (s'_j + s''_{ij}).$$

(ii) For $n = 6 + i$, we have $s_n = s'_i$ and $P_n = P'_i$ has

$$a = -\frac{1}{s'_i}, \quad b = s'_i + \frac{1}{s_i'^2} + \sum_{j \neq i} (s_j + s''_{ij}).$$

(iii) For $n > 12$, we have $s_n = s''_{ij}$ for some $i \neq j$, and $P_n = P''_{ij}$ has

$$a = -\frac{1}{s''_{ij}}, \quad b = s''_{ij} + \frac{1}{s_{ij}''^2} + (s_i + s'_i + s_j + s'_j) + \sum_{\{k,l\} \cap \{i,j\} = \emptyset} s''_{kl}.$$

The coefficients d, e of the y-coordinate $y = dt + e$ are given by

$$d = a^2 + \frac{1}{a}, \quad e = \left(2a - \frac{1}{a^2}\right) b - \left(a^3 + \frac{1}{a^3} + 2\right) + am_0 + p_1 + \frac{1}{a} q_2.$$

Algorithm 10.17 (*Algorithm* (M')) The input is the same as for Algorithm 10.15. The output consists of the new parameter λ', defined by (10.37), and of the 27 linear sections P_n determined by the above *(i)–(iii)*.

Remark 10.18 The two Weierstrass equations (10.35) and (10.16), are actually related by the coordinate transformation

$$x \mapsto x + \frac{m_0}{3}, \quad y \mapsto y - \frac{m_0}{6} t.$$

Thus one may argue that Algorithm (M') provides no essential improvement compared to Algorithm (M), *except* that the new formulas (10.37) work in every characteristic p different from 2; in particular, the algorithm works even in case $p = 3$

(and it does not rely on the specialization homomorphism sp'_∞ (which exists, extending sp^0_∞, by Lemma 10.10), since we only used the elimination theory part of the construction).

10.3 Non-degeneracy Condition and Vanishing Roots

10.3.1 Excellent Families and Non-degeneracy Condition

First we reformulate Theorem 10.8, esp. *(iii)*, on the multiplicative excellent families of type E_r, as a more geometric statement in which the parameters $\lambda = (\lambda_n)$ are not restricted to be generic over $\mathbb{Q}$ (similar to Lemma 9.35). This will be instrumental for the applications to follow momentarily.

Theorem 10.19 *Fix $r = 6, 7$ or 8. Consider the morphism ψ from the algebraic torus $\mathcal{T} := \mathbb{G}_m^r$ to the affine space $\mathbb{A}^r$,*

$$\psi : \mathcal{T} \longrightarrow \mathbb{A}^r, \quad \psi(s_1, \ldots, s_r) = (\lambda) := (p_i, q_j),$$

which is defined by the formulas (10.10) [esp. (10.30) for $r = 6$] using $W(E_r)$-invariant functions. Then ψ is a finite Galois covering with Galois group $W(E_r)$, unramified outside the discriminant locus $\{D^2 = 0\}$, where D is the basic anti-invariant (see 2.31), and ψ induces an isomorphism of the quotient variety $\mathcal{T}/W(E_r)$ onto $\mathbb{A}^r$.

Proof The assertion is just a restatement about the invariant ring of an affine ring under the action of a finite group and the quotient of the corresponding affine variety, except for the claim that the map ψ is unramified outside the discriminant locus $D^2 = 0$. This claim is a consequence of the invariant theory of the Weyl group recalled in Chap. 2, which states that the Jacobian determinant of the map ψ is equal to D^2 which factors as the product of $e^\alpha - 1$ with α running over the positive roots of E_r; see (2.31) and (2.32). In case $r = 6$, we can directly verify this claim by calculating the Jacobian determinant of the formulas of p_i, q_j in (10.30) with respect to $\{s_i (i < 6), r\}$ (free variables), and by checking that it is equal to (2.31) up to some monomial in the Laurent polynomial ring. □

We continue to elaborate on the above excellent families. For any $\lambda \in \mathbb{A}^r$, let X_λ denote the affine surface defined literally by the given Weierstrass Eqs. (10.4), (10.5) or (10.6). (See Theorem 10.22 for the additive excellent families.)

The surface X_λ can be called the *affine Weierstrass model.* The elliptic surface (or the Kodaira–Néron model) S_λ is obtained from the affine surface X_λ by

(1) blowing up the A, D, E-singular points (if any) of X_λ corresponding to the reducible fibres at $t \neq \infty$ (let $\tilde{X}_\lambda \to X_\lambda$ denote the blow-up), and then
(2) adjoin to $\tilde{X}_\lambda$ the zero section (O) and the special fibre $f^{-1}(\infty)$ at $t = \infty$.

Thus S_λ is isomorphic to $X_\lambda \cup (O) \cup f^{-1}(\infty)$ if and only if $\tilde{X}_\lambda = X_\lambda$, i.e. S_λ has no reducible fibres at $t \neq \infty$ (as in condition (9.13)).

Let $\{X_\xi\}$ denote the base-changed family of $\{X_\lambda\}$ by the map ψ. Namely, for each $\xi \in \mathcal{T}$, we set $\lambda = \psi(\xi)$ and

$$X_\xi := X_{\psi(\xi)} = X_\lambda, \quad S_\xi := S_{\psi(\xi)} = S_\lambda \quad (\xi \in \mathcal{T}). \tag{10.38}$$

Furthermore, denoting by $\mathcal{R}(L)$ the set of roots of the generic narrow Mordell–Weil lattice L (which presently is the root lattice E_r), we set

$$\Pi_\xi = \Pi(L)_\xi := \{(e^\alpha)|_\xi;\ \alpha \in \mathcal{R}(L)\} \subset \mathbb{Z}[\xi, \xi^{-1}], \tag{10.39}$$

where we let $(e^\alpha)|_\xi$ denote the monomial in $\mathbb{Z}[\xi, \xi^{-1}]$ corresponding to the root $\alpha \in L$, evaluated at ξ. For example, for $L = E_6$, $\Pi(L)_\xi$ is equal to the set Π given by (10.28). [Obviously, the set (10.39) is a multiplicative analogue of the set defined before in the additive setting in Definition 9.26.]

In order to treat the non-degeneracy condition appropriately, we insert an auxiliary lemma. Recall that we are presently studying the Mordell–Weil lattice $M_\lambda \simeq L^\vee$, $L \simeq E_r$ $(r = 6, 7, 8)$, endowed with a specialization homomorphism

$$sp'_\infty : M_\lambda \to k^\times$$

into the multiplicative group.

Lemma 10.20 *Suppose that $\{u_i\ (1 \le i \le r)\}$ is a $\mathbb{Z}$-basis of $L^\vee$, and let $\{P_i\ (1 \le i \le r)\}$ be the corresponding basis of M_λ for λ generic over $\mathbb{Q}$. Let $sp'_\infty(P_i) = s_i$. Then the s_i are algebraically independent (cf. the proof of Theorem 10.8), and if we set $\xi := (s_i)$, then the group ring $\mathbb{Z}[L^\vee]$ is isomorphic to the Laurent polynomial ring $\mathbb{Z}[\xi, \xi^{-1}] := \mathbb{Z}[s_i, s_i^{-1}]$ which contains the element $e^\alpha|_\xi$ for each root α of L. Similarly, let Q_α denote the corresponding root in M_λ. Then we have*

$$(e^\alpha)|_\xi = sp'_\infty(Q_\alpha). \tag{10.40}$$

Proof Each root α of L can be expressed as a $\mathbb{Z}$-linear combination $\alpha = \sum_i n_i u_i$. Hence we have

$$e^\alpha = \Pi_i (e^{u_i})^{n_i}.$$

Clearly, $(e^{u_i})|_\xi = s_i$ so

$$(e^\alpha)|_\xi = \Pi_i s_i^{n_i}.$$

On the other hand, since $Q_\alpha = \sum_i n_i P_i$, we have

$$sp'_\infty(Q_\alpha) = \Pi_i sp'_\infty(P_i)^{n_i} = \Pi_i s_i{}^{n_i},$$

which proves the lemma. □

We now draw the connection to the non-degeneracy condition (9.13). This will be extremely useful for practical applications, as we shall see shortly.

Theorem 10.21 (Non-degeneracy conditions) *In the above set-up, assume that S_λ generically has type $\{T, L, M\}$. Let $\xi \in \mathcal{T}$, $\lambda = \psi(\xi)$ and suppose that the trivial lattice of S_ξ has T^- as a direct factor, i.e.*

$$\mathrm{Triv}(S_\xi) \simeq U \oplus T^- \oplus T^-_{new}.$$

Then the following conditions are equivalent:

(0) $M_\lambda \simeq L^\vee$,
(i) S_ξ *has no new reducible fibres (at* $t \neq \infty$*), i.e.* $T_{new} = \{0\}$,
(ii) the affine surface $X_\lambda = X_\xi$ *is smooth,*
(iii) ξ *does not belong to the locus* $\{D^2 = 0\}$,
(iv) for any positive root $\alpha \in L$, $(e^\alpha)|_\xi \neq 1$,
(v) for any positive root $Q_\alpha \in M$, $sp'_\infty(Q_\alpha) \neq 1$,
(vi) the set Π_ξ *does not contain* 1.

Proof We know *(0), (i) and (ii)* are equivalent, as this is exactly what we discussed for rational elliptic surfaces, phrased in three different ways.

Next *(iii)* ⇔ *(iv)* since the locus $\{D^2 = 0\}$ is the union of the loci $\{(e^\alpha)|_\xi = 1\}$ for the positive roots $\alpha \in L$ by (2.31) and (2.32).

Then *(iv)* ⇔ *(v)* ⇔ *(vi)* by Lemma 10.20, since (10.40) holds for any ξ once it holds generically.

Now *(0)* ⇒ *(v)* by the height formula. Namely, for each root $Q \in M_\lambda$, $\langle Q, Q\rangle = 2 + (Q.O)$ implies $(Q.O) = 0$. Thus the section (Q) is disjoint from the zero section (O), and in particular, we have $sp'_\infty(Q) \neq sp'_\infty(O) = 1$, which shows *(v)*.

Finally, we claim *(v)* ⇒ *(i)*. Indeed, if *(i)* does not hold, there is a new reducible fibre. Take a non-identity component Θ', then it has the self-intersection number $(\Theta')^2 = -2$. It defines a negative root in the E_8-frame of $\mathrm{NS}(S)$ which is orthogonal to $U \oplus T^-$ (the generic trivial lattice). Hence its image Q in the quotient $\mathrm{NS}(S)/(U \oplus T^-) \simeq M$ must be a root. But then, by assumption, Q is equivalent to O in $\mathrm{NS}(S)/\,\mathrm{Triv}(S) \simeq M_\lambda$, so

$$sp'_\infty(Q) = sp'_\infty(O) = 1,$$

showing that *(v)* cannot hold either. □

The last argument in the above proof will be studied from the viewpoint of *vanishing roots* in the next subsection. It should be noted that the proof of the above Theorems 10.19 and 10.21 can be formally modified to the case of additive excellent families. We expatiate on the statement of the latter.

Theorem 10.22 (Non-degeneracy conditions, additive setting) *Assume that* $\{S_\lambda\}$ *is an excellent family of additive type* $\{T, L, M\}$. *Let* $\mathbf{u} \in \mathbb{A}^r$, $\lambda = \psi(\mathbf{u})$ *and suppose that the trivial lattice of* $S_{\mathbf{u}}$ *has* T^- *as a direct factor, i.e.*

$$\mathrm{Triv}(S_{\mathbf{u}}) \simeq U \oplus T^{-} \oplus T^{-}_{new}.$$

Then the following conditions are equivalent:

(0) $M_\lambda \simeq L^\vee$,
(i) $S_{\mathbf{u}}$ *has no new reducible fibres (at* $t \neq \infty$*), i.e.* $T_{new} = \{0\}$,
(ii) *the affine surface* $X_\lambda = X_{\mathbf{u}}$ *is smooth,*
(iii) $\mathbf{u}$ *does not belong to the locus* $\{D^2 = 0\}$,
(iv) *for any positive root* $\alpha \in L$, $\alpha|_{\mathbf{u}} \neq 0$,
(v) *for any positive root* $Q_\alpha \in M$, $sp'_\infty(Q_\alpha) \neq 0$,
(vi) *the set* $\Pi_{\mathbf{u}}$ *does not contain* 0.

Remark 10.23 To restate Theorem 9.23 more precisely, the induced family of rational elliptic surfaces $\{S_\xi\}$ parametrized by $\xi \in \mathcal{T}$ is a global analogue of the simultaneous resolution of E_r-singularities due to Brieskorn and Grothendieck (cf. [Bri68, Bri71, Art74b, Slo80]).

10.3.2 Vanishing Roots

We continue by extending the above consideration beyond the framework of excellent families. For this purpose, we will introduce the notion of vanishing roots.

In general, suppose that S is a rational elliptic surface with zero section O of type $\{T, L, M\}$ in the notation of Theorem 8.8. Recall that $T = \oplus_v T_v$, where v runs over the places of the reducible fibres $F_v = f^{-1}(v)$ and T_v^- is the $\mathbb{Z}$-span of the irreducible components $\Theta_{v,i} (i > 0)$ of F_v other than the identity component $\Theta_{v,0}$. The quotient group of NS(S) by the subgroup $Triv(S) := U \oplus T^-$ is isomorphic to the Mordell–Weil group MW(S), (where U is as usual the $\mathbb{Z}$-span of the zero section (O) and the fibre class F; see Theorem 6.5). Let π be the natural projection map

$$\pi : \mathrm{NS}(S) \to \mathrm{MW}(S).$$

Now consider the following set of *negative roots* (by which we mean the classes of (-2)-divisors) in the Néron–Severi lattice NS(S):

$$\mathcal{D}(S) := \{D \in \mathrm{NS}(S) \mid D^2 = -2, D \perp (O), F\} \subset U^\perp \subset \mathrm{NS}(S). \qquad (10.41)$$

This corresponds to the set of all roots of the frame E_8 (by Theorem 6.13), and hence the total number of negative roots on any S is equal to 240:

$$\#\mathcal{D}(S) = \tau(E_8) = 240. \qquad (10.42)$$

Consider the restriction of π to the set of negative roots:

$$\pi_S := \pi|_{\mathcal{D}(S)} : \ \mathcal{D}(S) \to \mathrm{MW}(S). \qquad (10.43)$$

Definition 10.24 A negative root $D \in \mathcal{D}(S)$ is called a *vanishing root on* S if the image $\pi_S(D)$ vanishes in $\mathrm{MW}(S)$, i.e. $\pi_S(D) = O$.

Proposition 10.25 *Each irreducible component* $\Theta_{v,i}$ $(i > 0)$ *(disjoint from the zero section* O*) is a vanishing root. The set of vanishing roots on* S *coincides with the set* $\mathcal{R}(T)$ *of roots in* $T = \oplus_v T_v$, *and hence the number of vanishing roots on* S *is equal to*

$$\tau(T) = \sum_v \tau(T_v),$$

where $\tau(T)$ *denotes the number of roots of a root lattice.*

Proof This is obvious since $\mathrm{Ker}(\pi) = \mathrm{Triv}(S) = U \oplus T^-$, so

$$\mathrm{Ker}(\pi) \cap U^\perp = T^- = \oplus_v T_v^-,$$

where each T_v is a root lattice of ADE-type. □

On the other hand, there are usually negative roots D which are not vanishing roots, i.e. $\pi_S(D) \neq O$. Recall, for instance, that for any integral section (cf. Definition 8.25) $P \in MW(S)$, the divisor class of

$$D_P := (P) - (O) - F \tag{10.44}$$

is a negative root in $\mathcal{D}(S)$ such that $\pi_S(D_P) = P \neq O$; cf. (6.7).

As an illustration of the concept of vanishing roots, let us look at some basic cases.

Proposition 10.26 *Consider the first three cases of Theorem 8.8, where* $T = \{0\}$, A_1 *or* A_2.

(i) First, in case $T = 0$, $L = M = E_8$, *there are no vanishing roots on* S, *and the map* π_S *defines a bijective correspondence*

$$D_P \longleftrightarrow P$$

from $\mathcal{D}(S)$ *to the set of 240 roots in* $MW(S) \simeq E_8$. *Conversely, such a bijection holds only in case* $T = 0$, $L = M = E_8$.

(ii) Next, in case $T = A_1$, $L = E_7$, $M = E_7^\vee$, *let* Θ_1 *denote the non-identity component of the unique reducible fibre with two irreducible components on* S. *Then the negative roots in* $\mathcal{D}(S)$ *are listed as follows:*

- *two vanishing roots:* $\pm\Theta_1$;
- *126* D_P *with* $\pi_S(D_P) = P$ *for 126 roots* P *of* $L = M^0 \simeq E_7$;
- *56* D_P *with* $\pi_S(D_P) = P$ *for 56 minimal sections* P *of height* $\frac{3}{2}$ *of* $M \simeq E_7^\vee$;
- *another 56 negative roots:* $D_P + \Theta_1$ *for the same* P *as above.*

(iii) Third, in case $T = A_2$, $L = E_6$, $M = E_6^\vee$, *let* Θ_1, Θ_2 *denote the non-identity components of the unique reducible fibre with three irreducible components on* S. *Then the negative roots in* $\mathcal{D}(S)$ *are listed as follows:*

- *six vanishing roots:* $\pm\{\Theta_1, \Theta_2, \Theta_1 + \Theta_2\}$;
- *72* D_P *with* $\pi_S(D_P) = P$ *for 72 roots* P *of* $L = M^0 \simeq E_6$;
- *54* D_P *with* $\pi_S(D_P) = P$ *for 54 minimal sections* P *of height* $\frac{4}{3}$ *of* $M \simeq E_6^\vee$;
- *a further 27 negative roots* $D_P + \Theta_1$ *for those* P *intersecting* Θ_1;
- *another 27 negative roots:* $D_P + \Theta_1 + \Theta_2$ *for the same* P *as above;*
- *analogously,* 2×27 *negative roots for the remaining 27 minimal sections* P *intersecting* Θ_2.

Proof It is elementary to check that all the given classes define negative roots in $U^\perp$, i.e. elements of $\mathcal{D}(S)$. It thus remains to compare the cardinalities.

(i) is obvious. In case *(ii)*, the given list contains

$$2 + 1 \times 126 + 2 \times 56 = 240$$

elements. By (10.42), it exhausts all the negative roots on S and the list is complete.

In case *(iii)*, the given lists contains

$$6 + 1 \times 72 + 3 \times 54 = 240$$

negative roots on S, which shows similarly that the list is complete. □

More generally, for any S of type $\{T, L, M\}$ with $T \neq \{O\}$, the map π_S is a surjective map from $\mathcal{D}(S)$ to $\{O\} \cup \mathcal{P}$, where $\mathcal{P}$ denotes the set of integral sections in S. The *multiplicity* $\mu(P)$ of an integral section P (such as the multiplicity $1, 2, 3$ appearing in the above Proposition 10.26) is defined by

$$\mu(P) = \#\pi_S^{-1}(P).$$

By definition, the following formula holds true:

$$240 = \tau(T) + \sum_{P\in\mathcal{P}} \mu(P),$$

which can be called the *conservation law* of the number of roots.

We shall now extend the above notion of vanishing roots slightly, in order to apply it to families of rational elliptic surfaces and their degenerations in the sequel. To this end, recall that given a type $\{T, L, M\}$ (not necessarily admissible), the narrow Mordell–Weil lattice L comes with a fixed embedding into E_8, or equivalently, $T \hookrightarrow E_8$ is given. Given any sublattice $N \hookrightarrow E_8$, consider the subset

$$\mathcal{D}(S, N) = \mathcal{D}(S) \cap (N^-)^\perp \subset \mathcal{D}(S).$$

Definition 10.27 A negative root $D \in \mathcal{D}(S, N)$ is called a *vanishing root relative to* N (or orthogonal to N) if $\pi_S(D) = O$.

Usually, we will apply this definition to capture the contribution from the reducible fibres by setting $N = T$:

$$L \supset \mathcal{D}(S, T) = \mathcal{D}(S) \cap (T^-)^\perp = \{D \in \mathrm{Triv}(S)^\perp \subset \mathrm{NS}(S) \mid D^2 = -2\}. \quad (10.45)$$

This set consists of the negative roots in the L-frame $\mathcal{D}(S, T)$, and is very well suited for studying degenerations in families.

Lemma 10.28 *The map $D \mapsto P = \pi_S(D)$ gives a bijective correspondence from $\mathcal{D}(S, T)$ to the set $\mathcal{R}(L)$ of all roots in L:*

$$\mathcal{R}(L) := \{P \in L \mid \langle P, P\rangle = 2\}.$$

The inverse correspondence is defined by $P \mapsto D_P$:

$$D_P = (P) - (O) - F.$$

Proof For any root P of the narrow Mordell–Weil lattice L, we have $(P.O) = 0$ by the height formula

$$\langle P, P\rangle = 2 + 2(P.O).$$

Hence $\varphi(P)^2 = -2$ and $\varphi(P) \in \mathcal{D}(S, T)$. Conversely, note that we defined in (6.16) an injective map with the same letter φ from $M = \mathrm{MW}(S)$ to $\mathrm{NS}(S) \otimes \mathbb{Q}$, in trying to split the homomorphism $\pi : \mathrm{NS}(S) \to M$. It is easy to check that φ induces a map from $L = M^0 \to \mathrm{NS}(S)$, and it sends any root $P \in \mathcal{R}(L)$ to $D = \varphi(P)$ defined above. □

Clearly, the lemma gives

$$\#\mathcal{D}(S, T) = \#\mathcal{R}(L) = \tau(L). \quad (10.46)$$

(When $T = 0$, A_1 resp. A_2 and $L = E_8$, E_7 resp. E_6, this reduces to the setting from Proposition 10.26.) For later reference, we also note the following immediate corollary:

Corollary 10.29 *The set $\mathcal{D}(S, T)$ contains no vanishing roots (for π).*

We will continue to discuss the notion of vanishing roots in the framework of families in the following sections. An application of vanishing roots to the degeneration of cubic surfaces will follow in Sect. 10.5.6.

10.3.2.1 Families of Rational Elliptic Surfaces

Our next aim is to apply the above set-up of vanishing roots to families of rational elliptic surfacesin order to compare specialization within the family against the degeneracy of the singular fibres (à la the generality condition (9.13), or as in Theorem 10.21).

To fix the notation, consider a family of rational elliptic surfaces with parameter λ which generically has the *admissible type* $\{T, L, M\}$ as before; here we impose the essential assumption that L is a root lattice (of positive rank; compare Corollary 8.24). Upon specializing the parameter to some λ^0 we obtain a rational elliptic surface S_0 of type $\{T_0, L_0, M_0\}$, say (where L_0 need not be a root lattice anymore). As in Theorem 10.21, we assume that T_0 is a direct sum

$$T_0 = T \oplus T_{new}. \tag{10.47}$$

Here T_{new} is a direct sum of the root lattices T_v at the new reducible fibres which occur upon specialization. Note that this induces embeddings

$$T \hookrightarrow T_0, \quad \mathcal{D}(S_0, T_0) \hookrightarrow \mathcal{D}(S_0, T) = \mathcal{D}(S, T); \tag{10.48}$$

in particular, the projection map

$$\pi_0 : \mathrm{NS}(S_0) \longrightarrow M_0 = \mathrm{MW}(S_0) \cong \mathrm{NS}(S_0)/\,\mathrm{Triv}(S_0) \cong E_8/T_0$$

factors through π. Using the concept of vanishing roots, we can easily characterize abstractly when there appear additional reducible fibres upon specialization (under the assumption (10.47) guaranteeing that the singular fibres encoded in T are preserved).

Theorem 10.30 *In the above set-up, the following conditions are equivalent:*

(i) the Mordell–Weil groups satisfy $M \cong M_0$;
(ii) the narrow Mordell–Weil groups satisfy $L \cong L_0$;
(iii) the lattices of the singular fibres satisfy $T \cong T_0$;
(iv) $T_{new} = 0$;
(v) in $\mathcal{D}(S_0, T)$ there are no vanishing roots for π_0;
(vi) the vanishing roots in $\mathcal{D}(S, T)$ (for π) and in $\mathcal{D}(S_0, T_0)$ (for π_0) are the same.

Remark 10.31 A more catchy way to phrase condition *(v)* would be:

In $\mathcal{D}(S, T)$ there are no vanishing roots for π_0.

However, that requires to interpret the statement via (10.48), since a priori the projection π_0 (on S_0) does not interact with $\mathcal{D}(S, T)$.

Proof The equivalence of conditions *(i)* through *(iv)* should be clear, same for *(v)* $\Leftrightarrow$ *(vi)*. The fact that *(ii)* $\Rightarrow$ *(v)* follows from Corollary 10.29 since with $L \cong L_0$ we can identify π and π_0 as well as $\mathcal{D}(S, T)$ and $\mathcal{D}(S_0, T_0)$. Finally, to see that *(v)* $\Rightarrow$ *(iv)*, assume that $T_{new} \neq 0$. Hence there is a root $\Theta \in T_{new}^- \cap \mathcal{D}(S_0, T)$ (an additional fibre component on S_0), so by construction, $\pi_0(\Theta) = 0$, i.e. Θ is a vanishing root for π_0 in $\mathcal{D}(S_0, T)$. □

10.3.2.2 Excellent Families

To round-up the preceding discussion, we shall now focus even more on the setting which prompted our whole investigation: an *excellent family* of rational elliptic surfaces $\{S_\lambda\}$ as in Definition 10.6 (both of additive and multiplicative type). Our aim is to gain more explicit control over the degeneration of the singular fibres within the excellent family than already provided by Theorem 10.30 (in the vein of Theorem 10.21).

In the given set-up, the generic member S has an *admissible type* $\{T, L, M\}$, with a root lattice L of positive rank, and the family is endowed with a specialization homomorphism

$$sp'_v : M \to k \quad \text{or} \quad k^\times,$$

extending the restriction sp^0_v from the narrow Mordell–Weil lattice L to M. Given a $\mathbb{Z}$-basis $\{P_i\ (1 \le i \le r)\}$ of the Mordell–Weil lattice M modulo torsion, we suppose that the splitting field $\mathcal{K}_\lambda$ of the Mordell–Weil group M over $k_0 = \mathbb{Q}(\lambda)$ is generated over the base field $\mathbb{Q}$ by the special values $s_i = sp'_v(P_i)$ (following notation from the multiplicative case). In this situation, any $\mathbb{Z}$-basis of L, say $\{\alpha_i\ (1 \le i \le r)\}$, and hence the set $\mathcal{R}(L) = \{\alpha_i\ (i = 1, \dots, \tau(L))\}$ of all roots in L is defined over $\mathcal{K}_\lambda = \mathbb{Q}(s_i\ (1 \le i \le r))$. Let $\xi = (s_1, \dots, s_r)$ determine $\lambda = \psi(\xi)$. We are interested in the specialization

$$\xi \to \xi^0 \in K \tag{10.49}$$

for some intermediate field $k_0 \subset K \subset k(\lambda)$. This results in a rational elliptic surface

$$S_0 = S_{\xi^0} = S_{\lambda^0}$$

as featured in Theorem 10.30. In order to make sense out of what follows, we will need the following quite natural assumption:

Assumption 10.32 The special fibre of S at v does not change its Kodaira type upon the specialization (10.49).

We point out that this non-degeneracy condition is automatically fulfilled for the excellent families, both additive and multiplicative, of type A_2, E_6, E_7, E_8 studied throughout the last two chapters (compare especially Theorem 10.21). Indeed, this shows once again that in those cases, the assumption

$$T_0 = T \oplus T_{new}$$

from Theorem 10.30 always holds true (though it need not persist for other excellent families, as it depends on the precise model of the chosen family), and that the generality condition (9.13) is equivalent to $T_{new} = 0$ (i.e. Theorem 10.30 *(iv)*).

Theorem 10.33 *In the above excellent set-up, Theorem 10.30 has another equivalent condition:*

(vii) *the special values* $sp'_v(\alpha_i)$ $(1 \le i \le \tau(L))$ *are distinct from the neutral element upon the specialization* (10.49).

Proof The proof follows closely (part of) the proof of Theorem 10.21. Indeed, the specialization homomorphisms sp'_v factor through π on S resp. through π_0 on S_0. Hence a vanishing root $\alpha \in \mathcal{D}(S_0, T)$ for π_0 (as in *(v)*) indeed yields the neutral element e under evaluation of the specialization homomorphism:

$$sp'_v(\alpha)(\xi^0) = e,$$

proving *(vii)* ⇒ *(v)*. Conversely, given a root (i.e. section) Q_0 in the narrow Mordell–Weil lattice L_0 of S_0, the height pairing implies that $(Q_0.O) = 0$, hence

$$sp'_v(Q_0) \neq e$$

(since (Q) always meets the same fibre component as (O)). On the other hand, by *(ii)*, Q_0 arises from a root $\alpha \in \mathcal{R}(L)$ upon specializing $\xi \to \xi^0$, so

$$sp'_v(\alpha)(\xi^0) = sp'_v(Q_0) \neq e.$$

This shows that *(ii)* ⇒ *(vii)* and completes the proof of Theorem 10.33. □

In the additive excellent setting, we have already seen and used analogues of the above two theorems around Theorem 10.21. Therefore we postpone the study of further examples and applications to Sect. 10.4.3.

10.4 Examples and Applications

Once a multiplicative excellent family is given, we have various applications of it to number theory and algebraic geometry as explored for the additive case in Sect. 9.5. In particular, we can systematically construct explicit examples of semi-stable rational elliptic surfaces over $\mathbb{Q}$ (i.e. without additive singular fibres) with interesting properties such as:

- big Galois over $\mathbb{Q}$,
- small Galois over $\mathbb{Q}$,
- controlled degeneration via "vanishing roots",
- polynomial sections

(compare Problem 9.6). The corresponding results using additive excellent families from Sect. 9.4.2 have been treated in Sect. 9.5, but the multiplicative family is strictly necessary for treating the semi-stable case (which is more flexible than the additive case and more general with respect to moduli, cf. Remark 9.17).

10.4.1 *Big Galois over* $\mathbb{Q}$

Fix $r = 6, 7$ or 8. By Hilbert's irreducibility theorem (cf. [Ser70]), Theorem 10.8 *(i)* implies that for most choices of $\lambda^0 \in \mathbb{Q}^r$, the polynomial $\Psi(X, \lambda^0) \in \mathbb{Q}[X]$ has the same Galois group $W(E_r)$ as the generic case. Conversely, the existence of a single example $\lambda^0 \in \mathbb{Q}^r$ with Galois group $W(E_r)$ will prove Theorem 10.8 *(i)* for generic λ, in view of [Ser70, 9.2, Proposition 2]. Let us exhibit such an explicit example of $\lambda^0 \in \mathbb{Q}^r$ in the case $r = 6$:

Example 10.34 Take $\lambda^0 = (1, \ldots, 1)$, i.e. $p_i = q_i = 1$ $(i = 0, 1, 2)$ in (10.4). Thus the elliptic surface S_{λ^0} (and later on the cubic surface V_{λ^0}, see Sect. 10.5) is defined by the Weierstrass equation:

$$y^2 + txy = x^3 + (1 + t + t^2)x + 1 + t + t^2 + t^3.$$

Then the polynomial $\Psi(X, \lambda^0)$ becomes

$$\begin{aligned}\Psi(X, \lambda^0) = X^{27} &+ 5X^{25} + 13X^{24} - X^{23} + 76X^{22} + 19X^{21} + 99X^{20} + 85X^{19} \\ &+ 122X^{18} + 133X^{17} + 222X^{16} + 232X^{15} + 450X^{14} + 340X^{13} \\ &+ 546X^{12} + 650X^{11} + 369X^{10} + 320X^9 + 287X^8 + 151X^7 \\ &+ 103X^6 + 82X^5 + 31X^4 + 13X^3 + 13X^2 + 6X + 1.\end{aligned}$$

This polynomial in $\mathbb{Q}[X]$ has Galois group $W(E_6)$, so we obtain the algebraic equation of degree 27 with Galois group $W(E_6)$.

Proof We can use the same argument as in the additive case (Example 9.42). Look at the factorization of $\Psi(X, \lambda^0)$ mod p into irreducible factors in $\mathbb{F}_p[X]$, and check that it has cycle type $(9)^3$ for $p = 23$, and $(2)(5)^3(10)$ for $p = 43$. Then the claim follows from Lemma 9.43. □

While the above proof is the same as for the additive case, the resulting $W(E_6)$-extension $\mathcal{K} = \mathcal{K}_{\lambda^0}$ is given with a "multiplicative" structure, that is, the 27 roots $\{a_i\}$ form a set of 27 units, stable under the Galois group $W(E_6)$. One could well ask what the structure of the unit group of $\mathcal{K}$ will be as $W(E_6)$-module.

For an explanation how to compute the zeta function of the rational elliptic surface S_{λ^0} from combinatorial data encoded in the polynomial Ψ, see Sect. 10.5.2. Especially Remark 10.52 will relate to Example 10.34.

Remark 10.35 For explicit multiplicative examples with Galois group $W(E_7)$ or $W(E_8)$, the reader is referred to [KumS13].

10.4.2 Small Galois over $\mathbb{Q}$

In a different direction, it is now easy to engineer rational elliptic surfaces over $\mathbb{Q}$ with Mordell–Weil lattice $E_6^\vee$ fully defined over $\mathbb{Q}$, as we can use the Algorithms 10.15 and 10.17 for this purpose.

Example 10.36 Take $\xi = (1, 2, 3, 4, -1, -\frac{1}{3}; 2)$. From Algorithm 10.17, we obtain

$$\lambda' = \left(-\frac{21}{4}, -\frac{175}{3}, \frac{68}{9}, \frac{105625}{324}, -\frac{175}{3}, -\frac{21}{4}\right)$$

so that the Weierstrass equation is given by

$$y^2 + txy = x^3 - \frac{21x^2}{4} + \left(\frac{68t}{9} - \frac{175}{3}\right)x + \frac{105625}{324} - \frac{175t}{3} - \frac{21t^2}{4} + t^3. \tag{10.50}$$

Let $E_{\lambda'}$ be the elliptic curve over $\mathbb{Q}(t)$ defined by (10.50). Then $E_{\lambda'}$ has Mordell–Weil lattice $E_{\lambda'}(\mathbb{Q}(t)) \cong E_6^\vee$. The 27 linear sections $P_n = (a_n t + b_n, d_n t + e_n)$ are given as follows. The first six:

$$P_1 = \left(-t, \frac{325}{18}\right), \quad P_2 = \left(\frac{19}{6} - \frac{t}{2}, \frac{395}{36} - \frac{7t}{4}\right),$$
$$P_3 = \left(\frac{40}{9} - \frac{t}{3}, \frac{385}{54} - \frac{26t}{9}\right), \quad P_4 = \left(\frac{85}{16} - \frac{t}{4}, \frac{2435}{576} - \frac{63t}{16}\right),$$
$$P_5 = \left(t + \frac{20}{3}, 2t - \frac{5}{18}\right), \quad P_6 = \left(3t + \frac{20}{3}, \frac{28t}{3} + \frac{5}{18}\right).$$

The second six:

$$P_7 = \left(\frac{19}{3} - 2t, \frac{7t}{2} - \frac{1}{9}\right), \quad P_8 = \left(\frac{25}{4} - t, -\frac{25}{36}\right),$$
$$P_9 = \left(\frac{55}{9} - \frac{2t}{3}, -\frac{19t}{18} - \frac{35}{27}\right), \quad P_{10} = \left(\frac{35}{6} - \frac{t}{2}, -\frac{7t}{4} - \frac{85}{36}\right),$$
$$P_{11} = \left(2t + 15, \frac{9t}{2} + \frac{365}{9}\right), \quad P_{12} = \left(6t + \frac{115}{3}, \frac{217t}{6} + \frac{1945}{9}\right).$$

The remaining fifteen:

$$P_{13} = \left(8 - t, -\frac{107}{18}\right), \quad P_{14} = \left(\frac{55}{6} - \frac{3t}{2}, \frac{19t}{12} - \frac{395}{36}\right),$$
$$P_{15} = \left(\frac{35}{3} - 2t, \frac{7t}{2} - \frac{205}{9}\right), \quad P_{16} = \left(\frac{t}{2} - \frac{15}{2}, \frac{9t}{4} + \frac{245}{36}\right),$$

$$P_{17}=\left(\frac{t}{6}-\frac{115}{18},\frac{217t}{36}-\frac{1615}{108}\right),\quad P_{18}=\left(\frac{40}{3}-3t,\frac{26t}{3}-\frac{565}{18}\right),$$

$$P_{19}=\left(\frac{85}{4}-4t,\frac{63t}{4}-\frac{715}{9}\right),\quad P_{20}=\left(t-\frac{20}{3},2t-\frac{245}{18}\right),$$

$$P_{21}=\left(\frac{t}{3}-\frac{20}{9},\frac{28t}{9}-\frac{1105}{54}\right),\quad P_{22}=\left(\frac{125}{3}-6t,\frac{215t}{6}-\frac{2225}{9}\right),$$

$$P_{23}=\left(\frac{3t}{2}-\frac{25}{6},\frac{35t}{12}-\frac{725}{36}\right),\quad P_{24}=\left(\frac{t}{2}+\frac{1}{2},\frac{9t}{4}-\frac{619}{36}\right),$$

$$P_{25}=\left(2t-1,\frac{9t}{2}-\frac{175}{9}\right),\quad P_{26}=\left(\frac{2t}{3}+\frac{25}{9},\frac{35t}{18}-\frac{325}{27}\right),$$

$$P_{27}=\left(\frac{125}{18}-\frac{t}{6},\frac{175}{108}-\frac{215t}{36}\right).$$

The first seven P_i $(1\le i\le 7)$ generate $M=E_{\mathcal{X}'}(\mathbb{Q}(t))$. Finally, the height 2 section (root) of M, $R=P_1-P_7$, is given by:

$$R=\left(2t^2+\frac{23t}{2}+\frac{125}{18},2t^3+\frac{115t^2}{6}+\frac{1525t}{36}+\frac{175}{108}\right).$$

The parameter was chosen to fit the non-degeneracy condition *(vi)* from Theorem 10.21 which thus implies that $E_{\mathcal{X}'}$ has Mordell–Weil lattice $E_6^\vee$. Given this, the example is an easy application of Algorithm 10.17.

10.4.3 Degenerations via Vanishing Roots

We now return to the idea of vanishing roots and apply it with a view towards degenerations (and how to control them) for our two key excellent families: those of type A_2 and E_6.

Example 10.37 (*Multiplicative hexagonal family*) Recall the multiplicative excellent hexagonal family from Sect. 10.1.2. Here generically the narrow Mordell–Weil lattice is $L\cong A_2$ with roots

$$\mathcal{R}(L)=\{P_i-P_j;\ 1\le i,j\le 3\}.$$

Since $sp^0_\infty(P_i-P_j)=\mu_i/\mu_j$ by (10.3), Theorem 10.33 shows that there are additional reducible fibres if and only if there is some equality $\mu_i=\mu_j$ $(i\ne j)$. Indeed we find

parameters	T_0	L_0	M_0	Thm. 8.8
$\mu_i=\mu_j\ne\mu_k$	A_5+2A_1	$\langle 6\rangle$	$\langle 1/6\rangle\times\mathbb{Z}/2\mathbb{Z}$	No. 53
$\mu_i=\mu_j=\mu_k$	$A_5+A_2+A_1$	0	$\mathbb{Z}/6\mathbb{Z}$	No. 66

We leave it to the reader to verify the additional reducible fibre at $t = \mu_k$ (in both cases) and the height pairing on the six base points of the cubic pencil (10.1).

10.4.3.1 Multiplicative Excellent Family of Type E_6

We now turn to the multiplicative excellent family S_λ of (admissible) type E_6 from (10.4). Recall that Assumption 10.32 as well as the assumption from Theorem 10.30 are automatically satisfied by the very definition of E_λ. Hence the conditions from Theorems 10.30 and 10.33 are equivalent to the non-degeneracy condition (9.13). These can be put into a very nice framework presently because of our good knowledge about the structure of the generic narrow Mordell–Weil lattice $L \cong E_6$, especially its (positive) roots and its symmetries.

Recall from Sect. 10.2.5 that we decided to investigate the excellent family using the extended parameter

$$\xi = (s_1, \ldots, s_6; r) \in \mathcal{T} = \{s_1 \cdots s_6 = r^3\} \subset \mathbb{G}_m^7,$$

the toric hypersurface from (10.33). Since we know the positive roots of E_6 from Sect. 10.2.3, we can spell out condition *(vii)* from Theorem 10.33 explicitly. To this end, we use the set corresponding to the 36 positive roots of E_6 from (10.28), with potential multiplicities of elements taken into account:

$$\Pi_\xi = \left\{ \frac{1}{r},\ \frac{s_i}{s_j}\ (i < j),\ \frac{r}{s_i s_j s_k}\ (i < j < k) \right\}. \tag{10.51}$$

Of course, $\xi \in \mathcal{T}$ is *non-degenerate* if and only if Π_ξ does not contain 1. Equivalently, the generality condition (9.13) holds, or either condition from Theorems 10.30 and 10.33 is satisfied.

Example 10.38 On the hypersurface $\{r = 1\} \subset \mathcal{T}$, there is generically one vanishing root in Π_ξ (i.e. $\nu(\xi) = 1$). Using that $2\nu(\xi)$ is equal to the number of roots in the new part T_{new}, it follows that S_λ will have a fibre of type I_2 at $t = \sum_{i=1}^6 s_i^{-1}$, thus realizing a rational elliptic surface of type $(A_2 + A_1, A_5, A_5^\vee)$ by Theorem 8.8.

We encourage the reader to check that this indeed gives a new excellent family:

Proposition 10.39 *The hypersurface $\{r = 1\} \subset \mathcal{T}$ induces a multiplicative excellent family of type A_5, parametrized by the maximal torus of $SL(6)$:*

$$\mathcal{T}' = \{(s_1, \ldots, s_6) \in \mathbb{G}_m^6 | s_1 \cdots s_6 = 1\}.$$

By the same method as above, we can also study further degenerations of S_λ under specialization of parameters, thus realizing a number of further excellent families and at the same time proving the existence of (excellent families of) rational elliptic surfaces for many types $\{T, L, M\}$ from Theorem 8.8. A few more examples may be extracted from Sect. 10.5.6 (where they appear as degenerations of cubic surfaces); we leave the details to the interested reader.

10.4.4 Application to Polynomial Sections

We conclude this section by observing the analogue of Lemma 9.35 for polynomial sections on specializations of excellent families of rational elliptic surfaces (in particular of type E_6, E_7, E_8).

Lemma 10.40 *Let E_λ be a multiplicative excellent family of rational elliptic surfaces over $k_0(\lambda)$ of admissible type $\{T, L, M\}$ (so L is a root lattice) with specialization homomorphism sp'_v. Assume that the Mordell–Weil group M admits no torsion. Specialize $\lambda \to \lambda^0$ in such a way to some field $K \supseteq k_0$ that the specialized values s_i^0 are multiplicatively independent. Then $E_{\lambda^0}(K(t)) \cong M$ does not contain any v-integral sections other than the integral ones.*

Proof First note that the assumption on multiplicative independence implies that Π_ξ does not contain 1. Hence $E_{\lambda^0}(K(t)) \cong M$ by Theorem 10.21 and its analogue for multiplicative excellent families of types other than E_r $(r = 6, 7, 8)$. Lemma 10.40 will then follow immediately from the following much stronger statement. □

Claim 10.41 *Any section in $E_{\lambda^0}(K(t))$ is integral at v.*

Proof Assume to the contrary that some section P intersects the zero section O non-trivially at v. Then

$$sp'_v(P) = sp'_v(O) = 1.$$

On the other hand, since the sections $P_i(\lambda^0)$ generate $E_{\lambda^0}(K(t))$, we know that there are integers $a_i \in \mathbb{Z}$ such that

$$P = \sum_{i=1}^{r} a_i P_i(\lambda^0) \quad (r = \mathrm{rank}(M)).$$

Hence we obtain a contradiction to the multiplicative independence from spelling out the specialization map:

$$1 = sp'_v(P) = \prod_{i=1}^{r} sp'_v(P_i(\lambda^0))^{a_i} = \prod_{i=1}^{r} (s_i^0)^{a_i}.$$

This contradiction completes the proof of Claim 10.41 and thus of Lemma 10.40. □

Remark 10.42 The above statement is much stronger than what we derived for the additive excellent case in Lemma 9.35. Notably, our assumptions allow for all s_i^0 to be contained in k_0 (for instance, in $\mathbb{Q}$); compare the additive case of type E_r $(r = 6, 7, 8)$ from Lemma 9.35, where in order to exclude polynomial sections, we had to consider extensions of $\mathbb{Q}$ which were either transcendental or of degree at least r over $\mathbb{Q}$.

10.5 Applications to Classical Topics

This section will discuss applications of the theory of Mordell–Weil lattices, and in particular excellent families, to some classical topics in algebraic geometry. Notably, we discuss the 27 lines on a smooth cubic surface in $\mathbb{P}^3$, both from an abstract and explicit viewpoint, and the 28 bitangents of a smooth plane quartic curve. With a more arithmetic flavour, we also comment on zeta functions in the vein of Weil's classical ICM notes [Wei54]. Finally, we also treat degenerations using the notion of vanishing roots.

10.5.1 The 27 Lines on a Cubic Surface

Suppose that the Weierstrass equation of an elliptic curve $E/k(t)$ is a polynomial of total degree at most 3 in x, y, t over k (like for (10.4)). Then, it defines an affine surface in $\mathbb{A}^3$ whose projective closure is a cubic surface, say V, in $\mathbb{P}^3$—one of the most classical objects of study in algebraic geometry. We briefly review the story before applying our ideas based on Mordell–Weil lattices to cubic surfaces.

A classical discovery in (algebraic) geometry, dating back to Cayley [Cay49] and Salmon [Salm49] in 1849, was that every smooth cubic surface $V \subset \mathbb{P}^3$ contains exactly 27 lines. Here, of course, Cayley and Salmon were only concerned with cubics over $\mathbb{C}$, but the result in fact holds over any algebraically closed field. Almost any textbook in algebraic geometry will contain a proof of this statement (at least over $\mathbb{C}$), so it may seem unnecessary to go into details, but obviously, the possible approaches vary greatly. Historically, one might have argued with a classical tool such as the flecnodal divisor (which is in fact very useful for studying surfaces of higher degree in $\mathbb{P}^3$ as well); from a more modern angle, one might want to pursue an approach using intersection theory or the incidence variety of (smooth) cubic surfaces and lines in $\mathbb{P}^3$ (or even degeneration techniques from tropical geometry). In fact, using the incidence approach, it is not terribly hard to show that the number of lines does not depend on the (smooth) surfaces chosen, and in fact that their configuration is stable, too. Then, working for instance with the Fermat cubic surface (outside characteristic 3), one can work out all the details explicitly and in particular determine that the group of symmetries of the 27 lines is exactly the Weyl group of E_6 (see [Mum76], for instance, for the detailed argument over $\mathbb{C}$)! Thus it comes very naturally, at least to us, to seek a connection with rational elliptic surfaces with narrow Mordell–Weil lattice E_6, or more precisely, to excellent families of type E_6. This will be achieved in the following paragraphs.

Lemma 10.43 *Let $E/k(t)$ be a non-constant elliptic curve which admits a polynomial Weierstrass equation of total degree at most 3 in x, y, t. Then the Kodaira–Néron model of E is a rational elliptic surface S with a singular fibre of Kodaira type* I_3, IV *or worse (i.e. with more than three components) at $t = \infty$.*

Conversely, if S is a rational elliptic surface with a singular fibre of Kodaira type I_3, IV *or worse, then its Weierstrass form can be normalized to a polynomial of total degree* 3.

Proof In the given set-up, the Kodaira–Néron model of E is a rational elliptic surface S by the degree bounds from Sect. 5.13. Generically, it has a single reducible fibre (of type I_3 or IV) at $t = \infty$ as Tate's algorithm in Sect. 5.8 shows.

In the converse direction, we work with a Weierstrass form (5.4) for S with each coefficient $a_i \in k[s]$ of degree at most i, again by Proposition 5.51. Apply a Möbius transformation to locate the special fibre at $s = 0$ for convenience, and a translation of x, y by constants so that the node of the Weierstrass model at $s = 0$ sits at $(0, 0)$. Then $s \mid a_3, a_4$ and $s^2 \mid a_6$ since the node is assumed to be a surface singularity. Write $a_{i,j} = a_i/s^j$. Blowing up, we consider the affine chart $s' = 1$ with $x = sx'$, $y = sy'$ and equation

$$S: \quad y'^2 + a_1x'y' + a_{3,1}y' = sx'^3 + a_2x'^2 + a_{4,1}x' + a_{6,2}.$$

The special fibre at $s = 0$ intersects this chart in the quadric

$$y'^2 + a_1(0)x'y' + a_{3,1}(0)y' = a_2(0)x'^2 + a_{4,1}(0)x' + a_{6,2}(0). \qquad (10.52)$$

By assumption, the difference of LHS and RHS of (10.52) factors into two linear polynomials which are distinct if and only if the Kodaira type is I_3 or IV. A linear transformation in x', y' takes the factors to

$$y(y + cx) \ \ (\text{type } \mathrm{I}_3), \quad y(y + c) \ \ (\text{type IV}), \quad y^2 \ \ (\text{else}) \qquad (c \neq 0).$$

In each case, this implies that s divides each of $a_2, a_{4,1}$ and $a_{6,2}$. Switching from the underlying Weierstrass form to the affine chart with coordinate $t = 1/s$ thus gives a cubic polynomial, using the standard admissible transformation $(x, y) \mapsto (x/t^2, y/t^3)$. □

By Lemma 10.43, cubic Weierstrass forms lead generically to rational elliptic surfaces with Mordell–Weil lattice $E_6^\vee$ as already hinted at (with linear sections etc. as explored in this and the previous chapter). On the other hand, in terms of homogeneous coordinates

$$[X, Y, T, Z] = [x, y, t, 1],$$

the elliptic curve E (or the associated elliptic surface $f : S \to \mathbb{P}^1$) defines a cubic surface, say V, in $\mathbb{P}^3$. Here the affine open set

$$V_{\mathrm{aff}} := V \cap \{Z \neq 0\}$$

of V can be identified with the affine surface defined by the Weierstrass equation in (x, y, t)-space $\mathbb{A}^3$. That is, V_{aff} can be identified with the affine elliptic surface

$$S_{\mathrm{aff}} := S - (O) - f^{-1}(\infty),$$

i.e. the complement of the zero section (O) and the fibre at $t = \infty$ in S. In fact, this gives a birational morphism from S to V inducing an isomorphism of the open subsets

$$S_{\mathrm{aff}} \to V_{\mathrm{aff}}.$$

Generically, the birational morphism can be given explicitly as follows. Assume that

$$f^{-1}(\infty) = \Theta_0 + \Theta_1 + \Theta_2,$$

where Θ_0 is the component met by (O) and Θ_1 is the unique component which all the linear sections intersect. Then V is obtained from S by blowing down first (O) (a (-1)-curve), secondly the image of Θ_0 (made a (-1)-curve by the first blow-down), and thirdly the image of Θ_2. The image of Θ_1 in V will be the cubic curve $V \cap \{Z = 0\}$, which is either a nodal or cuspidal curve according to the singular fibre $f^{-1}(\infty)$ being of type I_3 or IV. For an explicit instance of this in the framework of Weierstrass transformations, see Theorem 10.56.

For example, consider the Weierstrass Eq. (10.4) for the multiplicative excellent family of type E_6:

$$E_\lambda : \quad y^2 + txy = x^3 + x\left(\sum_{i=0}^{2} p_i t^i\right) + \sum_{i=0}^{2} q_i t^i + t^3.$$

It gives rise to the following cubic surface in $\mathbb{P}^3$,

$$\begin{aligned} V_\lambda : \quad Y^2Z + TXY = X^3 &+ X(p_0Z^2 + p_1TZ + p_2T^2) \\ &+ q_0Z^3 + q_1TZ^2 + q_2T^2Z + T^3 \end{aligned} \tag{10.53}$$

with parameter $\lambda = (p_0, p_1, p_2, q_0, q_1, q_2)$. By construction, the 27 linear sections of (10.4),

$$P_n : \ \mathrm{x} = a_nt + b_n, \quad \mathrm{y} = d_nt + e_n \quad (n = 1, \ldots, 27),$$

as worked out in Theorem 10.13, stay linear in $\mathbb{P}^3$. They correspond to the 27 lines in $\mathbb{P}^3$ given by

$$l_n : \quad X = a_nT + b_nZ, \quad Y = d_nT + e_nZ \quad (n = 1, \ldots, 27) \tag{10.54}$$

which lie on the cubic surface V_λ. We note that each line l_n, on the affine open set $V_\lambda \cap \{Z \neq 0\}$, defines the affine line defined by the same equation as the section P_n. Therefore we directly obtain the following consequence:

Theorem 10.44 *The* algebraic equation for the 27 lines on V_λ *is explicitly given by the same polynomial as that in (10.20):*

$$\Phi_\lambda(X) = 0.$$

(Recall that Φ has degree 27 with coefficients in $\mathbb{Z}[\lambda] = \mathbb{Z}[p_i, q_j]$.)

Remark 10.45 We point out that the above approach also suffices to prove that V_λ contains no more than 27 lines. Indeed, any line on V_λ gives a linear section of S_λ, except for lines not involving the coordinate X or Y, but these are easy to deal with.

10.5.1.1 Big Galois for the 27 Lines

By Theorem 10.8, the generic Galois group of the splitting field $\mathcal{K}$ of Φ_λ over $k_0 = \mathbb{Q}(\lambda)$ is equal to the Weyl group $W(E_6)$, of order 51840.

Moreover, Example 10.34 gives a numerical example where the coefficients satisfy $p_i = q_i = 1$ $(i = 0, 1, 2)$, so that the cubic surface is defined over $\mathbb{Q}$ by the equation:

$$V_1 : Y^2Z + TXY = X^3 + X(Z^2 + TZ + T^2) + Z^3 + TZ^2 + T^2Z + T^3. \quad (10.55)$$

In this case, $\Phi(X)$ becomes an integral polynomial of degree 27 (see Theorem 10.8) whose splitting field has Galois group $W(E_6)$ over $\mathbb{Q}$.

10.5.2 Zeta Functions and Cubic Surfaces

We want to draw some explicit consequences for the zeta functions of cubic surfaces. First we introduce the main terminology (see [Har77, Appendix C], for instance).

For any variety X defined over some finite field $\mathbb{F}_q$, we denote by N_r the number of $\mathbb{F}_{q^r}$-rational points for any $r \in \mathbb{N}$:

$$N_r = \#X(\mathbb{F}_{q^r}).$$

The zeta function of X is then defined as the following formal power series (which can be viewed as a kind of generating function):

$$Z(X, T) = \exp\left(\sum_{r=1}^{\infty} N_r \frac{T^r}{r}\right) \in \mathbb{Q}[[T]].$$

Example 10.46 If $X \cong \mathbb{A}^n$, then $N_r = q^{nr}$, and from the power series expansion of $\log(1-x)$ evaluated at $x = q^n T$, one reads off $Z(X, T) = 1/(1 - q^n T)$.

Example 10.47 If $X \cong \mathbb{P}^n$, then $N_r = 1 + q^r + \ldots + q^{nr}$, so

$$Z(X, T) = \frac{1}{(1-T)(1-qT)\cdots(1-q^n T)}.$$

If X is smooth (over $\bar{\mathbb{F}}_q$) projective of dimension n, then its zeta function has several remarkable properties which are generally known as the *Weil conjectures*. They were first proven in the one-dimensional case by Weil ([Wei48a]), and in general by Dwork [Dwo60], Grothendieck [Gro58, SGA4] and Deligne [Del74, Del80], building on important input from many other mathematicians. Here we point out the *rationality* (as in Examples 10.46, 10.47),

$$Z(X, T) = \frac{P_1(T)P_3(T)\cdots P_{2n-1}(T)}{P_0(T)P_2(T)\cdots P_{2n}(T)},$$

where the P_i are polynomials with rational coefficients. The *analogue of the Riemann hypothesis* makes more predictions on the P_i (in agreement with Example 10.47). Namely, one can take $P_0 = 1 - t$, $P_{2n} = 1 - q^n T$ and all other $P_i \in 1 + T\mathbb{Z}(T)$ as well. More precisely, we one has $\deg(P_i) = b_i(X)$ (the i-th Betti number), and factoring

$$P_i(T) = \prod_{j=1}^{b_i(X)} (1 - \alpha_{ij} T),$$

all reciprocal roots α_{ij} are algebraic integers of absolute value $|\alpha_{ij}| = q^{i/2}$. Note the remarkable consequence that it suffices to compute only a certain number of N_r to obtain $Z(X, T)$ and thus all N_r!

Example 10.48 Let E be a genus one curve over $\mathbb{F}_q$. Then

$$Z(E, T) = \frac{1 - aT \pm qT^2}{(1-T)(1-qT)}.$$

In particular, $Z(E, T)$ is determined by N_1, N_2.

A decisive step in proving the Weil conjectures consisted in developing a suitable cohomology theory (namely ℓ-adic étale cohomology, $\ell \neq p$), endowed with Galois representations and all desirable compatibilities and comparison theorems, such that the P_i are the reciprocal characteristic polynomials of the Frobenius morphism:

$$P_i(T) = \det(1 - T\,\mathrm{Frob}^*;\ H^i_{\text{ét}}(X \otimes \bar{\mathbb{F}}_q, \mathbb{Q}_\ell)).$$

Here the relation with the N_r comes from Lefschetz' fixed point formula:

$$N_r = \sum_{i=0}^{2n} (-1)^i \operatorname{trace}((\mathrm{Frob}^*)^r; H^i_{\text{ét}}(X \otimes \bar{\mathbb{F}}_q, \mathbb{Q}_\ell))$$

(which is independent of $\ell \neq p$, so the same will hold for the P_i, see below).

Example 10.49 In the notation of Example 10.48, one has

$$a = \operatorname{trace}(\mathrm{Frob}^*; H^1_{\text{ét}}(E \otimes \bar{\mathbb{F}}_q, \mathbb{Q}_\ell)).$$

This can also be computed in terms of the ℓ-adic Tate module of E, cf. Remark 9.1.

We point out that the even cohomology groups contain natural subspaces on which the action of Frobenius is easy to compute. These subspaces are generated by all algebraic cycles defined over $\bar{\mathbb{F}}_q$. Thus Frobenius permutes the cycles, and given a basis for the finite-dimensional subspaces, Frobenius acts through a finite group. For codimension one cycles, for instance, i.e. divisor classes inside H^2), this can be achieved by embedding $\mathrm{NS}(X \otimes \bar{\mathbb{F}}_q)$ into étale cohomology via the ℓ-adic cycle class map combined with a Tate twist, see (12.1).

Assume now that the smooth projective variety X of dimension n is defined over some number field K. After clearing denominators, we obtain defining equations in $\mathcal{O}_K$, so we can consider the reduction $X_\mathfrak{p}$ of X modulo any prime ideal $\mathfrak{p}$ of $\mathcal{O}_K$. For almost all primes, $X_\mathfrak{p}$ is smooth, leading to the local zeta function $\zeta(X_\mathfrak{p}, T)$ with all the above properties. In order to define a global zeta function of X, it remains to define the factors at the bad primes; generally speaking, it is advisable not to take the zeta function of the singular variety $X_\mathfrak{p}$ as input, which may depend on the chosen model of X, but rather the compatible system of Galois representations on each étale cohomology group:

$$\rho_{i,\ell} : \operatorname{Gal}(\bar{K}/K) \to \operatorname{GL}(H^i_{\text{ét}}(X \otimes \bar{K}, \mathbb{Q}_\ell)) \quad (i = 0, \ldots, 2n).$$

Each Galois representation $\rho_{i,\ell}$ is unramified outside the bad primes of X and the primes dividing ℓ. Here, a prime $\mathfrak{p}$ is unramified if and only if the inertia group $I_\mathfrak{p}$ acts trivially. In particular, this shows that the conjugacy class of the arithmetic Frobenius $\mathrm{Frob}_\mathfrak{p}$ of a good prime $\mathfrak{p}$ has a well-defined trace which, in fact, equals that of the geometric Frobenius Frob^* on $X_\mathfrak{p}$:

$$\operatorname{trace}(\rho_{i,\ell}(\mathrm{Frob}_\mathfrak{p})) = \operatorname{trace}(\mathrm{Frob}^*; H^i_{\text{ét}}(X_\mathfrak{p} \otimes \bar{\mathbb{F}}_\mathfrak{p}, \mathbb{Q}_\ell)).$$

At a bad prime $\mathfrak{p}$, we can adapt the above considerations by restricting to the subspace invariant under the inertia group $I_\mathfrak{p}$; indeed, the characteristic polynomial

$$\det\left(1 - T\rho_{i,\ell}(\mathrm{Frob}_\mathfrak{p}); H^i_{\text{ét}}(X \otimes \bar{K}, \mathbb{Q}_\ell)^{I_\mathfrak{p}}\right) \in 1 + T\mathbb{Z}[T]$$

is independent of ℓ (not divisible by $\mathfrak{p}$). The global zeta function is now obtained by piecing all local information together, using the norm map $N : K \to \mathbb{Q}$ (and tacitly picking at any given prime $\mathfrak{p}$ some suitable ℓ).

Definition 10.50 The *zeta function* of X over K is defined by

$$\zeta(X,s)=\prod_{\mathfrak{p}}\prod_{i=0}^{2n}\det\left(1-\rho_{i,\ell}(\mathrm{Frob}_{\mathfrak{p}})N(\mathfrak{p})^{-s};\,H^i_{\text{ét}}(X\otimes\bar{K},\mathbb{Q}_\ell)^{I_{\mathfrak{p}}}\right)^{(-1)^{i+1}}.$$

We note that the global zeta function contains all the local zeta functions at the good primes as factors. Often this involves the Riemann zeta function $\zeta(s)$ and its shifts by appropriate integers as in the next example.

Example 10.51 From Example 10.47, we obtain for $\mathbb{P}^n$ over $\mathbb{Q}$

$$\zeta(\mathbb{P}^n,s)=\zeta(s)\zeta(s-1)\cdots\zeta(s-n).$$

We shall now specialize to the case where $X=V$ is a smooth cubic surface in $\mathbb{P}^3$ defined over $\mathbb{Q}$. Then

$$\zeta(V,s)=\zeta(s)\zeta(s-2)\prod_p\frac{1}{\det\left(1-\rho_{2,\ell}(\mathrm{Frob}_p)p^{-s};\,H^2_{\text{ét}}(V\otimes\bar{\mathbb{Q}},\mathbb{Q}_\ell)^{I_p}\right)}. \tag{10.56}$$

Since $p_g(V)=0$, we have $H^2_{\text{ét}}(V\otimes\bar{\mathbb{Q}},\mathbb{Q}_\ell(1))=\mathrm{NS}(V)\otimes\mathbb{Q}_\ell$ with a Tate twist (tensoring the Galois representation with the inverse of the ℓ-adic cyclotomic character), so the Frobenii act through some finite group on *all* twisted cohomology groups. In other words, the infinite product in (10.56) equals the shifted L-function $L(s-1,\chi)$, where $L(s,\chi)$ is the Artin L-function corresponding to the character χ of the Galois representation on the Néron–Severi lattice $\mathrm{NS}(V)$.

For an interesting explicit example, we now consider the cubic surface V_1 from (10.55). Following Weil [Wei54] and Manin [Man86, Theorem 27.3], we recall that the factor $L(s-1,\chi)$ of the zeta function of V_1 over $\mathbb{Q}$ presently is non-abelian (corresponding to the finite extension $\mathcal{K}/\mathbb{Q}$ where $\mathcal{K}$ is the splitting field as before). Indeed, up to the class of the hyperplane section (which is Galois-invariant and thus contributes a factor $\zeta(s-1)$ to $L(s-1,\chi)$), the Néron–Severi lattice $\mathrm{NS}(V_1)$ equals the narrow Mordell–Weil lattice $E(K(t))\cong E_6$ (as a Galois representation, or as a lattice when considering the orthogonal complement of the hyperplane section).

We should like to emphasize that we can compute the local factors of $L(s,\chi)$ at a prime p of good reduction almost exclusively from the factorization of Φ in $\mathbb{F}_p[X]$. Namely, this factorization determines the permutation cycle type of (the conjugacy class) of Frobenius at p, and this in turn determines the trace by [Swi67] up to a little bit of ambiguity depending on the precise configuration of lines about which we will not go into the details here (cf. also [Shio92a]). For the reader's convenience, we reproduce the corresponding result from [Swi67] in Table 10.1, with one correction borrowed from [BFL19]. The table lists permutation cycle type, possible traces of Frobenius, and the precise local shape of $L(s,\chi)$ (up to the omnipresent factor corresponding to $\zeta(s)$). To this end, we denote by ϕ_n the minimal polynomial of the primitive n-th roots of unity with $x=p^{-s}$ substituted (thought reciprocally, i.e.

Table 10.1 Local factors of zeta function of cubic surface

Permutation type	Trace Frob_p^*	L_p/ϕ_1	p's
1^{27}	7	ϕ_1^6	127453
$1^7 2^{10}$	3	$\phi_1^4\phi_2^2$	2053
$1^3 2^{12}$	$\begin{cases}-1\\ 1\end{cases}$	$\begin{cases}\phi_1^2\phi_2^4\\ \phi_1^3\phi_2^3\end{cases}$	59, 439
$1^3 4^6$	3	$\phi_1^2\phi_4^2$	2111
$1\ 2^3 4^5$	1	$\phi_1^2\phi_2^2\phi_4$	19
$1^9 3^6$	4	$\phi_1^3\phi_3$	3461
$1^3 2^3 6^3$	2	$\phi_1^2\phi_2^2\phi_6$	31
$1\ 2^4 3^2 6^2$	0	$\phi_1\phi_2\phi_3\phi_6$	37
3^9	$\begin{cases}1\\ -2\end{cases}$	$\begin{cases}\phi_1^2\phi_3^2\\ \phi_3^3\end{cases}$	587,673
$3\ 6^4$	$\begin{cases}-1\\ 2\\ 1\end{cases}$	$\begin{cases}\phi_2^2\phi_3\phi_6\\ \phi_3\phi_6^2\\ \phi_1\phi_2\phi_3\phi_6\end{cases}$	5,73,79
$3\ 12^2$	0	$\phi_3\phi_{12}$	11,53,67,89
9^3	1	ϕ_9	23,97
$1^2 5^5$	2	$\phi_1^2\phi_5$	71
$1^{15} 2^6$	5	$\phi_1^5\phi_2$	1789
$1^5 2\ 4^5$	3	$\phi_1^3\phi_2\phi_4$	83
$1\ 2^3 4^5$	-1	$\phi_1\phi_2^3\phi_4$	41
$1\ 2\ 8^3$	1	$\phi_1^2\phi_2\phi_8$	3,7,17,29,47,61
$1^3 2^3 3^4 6$	2	$\phi_1^3\phi_2\phi_3$	709
$3^5 6^2$	-1	$\phi_1\phi_2\phi_3^4$	1187
$1\ 4^2 6\ 12$	2	$\phi_1\phi_2\phi_4\phi_6$	103
$2\ 5^3 10$	0	$\phi_1\phi_2\phi_5$	43

$\phi_1 = 1 - p^{-s}$ if $n = 1$). The table is enhanced by some exemplary primes p relative to the above choice of $p_i = q_j = 1$ from Example 10.34.

Remark 10.52 As announced, this also enables us to compute the zeta function of the corresponding elliptic surface S_1 from Example 10.34, as the surfaces only differ by blowing down three $\mathbb{Q}$-rational curves:

$$\zeta(S_1, s) = \zeta(s)\zeta(s-1)^3 L(s-1, \chi)\zeta(s-2).$$

A similar result applies for any cubic surface, i.e. independently of the precise Galois group encoded in $\Phi(X)$. The extreme case where all lines (or sections) are in fact defined over $\mathbb{Q}$ will be studied in the next section.

10.5.2.1 Small Galois for the 27 Lines

On the other hand, we can systematically construct cubic surfaces V such that all the 27 lines are defined over the field of definition of V. For example, Theorem 10.13, or either algorithm (M) or (M'), can be used for such a purpose.

Theorem 10.53 *Fix $\xi = (s_1, \ldots, s_6; r)$ such that $\Pi_{i=1}^6 s_i = r^3$, and define $\lambda = (p_i, q_j)$ by the formula (10.30). Then the 27 lines l_n on the cubic surface V_λ are given by* (10.54) *with coefficients given in Theorem 10.13.*

The cubic surface V_λ is smooth if and only if the non-degeneracy condition holds that the set Π_ξ from (10.51) does not contain 1.

For numerical examples, say with rational coefficients, we can just choose any rational point on the toric hypersurface, $\xi = (s_1, \ldots, s_6, r) \in \mathcal{T}(\mathbb{Q}^\times)$, i.e. to choose units $s_1, \ldots, s_6$ and r from $\mathbb{Q}^\times$ subject to the relation $\Pi_{i=1}^6 s_i = r^3$ and with the non-degeneracy condition that Π_ξ does not contain 1.

Example 10.54 Example 10.36 provides the 27 lines l_n on the cubic surface $V' = V_{\lambda'}$, where $\lambda' = (m_0, p_0, p_1, q_0, q_1, q_2)$ is given by the algorithm (M'). Namely, take the data $\xi = (1, 2, 3, 4, -1, -\frac{1}{3}; 2)$. Then we have

$$\lambda' = \left(-\frac{21}{4}, -\frac{175}{3}, \frac{68}{9}, \frac{105625}{324}, -\frac{175}{3}, -\frac{21}{4}\right).$$

Hence the cubic surface V' has the defining equation:

$$\begin{aligned} V' :\ Y^2Z + TXY = \ & X^3 - \frac{21}{4}X^2Z + \left(\frac{68}{9}TZ - \frac{175}{3}Z^2\right)X \\ & + \frac{105625}{324}Z^3 - \frac{175}{3}TZ^2 - \frac{21}{4}T^2Z + T^3. \end{aligned}$$

Then Example 10.36 gives the 27 lines l_n on V'. For instance, the first six lines are given as follows:

$$\begin{aligned} l_1 &: X = -T, & Y &= \frac{325}{18}Z, \\ l_2 &: X = -\frac{1}{2}T + \frac{19}{6}Z, & Y &= -\frac{7}{4}T + \frac{395}{36}Z, \\ l_3 &: X = -\frac{1}{3}T + \frac{40}{9}Z, & Y &= -\frac{26}{9}T + \frac{385}{54}Z, \\ l_4 &: X = -\frac{1}{4}T + \frac{85}{16}Z, & Y &= -\frac{63}{16}T + \frac{2435}{576}Z, \\ l_5 &: X = T + \frac{20}{3}Z, & Y &= 2T - \frac{5}{18}Z, \\ l_6 &: X = 3T + \frac{20}{3}Z, & Y &= \frac{28}{3}T + \frac{5}{18}Z. \end{aligned}$$

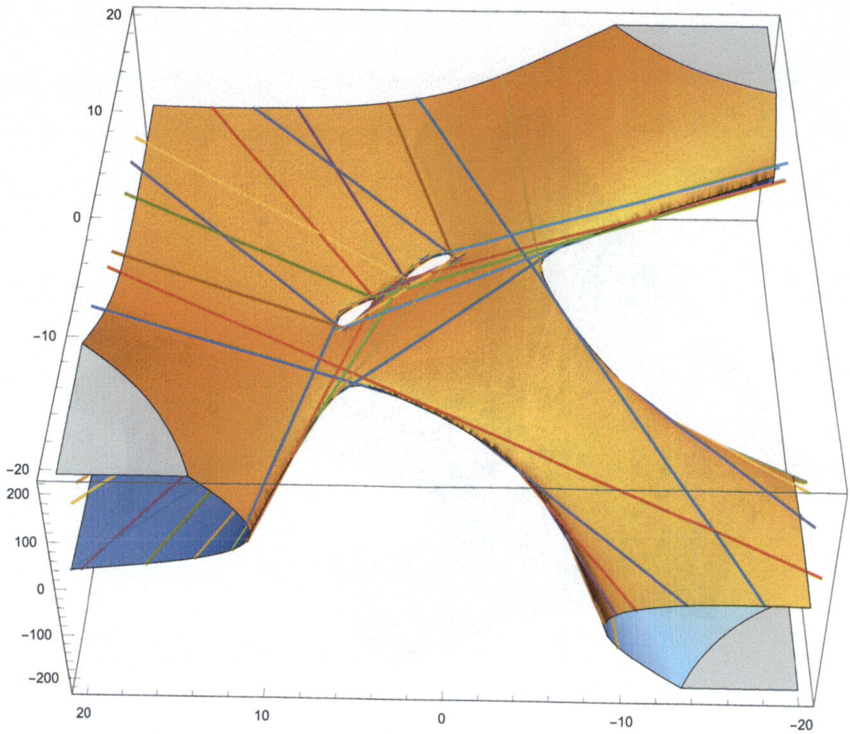

Fig. 10.1 Portion of V' with its 27 lines (from above) (© Shioda)

One can verify that these six lines are skew, i.e. disjoint from each other. We leave it to the reader as an easy exercise to write down the second six lines $l'_i = l_{i+6}$ ($i = 1, \ldots, 6$) and to verify that they are skew, too. Moreover, the following intersection properties hold:

$$l_i \cap l'_i = \emptyset, \quad l_i \cap l'_j \neq \emptyset \; (i \neq j).$$

Hence the 12 lines l_i, l'_i form a *double six* configuration in the sense of Schläfli.

Note that presently, all the 27 lines as well as the cubic surface itself have rational coefficients, and in particular, real coefficients. So we can draw the picture of the (real affine locus V_{aff} of) the cubic surface V' and its 27 lines l_n (Figs. 10.1 and 10.2).

10.5.3 Relation with Del Pezzo Surfaces

A del Pezzo surface X (at least over an algebraically closed field) is either $\mathbb{P}^1 \times \mathbb{P}^1$ or the blow-up of $\mathbb{P}^2$ in m points ($m = 1, \ldots, 8$) in general position. (In comparison, recall that a rational elliptic surface is the blow-up of $\mathbb{P}^2$ in nine points, but not in

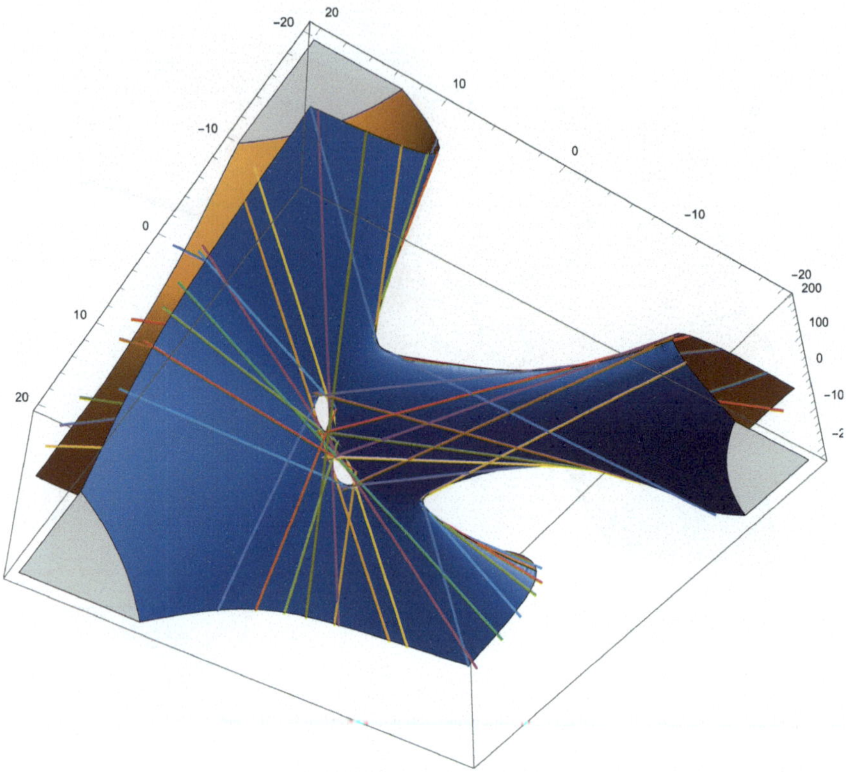

Fig. 10.2 Portion of V' with its 27 lines (from below) (© Shioda)

general position, since they are the base points of a cubic pencil.) The degree d of X is defined to be

$$d = K_X^2 = 9 - m.$$

For our purposes, the small degree cases $d = 1, 2, 3$ are of particular interest since they naturally give rise to the root lattice E_{9-d} by way of considering the orthogonal complement $K_X^\perp$ inside $\mathrm{NS}(X)$ (which is even by the adjunction formula from Theorem 4.11, and thus isomorphic to E_{9-d} for classification reasons). Indeed, one can obtain an elliptic surface from X by blowing up another d points. Here a word of care may be in order; namely, these d points are not at all independent of the others. For instance, if $d = 1$, then the eight points blown-up in $\mathbb{P}^2$ already determine a cubic pencil in $\mathbb{P}^2$. This pencil has one further base point determined by the given 8 which we thus have to blow-up in order to obtain a rational elliptic surface from X.

Here comes the easiest example of a del Pezzo surface which is not obviously the blow-up of $\mathbb{P}^2$: a smooth cubic surface V! Of course, it was classically known that V is, in fact, isomorphic to $\mathbb{P}^2$ blown-up in six points. Here is a rough idea how to see this: among the 27 lines on V, pick two skew lines L_1, L_2. Then any point $P \in V \setminus (L_1 \cup L_2)$ lies on a unique line $L \subset \mathbb{P}^3$ (generally not contained in V) connecting L_1 and L_2. The intersection points of L with L_1, L_2 thus determine a unique point on

$$L_1 \times L_2 \cong \mathbb{P}^1 \times \mathbb{P}^1.$$

This map extends to a morphism on all of V (by restricting to lines in the tangent plane at P if $P \in L_1$ or L_2), but it ceases to be injective on the five lines in V connecting L_1 and L_2. These five lines are blown down to five points in $\mathbb{P}^1 \times \mathbb{P}^1$. The birational inverse of this morphism has locus of indeterminacy exactly these five points, so V is seen to be isomorphic to the blow-up of $\mathbb{P}^1 \times \mathbb{P}^1$ in these five points. Since $\mathbb{P}^2$ blown up in two points is isomorphic to $\mathbb{P}^1 \times \mathbb{P}^1$ blown up in one point, we arrive at the claim.

In this picture, with the six points in general position, it is also easy to detect the 27 lines on V: next to the six exceptional divisors, they are given by the strict transforms of the 15 lines connecting any two of them in $\mathbb{P}^2$, and of the six conics through any five of them. Note that the six exceptional divisors and the strict transforms of the six conics exactly form Schläfli's double six configuration.

We note that this fits perfectly with our construction from Sect. 10.5.1, where we constructed a cubic surface V_λ from a rational elliptic surface with Mordell–Weil lattice E_6: here V_λ was naturally endowed with six disjoint lines $l_1, \dots, l_6$; blowing them down can only lead to $\mathbb{P}^2$. In the next section we will discuss a new approach to the converse problem building on the theory of Mordell–Weil lattices and excellent families.

10.5.4 Cubic Surfaces as the Plane Blown up in Six Points

We consider the following basic problem.

Problem 10.55 Given six points in the projective plane $\mathbb{P}^2$, write down the defining equations of the cubic surface V and the six lines $\{L_i\}$ arising from blowing up $\mathbb{P}^2$ in the six points.

There is an ad-hoc approach to Problem 10.55 by considering the map from $\mathbb{P}^2$ to $\mathbb{P}^3$ given by the system of cubics through the fixed 6 points. Then compute the image points of at least 20 points in $\mathbb{P}^2$ and find the cubic equation (unique up to scalars) satisfied by them, followed by determining the 27 lines. In the paragraphs to come, we propose a systematic solution to Problem 10.55 based on the theory of rational elliptic surfaces.

It is easy to see that given any six points in the plane, there always exists at least one nodal or cuspidal cubic curve passing through them. Assume the general situation that the six points lie on the smooth locus of the following affine nodal cubic (after some linear transformation of $\mathbb{P}^2$):

$$\Gamma: \; y^2 + xy = x^3. \tag{10.57}$$

As before, we parametrize the smooth points on Γ by

$$P(s) = \left(\frac{s}{(s-1)^2} : \frac{s}{(s-1)^3} : 1\right) \in \Gamma \subset \mathbb{P}^2 \quad (s \neq 0, 1).$$

Then our solution of Problem 10.55 is as follows.

Theorem 10.56 *Take six points $P_i = P(s_i)$ $(i = 1, ..., 6)$ on Γ such that*

$$s_i \neq s_j \ (i < j), \quad s_i s_j s_k \neq 1 \ (i < j < k), \quad s_1 \cdots s_6 \neq 1. \tag{10.58}$$

Fix a cube root s_0 of $s_1 \cdots s_6$:

$$s_0^3 = s_1 \cdots s_6.$$

Then the blown-up cubic surface V is defined by an equation of the form

$$Y^2 Z + TXY = X^3 + X(p_0 Z^2 + p_1 TZ + p_2 T^2) + Z(q_0 Z^2 + q_1 TZ + q_2 T^2) + T^3$$

and the six lines L_i (= blow-up of P_i) are given by the equations:

$$L_i : \ X = a_i T + b_i Z, \ \ Y = d_i T + e_i Z$$

such that

$$a_i = -s_i \cdot s_0 \ \ (1 \leq i \leq 6).$$

In other words, V is isomorphic to the cubic surface $V_{\mathbf{s}'}$, in the notation of Sect. 10.5.1, where

$$\mathbf{s}' = (s_0 \mathbf{s})^{-1} := \left(\frac{1}{s_0 s_1}, \ldots, \frac{1}{s_0 s_6}; \frac{1}{s_0^3}\right) \in \mathcal{T}.$$

Here V is smooth if and only if the condition (10.58) holds, i.e. if and only if the six points are in general position.

Proof The proof uses the idea of Weierstrass transformations (cf. [Shio95a]) and Mordell–Weil lattices.

(1) Find a linear pencil of plane cubics passing through the given six points P_i and the node $A : (x : y : 1) = (0 : 0 : 1)$ of the nodal curve (10.57).

(2) The generic member of the linear pencil, say Γ_t, is a smooth curve of genus 1 over the rational function field $k(t)$. Hence there is an isomorphism of elliptic curves over $k(t)$:

$$\Gamma_t \cong E, \ \ A \mapsto O,$$

where (E, O) (outside characteristics 2 and 3, say) is an elliptic curve defined by a Weierstrass equation:

$$E : \ \ \eta^2 = \xi^3 + p\xi + q \ \ (p, q \in k[t]), \quad O = (0 : 1 : 0).$$

(3) Check that the isomorphism of Γ_t to E extends to a birational transformation $\phi : \mathbb{P}^2 \to S$ from the ambient projective plane of Γ_t to the elliptic surface S which is the Kodaira–Néron model of $E/k(t)$. This is what we call a *Weierstrass transformation*; cf. [Shio95a].

(4) On the other hand, look at the base points $P_i \in \mathbb{P}^2$ and determine their image under the map ϕ. Since each base point P_i is at the same time a $k(t)$-rational point of Γ_t, the image of P_i, say Q_i, is an element of the Mordell–Weil group $E(k(t))$ which is identified with the group of sections of $S \to \mathbb{P}^1$.

(5) Check that the elliptic surface $S \to \mathbb{P}^1$ has a singular fibre of Kodaira type I_3 at $t = \infty$, and that p (or q) is a polynomial of degree 4 (or 6) in t.

(6) By a simple coordinate change of (t, ξ, η), we arrive at the Weierstrass equation of the form (10.4), and the sections Q_i of S (which form the image of the base points P_i) are defined by equations of the form

$$Q_i : \; x = a_i t + b_i, \; y = d_i t + e_i.$$

Furthermore, in passing to the homogeneous coordinates $(X : Y : T : Z) = (x : y : t : 1)$, the complement of the fibre at $t = \infty$ and the zero section in S is an open subset of S which is naturally isomorphic to the open set $V' = V \cap \{Z \neq 0\}$ of the cubic surface V defined by (10.4) (see 10.5.1), and the six sections Q_i become the six lines L_i on V. □

Remark 10.57 The case where the six points lie on a cuspidal cubic has been treated in detail in [Shio95a]. In that case the resulting rational elliptic surface has a singular fibre of type IV, and it can be realized as a cubic pencil with base points the six given points and a triple point accounting for the special fibre.

Remark 10.58 Examples and applications of this, such as the construction of smooth cubic surfaces over $\mathbb{R}$ for which the real locus in not connected, may be found in [PBT08].

10.5.5 The 28 Bitangents to a Plane Quartic

We continue to draw the connection to another classical topic in geometry, namely the 28 bitangents to a smooth plane quartic curve. From the cubic surface V above, defined by

$$Y^2 Z + TXY = X^3 + X(p_0 Z^2 + p_1 TZ + p_2 T^2) + Z(q_0 Z^2 + q_1 TZ + q_2 T^2) + T^3,$$

these are easily obtained as follows. Consider the projection from V to $\mathbb{P}^2$, i.e. the rational map of degree two given by

$$V \dashrightarrow \mathbb{P}^2$$

$$[T, X, Y, Z] \mapsto [T, X, Z].$$

Note that this rational map becomes a morphism after blowing up the point $[T, X, Y, Z] = [0, 0, 1, 0]$ on V, yielding a del Pezzo surface V' of degree two. Outside characteristic two (which is rather special and thus will be omitted here for simplicity, cf. [SV87]), the branch locus of the above rational map in $\mathbb{P}^2$ is exactly the plane quartic curve C defined affinely in $\mathbb{A}^2$ by

$$C: \quad x^3 + x\left(\sum_{i=0}^{2} p_i t^i\right) + \sum_{i=0}^{2} q_i t^i + t^3 + \frac{t^2x^2}{4} = 0.$$

Note that a line $L \subset \mathbb{A}^2$, say

$$L: \quad x = at + b,$$

splits on V into two components (smooth rational, so one automatically is a line for degree reasons) if and only if L meets C with multiplicity two each in two points, i.e. L is a bitangent to C. Hence 27 of the bitangents correspond to the lines on V, and the 28th bitangent, the line at ∞ given by $Z = 0$ in homogeneous coordinates $[T, X, Z] = [t, x, 1]$, has pre-image a nodal cubic in V given by the same equation (which is replaced by a smooth rational curve in V'). For a real picture (based on our equations with all bitangents defined over $\mathbb{Q}$), see Fig. 10.3.

Indeed, one can argue along the same lines directly for a multiplicative excellent family of type E_7 to obtain a morphism to $\mathbb{P}^2$ with branch curve a plane quartic curve whose bitangents correspond to half of the minimal sections of the Mordell–Weil lattice $E_7^\vee$. This construction via E_7 contains seven free parameters and thus has one more degree of freedom compared with six parameters for the previous E_6 case (for a thorough treatment of the moduli spaces of del Pezzo surfaces using root lattices, see [CvGL09]). We omit the details here, but we include two figures, one two-dimensional (Fig. 10.4 for the plane quartic) and one three-dimensional (Fig. 10.5 for the double cover del Pezzo surface V' of degree two).

10.5.6 Degenerations of Cubic Surfaces

To conclude this chapter, we sketch an application of the idea of vanishing roots to cubic surfaces. To this end, we take up again the case of the multiplicative excellent family (10.4) of type E_6 and consider the degenerations of the elliptic surfaces as in 10.4.3, which give at the same time the degeneration of affine surfaces or cubic surfaces. We look at $\xi = (s_1, \dots, s_6; r) \in \mathcal{T}$, with the associated 27-set

$$\Omega_\xi = \{s_n \mid 1 \le n \le 27\}$$

and 36-set

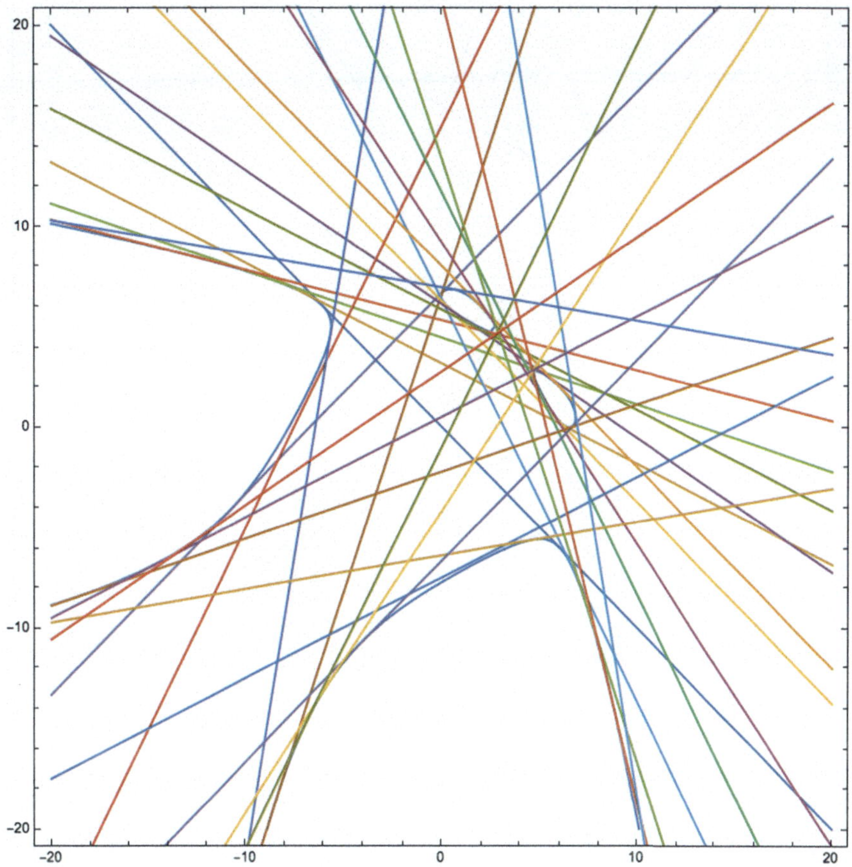

Fig. 10.3 Bitangents to a plane quartic curve (© Shioda)

$$\Pi_\xi = \left\{ \frac{1}{r},\ \alpha_{ij} = \frac{s_i}{s_j}\ (i < j),\ \beta_{ijk} = \frac{r}{s_i s_j s_k}\ (i < j, k) \right\}$$

as defined in (10.23) and (10.51). Below, the letter s_0 is used instead of r for the last coordinate of ξ, to avoid possible confusion with MW-rank.

First of all, there are exactly 21 types of triples $\{T, L, M\}$ in Theorem 8.8 such that the contribution T of the singular fibres to the trivial lattice contains A_2 as a direct factor:

$$T = T_{old} \oplus T_{new}, \quad T_{old} = T_\infty \simeq A_2, \quad T_{new} = \oplus_{v \neq \infty} T_v. \tag{10.59}$$

They are listed in the first three columns of Table 10.2, together with the structure of T_{new} and M (the Mordell–Weil lattice, employing the notation from Sect. 8.3).

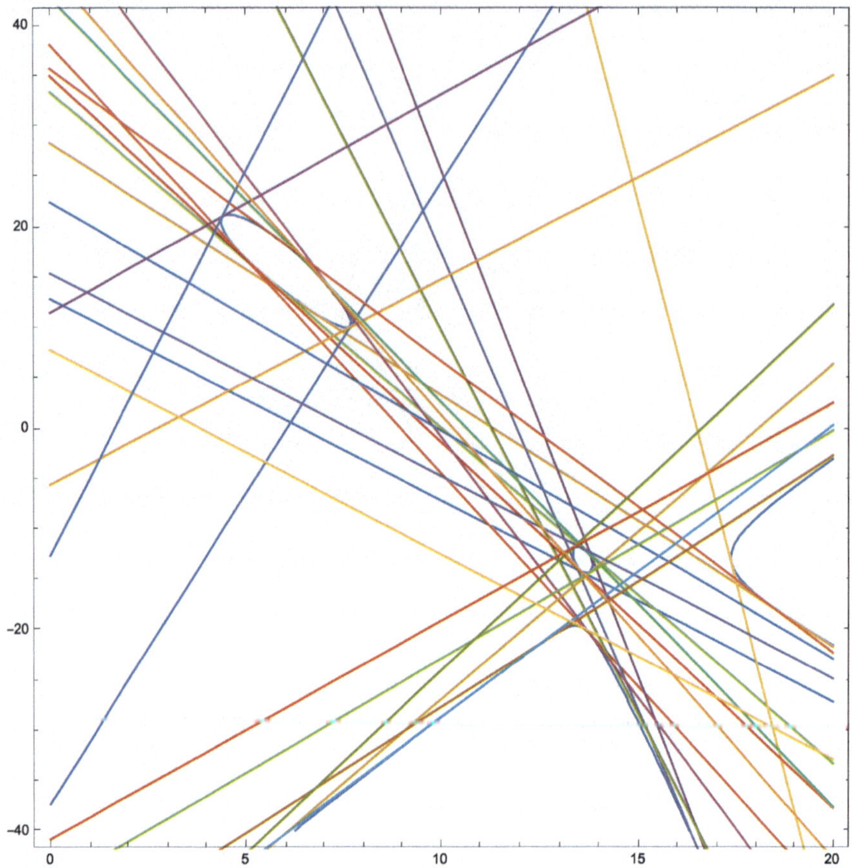

Fig. 10.4 Bitangents to a plane quartic curve (© Shioda)

The table should be read as follows: each of the data $\xi = \{s_1, \ldots, s_6; s_0\}$ in the fourth column satisfies

$$\prod_i s_i = s_0^3, \quad \text{i.e.} \quad \xi \in \mathcal{T}(K),$$

where $K = \mathbb{Q}(\xi) := \mathbb{Q}(s_1, \ldots, s_5, s_0)$. The fifth column computes the number of vanishing roots ν in the E_6-frame $\mathcal{D}(S_\xi, A_2)$, i.e. the number of times 1 occurs in Π_ξ. As in Example 10.38, 2ν is equal to the number of roots in the root lattice $\tau(T_{new})$ in the new part T_{new}. We know from Theorem 10.21 that $\nu = 0$ if and only if the elliptic surface S_ξ is non-degenerate, i.e. $M \simeq E_6^\vee$. Thus if $\nu = 1$, then $T_{new} = A_1$ as in Example 10.38; similarly, $\nu = 2$ implies $T_{new} = 2A_1$. One may continue along the same lines, but ν may not determine T_{new} as we hint at below. The last two columns of the table give the number ℓ of distinct linear sections of S_ξ (or the number of lines

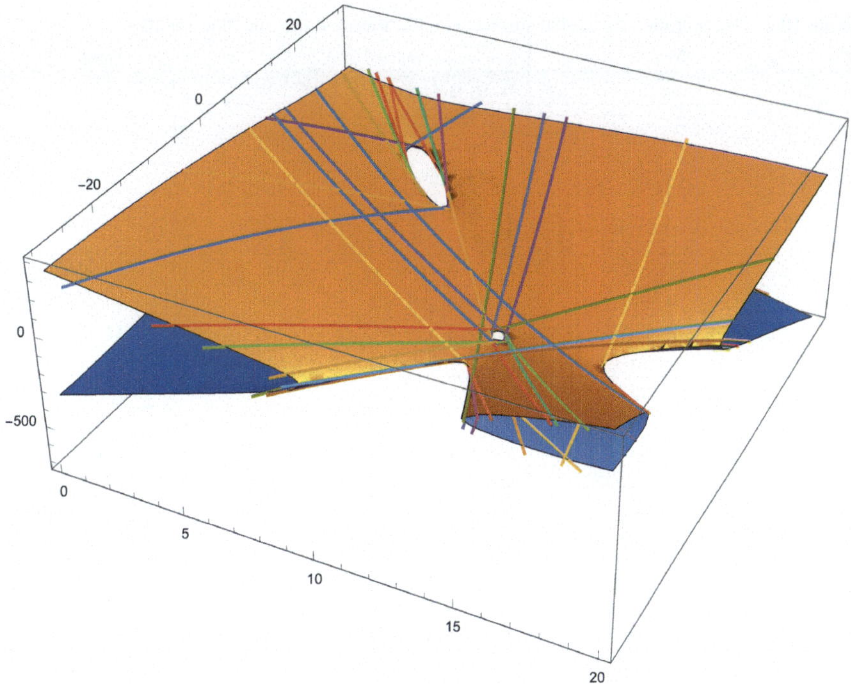

Fig. 10.5 Del Pezzo surface double covering of plane quartic (© Shioda)

on V_ξ) and their multiplicities; the symbol $\mu_1^{n_1}\mu_2^{n_2}\cdots$ in the last column means that there are n_1 lines with multiplicity μ_1, n_2 lines with multiplicity μ_2, etc. such that $\sum_i n_i\mu_i = 27$ and $\sum_i n_i = \ell$.

Note that the number of vanishing roots ν may not determine uniquely the type of root lattice T_{new}. For instance, if $\nu = 3$, there are two possibilities, as T_{new} is either A_2 or $3A_1$ (No. 11 or No. 23). These can be distinguished by the different types in the last two columns; for No. 11, there are 15 lines in total comprising nine lines of multiplicity 1 and six lines of multiplicity 3, while for No. 23, there are three lines with multiplicity 1, six lines with multiplicity 2, and three lines with multiplicity 4.

On the other hand, ξ determines the Weierstrass coefficients $\lambda = \psi(\xi) \in \mathbb{A}^6$ by algorithm (M) or (M'). Then computing the discriminant and the j-invariant by Tate's algorithm, one directly checks that the elliptic surface $S_\xi = S_\lambda$ has reducible fibres giving T_{new} as in the second column.

The Weierstrass Eq. (10.4) for each of the above data gives some explicit example of affine surface X_ξ in (x, y, t)-space which has precisely the ADE-singularities indicated by T_{new}. Moreover, all these singular points have coordinates in the given field $K = \mathbb{Q}(\xi)$ and their resolution can be achieved by blowing up only K-rational points (which are in fact rational over $\mathbb{Q}$ except for case No. 68, for the numerical choice of $u, v, \ldots \in \mathbb{Q}$). For a geometric account on the subject, compare [BW79].

Table 10.2 Degeneration of a cubic surface, singularities and the line multiplicity

No.	T_{new}	M	$\xi = \{s_1, \ldots, s_6; s_0\}$	ν	ℓ	*mult.*
3	0	$E_6^\vee$	$s_1, s_2, s_3, s_4, s_5, s_6; s_0$	0	27	1^{27}
6	A_1	$A_5^\vee$	$\begin{cases} s_1, s_2, s_3, s_4, s_5, s_6; 1 \\ u, u, s_3, s_4, s_5, s_6; s_0 \end{cases}$	1	21	$1^{15}2^6$
11	A_2	$(A_2^\vee)^2$	$u, u, u, s_4, s_5, s_6; s_0$	3	15	$1^9 3^6$
12	$2A_1$	$(A_5 : A_1)^\vee$	$\begin{cases} u, u, s_3, s_4, s_5, s_6; 1 \\ u, u, v, v, s_5, s_6; s_0 \end{cases}$	2	16	$1^7 2^8 4^1$
19	A_3	$(D_5 : A_2)^\vee$	$u, u, u, u, s_5, s_6; s_0$	6	10	$1^5 4^4 6^1$
20	$A_1 + A_2$	$A_2^\vee \oplus \langle 1/6 \rangle$	$u, u, u, s_4, s_5, s_6; 1$	4	11	$1^3 2^3 3^4 6^1$
23	$3A_1$	$A_1^\vee \oplus A_2^\vee(1/2)$	$u, u, v, v, s_5, 1/u^2v^2s_5; 1$	3	12	$1^3 2^6 4^3$
31	A_4	$\begin{pmatrix} 2/15 & 1/15 \\ 1/15 & 8/15 \end{pmatrix}$	$u, u, u, u, u, s_0^3/u^5; s_0$	10	6	$1^2 5^3 10^1$
32	D_4	$A_2^\vee(1/2)$	$u, u, v, v, 1/uv, 1/uv; 1$	12	6	$1^3 8^3$
37	$A_1 + A_3$	$A_1^\vee \oplus \langle 1/12 \rangle$	$u, u, u, u, v, 1/u^4v; 1$	7	7	$1^1 2^2 4^2 6^1 8^1$
39	$2A_2$	$A_2^\vee \oplus \mathbb{Z}/3\mathbb{Z}$	$u, u, u, v, v, v; uv$	6	7	$3^6 9^1$
40	$2A_1 + A_2$	$\langle 1/6 \rangle^2$	$u, u, v, v, v, 1/u^2v^3; 1$	5	8	$1^1 2^2 3^2 4^1 6^2$
41	$4A_1$	$A_2^\vee(1/2) \oplus \mathbb{Z}/2\mathbb{Z}$	$u, u, v, v, -1/uv, -1/uv; 1$	4	9	$1^3 4^6$
50	D_5	$\langle 1/12 \rangle$	$u, u, u, u, 1/u^2, 1/u^2; 1$	20	3	$1^1 10^1 16^1$
51	A_5	$A_1^\vee \oplus \mathbb{Z}/3\mathbb{Z}$	$u, u, u, u, u, u; u^2$	15	3	$6^2 15^1$
56	$A_1 + A_4$	$\langle 1/30 \rangle$	$u, u, u, u, u, 1/u^5; 1$	11	4	$2^1 5^1 10^2$
59	$2A_1 + A_3$	$\langle 1/12 \rangle \oplus \mathbb{Z}/2\mathbb{Z}$	$u, u, u, u, -1/u^2, -1/u^2; 1$	8	5	$1^1 4^1 6^1 8^2$
61	$A_1 + 2A_2$	$\langle 1/6 \rangle \oplus \mathbb{Z}/3\mathbb{Z}$	$u, u, u, 1/u, 1/u, 1/u; 1$	7	5	$3^2 6^2 9^1$
66	$A_1 + A_5$	$\mathbb{Z}/6\mathbb{Z}$	$-1, -1, -1, -1, -1, -1; 1$	16	2	$12^1 15^1$
68	$3A_2$	$(\mathbb{Z}/3\mathbb{Z})^2$	$\omega, \omega, \omega, \omega^2, \omega^2, \omega^2; 1$	9	3	9^3
69	E_6	$\mathbb{Z}/3\mathbb{Z}$	$1, 1, 1, 1, 1, 1; 1$	36	1	27^1

($s_6 = s_0^3/(s_1s_2s_3s_4s_5)$; ω : a primitive cube root of 1)

Proposition 10.59 *Fix any element ξ in the fourth column of the table such that the coordinates (other than s_6) are sufficiently general. Then the elliptic surface S_ξ has the properties described by other entries (No, T_{new}, M, ν, line multiplicities) in the same row as the chosen ξ. Moreover, the position of reducible fibres and the singular points of X_ξ are given by $K = \mathbb{Q}(\xi)$-rational points.*

Proof The number ν of vanishing roots can be read off from the 36-set Π_ξ. The number ℓ of lines and their multiplicities μ_i can be determined from the 27-set Ω_ξ when ξ is assumed to be sufficiently general, as the correspondence of the 27-set and the set of linear sections (lines) is then bijective. All other claims can be verified in the same way as the following example treating case No. 61, $\xi = \{u, u, u, 1/u, 1/u, 1/u; 1\}$. □

Example 10.60 Consider the case $\xi = \{u, u, u, 1/u, 1/u, 1/u; 1\}$, where u is a variable or any element such that $u^6 \neq 1$. In this case, it is clear that the following roots in the Dynkin diagram E_6 become vanishing roots:

$$\alpha_1 = s_1/s_2, \ \alpha_2 = s_2/s_3, \ \alpha_4 = s_4/s_5, \ \alpha_5 = s_5/s_6.$$

They define A_2^2 and, together with $s_0 = 1$, we have $\nu \geq 7$. By checking that $\nu = 7$ from Π_ξ, we have $T_{new} = A_1 \oplus A_2^2$. Then we see from Ω_ξ that $\ell = 5$ with multiplicity type $3^2 6^2 9^1$. This implies we are in case No. 61.

The two I_3-fibres are located at $t = (2 + 3u + u^3)/u$ and $t = (1 + 3u^2 + 2u^3)/u^2$, and one I_2-fibre at $t = 3(1 + u^2)/u$. The nodes in the Weierstrass model and the singular points of X_ξ have all $\mathbb{Q}(u)$-rational coordinates. The latter are given by the following two A_2-singular points (t, x, y):

$$((2 + 3u + u^3)/u, (1 + 3u^2 + 2u^3)/u^2, -(2 + 3u + u^3)(1 + 3u^2 + 2u^3)/(2u^3)),$$
$$((1 + 3u^2 + 2u^3)/u^2, (2 + 3u + u^3)/u, -(2 + 3u + u^3)(1 + 3u^2 + 2u^3)/(2u^3))$$

and one A_1-singular point

$$(t, x, y) = (3(1 + u^2)/u, 3(1 + u^2)/u, -9(1 + u^2)^2/(2u^2)).$$

Example 10.61 Specializing $u \to -1$ in Example 10.60 gives $\xi = \{-1, \ldots, -1; 1\}$. In this case, it is immediate that the number $\nu = 16$ and $T_{new} = A_5 + A_1$. In fact, we have $\lambda' = (-3, -12, 4, 36, -12, -3)$ so that the Weierstrass equation becomes:

$$E : y^2 + txy - 4tx = x^3 - 3x^2 - 12x + t^3 - 3t^2 - 12t + 36$$

and the discriminant is equal to $-(t-2)^6(t+6)^2(t-3)$. There is a new reducible fibre of type I_6 at $t = 2$ and one of type I_2 at $t = -6$.

The 27 linear sections reduce to the two distinct sections $P_1 = (t, 2t - 6)$ and $P_{13} = (-t + 4, -2)$ of mutiplicity 12 resp. 15. The Mordell–Weil group $M = E(k(t)) = E(\mathbb{Q}(t))$ is a cyclic group of order 6, generated by P_1. For the sake of completeness, we write down:

$$\begin{aligned} 2P_1 &= (-t + 4, t^2 - 4t + 2), \\ 3P_1 &= \left(-\frac{1}{4}t^2 + 3, \frac{1}{8}t^3 - \frac{3}{2}t\right), \\ 4P_1 &= (-t + 4, -2), \\ 5P_1 &= -P_1 = (t, -t^2 - 2t + 6). \end{aligned}$$

As it happens, this elliptic surface has turned up (in different coordinates) as a degeneration of the multiplicative hexagonal case in Example 10.37.

The corresponding cubic surface V_ξ has an A_5-singular point at $(t, x, y) = (2, 2, -2)$ and an A_1-singular point at $(t, x, y) = (-6, -6, -18)$.

Remark 10.62 As a by-product, Proposition 10.59 shows that all the types from Table 10.2 are realized by rational elliptic surfaces over $\mathbb{Q}$ (which, except for No. 68, can even be taken to be $\mathbb{Q}$-split). The same approach can be applied over other fields of characteristic $\neq 2$ (after ensuring that there are no higher multiplicities showing up in Π_ξ). It also works for other excellent families (see [KumS13]) to give a complete proof of Theorem 8.9 over $\mathbb{Q}$ (or other given fields).

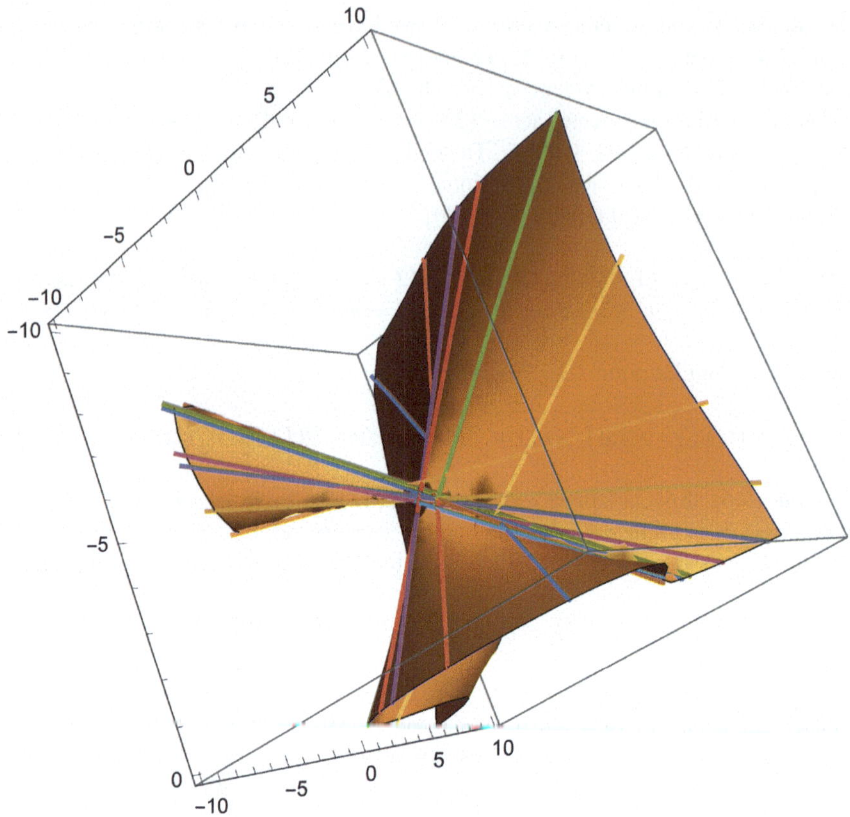

Fig. 10.6 Cubic surface with A_1 singularity (© Shioda)

Pictures. It should be evident from all that we have seen so far that our results can be visualized by drawing the picture of 27 lines (counted with multiplicities) on a cubic surface, since all the lines or the surface can be chosen to be $\mathbb{Q}$-rational, and in particular with real coefficients, for any $\mathbb{Q}$-split data $(s_1, \ldots, s_6, s_0)$. This holds true in the degenerate cases, too, as all the singular points then have rational (hence real) coordinates. We illustrate this below in Figs. 10.6 and 10.7.

Remark 10.63 It should be remarked that our method affords some concrete examples in commutative algebra which might be of some interest. Let $I = I_\lambda$ denote the ideal in the polynomial ring $A = \mathbb{Q}(\lambda)[a, b, c, d]$ generated by the four relations (10.17). Then it is a zero-dimensional ring with $\dim A/I = 27$ for any λ. The primary decomposition of I is given by

$$I = \cap_{i=1}^{m} \mathbf{q}_i, \quad \dim A/\mathbf{q}_i = \mu_i \tag{10.60}$$

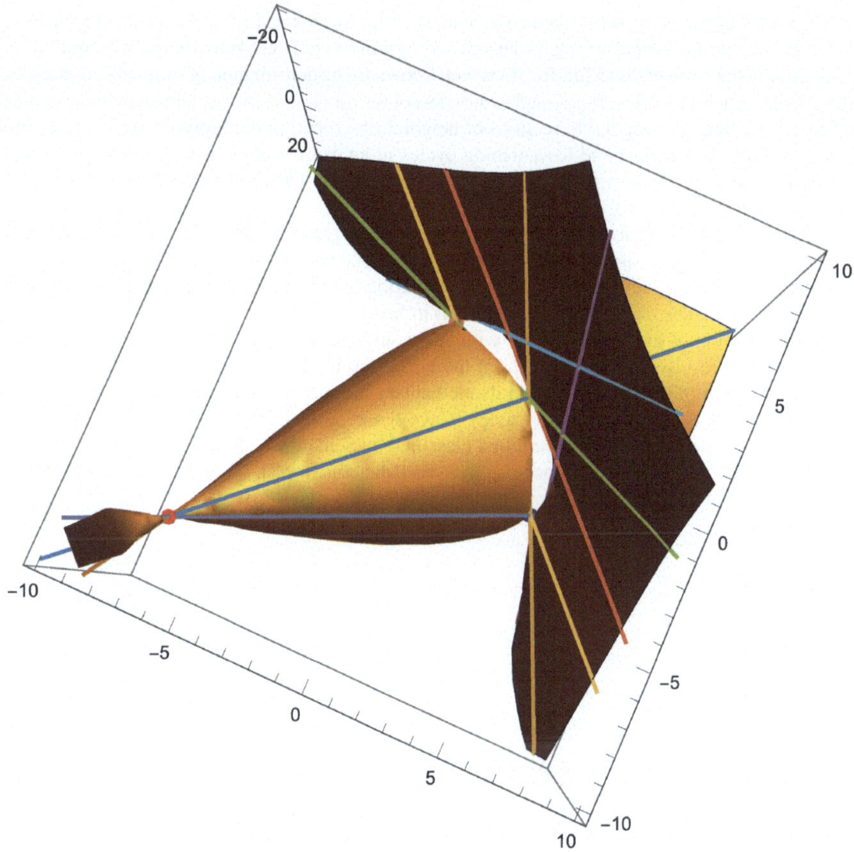

Fig. 10.7 Cubic surface with four A_1 singularites (© Shioda)

with primary ideals $\mathbf{q}_i$, where m, μ_i are the number of lines and multiplicities as in the table. (To verify this is a good exercise in Gröbner basis computation. For the corresponding fact for E_8, cf. [Shio10].)

Historical Notes

Almost twenty years passed between the introduction of the additive excellent families of types E_6, E_7, E_8 and that of their multiplicative analogues. This was for several reasons, from personal through theoretical to computational (as one may infer from [KumS13]). The multiplicative case affords much more flexibility in applications, as we have seen in Sect. 10.4. The resulting families are also more natural from the moduli-theoretic viewpoint: they cover the general rational elliptic surfaces with no reducible fibre, exactly one A_1 or A_2 fibre (according to whether $L = E_8$, E_7 or E_6)—these are the semi-stable ones from Theorem 8.39. The families also fit into the general picture of compactifying by filling in stable curves (as pioneered in [DM69]).

The notion of vanishing roots, defined in Sect. 10.3.2, is borrowed from the concept of vanishing cycles in singularity theory, originally introduced by Lefschetz [Lef24] to study the monodromy around a singular fibre within a family. As is well known in the deformation of rational singularities [Bri68], the vanishing cycles correspond to the roots of the root lattice. As explained in the examples in Sect. 10.3.2, they correspond to sections of height 2 (i.e. roots) in the Mordell–Weil lattice, and thus vanishing roots correspond to vanishing cycles in an explicit way. Indeed, in [Shio91b] we were still referring to "vanishing cycles" of the E_8-singularity. The algebraic version of vanishing cycles in terms of derived categories is treated in [SGA7.2].

Vanishing roots allow for a systematic study of the degenerations within a given (excellent) family of rational elliptic surfaces. For instance, they can be taken as a key input to explicitly work out concrete examples with any given fibre configuration over a fixed field (Theorem 8.9 which we proved much more abstractly (but over any field) in Sects. 8.8, 8.9).

The connection of excellent families with cubic surfaces, even though it was already alluded to in [Shio91d], received critical input from the idea of Weierstrass transformations in [Shio95a], and from the multiplicative set-up (with its elementary algorithm). Here we not only recover the classical result on the 27 lines due to Cayley [Cay49] and Salmon [Salm49], but also retain good control over degenerations (which were first classified by Schläfli [Schl63], see also [BW79]). For arithmetic questions, the key references seem to be [Man86, Swi67]. For the Galois action by the full Weyl group $W(E_6)$, cf. [Eke90].

Chapter 11
Elliptic K3 Surfaces—Basics

For rational elliptic surfaces we have developed a uniform theory of Mordell–Weil lattices that was basically characteristic-free. As soon as we allow the geometric genus of an elliptic surface to go up (or equivalently, the Euler–Poincaré characteristic of the surface, when fixing the (genus of the) base curve), the situation changes dramatically. In fact, one can argue that there is no really convincing structure theory beyond geometric genus one. In this and the next chapter, we thus concentrate on the case of geometric genus one, that is, elliptic K3 surfaces. These offer many new interesting phenomena, some of which we are going to discuss in detail.

K3 surfaces have been a central object of study in algebraic geometry and many neighbouring areas such as differential geometry, number theory and mathematical physics, in particular string theory, for the last half century. For good general references, see [BHPV04, B+85, Huy16]. Next to abelian surfaces, K3 surfaces may be considered the most approachable surfaces of non-negative Kodaira-dimension—and a prototype class for testing and verifying conjectures about algebraic surfaces. For instance, the Tate conjecture, one of the central open problems in algebraic geometry (cf. Conjecture 12.1) has recently been proved for K3 surfaces over finite fields, as we shall state in detail in Theorem 12.2. Historically the Tate conjecture was first proved for those K3 surfaces over finite fields of any characteristic, which admit a genus one fibration (with or without section) by the pioneering work of Artin and Swinnerton-Dyer [ASD73] on homogeneous spaces; this put a first spotlight on elliptic K3 surfaces.

We will discuss some of these specific features of K3 surfaces among other important properties in this and the next chapter.

11.1 Definition and First Examples

The theory of K3 surfaces was largely developed in the 1960s and 70s. K3 surfaces admit many rich structures, especially if it comes to moduli spaces, part of which we shall exploit in the sequel, see e.g. Sect. 11.3.4. Over the complex numbers, the

M. Schütt and T. Shioda, *Mordell–Weil Lattices*, Ergebnisse der Mathematik und ihrer Grenzgebiete. 3. Folge / A Series of Modern Surveys in Mathematics 70,
https://doi.org/10.1007/978-981-32-9301-4_11

moduli viewpoint is even more uniform if we allow non-projective K3 surfaces, but with elliptic surfaces with section in mind we shall limit ourselves to the algebraic K3 surfaces, considered over an algebraically closed field as usual.

Definition 11.1 A K3 surface X is an irreducible smooth projective surface with

$$\omega_X \cong \mathcal{O}_X, \quad h^1(X, \mathcal{O}_X) = 0.$$

Equivalently, a K3 surface has trivial canonical divisor $K_X = 0$ while being (algebraically) simply connected (so in particular $b_1(X) = 0$ regardless of the characteristic). As the canonical divisor is trivial, the geometric genus is one, so that the remaining invariants of X follow with Noether's formula:

$$p_g(X) = 1, \quad \chi(\mathcal{O}_X) = 2, \quad e(X) = 24, \quad b_2(X) = 22.$$

For a complex K3 surface, one also reads off the Hodge numbers

$$h^{2,0}(X) = h^{0,2}(X) = 1, \quad h^{1,1}(X) = 20.$$

Example 11.2 (*Quartics*) The classical example of a K3 surface is a smooth quartic in $\mathbb{P}^3$. Indeed, this has irregularity zero by Example 4.2 and trivial canonical divisor by the adjunction formula (4.3). Then, according to A. Weil's original definition (which later led to the equivalent Definition 11.1), a K3 surface is any (complex projective) surface with the same differentiable structure as a smooth quartic (compatible with the above definition due to Kodaira). One may also admit rational double point singularities and consider their minimal desingularization as this process does not take us out of the deformation class.

One of the prototype examples is the Fermat quartic surface

$$S_4 = \{x_0^4 + x_1^4 + x_2^4 + x_3^4 = 0\} \subset \mathbb{P}^3 \tag{11.1}$$

which is visibly smooth outside characteristic 2. In the sequel we will occasionally revisit this surface, mostly for geometric properties, for instance for the curves it contains. Notably there are 48 lines on S_4 which generate a sublattice of $\mathrm{NS}(S_4)$ of rank 20. From (4.15), it follows that the lines generate $\mathrm{NS}(S_4)$ in characteristic zero (a priori up to finite index, compare Example 13.22); the same holds in certain positive characteristics (compare Example 13.40). The Fermat quartic thus provides the first example of what we call a singular K3 surface:

Definition 11.3 A complex K3 surface X is called *singular*, if $\rho(X) = 20$. The determinant of $\mathrm{NS}(X)$ is called the *discriminant* of X.

Singular K3 surfaces carry rich arithmetic structures, as we shall exploit in Chap. 12. For instance, the zeta function of the Fermat quartic (or any singular K3 surface over $\mathbb{Q}$ by [Liv95]) leads to a certain modular form of weight 3; the zeta function was first studied in terms of Jacobi sums, as pioneered by A. Weil (more generally in any degree and dimension, see [Wei49]).

In a way, the definition of a K3 surface as well as the above classical example parallel elliptic curves (trivial canonical divisor, cubics in $\mathbb{P}^2$). However, while any elliptic curve admits a model as a cubic in $\mathbb{P}^2$ as a consequence of Riemann–Roch (see Theorem 3.2), for K3 surfaces the situation is completely different. The information in which projective space $\mathbb{P}^N$ a given (abstract) K3 surface X can be embedded, is captured by the class of the hyperplane section of $\mathbb{P}^N$, considered as a (very ample) divisor on X. This will lead the concept of polarization, see Sect. 11.3. Here we give but a few more examples:

Example 11.4 (*Double sextics*) Another way of mimicking elliptic curves starts from double covers of $\mathbb{P}^1$ branched along four points (outside characteristic 2). In analogy, double covers of $\mathbb{P}^2$ branched along a sextic curve (smooth or with no worse than ADE-singularities), written affinely as

$$w^2 = f(x, y), \quad f \in k[x, y], \quad \deg(f) = 5, 6. \tag{11.2}$$

give rise to K3 surfaces (which come in a 19-dimensional family up to the linear transformations induced from $\mathbb{P}^2$, cf. [Per85]).

Remark 11.5 (*Characteristic two*) The double sextic K3 surfaces from Example 11.4 can be adapted to characteristic 2 by introducing an auxiliary summand $g \in k[x, y]$ of degree at most 3, just as in the generalized Weierstrass form:

$$w^2 + g(x, y)w = f(x, y), \quad f, g \in k[x, y], \quad \deg(f) \leq 6, \quad \deg(g) \leq 3.$$

Here one requires $\deg(g) > 2$ or $\deg(f) > 4$. Indeed, if $g \not\equiv 0$, then this defines a separable cover of $\mathbb{P}^2$ with the same invariants and total number of parameters (up to linear transformations) as before. Note, however, that the branch locus now is the cubic curve $Z(g)$. In comparison, if $g \equiv 0$, then the double cover becomes purely inseparable, and (11.2) automatically has some singularities, but in the ADE-case, the resolution still is a K3 surface (later to be called *supersingular* in Sect. 12.4). As a side-effect, there is no branch locus, and the number of parameters drops to 9.

Example 11.6 (*Kummer surfaces*) Yet another classical class of examples is formed by *Kummer surfaces*. Here we start from an abelian surface A and divide out by the inversion map ι with respect to the group structure. Outside characteristic 2, this involution has 16 isolated fixed points, resulting in A_1 singularities on the quotient which are resolved to a K3 surface denoted by $\mathrm{Km}(A)$. This implies that

$$\rho(\mathrm{Km}(A)) = \rho(A) + 16 \geq 17. \tag{11.3}$$

There is an easy case where we can bring the above two examples together: **Kummer surfaces of product type**. In the above construction, take A to be of product type, i.e. decomposable into two elliptic curves E, E':

$$A \cong E \times E'.$$

Assume that both elliptic curves are given in affine (extended) Weierstrass form

$$E: \; y^2 = f(x), \quad E': \; y'^2 = f'(x').$$

Then the involution ι on $E \times E'$ is defined as

$$\iota: \; (x, y, x', y') \mapsto (x, -y, x', -y').$$

The ring of invariants is generated by $x, x', w = yy'$. A model of the Kummer surface $\mathrm{Km}(E \times E')$ is thus given as (singular affine) double sextic

$$\mathrm{Km}(E \times E'): \quad w^2 = f(x)\, f'(x'). \tag{11.4}$$

Let us comment on the Picard number of the desingularization. An abelian surface has generically $\rho(A) = 1$. Presently, the product structure equips $E \times E'$ with two algebraically independent divisor classes, the fibres of the two projections. It follows that $\rho(E \times E') \geq 2$, and from (11.3), $\rho(\mathrm{Km}(E \times E')) \geq 18$ (with equality unless E and E' are isogenous, see (12.12) for a general formula for the Picard number). These divisors are also visible on the double sextic model: the RHS of (11.4) gives six lines, or in fact, two triplets of parallel lines forming the grid displayed in Fig. 11.1. The lines intersect in nine ordinary double points (type A_1) in the affine plane and two D_4-singularities outside the given chart. Together with the pull-back of the hyperplane class from $\mathbb{P}^2$, this produces 18 independent algebraic curves on the minimal resolution $\mathrm{Km}(E \times E')$.

We should point out that in practice, proving that a given K3 surface is a Kummer surface can be a highly non-trivial task. For instance, one can deduce in the context of singular K3 surfaces that the Fermat quartic is a Kummer surface of product type (see Example 12.5). However, the reasoning is fairly abstract as it relies on lattices and the Torelli theorem, and it is not so immediate how to derive the elliptic curves explicitly from Eq. (11.1).

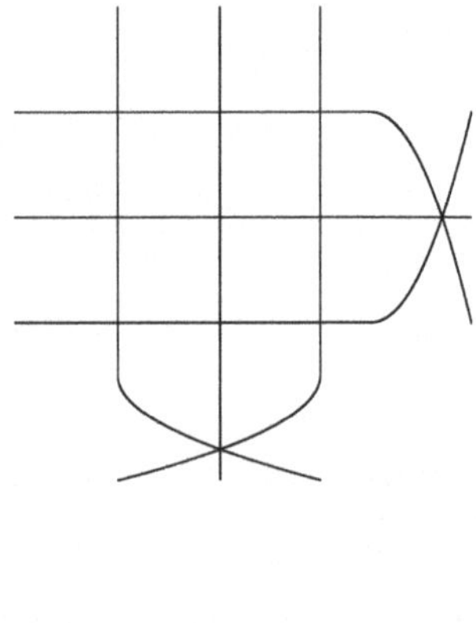

Fig. 11.1 Lines on double sextic

We conclude this section by emphasizing why there is no harm in passing from one model of a K3 surface to another, say affinely or a priori birationally. The reason comes from general surface theory because K3 surfaces are minimal by definition: a birational map between K3 surfaces is automatically an isomorphism because the canonical divisor is nef.

11.2 Elliptic K3 Surfaces

We shall now introduce the key players of this chapter: elliptic K3 surfaces. We first recall some easy facts from previous sections.

Assume that the K3 surface X admits a genus one fibration. Since X is simply connected by definition, the base curve is necessarily $\mathbb{P}^1$ as any regular 1-form on the base pulls back to a regular 1-form on X. (This can also be regarded as a consequence of the functoriality of the Albanese map.) In the following, we assume that the genus one fibration

$$X \to \mathbb{P}^1$$

has a section; i.e. the fibration is elliptic. The corresponding smooth rational curve on X is a (-2)-curve by the adjunction formula (Theorem 4.11). Thanks to the section, the K3 surface X admits a globally minimal Weierstrass form as in (5.4):

$$X: \quad y^2 + a_1xy + a_3y = x^3 + a_2x^2 + a_4x + a_6. \tag{11.5}$$

We can rescale x, y so that the coefficients are all polynomials in $k[t]$. For the canonical divisor to be trivial, the degrees ought to satisfy $\deg a_i \leq 2i$ with some i such that $\deg a_i > i$ (and no common divisibilities after admissible transformations in x, y, see Proposition 5.51).

11.2.1 *Kummer Surfaces of Product Type*

We continue to discuss an easy, but instructive example which will allow us to experience already some of the special features of elliptic K3 surfaces: Kummer surfaces of product type. Geometrically, two elliptic fibrations are easily derived; they are induced from the projections of the product to its single factors:

$$\begin{array}{ccc} E & \longrightarrow & \mathbb{P}^1 \\ \uparrow & & \uparrow \pi \\ E \times E' & \dashrightarrow & \mathrm{Km}(E \times E') \\ \downarrow & & \downarrow \pi' \\ E' & \longrightarrow & \mathbb{P}^1 \end{array}$$

By construction, either fibration is isotrivial with generic fibre E' resp. E (the complementary factor of the product). Moreover, one sees that the 16 singularities resulting from the Kummer quotient sit in the four fibres above the points where the double cover $E \to \mathbb{P}^1$ (resp. $E' \to \mathbb{P}^1$) ramifies, as they come from the 2-torsion points. In view of Kodaira's classification of the singular fibres (combined with Euler number considerations), this provides an ad-hoc way to see that the singular fibres can only have type I_0^*, and there cannot be any further singular fibres.

In terms of equations, we can explicitly write down models of these fibrations by multiplying one cubic polynomial from the RHS of (11.4) to the LHS. For instance, we obtain with $t = x'$ the family of cubics in x, y

$$\mathrm{Km}(E \times E') : \quad f'(t)\, y^2 = f(x). \tag{11.6}$$

Following [KuwS08], we call the rational function t on $\mathrm{Km}(E \times E')$, defining the elliptic fibration

$$\mathrm{Km}(E \times E') \xrightarrow{\pi'} \mathbb{P}^1,$$

the *elliptic parameter*. The cubic equation (11.6) is easily transformed to Weierstrass form, but the present equation displays the isotriviality better. The discriminant turns out to be a constant times $f'(t)^6$ which gives the four singular fibres of Kodaira type I_0^* as stated above. Moreover, we read off that the Mordell–Weil group contains full 2-torsion over $\bar{k}$, given exactly by $E[2]$ (i.e. the non-trivial 2-torsion points are $(\alpha_i, 0)$ for the zeroes α_i of $f(x)$, just like on E). Note that the formula for the Picard number from Corollary 6.7 returns

$$\rho(\mathrm{Km}(E \times E'), \pi') = 2 + 4 \times 4 + \mathrm{rk}\, \mathrm{MW}(\mathrm{Km}(E \times E'), \pi') \geq 18$$

in agreement with the considerations in the previous section. For the remainder of this section we shall make the following generality assumption for simplicity:

Assumption 11.7 The elliptic curves E, E' are not isogenous.

Otherwise there will be additional algebraic cycles on $E \times E'$ and on $\mathrm{Km}(E \times E')$ such that $\rho(\mathrm{Km}(E \times E')) > 18$. These curves may even appear as additional reducible fibres for some of the elliptic fibrations to come (or induce additional sections by Theorem 6.5). To ease the exposition, we will rule this out for a while and restrict to the very general situation. For the remaining cases, see Sect. 12.2.4, especially (12.12).

Under Assumption 11.7, we infer from $\rho(E \times E') = 2$ and (11.3) that $\rho(\mathrm{Km}(E \times E')) = 18$ and thus the Mordell–Weil group of the fibration π' has rank zero. But then the 2-torsion sections $(\alpha_i, 0)$ over $\bar{k}$, compared against the singular fibres, imply by the discussion in Sect. 5.6 that

$$\mathrm{MW}(\mathrm{Km}(E \times E'), \pi') \cong (\mathbb{Z}/2\mathbb{Z})^2,$$

and by symmetry the same holds for the elliptic fibration π and $MW(Km(E \times E'), \pi)$. By the formula for the determinant (6.23), we obtain

$$\det NS(Km(E \times E')) = -2^4. \tag{11.7}$$

This fact was not apparent at all from the previous models. We shall see a lattice theoretic way to deduce (11.7) in Sect. 12.2.2.

With the two inequivalent elliptic fibrations at hand, the reader might wonder whether there are any more, and if so, how many. This problem was solved in the generic situation of non-isogenous elliptic curves by Oguiso [Ogui89] by a genuinely geometric argument. We shall discuss an alternative lattice theoretic approach going back to Nishiyama (and tracing back to Kneser) in Sect. 12.3.2. In this section, we content ourselves with elaborating on the two further more or less obvious elliptic fibrations on $Km(E \times E')$.

11.2.2 First Alternative Elliptic Fibration

One more elliptic fibration, say π_1, follows immediately from (11.6) by taking y, rather than t, as the elliptic parameter (i.e. the parameter of the base curve $\mathbb{P}^1$). Then the generic fibre is a plane cubic in x, t. More precisely, $Km(E \times E')$ is the quadratic base change by $w = y^2$ of the rational elliptic surface $S \to \mathbb{P}^1_w$ defined by

$$S: \quad f'(t)\, w = f(x). \tag{11.8}$$

The base points (α_i, α'_j) are given exactly by the x- resp. x'-coordinates α_i, α'_j of the non-trivial 2-torsion points $E[2], E'[2]$, so that over $\bar{k}$ these elliptic K3 surfaces have nine sections right away.

Keeping the assumption that E and E' are not isogenous, the only reducible singular fibres of S sit at $w = 0, \infty$. They are formed by the three parallel lines, meeting in a single point outside the given affine chart of $\mathbb{P}^2$ (see Fig. 11.2). Thus the fibres have Kodaira type IV.

By Theorem 8.8 the Mordell–Weil group of S has rank 4, and in fact the shape of the Mordell–Weil lattice is not affected by the degeneracy of the fibre type for the corresponding root lattices A_2 (having Kodaira type IV as opposed to the usual I_3):

$$MWL(S) = A_2^\vee \oplus A_2^\vee.$$

Presently, one easily confirms that the base points of the cubic pencil (with one chosen as zero section) generate the full MWL (in agreement with [Pas12], but contrary to the general situation from Theorem 7.26). To do this, one simply works out the matrix with the entries of the height pairing.

Fig. 11.2 Singular fibres of first alternative fibration

Now we switch back to the elliptic K3 surface $X = \mathrm{Km}(E \times E')$. From Table 5.2 we read off the singular fibres of type IV* at $0, \infty$ resulting from the quadratic base change. Moreover, by Proposition 6.53 the induced sections form a sublattice

$$\mathrm{MWL}(S)(2) = A_2^\vee(2) \oplus A_2^\vee(2) \subseteq \mathrm{MWL}(X, \pi_1),$$

where as usual intersection pairings on the lattices are scaled by the numbers in brackets. By inspection, these sections together with the fibre components form a finite index sublattice of $\mathrm{NS}(X)$ (rank 18). Computing determinants and comparing with (11.7), we find that in fact, equality holds:

$$\mathrm{MWL}(X, \pi_1) = \mathrm{MWL}(S)(2) = A_2^\vee(2) \oplus A_2^\vee(2).$$

In particular, this shows that $\mathrm{MWL}(X, \pi_1)$ is still generated by the base points of the cubic pencil (11.6) (under Assumption 11.7).

11.2.3 Second Alternative Elliptic Fibration

There is another way to convert (11.4) into a plane cubic over $\bar{k}(t)$. Here we first factorize the polynomials $f(x), f'(x')$ over $\bar{k}$ and then pair up two factors of one polynomial with one of the other. Eventually, this results in cubics such as

$$X: \quad (x-\alpha)(x-\beta)(x'-\gamma')\, y^2 = (x'-\alpha')(x'-\beta')(x-\gamma). \qquad (11.9)$$

Again we observe the quadratic base change from a rational elliptic surface, this time given by the cubic pencil

$$S': \quad (x-\alpha)(x-\beta)(x'-\gamma')\, t = (x'-\alpha')(x'-\beta')(x-\gamma).$$

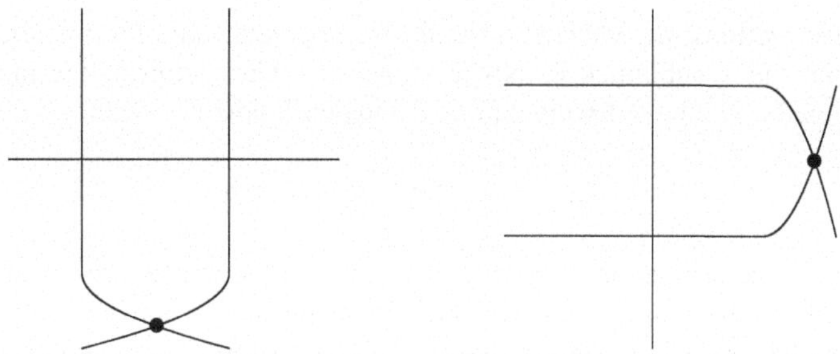

Fig. 11.3 Singular fibres with singularity at double base point

Contrary to the previous case, the given model of S' is not smooth yet. There are two rational double point singularities at the double base points; in partly homogeneous coordinates (i.e. homogeneous on $\mathbb{P}^2$, but not on $\mathbb{P}^1$), they are given as

$$([x, x', v], t) = ([1, 0, 0], 0), ([0, 1, 0], \infty).$$

The resolution of the two A_1 singularities results in singular fibres of type I_4 (see Fig. 11.3).

Then the two double base points can be interpreted as the zero section and a 2-torsion section on the resolved rational elliptic surface S'. This follows directly from the height pairing as the sections meet opposite components in both singular fibres (see Notation 5.13). Indeed, picking $[0, 1, 0]$ as zero section as usual, we compute the height of the section $P = [1, 0, 0]$ as

$$h(P) = 2 - 1 - 1 = 0.$$

Since generally there are no further reducible fibres (as one easily checks using the discriminant), it follows from Theorem 8.8 that

$$\mathrm{MWL}(S') = A_1^\vee \oplus A_1^\vee.$$

Generators corresponding to this direct sum are easily found in terms of the base points $[\alpha, \alpha', 1]$ and $[\beta, \alpha', 1]$, for instance. Thus the base points of the cubic pencil generate the full Mordell–Weil group (contrary to the general situation from Sect. 7.6).

On the K3 surface X with the elliptic fibration π_2 from (11.9) given as projections onto the projective y-line, we derive singular fibres of type I_8 at 0 and ∞ (see also [KuwS08]). Base change induces a sublattice

$$\mathrm{MWL}(S')(2) = A_1^\vee(2) \oplus A_1^\vee(2) \cong A_1 \oplus A_1 \subseteq \mathrm{MWL}(X, \pi_2).$$

Again, these sections, together with the fibre components and the torsion section, form a finite index sublattice of $\mathrm{NS}(X)$ (rank 18). Computing determinants and comparing with (11.7), we find that in fact, equality holds:

$$\mathrm{MWL}(X, \pi_2) = \mathrm{MWL}(S')(2) = \langle 1 \rangle^2.$$

Remark 11.8 There is another notable difference in the above two constructions: while the elliptic fibration π_1 in Sect. 11.2.2 was derived in a canonical way over k, exhibiting the second fibration π_2 above required factoring the polynomials f, f' over $\bar{k}$, and so to say, specifying 2-torsion points of the elliptic curves E, E'. In general this can be done in 3×3 different ways, so we obtain nine inequivalent elliptic fibrations on $\mathrm{Km}(E \times E')$ with the same configuration of singular fibres (compare Sect. 12.3.1).

11.3 Lattice Theory for K3 Surfaces

In this section we work over $\mathbb{C}$, although some of the results, notably those concerning moduli, carry over to positive characteristic (see Sect. 11.3.7).

11.3.1 K3 Lattice

We have already computed that a K3 surface X has $b_2(X) = 22$. We now consider $H^2(X, \mathbb{Z})$ and recall some basic properties from [BHPV04], for instance. Since $H_1(X, \mathbb{Z})$ is free (in fact, it is zero), so is $H^2(X, \mathbb{Z})$; we view it as a lattice endowed with cup-product. By Poincaré duality, this lattice is unimodular. Evenness follows from Wu's formula ([BHPV04, p. 310]). Then the signature (3, 19) (computed through the topological index theorem) determines $H^2(X, \mathbb{Z})$ uniquely up to isometries as

$$H^2(X, \mathbb{Z}) \cong U^3 \oplus (E_8^-)^2 =: \Lambda.$$

We call Λ the K3 lattice.

11.3.2 Polarized K3 Surfaces

Since we assumed X to be projective, the ambient projective space provides it with a very ample divisor (the class of the hyperplane section). This divisor class need not be primitive in $\mathrm{NS}(X)$, but there a unique primitive ample divisor class H such that the hyperplane section is a multiple of H. Via the first Chern map, H gives an element of Λ. This is called a *polarization* of X of degree $d = H^2 \in 2\mathbb{N}$.

Example 11.9 On a smooth double sextic (as in Example 11.4), the pull-back of the hyperplane class in $\mathbb{P}^2$ gives a polarization of degree 2. Similarly, a smooth quartic in $\mathbb{P}^3$ has a polarization of degree 4, and conversely, a very ample divisor H of degree 4 on a K3 surface gives such a model via the linear system $|H|$.

Here comes a striking difference to elliptic curves: for any even positive d, there is a 19-dimensional moduli space of K3 surfaces with polarization of degree d. Note the generic K3 surface in such a moduli space will have Picard number one with $\text{Pic} \cong \mathbb{Z} = \langle d \rangle$, so in particular the intersection of two moduli spaces of different polarizations has codimension one in either moduli space.

Continuing the above examples, one could consider complete intersections of type $(2, 3)$ in $\mathbb{P}^4$ or $(2, 2, 2)$ in $\mathbb{P}^5$ (of degree 6 reps. 8), but for higher-degree polarizations more sophisticated ideas are required for explicit constructions of polarized K3 surfaces (see work of Mukai, for instance) until very soon one runs out of generic constructions. The geometry of the moduli spaces of polarized K3 surfaces has been an area of intense research for the last decades, and by now we know that very soon they tend to be of general type (work of Kondō [Kon93, Kon99], Gritsenko-Hulek-Sankaran [GHS07]). Notice that an explicit construction of projective models of all K3 surfaces with a given polarization often implies that the corresponding moduli space will be uniruled, unirational or even rational (a phenomenon already observed for moduli spaces of curves when the genus g grows).

11.3.3 Hodge Theory and Transcendental Lattice

The fundamental problem in the theory of K3 surfaces goes under the name of the Torelli problem:

Problem 11.10 What additional condition can be imposed on an isometry

$$\imath : \; H^2(X', \mathbb{Z}) \cong H^2(X, \mathbb{Z})$$

for two complex K3 surfaces X, X' to imply that X, X' are isomorphic? Given this, how does it determine the isomorphism?

The answer to this question is provided by *Hodge theory*. We have already discussed that $\text{NS}(X)$ (or $\text{Pic}(X)$, for both are the same for a K3 surface) embeds into $H^2(X, \mathbb{Z})$ via the first Chern map.Then we consider its orthogonal complement with respect to cup-product:

Definition 11.11 The transcendental lattice of a K3 surface X is defined as

$$T(X) = \text{NS}(X)^{\perp} \subset H^2(X, \mathbb{Z}).$$

Remark 11.12 In the context of rational elliptic surfaces and their Mordell–Weil lattices, especially in the classification of Theorem 8.8, we reserved the letter T for the fibre contribution to the trivial lattice, or to the frame. Still we decided to use the same letter here for the transcendental lattice, since in either context this is the standard notation. It is our hope that the reader will excuse this as throughout this book there will be little room for ambiguity.

Since NS(X) has signature $(1, \rho(X) - 1)$ by the Hodge index theorem, the transcendental lattice $T(X)$ has signature $(2, 20 - \rho(X))$. Its determinant agrees with that of NS(X) up to sign, and in fact, as orthogonal complements in the unimodular lattice Λ, the discriminant forms of $T(X)$ and NS(X) are opposite to each other:

$$q_{T(X)} = -q_{\mathrm{NS}(X)}.$$

Example 11.13 The Fermat quartic surface S_4 has transcendental lattice

$$T(S_4) = \begin{pmatrix} 8 & 0 \\ 0 & 8 \end{pmatrix}.$$

For the history of the proof of this, see the discussion in [SSvL10].

Of course, there are other ways of viewing the transcendental lattice $T(X)$. For instance, it gives rise to a compatible system of Galois representations if X is defined over some number field (thus lending itself to questions concerning modularity or automorphy). We shall now interpret the transcendental lattice in yet another way. Recall that $H^2(X, \mathbb{C})$ is endowed with a Hodge structure:

$$H^2(X, \mathbb{C}) = H^{2,0}(X) \oplus H^{1,1}(X) \oplus H^{0,2}(X).$$

Via $H^2(X, \mathbb{C}) = H^2(X, \mathbb{Z}) \otimes \mathbb{C}$, we can view $H^2(X, \mathbb{Z})$ as sitting inside $H^2(X, \mathbb{C})$. Then Lefschetz' theorem states that NS(X) has Hodge type (1, 1), i.e. it fully lies inside $H^{1,1}(X)$ (see (4.15)). In turn, $T(X)$ corresponds to the smallest Hodge substructure of $H^2(X, \mathbb{Z})$ containing $H^{2,0}$ after complexifying. Since $h^{2,0}(X) = 1$ for K3 surfaces, the Hodge structure $T(X)$ is thus automatically irreducible over $\mathbb{Z}$ (and also over $\mathbb{Q}$).

Let us return to the Torelli problem. Clearly, if X and X' are isomorphic, then the induced isometry on the K3 lattices will respect the Hodge structures. Any isometry respecting the Hodge structures is called a *Hodge isometry*. Conversely, the Torelli theorem states that this condiction is in fact sufficient:

Theorem 11.14 ((Weak) Torelli theorem) *Two K3 surfaces X, X' are isomorphic if and only if there exists a Hodge isometry*

$$\imath : \; H^2(X', \mathbb{Z}) \cong H^2(X, \mathbb{Z}).$$

The strong version of the Torelli theorem specifies whether a Hodge isometry ι is indeed induced by an isomorphism $X \cong X'$. This holds true if and only if ι is effective, i.e. preserves the positive cones and induces a bijection between effective classes.

11.3.4 Lattice Polarized K3 Surfaces

We can also ask for K3 surfaces with more structure than simply a polarization. For instance, one could think of automorphisms and/or higher Picard number—or of an elliptic fibration, but we will soon see how to interpret elliptic fibrations (and more generally genus one fibrations) in terms of the Néron–Severi lattice (Theorem 11.24, Proposition 11.31). Indeed, given a K3 surface X, one may wonder about the deformations of X preserving $\mathrm{NS}(X)$. In order to allow for higher Picard number than $\rho(X)$, one may weaken this to require that $\mathrm{NS}(X)$ embeds primitively into $\mathrm{NS}(X')$ for the deformation X'. Ultimately, this leads to the concept of lattice polarization, which we introduce abstractly below.

What are the requirements for a lattice

$$M \hookrightarrow \Lambda \tag{11.10}$$

to occur as Néron–Severi lattice of some algebraic K3 surface? For sure, M ought to have signature $(1, \mathrm{rk}(M) - 1)$ and the embedding (11.10) ought to be primitive. It turns out that this condition is sufficient (see Proposition 11.15), but in order to give a more precise answer and setup a moduli space of the corresponding K3 surfaces, we have to fix a marking $H^2(X, \mathbb{Z}) \cong \Lambda$ for every K3 surface X. This allows us to speak about ample classes in Λ, and it appears natural to require that the image of M in Λ contains an ample class. We call such a (marked) K3 surface M-polarized.

Proposition 11.15 ([Morr84, Corollary 1.9]) *Let M be an even integral lattice of signature $(1, r)$, $r \in \mathbb{N}_0$. If M has a primitive embedding into Λ, then M-polarized complex K3 surfaces admit a local moduli space of dimension $19 - r$.*

We remark that each inequivalent primitive embedding $M \hookrightarrow \Lambda$ gives rise to a different moduli space of M-polarized K3 surfaces. Here is an interesting twist of this story: in low moduli dimensions, the K3 surfaces themselves can be investigated in order to collect information, possibly even explicit equations about the moduli spaces which turn out to be interesting objects themselves. For instance, moduli spaces of K3 surfaces of Picard number $\rho \geq 19$ are related to modular curves and Shimura curves; in higher dimension we find Humbert surfaces, Hilbert modular surfaces and Shimura varieties, compare [Elk08, EK14, GS12]. We will briefly come back to these issues later in Sect. 13.1.4.

11.3.5 Examples

A. The simplest lattice of rank greater than one with a primitive embedding into Λ is the hyperbolic plane U. The embedding is seen to be unique up to isometries of Λ, so there is a unique moduli space of U-polarized K3 surfaces. Note that we indeed have suitable candidates at hand: elliptic K3 surfaces! Here zero section and general fibre generate the rank 2 lattice

$$\begin{pmatrix} -2 & 1 \\ 1 & 0 \end{pmatrix} \cong \begin{pmatrix} 0 & 1 \\ 1 & 0 \end{pmatrix} = U.$$

We give a heuristic argument that all U-polarized K3 surfaces are elliptic with section. For this purpose we simply consider the moduli dimension of elliptic surfaces with Weierstrass form (11.5). As discussed, this corresponds to the $d = 2$ case in Theorem 5.48, yielding 18 moduli as expected for a K3 surface with $\rho \geq 2$. A precise argument (without reference to moduli) will be given in Theorem 11.24 and Proposition 11.31 in Sect. 11.4.

B. From the geometric viewpoint we can also enhance K3 surfaces with specific curves. For instance, consider a smooth quartic surface S containing a line. Denoting the line by ℓ, we have $\ell^2 = -2$ by the adjunction formula (Theorem 4.11). Together with the hyperplane section H, this gives the following sublattice of $\mathrm{NS}(S)$:

$$\langle H, \ell \rangle = \begin{pmatrix} 4 & 1 \\ 1 & -2 \end{pmatrix} =: M \hookrightarrow \mathrm{NS}(S).$$

On the other hand, containing a line clearly is a codimension one condition on the moduli space of quartic surfaces (aka K3 surfaces of degree 4, cf. Example 11.9) as the quartics containing a line can easily be parametrized. Taking the uniqueness of the embedding $M \hookrightarrow \Lambda$ for granted, we deduce that a very general M-polarized K3 surface is quartic containing a line (cf. [S-D74]). In fact, these quartics come with genus one fibrations, i.e. generically without section, see Example 11.30.

C. In order to consider an example of higher rank, let us return to the Kummer surfaces of product type (see (11.4) and Sect. 11.2.1). There $\mathrm{NS}(E \times E')$ is generically a lattice M of rank 18 and determinant -16 by (11.7). Even without having determined the precise shape of M explicitly (we will find a convenient description in Sect. 12.2.2), we can deduce that its embedding into Λ is unique as follows: the isotrivial elliptic fibration with four fibres of type I_0^* gives a sublattice of index 4,

$$U \oplus D_4^4 \subset M.$$

With the sublattice, also M is 2-elementary, i.e. its discriminant group is only 2-torsion. But then the primitive embedding $M \hookrightarrow \Lambda$ (exists and) is unique by Theorem 2.42. As a consequence, the moduli space of M-polarized K3 surfaces is irreducible of dimension two by Proposition 11.15. Since the Kummer surfaces at hand have two independent parameters coming from the two elliptic factors of the abelian surface, we find that all M-polarized K3 surfaces, possibly up to some boundary points, are Kummer surfaces of product type.

11.3.6 (−2)-Curves on K3 Surfaces

We have to mention the main obstruction against global moduli of K3 surfaces (or generally algebraic surfaces), namely (-2)-*curves*, i.e. smooth rational curves C of self-intersection $C^2 = -2$. We have seen such curves in several instances, most notably as components of reducible fibres of genus one fibrations, but also (related) as exceptional curves on resolutions of isolated ADE-singularities (as occurring on the Weierstrass models)—or as sections of elliptic K3 surfaces (by inspection of the Euler characteristic).

To illustrate how (-2)-curves compete with a polarization, consider a family of smooth quartic surfaces $\mathcal{X}$, thus carrying a polarization of degree 4 given by a hyperplane section H. Assume that $\mathcal{X}$ degenerates to a quartic $X \subset \mathbb{P}^3$ with an isolated ADE-singularity. Then the minimal resolution $\tilde{X}$ will again be a K3 surface (deformation equivalent to the smooth members of $\mathcal{X}$), but H does not pull back to an ample divisor on $\tilde{X}$ since it is perpendicular to all exceptional curves. In terms of moduli of polarized K3 surfaces as in Proposition 11.15, the degeneration X (or $\tilde{X}$) will correspond to a point on the boundary of the given moduli space.

A special feature of K3 surfaces is that (-2)-curves can be detected completely on the level of lattices. Since this will prove extremely useful in the following, we sketch the two most important properties in Lemmas 11.16 and 11.17. We will call a divisor $D \in \mathrm{Div}(X)$ a (-2)-*divisor*, if $D^2 = -2$.

Lemma 11.16 *Any (-2) divisor on a K3 surface is either effective or anti-effective.*

Proof Apply Riemann–Roch as in (4.9). □

Recall from (2.11) that for any negative root $\alpha \in \mathrm{NS}(X)$, the reflection

$$s_\alpha : \quad x \mapsto x + \langle x, \alpha \rangle\, \alpha$$

defines an isometry of $\mathrm{NS}(X)$. Here we will only use reflections in (-2)-curves to obtain the following structure result about (-2)-divisors:

Lemma 11.17 *Let D be a (-2)-divisor class on a K3 surface X. Pick the sign such that $\pm D > 0$. Then $\pm D$ is supported on (-2)-curves, i.e. there are (-2)-curves $C_i \subset X$ and $m_i \in \mathbb{N} (i = 1, \dots, r)$ such that*

$$D = \sum_{i=1}^{r} m_i C_i.$$

Proof Let $D \in \mathrm{Div}(X)$ with $D^2 = -2$. After changing sign, if necessary, we may assume that $D > 0$ by Lemma 11.16. Write D as a finite sum

$$D = \sum_C m_C C \quad (m_C \in \mathbb{N})$$

for distinct irreducible curves $C \subset X$. Here $D^2 = D.(\sum_C m_C C) = -2 < 0$ not only implies that there is some C_1 such that $D.C_1 < 0$, but writing $D.C_1 = (\sum_C m_C C).C_1$ one even deduces that $C_1^2 < 0$ (so that by adjunction (Theorem 4.11), C_1 is a (-2)-curve). The reflection in C_1 thus results in the (-2)-divisor

$$D_1 = s_{C_1}(D) < D.$$

In particular, the degree (relative to any fixed polarization) drops. At the same time D_1 is still effective or anti-effective by Lemma 11.16. Note that we can continue reflecting in (-2)-curves $C_2, \ldots, C_n$ to obtain (-2)-divisors $D_2, \ldots, D_n$ of smaller and smaller degree until the degree switches from positive to negative, and so does the property of being effective or anti-effective:

$$0 < D_{n-1} \mapsto D_n = s_{C_n}(D_{n-1}) < 0.$$

For the degrees, we will use the inequality

$$\deg(-D_n) \leq \deg(D_{n-1}). \tag{11.11}$$

To see this, simply write out

$$D_{n-1} = \sum_C m'_C C \quad \text{with} \quad \deg(D_{n-1}) = \sum_C m'_C \deg(C) \quad (m'_C \in \mathbb{N}),$$

as before, and compare with

$$\deg(-D_n) = \left(m'_{C_n} - \sum_{C \neq C_n} m'_C (C.C_n) \right) \deg(C_n) - \sum_{C \neq C_n} m'_C \deg(C).$$

In particular, equality in (11.11) holds if and only if $m'_C = 0$ for all $C \neq C_n$. But then the same argument applies to the effective (-2)-divisor $-D_n$ to show that

$$\deg(D_{n-1}) = \deg(-s_{C_n}(-D_n)) \leq \deg(-D_n).$$

By symmetry we infer that equality holds in (11.11), so $D_{n-1} = m'_{C_n} C_n$, and $D_{n-1}^2 = -2$ implies $m'_{C_n} = 1$. To conclude, we recover D as the effective divisor

$$D = C_n + \sum_{i=1}^{n-1} (D_{i+1}.C_i)C_n,$$

which obviously is supported on the (−2)-curves $C_1, \ldots, C_n$. □

Remark 11.18 From basic properties of root lattices, one can also deduce the following useful property (see [Schü18]). Let $R \subset \mathrm{NS}(X)$ be a negative-definite root lattice generated by (−2)-curves and consider the primitive closure $R' \subset \mathrm{NS}(X)$. Then

$$\text{any divisor } D \in R' \setminus R \text{ satisfies } D^2 \leq -4. \tag{11.12}$$

Note that this agrees precisely with our results for the Mordell–Weil group: the frame W of an elliptic fibration contains the root lattice W_{root} generated by fibre components; this may not be primitive in W as

$$(W_{\mathrm{root}})'/W_{\mathrm{root}} \cong \mathrm{MW}(X)_{\mathrm{tors}},$$

by Proposition 6.42, but all non-zero sections, in particular those accounting for the torsion, are represented by vectors $w \in W$ of square $w^2 \geq 2\chi = 4$ as in (11.12).

11.3.7 Positive Characteristic

In positive characteristic, the moduli question has a rather different flavour. For one thing, the Lefschetz bound

$$\rho(X) \leq h^{1,1}(X) = b_2(X) - 2p_g(X) \tag{11.13}$$

collapses; in fact, there are so-called *supersingular* K3 surfaces with $\rho = 22$, as was first shown by Tate [Tat65]. Systematically, one can study K3 surfaces over an algebraically closed field k of characteristic p in terms of the *formal Brauer group* $\widehat{\mathrm{Br}}(X)$, following Artin and Mazur [AM77]. This is a smooth formal group which pro-represents the following functor on the category of finite local k-algebras with residue field k:

$$A \mapsto \widehat{\mathrm{Br}}(A) = \ker(H^2_{\text{ét}}(X \otimes \mathrm{Spec}(A), \mathbb{G}_m) \to H^2_{\text{ét}}(X, \mathbb{G}_m)).$$

With X a K3 surface, $\widehat{\mathrm{Br}}(X)$ has dimension one $(= h^{0,2}(X))$ and is therefore classified by its height h by [Laz55] which a priori may assume any value in $\mathbb{N}$ and ∞. The values in $\mathbb{N}$, however, are presently limited by the inequality

$$\rho(X) \leq b_2(X) - 2h = 22 - 2h \quad (h < \infty) \tag{11.14}$$

from [AM77] which may be regarded as a substitute for (11.13). That is, if $h < \infty$, then $h \leq 10$. We recall the following standard terminology, part of which was introduced in [Art74a]. It reflects the terminology for elliptic curves reviewed at the end of Sect. 3.2 (the connections become clear, for instance, in the framework of Kummer surfaces of product type, see Sect. 12.2.4).

Definition 11.19 A K3 surface X over k is said to be:

1. *ordinary* if $h = 1$;
2. *of finite height* if $h < \infty$;
3. *supersingular* if $h = \infty$.

Remark 11.20 There is a second notion of supersingularity, requiring

$$\rho(X) = b_2(X) = 22,$$

as mentioned just below (11.13). By (11.14), supersingularity by Picard number implies supersingularity by height, but the equivalence of these two notions over finite fields was only recently proved as part of the Tate conjecture for K3 surfaces, as we shall explain in Sect. 12.1.

Supersingular K3 surfaces themselves admit a rich moduli theory which was discovered by Artin in [Art74a] and greatly extended by Ogus [Ogus79] (see also [RS78] for an explicit approach in characteristic 2).

Quite remarkably, there are also Torelli theorems for K3 surfaces in positive characteristic. This has been proved in the supersingular case by Ogus [Ogus83] and in the ordinary case by Nygaard [Nyg83b]. Proposition 11.15 has a natural analogue for ordinary K3 surfaces in positive characteristic in the context of canonical liftings, see [Del81, Nyg83a]. With a view towards elliptic fibrations, we will consider supersingular K3 surfaces in more detail in Sect. 12.4.

Remark 11.21 If X is defined over a finite field k, then the height is visible in the Newton polygon of the characteristic polynomial of the induced action of Frobenius on $H^2_{\text{ét}}(X \otimes \bar{k}, \mathbb{Q}_\ell)$: the height is ∞ if and only if the Newton polygon has slope 1, while in case $h < \infty$, there are exactly h segments of slope $(h-1)/h$ and $(h+1)/h$ each, all others having slope one. In particular, X is ordinary if and only if there is a segment of slope one, i.e. the characteristic does not divide the trace of Frobenius.

11.4 Elliptic Fibrations

In the last section, we have seen that K3 surfaces are governed to a large extent by lattices. This leads to the question whether it is possible to read off an elliptic fibration from the Néron–Severi lattice—or in the absence of a section, *a genus one fibration*. The answer is surprisingly simple, much as in the examples in Sect. 11.3.5 over $\mathbb{C}$, and it is independent of the characteristic which we allow again to be positive from this section on. This happens at a small price that we have to pay in characteristic 2 and 3; it concerns the existence of *quasi-elliptic fibrations* whose importance was first recorded in the framework of surface classification in [BM76/77] (see Definition 7.21 and the discussion around there).

Convention 11.22 *In characteristic 2 and 3 we let the notion "elliptic K3 surface" include "quasi-elliptic K3 fibrations" with section.*

This weakening of terminology will not really be essential to our issues in the sequel. Mostly, this is because quasi-elliptic fibrations are easy to detect for several reasons. For instance, they come automatically with a group structure on the smooth locus of the generic fibre, but they have finite Mordell–Weil groups, thus no Mordell–Weil lattices by Corollary 7.22. Moreover, the configuration of reducible singular fibres of a quasi-elliptic surface is often impossible for a proper elliptic surface. In most instances, this is immediate from Euler number considerations. To this end, note that the corrected Euler number formula for a quasi-elliptic surface $S \to C$ is

$$e(S) = 2e(C) + \sum_{v \in C} (e(F_v) - 2),$$

which is different compared to Theorem 5.47. For instance, a general quasi-elliptic K3 surface (with or without section) has 20 reducible fibres of Kodaira type III in characteristic 2, and 10 reducible fibres of Kodaira type IV in characteristic 3. Finally, quasi-elliptic surfaces over $\mathbb{P}^1$ with section (such as quasi-elliptic K3 surfaces) can be shown to be unirational (cf. [Miy76]).

Example 11.23 (*Fermat quartic in characteristic 3*) Consider the Fermat quartic surface S_4 over $\bar{\mathbb{F}}_3$. Then any line on S_4 induces a quasi-elliptic fibration, with 10 fibres of Kodaira type IV. This special situation is related to the fact that S_4 contains 112 lines over $\mathbb{F}_9$ (the maximum on a smooth quartic by [RS15b]), while in any other characteristic there cannot be more than 64 lines on a smooth quartic in $\mathbb{P}^3$ ([RS15a, RS18], see also [Deg16]).

For working with quasi-elliptic fibrations (with section), one can simply write down Weierstrass forms, for instance an extended or normal Weierstrass form

$$y^2 = x^3 + a_2x^2 + a_4x + a_6$$

in characteristic 2, and then either resolve singularities (by Tate's algorithm 5.8 just like for elliptic surfaces) or appeal to an appropriate replacement of the discriminant (see [CD89, Sect. 5]). Indeed, in this way one can parametrize supersingular K3 surfaces of any Artin invariant in characteristic 2 by [RS78] (see Sect. 12.4).

Taking quasi-elliptic K3 surfaces tacitly into account, we can formulate the following characterization of a genus one fibration on a K3 surface:

Theorem 11.24 ([PSS71, Sect. 3, Theorem 1]) *A K3 surface X admits a genus one fibrationif and only if there is a divisor* $0 \neq D \in \mathrm{NS}(X)$ *with* $D^2 = 0$.

Proof We use Riemann–Roch for the isotropic divisor D on X, given by (4.9), to see that either D or $-D$ is effective, say D. One may as well assume that D is primitive (in $\mathrm{NS}(X)$). The linear system $|D|$ associated to this isotropic divisor may still contain a non-empty base locus, but this exclusively consists of a finite number of (-2) curves (cf. the reasoning in the proof of Lemma 11.17). Hence these can be eliminated successively by reflections on the lattice level. Here the image of D remains isotropic, while its degree (with respect to any given polarization) drops at each reflection, so this procedure terminates after finitely many steps. The resulting effective divisor D' is isotropic and effective such that $|D'|$ is base point-free. It follows that the linear system $|D'|$ induces a genus one fibration on X. Since we made sure that D (and thus D') is primitive, the divisor D' will even feature as a singular fibre of the fibration. □

In the sequel, we will also say the the isotropic divisor D *induces* the genus one fibration on X given by $|D'|$ in the notation of the proof of Theorem 11.24. The above reasoning also gives the following characterization of a fibre class of a genus one fibration:

Lemma 11.25 *A non-zero isotropic divisor* $D \in \mathrm{NS}(X)$ *is the fibre class of a genus one fibration on X if and only if D is primitive and nef.*

Example 11.26 Let X be the minimal resolution of a double sextic, where the curve is endowed with a single node. Then X contains two linearly independent divisors: the pull-back H of a hyperplane class with $H^2 = 2$ and the exceptional curve C with $C^2 = -2$ above the node. Hence $D = H + C$ is isotropic and effective; however, D is not nef, but the reflection in C leads to the nef isotropic vector $D' = H - C$. The linear system $|D'|$ defining the genus one fibration is the pencil of strict transforms of lines through the node, each of which indeed defines a genus one curve on X. Note that C gives a bisection of the fibration.

We note two immediate consequences. For this purpose, we introduce the notion of a *divisor of Kodaira type*.

Definition 11.27 A divisor D on an algebraic surface is said to be of *Kodaira type* if the support of D consists of smooth rational curves of self-intersection -2 which are arranged according to one of the fibres of Kodaira's classification (Theorem 5.12).

Corollary 11.28 *If D is an irreducible curve of arithmetic genus one on a K3 surface X, or a divisor of Kodaira type, then* $|D|$ *induces a genus one fibration* $X \to \mathbb{P}^1$ *which has D as fibre.*

Proof This follows from Theorem 11.24 and its proof, since the divisor D contains no components to subtract. □

Corollary 11.29 *Let X be a K3 surface with Picard number $\rho(X) \geq 5$. Then X admits a genus one fibration.*

Proof Since $\mathrm{NS}(X)$ has rank at least five and signature $(1, \rho(X) - 1)$, the underlying quadratic form represents zero as a consequence of the Hasse–Minkowski theorem (cf. [Ser70, IV. 3.2]). Hence we can apply Theorem 11.24 to deduce the claim. □

For the question of an elliptic fibration, i.e. a genus one fibration *with* section according to our convention, we will offer two answers below: Proposition 11.31 will provide an explicit argument for a given K3 surface, and Lemma 11.34 develops a systematic description in the vein of the lattice-theoretic Corollary 11.29 (but valid only over $\mathbb{C}$).

Example 11.30 Consider a smooth quartic surface $X \subset \mathbb{P}^3$ containing a line ℓ as in Sect. 11.3.5.B. Letting $D = H - \ell$, we calculate $D^2 = 0$. By Theorem 11.24, the K3 surface X thus admits a genus one fibration. Geometrically, this is realized by the pencil of hyperplanes in $\mathbb{P}^3$ through ℓ (which indeed corresponds to the linear system $|H - \ell|$). Each such hyperplane intersects X in ℓ and a residual cubic which gives a curve in $\mathbb{P}^2$ of arithmetic genus one by adjunction (Theorem 4.11). Note that ℓ induces a trisection of the fibration as it meets each fibre in three points.

We highlight that one could also start with the isotropic vector $D' = H + 2\ell$ instead. Then the line ℓ is contained in the base locus of $|D'|$, but reflecting in ℓ precisely leads to D as above.

The above genus one fibration does generically not admit a section. To see this, we argue with the generic Néron–Severi lattice spanned by D and ℓ,

$$M = \begin{pmatrix} 0 & 3 \\ 3 & -2 \end{pmatrix}.$$

To show that M is primitive in the present situation, we compute the discriminant group $A_M = M^\vee/M \cong \mathbb{Z}/9\mathbb{Z}$. It follows that the only divisible divisor could be D itself (up to sign and adding elements from $3M$), so assume that $D \sim 3D'$ for some $D' \in \mathrm{Div}(X)$. Clearly $(D')^2 = 0$, so either D' or $-D'$ is effective by Riemann–Roch (as in the proof of Theorem 11.24). But then $D \geq 0$ implies $D' \geq 0$. On the other hand, since D has degree three in $\mathbb{P}^3$ (computed as $(D.H)$), we infer that D' has degree one. I.e. the effective divisor D' is represented by a line, and as such, it is smooth rational, so $(D')^2 = -2$ by adjunction, contradiction. Hence D was primitive to start with. (Alternatively, one can use the smoothness of X to infer that any hyperplane section is reduced.)

This observation initiates the problem to decide whether a genus one fibration derived from an isotropic divisor as in Theorem 11.24 admits a section. The solution again is very simple:

Proposition 11.31 ([PSS71, Sect. 3, Theorem 1]) *Let X be a K3 surface with a genus one fibration induced by a divisor D with $D^2 = 0$ (as in the proof of Theorem*

11.24). Then any divisor $E \in \mathrm{Div}(X)$ with $(D.E) = 1$ gives a section of the fibration, i.e. the fibration is indeed elliptic.

Proof For starters, note that D is automatically primitive, for otherwise $D.E \neq 1$. After changing signs, if necessary, we may assume that $D \geq 0$ as before by Riemann–Roch. We proceed as in the proof of Theorem 11.24 to apply successive reflections to subtract the base locus. We arrive at the isotropic divisor $D' = s(D)$ whose linear system $|D'|$ induces the genus one fibration in question. In particular, D' is nef (as any fibre is).

Let $E^2 = (s(E))^2 = n \in 2\mathbb{Z}$ and define the auxiliary divisor

$$E' = s(E) - \left(\frac{n}{2} + 1\right) D' \quad \text{with} \quad (E')^2 = -2, \quad D'.E' = 1.$$

By Lemma 11.16, E' is either effective or anti-effective, so $D'.E' = 1$ determines the first alternative (since D' is nef). By Lemma 11.17, there is a linear equivalence

$$E' \sim \sum_i n_i C_i, \quad n_i \in \mathbb{N},$$

where the C_i are distinct (-2)-curves. Again, since D' is nef, we deduce from $(D'.E') = 1$ that $(D.C_i) = 0$ for all but one index i_0, whence $(D.C_{i_0}) = 1$ (and $n_{i_0} = 1$). We conclude that the genus one fibration given by $|D'|$ admits a section given by C_{i_0}. □

Remark 11.32 Note that even if the divisor E in Proposition 11.31 equals a (-2)-curve C, this does not mean that C will be the precise section of the induced fibration. For instance, let $D = F + (O)$ for general fibre and some section, and $C = \Theta$ for some smooth rational fibre component met by (O). Then D is isotropic and $D.C = 1$ as required. The base locus of $|D|$ consist of (O). Reflection in (O) yields

$$D' = s_{(O)}(D) = F, \quad s_{(O)}(C) = \Theta + (O),$$

and the section (O) is recovered by applying a reflection in Θ. In comparison, if D is an irreducible curve, or a divisor of Kodaira type as in Corollary 11.28, then indeed a (-2)-curve C with $(C.D) = 1$ provides a section of the elliptic fibration given by $|D|$.

Note that D and E as in Proposition 11.31 will automatically span the hyperbolic plane U. Thus we have derived the following lattice-theoretic extension of Theorem 11.24.

Proposition 11.33 *Any U-polarized K3 surface admits an elliptic fibration (with section).*

Consider a U-polarized K3 surface X with elliptic fibration as in Proposition 11.33. Writing $\mathrm{NS}(X) = U \oplus W^-$, there is the following relation between the frame

W and the induced elliptic fibration by ([Kon92, Lemma 2.1]): letting F denote any fibre, we have

$$W^- \cong F^\perp/\langle F\rangle \subset \mathrm{NS}(X).$$

More specifically, if W decomposes into an orthogonal sum of root lattices associated to Dynkin diagrams, then these correspond to singular fibres of the fibration.

Given a K3 surface X it seems natural to ask the general question whether X admits any elliptic fibration at all; by the above reasoning, this is equivalent to the hyperbolic plane U embedding into $\mathrm{NS}(X)$. Over $\mathbb{C}$ there is a uniform answer if the Picard number is big enough:

Lemma 11.34 *Every complex K3 surface of Picard number at least* 12 *admits an elliptic fibration.*

Proof For $\rho \geq 13$, the claim follows from a result of Nikulin ([Nik80, Corollary 1.13.5]). In detail, Nikulin derives the following implication relating the rank of an indefinite even integral lattice L with its length, i.e. the minimum number of generators of the discriminant group $A_L = L^\vee/L$:

$$\mathrm{rank}(L) \geq \mathrm{length}(L) + 3 \implies U \hookrightarrow L. \tag{11.15}$$

In the present situation, the length of $\mathrm{NS}(X)$ is bounded by both the rank of $\mathrm{NS}(X)$ and the rank of the transcendental lattice $T(X)$ by Proposition 2.41. As the latter is $22 - \rho(X) \leq 9$ by assumption, the inequality (11.15) of rank against length holds for $\mathrm{NS}(X)$. Hence (11.15) gives an embedding of U into $\mathrm{NS}(X)$ which in turn guarantees an elliptic fibration on X by Proposition 11.33.

For $\rho = 12$, the argument that $U \hookrightarrow \mathrm{NS}(X)$ is a little more delicate and can be found in Huy ([Huy16, Chap. 14, Corollary 3.8]), partly based on [Kov94]. □

We point out that Lemma 11.34 does not have an analogue in positive characteristic. In fact, we will see in Corollary 12.45 that certain supersingular K3 surfaces do not admit any elliptic fibration at all (despite the Picard number being maximal, i.e. $\rho = 22$, so the presence of genus one fibrations is guaranteed by Corollary 11.29). Yet there are abstract lattice theoretic criteria that can be applied in any characteristic (see Theorem 12.40, Remark 12.41), and certainly Lemma 11.34 applies to all K3 surfaces of finite height (compare Sect. 10.5.3).

11.4.1 Explicit Divisors and Elliptic Fibrations

In practice, in order to exhibit an elliptic fibration on a K3 surface X, one often looks for a divisor D of Kodaira type right away, assuming one knows enough smooth rational curves on X. Then X admits a unique elliptic fibration with D as singular fibre by Corollary 11.28. We illustrate this by revisiting the Kummer surfaces of

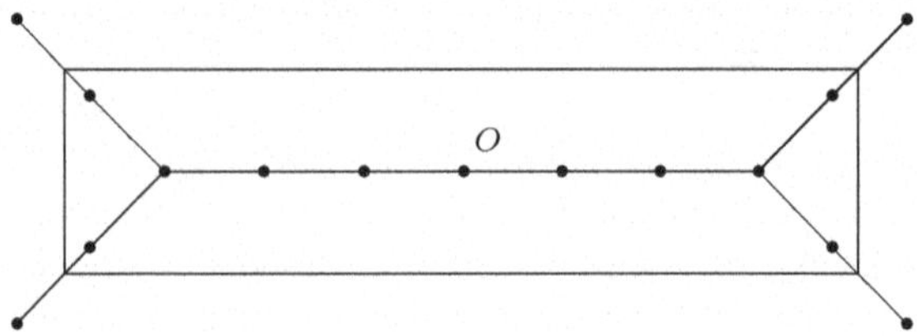

Fig. 11.4 $\tilde{D}_{10}$ divisor supported on fibres of type IV* connected by zero section

product type. Consider the elliptic fibration π_1 from Sect. 11.2.2 with two singular fibres of type IV*. Together with the zero section O, the fibre components generate the root system depicted in Fig. 11.4. Indicated there also is a divisor D of Kodaira type $\tilde{D}_{10}$ (with appropriate multiplicities).

By Theorem 11.24 the divisor D induces an elliptic fibration

$$\pi_3 : \ X \to \mathbb{P}^1$$

with singular fibre of type I_6^* given by D. Then Proposition 11.31 implies that each outer (-2)-curve in the figure is a section of the fibration. Upon closer inspection, one can detect further singular fibres. Namely, the cubic pencil (11.8) has exactly four base points which meet each IV fibre in a non-identity component. The induced sections of π_1 meet only the outer (-2)-curves in Fig. 11.4. As these rational curves are perpendicular to D, they have to be contained in some other singular fibres. Generically, these will have type I_2 (four times). In fact, the second fibre component will be given by the inverse of such a section P: by construction, $(-P)$ will again meet both IV* fibres in non-identity components (but in the opposite components compared with those met by (P); here we use that the non-identity components of the IV* fibre of π_1 correspond to the non-identity components of the IV fibres of the cubic pencil (11.8)). From the height pairing we obtain $h(P) = 4 - 4/3 - 4/3 = 4/3$, so

$$-4/3 = \langle P, -P \rangle = 2 - (P.(-P)) - 2/3 - 2/3 \Longrightarrow (P.(-P)) = 2.$$

That is, the sections $(P), (-P)$ form the configuration of an I_2 fibre as claimed (or III a priori, but this will presently be ruled out by Euler–Poincaré characteristic considerations). Note that $-P$ cannot be induced from a base point of the original cubic pencil (11.8), as (P) and $(-P)$ are not perpendicular.

We claim that the fibration π_3 will have two more singular fibres of type I_2 in addition to the four divisors of shape $(P) + (-P)$. An abstract argument starts from two base points P, Q of the cubic pencil (11.8) as above, meeting exactly opposite fibre components of the IV fibres. In other words, (P) and $(-Q)$ meet the same non-identity components. By the height formula, we find

$$\langle P, Q \rangle = 2 - 2/3 - 2/3 = 2/3 \quad \text{and} \quad \langle P, -Q \rangle = -2/3.$$

This directly shows that the section $P - Q$ has height $4/3$, which is small enough to imply that $(P - Q)$ also meets both IV fibres at non-identity components downstairs on the cubic pencil, and same with the IV* fibres upstairs on X. Hence the sections $(P - Q), (Q - P)$ of the old fibration π_1 form another I_2 fibre of the new fibration π_3, and this game can be played in two "orthogonal" ways (choosing the pair (P, Q)). In summary, we find one fibre of type I_6^* and six I_2's for the elliptic fibration π_3. As hinted, the Euler–Poincaré characteristic excludes further singular fibres or degenerations to type III.

As a consequence, the trivial lattice has rank 18, equalling the Picard number of X. Hence there can only be torsion in MW, or to be precise, only 2-torsion due to the I_6^* fibre by (5.5). We conclude that MW consists of the four sections depicted as the outer corners in Fig. 11.4, accounting for full 2-torsion $(\mathbb{Z}/2\mathbb{Z})^2$.

Finding the new elliptic fibration explictly amounts to working out a basis of $H^0(\mathcal{O}(D))$. For this purpose, let us first convert the equation of cubics (11.6) to Weierstrass form (over $\bar{k}$). For convenience, we use the standard notation with coordinates x, y, t, although these are different from the coordinates from (11.6). Neglecting the precise coefficients of the Weierstrass form, we use Tate's algorithm to deduce the shape,

$$X: \quad y^2 = x^3 + At^3x + Bt^4, \tag{11.16}$$

with polynomials $A, B \in k[t]$ satisfying $\deg(A) \leq 2$, $\deg(B) = 4$. We now use some concrete input from our previous considerations concerning involutions on K3 surfaces. We distinguish involutions according to their action on a regular non-zero 2-form ω_X, calling $\imath$ *symplectic* if $\imath^*\omega_X = \omega_X$, and *non-symplectic* if $\imath^*\omega_X = -\omega_X$. For elliptic surfaces over $\mathbb{P}^1$, the action on the regular 2-forms can be computed using Sect. 5.13. Presently, X admits the non-symplectic involution

$$\imath: \quad (x, y, t) \mapsto (x, y, \quad t)$$

(induced from the corresponding automorphism on the base $\mathbb{P}^1$); this was perfectly visible in (11.6) (only that in those coordinates, it would have been expressed by the elliptic parameter $y \mapsto -y$), and we have used it in Sect. 11.2.2 in order to relate the fibration to a the cubic pencil (11.8), i.e. a rational elliptic surface.

The involution $\imath$ implies that $A = at$ for some constant $a \in k$, while $B(t) = b(t^2)$ for a degree 2 polynomial $b \in k[t]$. We proceed to work out a basis of the linear system $H^0(\mathcal{O}(D))$ which we claim to be given by $1, x/t^2$. Indeed, affinely these functions are easily checked to lie in $H^0(\mathcal{O}(D))$, and by symmetry the same holds at ∞. Similarly, one deals with their regularity at O. Hence we can take

$$u = x/t^2$$

as the elliptic parameter of the new fibration π_3. We substitute in the equation and simplify—together this amounts to the double blow-up at $(0, 0, 0)$ in the chart $t' = 1$—to obtain

$$X: \quad v^2 = t^2u^3 + at^2u + b(t^2).$$

As the RHS is a quartic in t, this gives an elliptic fibration over $\mathbb{P}^1_u$. Indeed there are four sections at $t = 0$ and $t = \infty$ (defined over k if and only if the original IV* fibres are split, i.e. all components are k-rational). The discriminant of the RHS reveals six fibres of type I_2, as predicted. The degree 6 polynomial in $k[u]$ describing the location of the I_2 fibres exactly encodes the x-coordinates $ut^2 (u \in \bar{k})$ of sections of (11.16) which meet both IV* fibres non-trivially, but are perpendicular to O. (Thanks to the symmetry in t, the discriminant of the RHS is zero if and only if the RHS is already a square.)

Remark 11.35 The precise coefficients in (11.16) can be determined explicitly in terms of the elliptic curves and their j-invariants; this goes back to Inose and will be discussed in the context of Shioda–Inose structures in Sect. 12.2.

11.5 Classification of Complex Elliptic K3 Surfaces

Continuing this section, we turn to the problem of (roughly) classifying the elliptic fibrations on *all* (complex) K3 surfaces (with a section as usual). Recall from Sect. 11.3.5.A and Proposition 11.33 that an elliptic fibration on a K3 surface X induces and is determined by a specific primitive embedding

$$U \hookrightarrow \mathrm{NS}(X).$$

As in (6.25), this induces a decomposition

$$\mathrm{NS}(X) = U \oplus W^-. \tag{11.17}$$

Here W is the frame of the elliptic fibration (as introduced in Sect. 6.4) which by Theorem 6.13 is even and positive-definite. Recall from Proposition 6.42 how one can retrieve the singular fibres and the Mordell–Weil lattice of the elliptic K3 surface X from the frame:

$$\sum_v T_v = \mathcal{R}(W), \quad \mathrm{MWL}(X) \cong W/\mathcal{R}(W)',$$

where $\mathcal{R}(W)'$ denotes the primitive closure of the root lattice $\mathcal{R}(W)$ in W.

The decomposition (11.17) depends on the chosen elliptic fibration. By the canonical bundle formula, this is a special feature of elliptic K3 surfaces. Indeed, each abstract decomposition (11.17) induces an elliptic fibration by Proposition 11.33, and these may well be inequivalent as we have seen before. This can be used to determine all elliptic fibrations (with section) on a given K3 surface as we shall exploit in Sect. 12.3.2.

For the remainder of this section we limit ourselves to complex K3 surfaces. In positive characteristic some extra care is required in order to take into account special issues such as extra cases related to supersingular K3 surfaces and inseparable base change, but also non-existing configurations (for instance, those ruled out by Proposition 6.33(vi)). Recall from Sect. 11.3.4 that complex K3 surfaces are governed by lattice theory by virtue of primitive embeddings

$$\mathrm{NS}(X) \hookrightarrow \Lambda .$$

In consequence, elliptic fibrations on complex K3 surfaces are fully governed by lattice theory. In principle, this enables one to determine whether a given configuration of singular fibres and sections occurs on a (family of) complex K3 surface.

Theorem 11.36 (Shimada [Shima00]) *Complex elliptic K3 surfaces admit exactly 3693 distinct pairs of possible configurations,*

$$\left(\sum_v T_v^-, \mathrm{MW}_{\mathrm{tors}}\right) = (R, G),$$

where $R = \sum_i R_i$ is a sum of root lattices R_i (ADE-type) and G is a finite abelian group.

Using Tate's algorithm (or simply the discriminant) one can work out explicit equations for many configurations. For instance, single singular fibres or configurations with small Euler number can often be parametrized fairly easily using extended Weierstrass forms. Sometimes it is even possible to include a section. As an illustration, we consider two examples.

Example 11.37 Start with $R = E_8$. Then there is a singular fibre of type II* which we may locate at ∞. Tate's algorithm prescribes the degrees in the extended Weierstrass form as

$$y^2 = x^3 + a_2x^2 + a_4x + a_6, \quad \deg(a_2) \leq 2, \quad \deg(a_4) \leq 4, \quad \deg(a_6) = 7. \tag{11.18}$$

(In fact, this works in any characteristic other than 2.) We can impose another summand $A_n (n \leq 5)$ in R by letting

$$t^{\lceil \frac{n+1}{2} \rceil} \mid a_4, \quad t^{n+1} \mid a_6$$

(compare the outline of Tate's algorithm in the multiplicative case in Sect. 5.8.1). On the other hand, we can also parametrize those Weierstrass forms generically admitting a section (P), say of height 4 (that is, disjoint from (O)). Such a section takes the shape $P = (\mathrm{x}, \mathrm{y})$ with polynomials x, y of degree 4 resp. 6. Upon substituting, (11.18) lets us read off y up to sign from the coefficients of $t^6, \dots, t^{12}$ on the RHS.

But then we can choose the coefficients of a_6 in degrees $0, 1, \dots, 5$ to accommodate the section P. Thus the elliptic K3 surfaces with a fibre of type II* and a section of height 4, i.e. with $U \oplus E_8^- \oplus \langle -4 \rangle \hookrightarrow \text{NS}$, are parametrized by a_2, a_4, x and the top two coefficient of a_6 (so up to normalizations, we obtain nine parameters as expected).

Example 11.38 Let $R = E_8^2$ and locate the corresponding fibres of type II* at $0, \infty$, Then Tate's algorithm prescribes the Weierstrass form as

$$X: \quad y^2 = x^3 + At^2x^2 + Bt^4x + Ct^5, \quad A, B \in k, \quad \deg(C) = 2, \quad (11.19)$$

again over fields of characteristic $\neq 2$. Abstractly, we can enhance the resulting generic trivial lattice

$$L = U \oplus (E_8^-)^2 \hookrightarrow \text{NS}(X)$$

by any even negative-definite lattice M of rank two, since any such lattice is supported on the remaining two copies of U in the K3 lattice Λ (perpendicular to L). By Proposition 6.42, the lattice M will encode the Mordell–Weil lattice of X (and possible additional reducible fibres of type I_2 or IV). However, working out explicit equations for such an enhanced K3 surface is a non-trivial task. We will address it in the next section in the context of Shioda–Inose structures.

We have seen in the above examples how one can construct K3 surfaces of Mordell–Weil rank 1 or 2 abstractly. It is not too hard to cover the whole possible range of ranks based on the theory of lattice-polarized K3 surfaces:

Lemma 11.39 (Cox) *Any Mordell–Weil rank* $r = 0, 1, \dots, 18$ *is attained by some complex elliptic K3 surface.*

This result was first proved by Cox [Cox82] as a consequence of the surjectivity of the period map (see [BHPV04, VIII. 14], [Huy16, Chap. 7]); however, Cox did not give any explicit examples. Indeed, spelling out explicit examples with given Mordell–Weil rank is a completely different task. Surprisingly, there is a systematic answer that takes us back to Kummer surfaces of product type. We will discuss it in the next chapter, esp. Sect. 12.2.6, and also derive the analogue of Lemma 11.39 in positive characteristic (Corollary 12.15).

We end this chapter by commenting on the hierarchy of Mordell–Weil lattices of complex elliptic K3 surfaces, following Oguiso [Ogui03]. It is an immediate consequence of the Torelli theorem that for any integer $0 < r \leq 18$ there are infinitely many different lattices of rank r which occur as Mordell–Weil lattices of complex elliptic K3 surfaces. Hence there cannot be any strict hierarchy as in the case of rational elliptic surfaces (Chap. 8). However, any given complex elliptic K3 surface can be placed in an analogous hierarchy:

Theorem 11.40 (Oguiso) *Let* X *be a complex elliptic K3 surface of Mordell–Weil rank* r_0. *Then there is a sequence of K3 surfaces* $\{X_r\}_{r_0 < r \leq 18}$ *such that*

- *each* X_r *has the same configuration of singular fibres as* X,
- $\operatorname{rk} \mathrm{MW}(X_r) = r$, *and*
- *the narrow Mordell–Weil lattices admit a chain of isometric embeddings*

$$\mathrm{MW}^0(X) \hookrightarrow \mathrm{MW}^0(X_{r_0+1}) \hookrightarrow \ldots \hookrightarrow \mathrm{MW}^0(X_{18}).$$

Historical Notes

The study of K3 surfaces was sparked by Andreotti and Weil, starting from smooth quartics in $\mathbb{P}^3$ (see [And57, Wei58]). After proving a local Torelli-type theorem, they stated the following famous conjectures which shaped the research on complex analytic K3 surfaces (not necessarily algebraic) for more than 20 years:

1. all K3 surfaces form a connected family (proved by Kodaira ([Kod64+, I, Theorem 17]));
2. all K3 surfaces are Kähler (established by Siu [Siu83]);
3. the period map is surjective (due to Kulikov [Kul77], Persson–Pinkham [PP81], see also [Loo81]);
4. the global Torelli theorem (in some form, for algebraic K3 surfaces mostly due to Piatetski-Shapiro and Shafarevich [PSS71], but see also [BR75, Shio78, LP81, Fri84]).

The complex theory of K3 surfaces, both algebraic and analytical, was almost completely settled by Kodaira on one side and the Shafarevich-school on the other. That is, in a way, until their moduli got within reach, starting from [BR75] or even earlier. For the first few polarizations, of degrees $2, 4, 6, 8$, it is elementary to work out K3 surfaces as smooth double sextics, quartics in $\mathbb{P}^3$ and complete intersections in $\mathbb{P}^4$ or $\mathbb{P}^5$. The underlying moduli spaces are thus unirational. This result can be extended up to polarizations of degree 24 (and for degrees $30, 34, 38$), but this relies on deep work of Mukai (see [Muk06, Muk16] for the latest results). At the other end of the scale, from polarizations of degree 124 on, the moduli spaces have been proven to be of general type by [GHS07] (see also [Kon93, Kon99, Pet15]).

Elliptic fibrations on K3 surfaces must have already been encountered by the Italian school, but the first systematic treatment in terms of lattice theory, next to Kodaira's complex analytic theory reaching way beyond K3 surfaces, seems to be due to Piatetski-Shapiro and Shafarevich in the course of their work towards the Torelli theorem for K3 surfaces (see Theorem 11.24). This was to have immediate consequences, for instance for the Tate conjecture in [ASD73] and for the study of supersingular K3 surfaces [Art74a], as we shall recall in the next chapter. Later, elliptic fibrations played a key role in Kondo's works (e.g. [Kon92]) and in Nikulin's papers (culminating in [Nik14]).

The fundamental role of smooth rational curves (or (-2)-curves) for the understanding of K3 surfaces cannot be overestimated, since they govern much of the structure of the surfaces. Notably, this applies to elliptic fibrations, but more generally also to automorphisms (see [Nik79, Nik83]). Note, however, that a very general K3 surface does not contain a smooth rational curve. In comparison, it was proved by Mumford that every complex K3 surface contains a (possibly singular) rational curve (cf. [MM83]). Recently, it was even shown that there are infinitely many rational curves on every K3 surface ([BHT11, LL12]).

A classical instance where interesting configurations of (-2)-curves and elliptic fibrations come together consists in Kummer surfaces. Indeed, they contain 16 disjoint (-2)-curves by construction, and equipping them with elliptic fibrations can be achieved in a canonical way not only in the product case (as in Sect. 11.2), but also in general (cf. [Dol04, Kum08], e.g.). A good account of the classical theory of Kummer surfaces can be found in [GH78].

We note that despite the theoretical progress, explicit calculations on K3 surfaces can still be quite delicate. For instance, one can infer this from the computation of the transcendental lattice of the Fermat quartic surface, which took quite a while (see Example 11.13, or consult [SSvL10] for a brief account).

Chapter 12
Elliptic K3 Surfaces—Special Topics

In this chapter, we discuss more specific topics from the theory of elliptic K3 surfaces which often have a more arithmetic flavour. Our focus lies especially on three subjects:

- Shioda–Inose structures and their applications, particularly to Mordell–Weil groups and their ranks,
- the problem of classifying all elliptic fibrations on a given K3 surface, and
- supersingular K3 surfaces.

Before going into the details, we give a brief recapitulation of the Tate conjecture, one of the most fundamental problems at the interface of algebraic geometry and number theory which was recently solved to large extent for K3 surfaces, for instance for all K3 surfaces over finite fields. This conjecture will also play a prominent role in the next chapter.

12.1 Tate Conjecture

Let X be a (geometrically irreducible) smooth projective variety defined over some finite field $k = \mathbb{F}_q$. Then we can consider the subgroup of the geometric Néron–Severi group $\mathrm{NS}(X \otimes \bar{k})$, which is generated by divisor classes defined over k; for clarity we denote this by $\mathrm{NS}(X/k)$. Fixing some prime ℓ relatively prime to q, there is an ℓ-adic cycle class map

$$\mathrm{NS}(X/k) \otimes \mathbb{Q}_\ell \hookrightarrow \mathrm{NS}(X \otimes \bar{k}) \otimes \mathbb{Q}_\ell \hookrightarrow H^2_{\text{ét}}(X \otimes \bar{k}, \mathbb{Q}_\ell(1)). \qquad (12.1)$$

Here we have applied a Tate twist to the second ℓ-adic cohomology in order to make the embedding compatible with the natural Galois action on the Néron–Severi group. More precisely, the Frobenius morphism Frob_q acts trivially on $\mathrm{NS}(X/k)$, and there are induced actions

M. Schütt and T. Shioda, *Mordell–Weil Lattices*, Ergebnisse der Mathematik und ihrer Grenzgebiete. 3. Folge / A Series of Modern Surveys in Mathematics 70,
https://doi.org/10.1007/978-981-32-9301-4_12

- on $\mathrm{NS}(X \otimes \bar{k})$ (which factors through a finite group, since $\mathrm{NS}(X \otimes \bar{k})$ is always generated by divisor classes defined over some finite extension of the ground field), and
- on $H^2_{ét}(X \otimes \bar{k}, \mathbb{Q}_\ell)$ (where Frob_q^* has eigenvalues of absolute value q by the Weil conjectures, as proved by Deligne for K3 surfaces originally in [Del72], see also Sect. 10.5.2).

The Tate twist multiplies eigenvalues by q^{-1} by tensoring with the inverse of the ℓ-adic cyclotomic character, thus enforcing the required compatibility. In particular, we have

$$\rho(X/k) = \mathrm{rank}\mathrm{NS}(X/k) \leq \dim_{\mathbb{Q}_\ell} H^2_{ét}(X \otimes \bar{k}, \mathbb{Q}_\ell(1))^{\mathrm{Frob}_q^*}, \tag{12.2}$$

and since $\mathrm{NS}(X \otimes \bar{k})$ admits a basis of divisor classes defined over some finite extension of k, as we have pointed out above,

$$\rho(X \otimes \bar{k}) \leq \sum_{\zeta \in \mu_\infty} \dim_{\bar{\mathbb{Q}}_\ell} H^2_{ét}(X \otimes \bar{k}, \bar{\mathbb{Q}}_\ell(1))^{\mathrm{Frob}_q^*=\zeta}. \tag{12.3}$$

Note that the last sum, involving a priori all roots of unity $\mu_\infty \subset \bar{\mathbb{Q}}_\ell$, amounts to a finite computation since the characteristic polynomial of Frob_q^* is defined over $\mathbb{Q}$, so there can only be roots of unity appearing whose order N satifies $\phi(N) \leq b_2(X)$ (or in fact, $\phi(N) \leq b_2(X) - 1$, since a hyperplane class is trivially Galois-invariant).

The Tate conjecture states that the divisor classes defined over k fill out all of the invariant second cohomology, that is, the inequalities (12.2), (12.3) are in fact equalities:

Conjecture 12.1 (Tate conjecture [Tat65]) *For any smooth projective variety over* $\mathbb{F}_q$*, one has*

$$\mathrm{NS}(X/\mathbb{F}_q) = \dim_{\mathbb{Q}_\ell} H^2_{ét}(X \otimes \bar{\mathbb{F}}_q, \mathbb{Q}_\ell(1))^{\mathrm{Frob}_p^*}.$$

There are also natural variants for algebraic cycles of higher codimension, but those will not be relevant for our issues dealing with algebraic surfaces.

The characteristic polynomial of Frob_q^* can, at least in principle, be computed explicitly using Lefschetz' fixed point formula (without Tate twist!),

$$\#X(\mathbb{F}_q) = \sum_{i=0}^{2\dim X} (-1)^i \,\mathrm{trace}\, \mathrm{Frob}_q^*(H^2_{ét}(X \otimes \bar{\mathbb{F}}_q, \mathbb{Q}_\ell)), \tag{12.4}$$

after counting points over sufficiently large extensions of $\mathbb{F}_q$. In practice, of course, this may be far from realistic, even for surfaces, although recently there has been substantial progress in p-adic cohomology theories to make extensive computations feasible for K3 surfaces, for instance (cf. [CT14]). It may be instructive to note that many numerical calculations do not a priori require the validity of the Tate conjecture (which, in fact, often follows as a side-product for the given surface a posteriori, cf.

Example 12.18). For K3 surfaces, however, the Tate conjecture has been proven to hold true very recently:

Theorem 12.2 *The Tate conjecture holds true for K3 surfaces over finite fields.*

Historically, this was first proved for elliptic K3 surfaces in any characteristic (with or without section) by Artin and Swinnerton-Dyer using homogeneous spaces [ASD73]. Then Nygaard gave a proof of the Tate conjecture for ordinary K3 surfaces, again in arbitrary characteristic [Nyg83a], and together with Ogus for K3 surfaces of finite height in characteristic $\neq 2, 3$ [NO85]. Only very recently, a full proof in odd characteristic was achieved by Maulik [Mau14], Charles [Cha13] and Madapusi Pera [Mad15], supplemented by a paper by Kim and Madapusi Pera [KM16] concerning characteristic 2.

12.2 Isogeny Structures Between K3 Surfaces

Soon, in Sect. 12.2.6, we will turn to the problem of determining which ranks are assumed by the Mordell–Weil groups of (complex) elliptic K3 surfaces. Compared with the abstract answer from Lemma 11.39, based on the Torelli theorem, we will, in fact, aim for an explicit answer (which will also prepare for the general Mordell–Weil rank discussion in the next chapter). The key input will be provided by a suitable notion of isogeny on K3 surfaces and its impact on Picard numbers, due to Inose (Theorem 12.9). Before discussing the details, let us go back to our prototype example of Kummer surfaces of product type. This will form a good motivation for the isogeny notion to follow since it admits an instructive sandwich structure which resembles a little the interplay of isogenies and dual isogenies on elliptic curves.

12.2.1 Sandwich Structure

For Kummer surfaces of product type, we have worked out the elliptic fibration π_1 with two fibres of type IV*. The Weierstrass form from (11.16) can be spelled out as

$$\mathrm{Km}(E \times E'): \quad y^2 = x^3 + ax + b_2 t^2 + b_1 + b_0 t^{-2}, \quad a, b_i \in k. \qquad (12.5)$$

This is related to the elliptic K3 surfaces X with two II* fibres from Example 11.38 as follows: the above equation arises from (11.19) by applying first the coordinate change $(x, y) \mapsto (t^2(x - At^2/3), t^3 y)$, so that X is given by

$$X: \quad y^2 = x^3 + ax + b_2 t + b_1 + b_0 t^{-1}, \qquad (12.6)$$

and then the quadratic base change $t \mapsto t^2$.

An intriguing aspect of the above situation is that X is in fact sandwiched by $\mathrm{Km}(E \times E')$: on the one hand, X is the minimal resolution of the quotient of $\mathrm{Km}(E \times E')$ by the symplectic involution

$$\iota : \quad (x, y, t) \mapsto (x, y, -t);$$

on the other hand, $\mathrm{Km}(E \times E')$ can be recovered as a quotient of X by extending the involution of the base

$$t \mapsto \frac{b_0}{b_2 t}$$

to a symplectic involution of X. The quotient inherits yet another elliptic fibration π_4 from X whose Weierstrass equation is computed in terms of the invariant affine coordinate s on the base curve as

$$\mathrm{Km}(E \times E') : \quad y^2 = x^3 + a(s^2 - 2b_2 b_0)^2 x + (s^2 - 2b_2 b_0)^3 (s + b_1).$$

Generally there are reducible fibres of type II^* (at ∞, the image of 0 and ∞ under the degree 2 map $t \mapsto s = t + \frac{b_0}{b_2 t}$) and twice I_0^* (at the branch points $\pm 2\sqrt{b_2 b_0}$ of the above map). Here the fibre types are as stated under Assumption 11.7, or in full generality if $E \not\cong E'$, cf. Observation 12.6.

Details on the sandwich structure can be found in [Nik91, Shio06]; for extensions to K3 surfaces of Picard number 17, see [Ma12, Koi12, Schü13]. Here we content ourselves with drawing the schematic diagram:

(12.7)

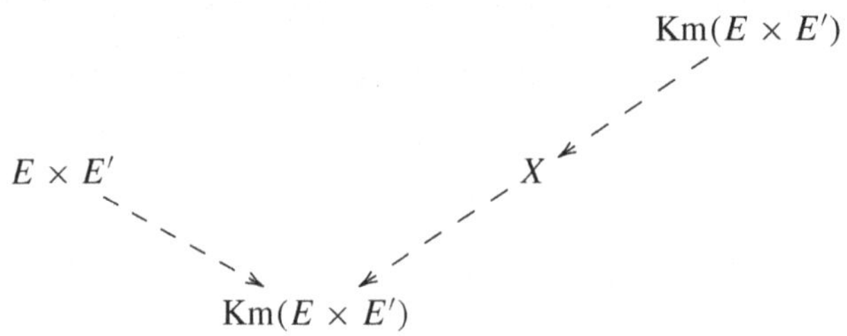

12.2.2 Shioda–Inose Structure

Consider the lower part of the diagram (12.7). Between the transcendental lattices of an abelian surface A and of the Kummer surface $\mathrm{Km}(A)$, there is the general relation

$$T(\mathrm{Km}(A)) = T(A)(2) \tag{12.8}$$

by [Morr84, Proposition 4.2] (see also [Nik75, Remark 2]). In particular, this allows us to derive the transcendental lattice and (determinant of) Néron–Severi lattice of a very general Kummer surface of product type (for instance, it gives an alternative proof for (11.7)).

In [SI77] it is proved (using the elliptic fibration π_4 on $\mathrm{Km}(E \times E')$) that the K3 surface X recovers the transcendental lattice of the abelian surface $E \times E'$:

$$T(X) \cong T(E \times E'). \tag{12.9}$$

This construction extends to (certain) K3 surfaces of Picard number 19, 18, and 17. This was analysed by Morrison in [Morr84], where also the following notion was introduced.

Definition 12.3 A complex K3 surface X admits a *Shioda–Inose structure* if it admits a rational map of degree two to some Kummer surface $\mathrm{Km}(A)$ such that $T(X) \cong T(A)$.

Remark 12.4 The above result from [SI77] was originally geared towards complex *singular K3 surfaces*, i.e. with maximal Picard number 20. Here the surjectivity of the period map would now follow from the corresponding statement for singular abelian surfaces (proven previously by an explicit construction in [SM74]) and from the above construction recovering transcendental lattices; indeed, by the Torelli theorem, isomorphism classes of singular K3 surfaces (or of singular abelian surfaces) are in bijective correspondence with the oriented positive definite quadratic forms representing their transcendental lattices.

Example 12.5 The Fermat quartic S_4 (see Example 11.13, where also the transcendental lattice $T(S_4) = A_1(4)^2$ is given) can be related by several constructions to elliptic curves with complex multiplication in $\mathbb{Q}(i)$, where, as before, $i = \sqrt{-1}$. First, $S_4 \cong \mathrm{Km}(E_i \times E_{2i})$. Here the Kummer surface of product type for the elliptic curves E_i with endomorphism ring $\mathbb{Z}[i]$ (and j-invariant 12^3) and E_{2i} with $\mathrm{End}(E_{2i}) = \mathbb{Z}[2i]$ has transcendental lattice $T(E_i \times E_{2i}) = A_1(2)^2$ by [SM74], so the stated isomorphism follows from (12.8) and Remark 12.4.

Second, S_4 arises through a Shioda–Inose structure from $E_i \times E$ for either elliptic curve E with endomorphism ring $\mathbb{Z}[4i]$ by the same reasoning as above. (There are two non-isomorphic such curves, but their products with E_i lead to isomorphic abelian surfaces and thus also to isomorphic singular K3 surfaces.)

12.2.3 Inose's Fibration

Explicit formulas for the above Weierstrass equations for $\mathrm{Km}(E \times E')$ and X were first derived by Inose in [Inos78]. Consider the j-invariants of E and E'; since this simplifies the formulas greatly, we shall normalize them here and only here in the classical way such that the elliptic curve with an automorphism of order 4 has $\mathrm{j} = 1$ (compare Remark 3.3):

$$\mathrm{j} = j(E)/12^3, \quad \mathrm{j}' = j(E')/12^3.$$

Let $a, b \in k$ such that

$$a^3 = \mathrm{j}\mathrm{j}', \quad b^2 = (1 - \mathrm{j})\,(1 - \mathrm{j}'). \tag{12.10}$$

Then the Weierstrass equation (11.19) of X can be normalized to take the shape

$$X: \quad y^2 = x^3 - 3at^4x + t^5(t^2 - 2bt + 1) \tag{12.11}$$

with discriminant $\Delta = -27t^{10}(t^4 - 4bt^3 + (4b^2 - 4a^3 + 2)t^2 - 4bt + 1)$. This fibration is often called *Inose's fibration* or *Inose's pencil.* Note that when working over $\mathbb{C}$, the coefficients a, b can be seen as products of Weber functions, thus of natural automorphic origin (just like the j-invariant itself, see [HS12, Lemma 6.12] for a discussion). In [Schü07b, Proposition 4.1], an easy twist of (12.11) was carried out to achieve a model of Inose's fibration defined over $k(j(E), j(E'))$.

Observation 12.6 *Inose's fibration degenerates, i.e. there is an additional reducible fibre, if and only if the elliptic curves are isomorphic (i.e. $j(E) = j(E')$).*

12.2.4 Picard Numbers

In this subsection we study the Picard numbers of Kummer surfaces (outside characteristic 2) – and in fact, of many K3 surfaces related to them by rational maps, see Theorem 12.9. This will eventually allow us to answer the problem of finding explicit elliptic K3 surfaces with given Mordell–Weil rank (solved abstractly over $\mathbb{C}$ in Lemma 11.39).

Let $A = E \times E'$ denote an abelian surface of product type as before. The Picard number of A depends on whether E, E' are isogenous and have complex multiplication (CM), or even are supersingular:

$$\rho(A) = 2 + \text{rank}\,\text{Hom}(E, E') = \begin{cases} 2, & \text{if } E \not\sim E', i.e. E, E' \text{ are not isogenous;} \\ 3, & \text{if } E \sim E' \text{ without CM;} \\ 4, & \text{if } E \sim E' \text{ with CM,} \\ & \qquad \text{but not supersingular;} \\ 6, & \text{if } E \sim E' \text{ are supersingular.} \end{cases} \tag{12.12}$$

Note that this classification perfectly agrees with the geometric picture where additional curves arise as graphs of the isogenies—as these can be composed with the endomorphisms of the curves, we obtain exactly the ranks anticipated at the end of Sect. 3.2 and also around (11.3).

Since the Picard number of the Kummer surface $\text{Km}(E \times E')$ equals

$$\rho(\mathrm{Km}(E \times E')) = \rho(E \times E') + 16,$$

we find the following Picard numbers for the Kummer surface:

$$\rho(\mathrm{Km}(E \times E')) = \begin{cases} 18, & \text{if } E, E' \text{ are not isogenous } (E \not\sim E'); \\ 19, & \text{if } E \sim E' \text{ without CM;} \\ 20, & \text{if } E \sim E' \text{ with CM, but not supersingular;} \\ 22, & \text{if } E \sim E' \text{ are supersingular.} \end{cases} \tag{12.13}$$

As a follow-up question, the reader might wonder about the Picard number of X, the elliptic K3 surface sandwiched by $\mathrm{Km}(E \times E')$ as in Sect. 12.2.1. Presently, there would be several ways to argue that in fact,

$$\rho(X) = \rho(\mathrm{Km}(E \times E')). \tag{12.14}$$

For instance, one could argue with the sandwich structure, enriched by the induced elliptic fibrations. We sketch the situation in the following diagram where the squares are quadratic base changes.

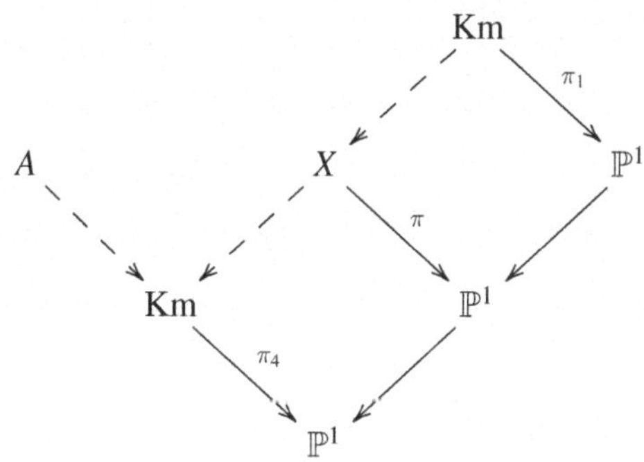

Here $A = E \times E'$ and $\mathrm{Km} = \mathrm{Km}(A)$ while the elliptic fibrations are:

$\pi_1 =$ elliptic fibration on Km from (12.5) with 2 IV^* (see 11.22);
$\pi_4 =$ elliptic fibration on X with 2 II^* from (12.11) (see Example 11.38);
$\pi_4 =$ elliptic fibration on Km from Sect. 12.2.1 with II^* and 2 I_0^*.

By way of base change and functoriality (Theorem 6.53), there are embeddings of scaled Mordell–Weil lattices

$$\mathrm{MWL}(\pi_4)(4) \hookrightarrow \mathrm{MWL}(\pi)(2) \hookrightarrow \mathrm{MWL}(\pi_1).$$

Note that $\mathrm{MWL}(\pi_1)$ comes with a contribution from the cubic pencil (11.8) since π_1 arises from the corresponding rational elliptic surface S via quadratic base change. In fact, this means that π and S are quadratic twists of each other. Equivalently, there are two involutions on Km,

$$\iota : (x, y, t) \mapsto (x, y, -t) \quad \text{and} \quad (x, y, t) \mapsto (x, -y, -t),$$

such that the quotient of π_4 by ι gives π as before and by the other involution gives S. Since the involutions are obtained from each other by composing with inversion on the fibres, the pull-backs of sections from π and S to π_1 are orthogonal with respect to the height pairing; that is

$$\text{MWL}(\pi) \hookrightarrow \text{MWL}(\pi_1)^{\iota^*=1} \perp \text{MWL}(\pi_1)^{\iota^*=-1} \hookleftarrow \text{MWL}(S). \qquad (12.15)$$

More precisely, it can be seen that both embeddings become equalities after tensoring with $\mathbb{Q}$, as does the embedding

$$\text{MWL}(\pi_1)^{\iota^*=1} \perp \text{MWL}(\pi_1)^{\iota^*=-1} \hookrightarrow \text{MWL}(\pi_1), \qquad (12.16)$$

which would provide yet another proof for (12.14) (cf. Remark 5.37). We shall now piece all this information together in the non-degenerate case where $E \not\cong E'$. Then Observation 12.6 can be used to show that neither of the above elliptic fibrations has other reducible fibres than the given ones, since with the ramification happening over singular fibres, fibre type changes for either fibration would always be visible on π. This implies that S has Mordell–Weil rank 4 by Theorem 8.8, so (12.15) and (12.16) give

$$\text{rk}\,\text{MWL}(\pi_1) = 4 + \text{rk}\,\text{MWL}(\pi).$$

Thus we find, using Corollary 6.7 and functoriality for the base change from π_4 to π,

$$\rho(X) = 18 + \text{rk}\,\text{MWL}(\pi) = 14 + \text{rk}\,\text{MWL}(\pi_1) = \rho(\text{Km}), \qquad (12.17)$$

so the claimed equality (12.14) follows.

Remark 12.7 Alternatively, we could have omitted the discussion of (12.16) and instead argued by extending (12.17) to the left using the sandwich structure:

$$\rho(\text{Km}) = 18 + \text{rk}\,\text{MWL}(\pi_4) \leq 18 + \text{rk}\,\text{MWL}(\pi) = 14 + \text{rk}\,\text{MWL}(\pi_1) = \rho(\text{Km}).$$

12.2.5 Isogeny Notion

In this context, we cannot possibly continue without briefly explaining how the above equality (12.14) of Picard numbers fits into the greater picture. Following Inose, we introduce a notion of isogeny of K3 surfaces (regardless of the presence of a sandwich structure):

Definition 12.8 Two complex K3 surfaces X, X' are called isogenous if they admit a dominant rational map $X \dashrightarrow X'$.

Theorem 12.9 *Let X, X' be isogenous complex K3 surfaces. Then $\rho(X) = \rho(X')$.*

Proof The proof of Theorem 12.9 amounts to a nice exercise using Hodge structures. Let $\tilde{X}$ be the blow-up of X in the locus of indeterminacy of the rational map $X \dashrightarrow X'$. Then we have a commutative diagram

$$\begin{array}{ccc} \tilde{X} & & \\ \downarrow & \searrow & \\ X & \dashrightarrow & X' \end{array}$$

Via the vertical morphisms we can pull back both transcendental Hodge structures (for which we use the same letter as for the transcendental lattice), $T(X)$ on X and $T(X')$ on X', to $\tilde{X}$. Since $h^{2,0}(\tilde{X}) = 1$ as blowing up only affects $H^{1,1}$, both $\mathbb{Q}$-Hodge structures on $\tilde{X}$, pulled back from X and from X', contain $H^{2,0}(\tilde{X})$ upon complexifying. But then they are irreducible (in fact, purely transcendental), so they agree with the smallest Hodge structure on $\tilde{X}$ containing $H^{2,0}(\tilde{X})$, that is, with $T(\tilde{X})$. Thus we have

$$T(X) \cong T(\tilde{X}) \cong T(X')$$

as $\mathbb{Q}$-Hodge structures. In particular, $T(X)$ and $T(X')$ have the same dimension, so the complementary Picard numbers agree as well:

$$\rho(X) = 22 - \operatorname{rank} T(X) = 22 - \operatorname{rank} T(X') = \rho(X').$$

This proves Theorem 12.9. □

Remark 12.10 The above proof carries over to any birational map between two surfaces. Thus it hints towards the general fact that the Lefschetz number $\lambda = b_2 - \rho$ (i.e. the rank of the transcendental Hodge structure) is a birational invariant of an algebraic surface (in any characteristic, in fact).

Remark 12.11 Theorem 12.9 does not hold in positive characteristic in general. Indeed, take a supersingular Kummer surface X Then X is unirational by [Shio77]. Since any K3 surface dominates $\mathbb{P}^2$, we infer that in fact any K3 surface dominates X, but, of course, the Picard numbers vary.

However, if we restrict to K3 surfaces of finite height (one of the cases where the Tate conjecture has been known for a while, see Sect. 11.3.7), or to separable rational maps (in particular to those of degree relatively prime to the characteristic), then work of [Maz72] proves an analogue of Theorem 12.9 using the Newton polygon.

12.2.6 Explicit Mordell–Weil Ranks

We return to the problem of exhibiting explicit K3 surfaces (preferably over $\mathbb{Q}$ or $\bar{\mathbb{Q}}$) with given Mordell–Weil rank. [Kuw00] pioneered the idea to apply cyclic base

Table 12.1 Mordell–Weil ranks of cyclic base changes $X^{(n)}$ of Inose's fibration ($j, j' \neq 0, 12^3$)

n	$E \cong E'$		$E \not\cong E'$	$E \sim E'$	$E \nsim E'$
	config.	MW-rank	config.	MW-rank	MW-rank
1	$2\mathrm{II}^*, \mathrm{I}_2, 2\mathrm{I}_1$	$\begin{cases} 1 \\ 0 \end{cases}$	$2\mathrm{II}^*, 4\mathrm{I}_1$	$\begin{cases} 2 \\ 1 \end{cases}$	0
2	$2\mathrm{IV}^*, 2\mathrm{I}_2, 4\,\mathrm{I}_1$	$\begin{cases} 4 \\ 3 \end{cases}$	$2\mathrm{IV}^*, 8\mathrm{I}_1$	$\begin{cases} 6 \\ 5 \end{cases}$	4
3	$2\mathrm{I}_0^*, 3\mathrm{I}_2, 6\mathrm{I}_1$	$\begin{cases} 7 \\ 6 \end{cases}$	$2\mathrm{I}_0^*, 12\,\mathrm{I}_1$	$\begin{cases} 10 \\ 9 \end{cases}$	8
4	$2\mathrm{IV}, 4\mathrm{I}_2, 8\mathrm{I}_1$	$\begin{cases} 10 \\ 9 \end{cases}$	$2\mathrm{IV}, 16\mathrm{I}_1$	$\begin{cases} 14 \\ 13 \end{cases}$	12
5	$2\mathrm{II}, 5\mathrm{I}_2, 10\mathrm{I}_1$	$\begin{cases} 13 \\ 12 \end{cases}$	$2\mathrm{II}, 20\mathrm{I}_1$	$\begin{cases} 18 \\ 17 \end{cases}$	16
6	$6\mathrm{I}_2, 12\mathrm{I}_1$	$\begin{cases} 12 \\ 11 \end{cases}$	$24\mathrm{I}_1$	$\begin{cases} 18 \\ 17 \end{cases}$	16

changes to Inose's fibration X from (12.11). With ramification at $0, \infty$, that is, base changes

$$t \mapsto t^n \quad (n \in \mathbb{N}),$$

we obtain another K3 surface $X^{(n)}$ if and only if $n \leq 6$ (by Euler number considerations, based on Table 5.2 from Sect. 5.9.2, or using Proposition 5.51 after minimalizing). Later we will use the notation $X^{(n)}(j, j')$ indicating also the j-invariants of the elliptic curves involved. For $n = 2$, this will exactly be a Kummer surface of product type with induced elliptic fibration π_1 as detailed for the sandwich structure in Sect. 12.2.1. In any case, the K3 surfaces X and $X^{(n)}$ have the same Picard number by Theorem 12.9; since the trivial lattices of the elliptic fibrations differ drastically, this results in a plentitude of Mordell–Weil ranks. These can be determined without difficulty using Corollary 6.7 and our knowledge about $\rho(X^{(n)})$ from (12.13) and Theorem 12.9, compared against the singular fibres resulting from the base change following Table 5.2. We depict the ranks in Table 12.1, where the two alternatives refer to the CM and non-CM case (with bigger rank, of course, corresponding to CM). The configurations are valid under the assumptions that in the first case $j(E) \neq 0, 12^3$, and in the second case $j(E)\,j(E') \neq 0$. These special cases imply different fibre types (more fibres of Kodaira type I_2 or II, IV), but can be analysed immediately along the same lines. We only omitted them for ease of presentation.

The rank formula can be summarized in the following form, which is extended to cover the case of positive characteristic p ([Shio13]). In case $p > 0$, assume that

$$p > 3 \text{ (for } n \leq 4 \text{ or } n = 6) \quad \text{and} \quad p > 5 \text{ (for } n = 5).$$

Theorem 12.12 *For any $n \leq 6$, the rank of the Mordell–Weil lattice* $\mathrm{MWL}(X^{(n)})$ *is given by the following formula:*

$$r^{(n)} = h + \min\{4(n-1), 16\} - \begin{cases} 0 & \text{if } j_1 \neq j_2, \\ n & \text{if } j_1 = j_2 \neq 0, 12^3, \\ 2n & \text{if } j_1 = j_2 = 0 \text{ or } 12^3. \end{cases} \tag{12.18}$$

where $j_1 = j(E)$, $j_2 = j(E')$ and h denotes the rank of the group of homomorphisms from (12.12)*:*

$$h = \mathrm{rk}\,\mathrm{Hom}(E, E') \in \{0, 1, 2, 4\}.$$

Remark 12.13 Over $\mathbb{C}$, one can also compute the transcendental lattices of $X^{(n)}$. Indeed, by [Shio08], we have

$$T(X^{(n)}) = T(X)(n)$$

(regardless of E, E' being isogenous or not).

Remark 12.14 Shioda–Inose structures can also be used to prove that any singular K3 surface, i.e. with Picard number 20, has infinite automorphism group, as originally shown in [SI77]. In this direction, note that the MW-ranks in Table 12.1 are always positive in the CM case, thus yielding automorphisms of infinite order. It then remains to cover the cases with $j = 0$ and 12^3.

As a byproduct, we note the following consequence of Theorem 12.12 which forms an analogue of Lemma 11.39:

Corollary 12.15 *Let k be an algebraically closed field of characteristic $p > 3$. Then there exists an elliptic K3 surfaces of Mordell–Weil rank* 20 *over k.*

For the remaining characteristics, see the discussion in Sect. 12.4.5 leading to Theorem 12.46.

12.2.7 *Mordell–Weil Rank* 15

There is only one possible rank missing from Table 12.1: Mordell–Weil rank 15. This gap was filled by Kloosterman in [Klo07b]. Here we review his construction and explain how to single out an elliptic K3 surface of Mordell–Weil rank 15. Our method is a slight improvement compared to Kloosterman's argument, and also to the subsequent argument in [TdZ09]. It builds on recent work of Elsenhans and Jahnel [EJ11] and was developed for K3 surfaces, but applied to quintic surfaces which are suitable deformations of Delsarte surfaces (see Sect. 13.2.1) in [Schü15]. (Recently, there was also a new construction in the spirit of Kuwata's approach, as reviewed above, in [KumK18].)

Consider the family of K3 surfaces X which are given in terms of an elliptic fibration with singular fibres of type III* and twice I_0^*. The trivial lattice

$$T(X) = U \oplus E_7^- \oplus 2D_4^-$$

has generically rank 17, so these K3 surfaces come in a 3-dimensional family by Proposition 11.5. Using Tate's algorithm (see Sect. 5.8), we easily find the following extended Weierstrass form with III* fibre at $t = 0$ and I_0^*'s at $t = -1, \infty$:

$$X: \quad y^2 = x^3 + at^2(t+1)x^2 + t^3(t+1)^2 l_1 x + t^5(t+1)^3 l_0 \tag{12.19}$$

for linear forms $l_0, l_1 \in k[t]$ and a constant $a \in k$. By definition, the generic member of this family has Mordell–Weil rank zero, but we can subsequently raise the Mordell–Weil rank by switching to the elliptic K3 surfaces resulting from the following base changes:

1. The cyclic base change $t \mapsto t^2 + 2t$ results in an elliptic K3 surface X' with two fibres of type III* generically as the only reducible fibres (since the I_0^* fibres are replaced after minimalization by smooth fibres by Table 5.2). Hence X' has trivial lattice $U \oplus (E_7^-)^2$, but $\rho(X') = 17$ by Theorem 12.9, so generically X has Mordell–Weil rank one.
2. Next, the cyclic base change of degree 4 which ramifies exactly at the two III* fibres of X' gives an elliptic K3 surface X'' with generically only I_1 fibres (again by Table 5.2), and thus Mordell–Weil rank 15 by the same argument as before.

It remains to determine a very general member of the family of elliptic K3 surfaces X'' (over $\mathbb{Q}$!). In fact, this can be achieved on the level of the simpler family X (again by Theorem 12.9). Kloosterman proceeds along the lines introduced by van Luijk in [vLui07] in order to exhibit quartics in $\mathbb{P}^3$ with Picard number $\rho = 1$. He picks some specific member and compares reductions modulo two primes where the characteristic polynomial of Frobenius on $H^2_{\text{ét}}(X \otimes \bar{\mathbb{Q}}, \mathbb{Q}_\ell)$ (obtained from point counting over $\mathbb{F}_p, \mathbb{F}_{p^2}, \mathbb{F}_{p^3}$ via Lefschetz' fixed point formula (12.4)) implies $\rho \leq 18$ by (12.3); then he proceeds to check that the rank 18 Néron–Severi lattices of the reductions cannot be compatible, so by way of the specialization embedding

$$\mathrm{NS}(X \otimes \bar{\mathbb{Q}}) \hookrightarrow \mathrm{NS}(X \otimes \bar{\mathbb{F}}_p) \tag{12.20}$$

the original K3 surface can only have $\rho = 17$.

Here we pursue a slightly differently approach which requires less machine assistance. We start by spelling out how the original family of K3 surfaces X may admit a section P over k (or some quadratic extension of k) of minimal height 3/2. This requires that the section meets the III* fibre and exactly one I_0^* fibre non-trivially. Without loss of generality, we can assume that the latter fibre is located at $t = \infty$.

Automatically, the section over k implies that the fibre components met by P are all defined over k. Then a translation $x \mapsto x + \alpha t^2(t+1)$ ($\alpha \in \bar{k}$) normalizes this component in the I_0^* fibre to be $x' = 0$ in the suitable chart of the blow-up at ∞ while preserving the overall shape of (12.19). The Weierstrass form thus has an extra zero at ∞ in the constant coefficient:

$$X: \quad y^2 = x^3 + at^2(t+1)x^2 + t^3(t+1)^2 l_1 x + t^5(t+1)^3 c, \qquad (12.21)$$

with $a, c \in k$ and $l_1 = b_1 t + b_0$ of degree at most one in $k[t]$; note that $b_0 \neq 0$ for otherwise the fibre type at $t = 0$ would be (at least) II^*, and $c \neq 0$ for otherwise we get a fibre of type I_2 at $l_1 = 0$ (or further degeneracy). In this situation, the integral section P meeting III^* non-trivially and the distinguished component of I_0^* at ∞ takes the shape

$$P = (u_0 t^2, (v_1 t + v_0)t^3).$$

Upon substituting into (12.21), the lowest-order coefficient appears at t^5, but only on the RHS. Thus its vanishing gives

$$u_0 = -c/b_0.$$

As a consequence, this leads to the problem whether the quadratic polynomial

$$q(t) = -(c/b_0)^3 + a(t+1)(c/b_0)^2 - (t+2)l_1 c/b_0 + (t^2 + 3t + 3)c$$

is a square in $k[t]$ (in fact, $(v_1 t + v_0)^2$). Equivalently (over $\bar{k}$), the discriminant vanishes; eliminating non-zero factors, one obtains that X admits a section P of height $3/2$ meeting the distinguished non-identity component of the I_0^* fibre if and only if

$$a^2 + 4(b_0 - b_1) = 0. \qquad (12.22)$$

We continue by exhibiting a K3 surface X over $\mathbb{Q}$ of Picard number 17. For this purpose we proceed by fixing some prime p and choose $a, b_0, b_1 \in \mathbb{Q} \cap \mathbb{Z}_p$ (i.e. p-adically integral) in such a way that (12.22) is zero modulo p, but not in $\mathbb{Q}$, while X has good reduction at p and the singular fibres do not degenerate. Then, possibly after adjoining a square root, we have by construction:

Lemma 12.16 *Let X be as above. Then $X \otimes \mathbb{F}_{p^2}$ admits a section P of height $3/2$ as above, but $X \otimes \bar{\mathbb{Q}}$ does not.*

Next, we use a recent result of [EJ11, Theorem 1.4], building on work of Raynaud [Ray79], which states that the specialization embedding (12.20) is in fact primitive in suitable situations, for instance if X is defined over $\mathbb{Q}$ and $p > 2$. In other words, if $\rho(X \otimes \bar{\mathbb{Q}}) = \rho(X \otimes \bar{\mathbb{F}}_p)$, then every divisor class lifts from characteristic p to characteristic zero (as opposed to some multiple of the divisor lifting). Presently, this would imply that P lifts to a divisor D whose reduction modulo p is equivalent to P. But then, by Theorem 6.5, D induces a section P_0 in characteristic zero which thus lifts P directly. However, by construction, P does not lift to $X \otimes \bar{\mathbb{Q}}$. Hence the primitivity of (12.20) implies:

Corollary 12.17 *In the above set-up,* $\rho(X \otimes \bar{\mathbb{Q}}) < \rho(X \otimes \bar{\mathbb{F}}_p)$.

In order to achieve that $\rho(X \otimes \bar{\mathbb{Q}}) = 17$ (the minimum possible due to the singular fibres), it thus suffices to choose the coefficients in such a way that $\rho(X \otimes \bar{\mathbb{F}}_p) = 18$. The latter condition can be checked explicitly using (12.3). This has as input the characteristic polynomial of Frobenius on $H^2_{\acute{e}t}(X \otimes \bar{\mathbb{F}}_p, \mathbb{Q}_\ell)$; since we know an 18-dimensional Galois-submodule of $H^2_{\acute{e}t}(X \otimes \bar{\mathbb{F}}_p, \mathbb{Q}_\ell)$ (the sublattice N of $\mathrm{NS}(X \otimes \bar{\mathbb{F}}_p)$ generated by fibre components and the sections (O), (P), embedded via the ℓ-adic cycle class map), the characteristic polynomial essentially only depends on a degree 4 factor $\chi(T)$ corresponding to the four-dimensional subspace $V = N^\perp \subset H^2_{\acute{e}t}(X \otimes \bar{\mathbb{F}}_p, \mathbb{Q}_\ell)$. In turn, this is determined by point counting over $\mathbb{F}_p$ and $\mathbb{F}_{p^2}$ using the Lefschetz fixed point formula (see Sect. 10.5.2).

Example 12.18 Let $b_0 = b_1 = c = 1$ and $a = p$ for some prime to be determined. Then for $p = 3$ the elliptic fibration on X degenerates modulo p, and for $p = 5$ the characteristic polynomial of Frob_p^* on $H^2_{\acute{e}t}(X \otimes \bar{\mathbb{F}}_p, \mathbb{Q}_\ell)$ has the essential factor

$$\chi(T) = (T + p)^2(T^2 + p^2).$$

By the Tate conjecture (Theorem 12.2), $X \otimes \mathbb{F}_{p^4}$ is thus supersingular ($\rho = 22$). For $p = 7$, however, the section P is defined over $\mathbb{F}_{p^2}$, and the characteristic polynomial of Frob_p^* on $V \subset H^2_{\acute{e}t}(X \otimes \bar{\mathbb{F}}_p, \mathbb{Q}_\ell)$ is

$$\chi(T) = T^4 - 3T^3 + 20T^2 - 3p^2T + p^4.$$

This polynomial is irreducible over $\mathbb{Q}$ while $p^4\chi(T/p)$ is not a cyclotomic polynomial. Hence $\rho(X \otimes \bar{\mathbb{F}}_p) \leq 18$ by (12.3) (and X is seen to be ordinary, cf. Sect. 11.3.7). On the other hand, $\rho(X \otimes \bar{\mathbb{F}}_p) \geq \operatorname{rank} N = 18$, so in fact, $\rho(X \otimes \bar{\mathbb{F}}_p) = 18$. By Corollary 12.17, we deduce that $\rho(X \otimes \bar{\mathbb{Q}}) = 17$, equalling the rank of the trivial lattice (i.e. $\mathrm{MW}(X \otimes \bar{\mathbb{Q}}) = 0$). As a consequence, X'' is equipped with an elliptic fibration of Mordell–Weil rank 15 over $\bar{\mathbb{Q}}$.

We emphasize that the above proof does not a priori require the validity of the Tate conjecture (Theorem 12.2) since we have exhibited an explicit extra divisor on the reduction, thus a posteriori verifying the Tate conjecture for $X \otimes \mathbb{F}_{7^r}$ for any $r \in \mathbb{N}$. In any case, the Tate conjecture has been known to be valid for elliptic K3 surfaces ever since [ASD73].

To end this section, we would like to point out again which particular property of elliptic surfaces with section enables us to simplify Kloosterman's argument so significantly (and in a fairly conceptual and constructive way): by [EJ11] we know that the section P on $X \otimes \mathbb{F}_{p^2}$ lifts to some divisor D in characteristic zero on $X \otimes \bar{\mathbb{Q}}$; a priori this need not be effective etc. It is only the Structure Theorem 6.5 for elliptic surfaces which allows us to infer that D is represented by a section on X, and more precisely exactly of the same shape as P. This line of argument (which was applied to certain quintics in $\mathbb{P}^3$ in [Schü15]) works indeed for any elliptic surface, regardless of being K3.

12.3 Classification of Elliptic Fibrations on a Given K3 Surface

We have already mentioned that a K3 surface may admit more than one elliptic fibration. As an example we have discussed a few elliptic fibrations on the Kummer surface of product type in Sects. 11.2.1–11.2.3. Before revisiting them again in Sect. 12.3.1, we have to spell out the different ways to classify the different elliptic fibrations on a given K3 surface X:

1. by the types of the singular fibres—either in terms of the ADE lattices corresponding to the singular fibres as in Theorem 11.36, or even
2. distinguishing between fibres of types I_n $(n = 1, 2, 3)$ and II, III, IV;
3. by the frame W;
4. up to isometries of $\mathrm{NS}(X)$, i.e. lattice equivalence classes of embeddings $U \hookrightarrow \mathrm{NS}(X)$ modulo $O^+(\mathrm{NS}(X))$ (following Proposition 11.33);
5. up to automorphism of X, i.e. isomorphism classes of elliptic fibrations modulo $\mathrm{Aut}(X)$.

Except for the frame, these notions extend to genus one fibrations, as we shall exploit to some extent in Sect. 12.4.5. Each classification is finite in number (a consequence of work of [Ste85] in the complex case, and of [LM18] in odd characteristic), and the classifications are in hierarchical order, i.e. any classification is finer than each of the preceding ones, except that the third and fourth classification may not detect the differences from the second. For instance, the frame allows us to read off the ADE types by Proposition 6.42, but not the precise singular fibres. Note, however, that the frame does encode relevant information how the singular fibres are met by the sections which is not at all contained in either of the preceding data.

In what follows, we will mostly concentrate on the third classification since this fits very well with our preparations as it is based on a lattice theoretic technique going back to Kneser and Nishiyama. A nice overview about all classifications can be found in [BKW13] which in particular gives a good account how to apply the so-called Borcherds' method to the last two classifications (following Borcherds [Bor87, Bor98, Kon98b]). The fourth classification will briefly reappear in Sect. 12.4.5.

12.3.1 Elliptic Fibrations on a Kummer Surface of Product Type

In [Ogui89], Oguiso pioneered a geometric technique to work out all elliptic fibrations up to isomorphism on a very general complex Kummer surface of product type. Recall from Assumption 11.7 that by very general (in moduli theoretic terms) we simply mean that

$$S = \text{Km}(E \times E') \quad \text{for non-isogenous complex elliptic curves } E, E';$$

equivalently, $\varrho(S) = 18$. More precisely, the transcendental lattice and the Néron–Severi lattice of S can be written as

$$T(S) = T(E \times E')(2) = U(2)^2, \quad \text{NS}(S) = U \oplus E_8^- \oplus 2D_4^-.$$

In the next theorem, we reproduce Oguiso's result. We also include the multiplicities, i.e. the number of non-isomorphic fibrations of the same type, but for simplicity we only list the case where neither E nor E' has extra automorphisms, i.e. for $j, j' \neq 0, 12^3$. Mostly, the multiplicity can be explained by the freedom of choice of 2-torsion point(s) on E or E' (cf. Remark 11.8). Another case is No. 4, where the two inequivalent fibrations π, π' are induced by the projections from $E \times E'$ to either factor (see Sect. 11.2.1). All types with a unique elliptic fibration up to Aut(S) (No. 3, 5 and 9) have already been encountered before.

Theorem 12.19 (Oguiso) *Let S be a very general complex Kummer surface of product type. Then S admits 11 different types of elliptic fibrations as tabulated below:*

No.	1	2	3	4	5	6
Reducible fibres	$2\,\text{I}_8 + 8\,\text{I}_1$	$\text{I}_{12} + \text{I}_4$	$2\,\text{IV}^*$	$4\,\text{I}_0^*$	$\text{I}_6^* + 6\,\text{I}_2$	$2\,\text{I}_2^* + 4\,\text{I}_2$
torsion in MW	$\mathbb{Z}/2\mathbb{Z}$	$\mathbb{Z}/2\mathbb{Z}$	0	$(\mathbb{Z}/2\mathbb{Z})^2$	$(\mathbb{Z}/2\mathbb{Z})^2$	$(\mathbb{Z}/2\mathbb{Z})^2$
rkMW	2	2	4	0	0	0
multiplicity	9	6	1	2	1	9

No.	7	8	9	10	11
Reducible fibres	$\text{I}_4^* + 2\,\text{I}_0^*$	$\text{III}^* + \text{I}_2^* + 3\,\text{I}_2/2\,\text{I}_2 + \text{III}$	$\text{II}^* + 2\,\text{I}_0^*$	$\text{I}_8^* + \text{I}_0^*$	$2\,\text{I}_4^*$
torsion in MW	$\mathbb{Z}/2\mathbb{Z}$	$\mathbb{Z}/2\mathbb{Z}$	0	0	0
rkMW	0	0	0	0	0
multiplicity	9	6	1	6	9

We note that the above classification carries over to algebraically closed fields of odd characteristic under Assumption 11.7 ($E \not\sim E'$), provided that still $j, j' \neq 0, 12^3$. In characteristic 2, however, there are several subtleties occurring, notably because the Kummer construction degenerates and because there can be quasi-elliptic fibrations.

The technique pursued by Oguiso to prove Theorem 12.19 built on finiteness arguments due to [Ste85] which in particular showed that any complex K3 surface X has only finitely many elliptic fibrations (or genus one fibrations) up to $\mathrm{Aut}(X)$.

Explicit equations for all these fibrations have subsequently been worked out by Kuwata and one of us in [KuwS08]. For simplicity, this paper considered the situation where both elliptic curves E and E' have all 2-torsion points k-rational (i.e. E, E' are given in Legendre form). A special feature of [KuwS08] is the determination of the elliptic parameter for each fibration (cf. Sect. 11.2.1).

12.3.2 Kneser–Nishiyama Method

We shall now start discussing another approach towards the problem of classifying all elliptic fibrations on a given K3 surface. This approach goes back to Nishiyama [Nis96]. It is based on lattice theoretic ideas which partly can be traced back to gluing techniques due to Kneser and Witt [Kne56]. Moreover, it builds on Niemeier's classification of definite unimodular lattices of rank 24 in [Nie73] and Nikulin's converse of Proposition 2.41 which we review below. We start with the latter:

Theorem 12.20 ([Nik80, Corollary 1.6]) *Let L, M be even integral lattices such that*

$$L^\vee/L \cong M^\vee/M, \quad q_L = -q_M.$$

Then there exists an even unimodular overlattice N of $L \oplus M$ such that both L and M embed primitively into N as each other's orthogonal complement:

$$L = M^\perp \subset N, \quad M = L^\perp \subset N.$$

Obviously, we can apply Theorem 12.20 to the Néron–Severi lattice and the transcendental lattice of any complex K3 surface X. What makes elliptic fibrations (with section!) special are the following intrinsic properties of the frame $W(X)$:

(∗) it is even and positive-definite of rank $\rho(X) - 2$ by Theorem 6.13, and
(∗∗) its discriminant form equals that of $\mathrm{NS}(X)$ up to sign.

Hence it makes sense to aim for embedding the frame into an even positive-definite unimodular lattice. It turns out that positive-definite lattices of small rank have been fully classified. In the complex case, $W(X)$ has rank at most 18 (since $\rho(X) \leq 20$ by (4.15)), and modest discriminant group, so one could try to embed $W(X)$ into an even unimodular positive-definite lattice of rank 24 (which will indeed turn out to be possible, see Lemma 12.25). These so-called *Niemeier lattices* admit a instructive classification in terms of their root lattices.

Theorem 12.21 ([Nie73, Satz 8.3]) *A Niemeier lattice N (positive-definite even unimodular of rank 24) is determined by its root lattice $\mathcal{R}(N)$ up to isometry. Among these, the Leech lattice Λ_{24} has no roots at all. Other than that, there are the following 23 non-zero possibilities for $\mathcal{R}(N)$:*

$$A_1^{24}, A_2^{12}, A_3^8, A_4^6, A_5^4 \oplus D_4, D_4^6, A_6^4, A_7^2 \oplus D_5^2, A_8^3, A_9^2 \oplus D_6, D_6^4, E_6^4, A_{11} \oplus D_7 \oplus E_6,$$
$$A_{12}^2, D_8^3, A_{15} \oplus D_9, A_{17} \oplus E_7, D_{10} \oplus E_7^2, D_{12}^2, A_{24}, D_{16} \oplus E_8, E_8^3, D_{24}.$$

Niemeier lattices play a central role for several aspects of K3 surfaces, not only elliptic fibrations. To name but the two most prominent ones:

- Mukai's groundbreaking classification of finite symplectic group actions on K3 surfaces [Muk88] has a neat interpretation in terms of Niemeier lattices and Mathieu groups due to Kondō [Kon98a].
- The expansion of the K3 elliptic genus leading to Mathieu moonshine due to [EOT11] was subsequently generalized by Cheng–Duncan–Harvey to the umbral moonshine conjecture in [CDH14], relating Niemeier lattices and certain modular functions (proved in [DGO15]).

12.3.3 Partner Lattice

Given an elliptic K3 surface X, we aim at embedding the frames of all elliptic fibrations on X into Niemeier lattices (as classified in Theorem 12.21) following Kneser and Nishiyama. For this purpose we start by determining a partner lattice for $\mathrm{NS}(X)$.

Definition 12.22 An even positive-definite lattice M is called a *partner lattice* of $\mathrm{NS}(X)$ if the following two equalities are satisfied:

$$q_M = q_{\mathrm{NS}(X)}, \quad \mathrm{rank}(M) + \rho(X) = 26.$$

Note how the definition of a partner lattice resembles that of a companion in Definition 8.16, but it is not limited to root lattices, nor does it allow for imprimitive partner lattices.

Given a partner lattice M, we can use Theorem 12.20 to find the following relation with each frame W of an elliptic fibration on X:

Corollary 12.23 *Let W be a frame of an elliptic fibration on X and M be a partner lattice of* $\mathrm{NS}(X)$. *Then $M \oplus W$ has a Niemeier lattice as an overlattice.*

Conversely, we can thus determine a partner lattice of $\mathrm{NS}(X)$ by embedding the frame W^- of any elliptic fibration on X primitively into any Niemeier lattice and taking M^- to be the orthogonal complement.

Example 12.24 If $\mathrm{NS}(X) = U \oplus E_8^-$, then the possible partner lattices are E_8^2 and D_{16}', the index two even unimodular overlattice of D_{16}.

Similarly, for $\mathrm{NS}(X) = U \oplus 2E_8^-$, there is only one partner lattice, namely $M = E_8$.

In order to classify all types of elliptic fibrations on X, Corollary 12.23 prompts us to determine the inequivalent primitive embeddings of the partner lattice M of X into the Niemeier lattices from Theorem 12.21. In many situations, the main step to achieve this consists in comparing the root lattices involved; that is, we first embed

$$\mathcal{R}(M) \hookrightarrow \mathcal{R}(N)$$

and then extend this to an embedding

$$M \overset{\iota}{\hookrightarrow} N \tag{12.23}$$

for some Niemeier lattice N. In each case, the orthogonal complement of M in the respective choice N of Niemeier lattice gives the corresponding frame:

$$W = \iota(M)^\perp \subset N.$$

By Proposition 6.42 one can read off the Mordell–Weil lattice (plus the torsion in MW) and the singular fibres (or rather the associated restricted root lattices T_v) of the elliptic fibration associated to the embedding ι in (12.23) from the frame W by

$$\bigoplus_{v \in \mathbb{P}^1} T_v = \mathcal{R}(W), \quad \mathrm{MW}(X) \cong W/\mathcal{R}(W).$$

The Kneser–Nishiyama method, as sketched above, is purely lattice theoretic in nature. Hence it works in any characteristic as soon as the frame W has a partner lattice M as in Definition 12.22. In positive characteristic, such a partner need not exist (for the instance of supersingular K3 surfaces, see Sect. 12.4.4), but over $\mathbb{C}$ there is a short argument which we briefly explain next.

Lemma 12.25 ([Nis97]) *Assume that X is a complex K3 surface. Then* $\mathrm{NS}(X)$ *has a partner lattice.*

Proof Recall the transcendental lattice $T(X)$ which is an even lattice of rank $r = 22 - \rho(X)$ and signature $(2, 20 - \rho(X))$. Let $t = r - 2$. By [Nik80, Theorem 1.12.4], the transcendental lattice with reversed sign, $T(X)^-$, admits a primitive embedding into the following indefinite even unimodular lattice:

$$T(X)^- \hookrightarrow U^t \oplus E_8.$$

Define M as the orthogonal complement of $T(X)^-$ in $U^t \oplus E_8$. By construction, this lattice is negative-definite of rank $t + 6 = r + 4 = 26 - \rho(X)$. The discriminant form of M satisfies

$$q_M = -q_{T(X)^-} = q_{T(X)} = -q_{\mathrm{NS}(X)},$$

where the last equality follows from Proposition 2.41. Hence M is a partner lattice of $\mathrm{NS}(X)$. □

Remark 12.26 For K3 surfaces of finite height in odd characteristic, one can apply a lifting argument (with full Néron–Severi lattice) to use the results above (see [Del81], [NO85, p. 505], [LM18, Corollary 4.2]).

12.3.4 Elliptic Fibrations from the Partner Lattice

We conclude our discussion of the classification of elliptic fibration by frames by clarifying when the above technique guarantees the existence of an elliptic fibration with frame W on a fixed K3 surface X, given a lattice W satisfying $(*)$ and $(**)$ from page xxx.

For starters, it is instructive to consider the complex case again. Note that in any case there is some K3 surface with an elliptic fibration with frame W. To see this, apply Theorem 12.20 to $T(X)$ and $U \oplus W$ to derive a primitive embedding of $U \oplus W$ into the K3 lattice Λ. By Proposition 11.5, there is a family of K3 surfaces with Néron–Severi group $\mathrm{NS} = U \oplus W$. In particular, each of these K3 surfaces admits an elliptic fibration with section and frame W by Proposition 11.33.

Now let us turn to the specific K3 surface X. Assume that we have found a candidate lattice W for the frame of an elliptic fibration on X, for instance by Nishiyama's technique. By construction, $\mathrm{NS}(X)$ and $U \oplus W$ have the same signature and discriminant form. Due to the hyperbolic plane in $U \oplus W$, the latter lattice, and hence both lattices satisfy the conditions of Theorem 2.40. Hence they are isometric. In particular, the reasoning from the complex case applies to show that X admits an elliptic fibration with frame W by Proposition 11.33.

12.3.5 Applications to Complex K3 Surfaces

Nishiyama applied this technique to several complex K3 surfaces. For instance, he determined lattice-theoretically all elliptic fibrations on certain Kummer surfaces of product type (see Sect. 11.2.1). This is in perfect agreement with Oguiso's result from Theorem 12.19. Below we only give the essential information on Nishiyama's findings:

E, E'	$\rho(\mathrm{Km}(E \times E'))$	no. of fibrations
$E \cong E', j(E) = 0$	20	30
$E \cong E', j(E) = 12^3$	20	25
$E \cong E'$ without CM	19	34
$E \not\sim E'$	18	11

We omit the details here since they would go beyond the scope of this section. Instead we focus on two examples which are very easy to handle. Several other complex K3 surfaces have been analysed along similar lines in recent years. We mention [ElS08, BL13, Kum14] the latter of which also develops important additional techniques geared towards the given problem (and also for non-jacobian elliptic fibrations).

We conclude this section with two examples which are easy to understand using the above methods. A more complicated example, together with detailed computations, will appear in relation with the rank problem for elliptic curves in Sect. 13.1.4.

12.3.5.1 First Example

To start with, consider the K3 surfaces X arising from a pair of non-isogenous elliptic curves E, E' through the Shioda–Inose structure Sect. 12.2.2. Directly, we derive that $\mathrm{NS}(X)$ equals the trivial lattice,

$$\mathrm{NS}(X) = U \oplus 2E_8^-, \tag{12.24}$$

from the presence of two singular fibres of type II* (using the non-isogeny assumption), and in the complex case, $T(X) = U^2$ like for the abelian surface $E \times E'$ from (12.9). Thus a partner lattice for $\mathrm{NS}(X)$ would be $M = E_8^-$ as in Example 12.24. We use that E_r admits no embeddings into A_n, D_m (and similarly, D_m does not embed into A_n, see [Nis96], for instance). Up to isometries, M therefore has exactly two embeddings into a Niemeier lattice:

1. $N = \mathcal{R}(N) = 3E_8^-$ gives back the representation (12.24) of $\mathrm{NS}(X)$; this corresponds to Inose's fibration (12.11).
2. $\mathcal{R}(N) = E_8^- \oplus D_{16}^-$ (and $N = E_8 \oplus (D_{16}')^-$, involving the second even unimodular negative-definite lattice of rank 16) yields an elliptic fibration on X with a fibre of Kodaira type I_{12}^* and a 2-torsion section (accounting for $N/\mathcal{R}(N) \cong M^\perp/\mathcal{R}((M^\perp)) \cong D_{16}'/D_{16} \cong \mathbb{Z}/2\mathbb{Z}$).

To connect the two fibrations, we start with Inose's fibration (12.11) with two II* fibres on X. It is easy to identify a divisor of Kodaira type I_{12}^* in the intersection graph consisting of zero section and II*'s: simply omit the vertices corresponding to the far double components of the II* fibre, as sketched in Fig. 12.1.

By Corollary 11.28, the divisor D indicated in Fig. 12.1 induces an elliptic fibration on X which corresponds to the second case above. Here the two sections are given by

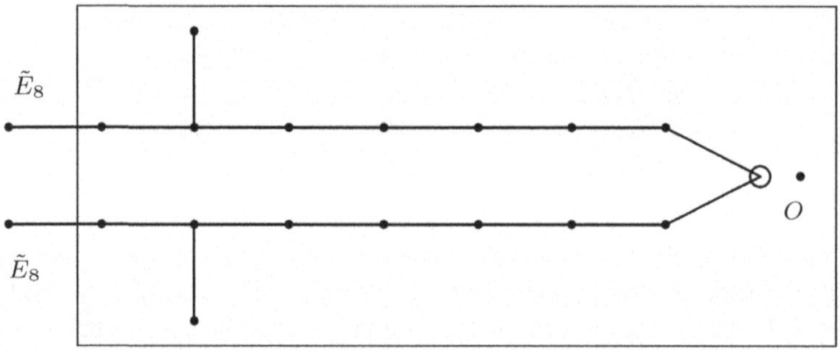

Fig. 12.1 A I_{12}^* divisor supported on zero section and fibre components

the far components omitted in the indicated divisor by Proposition 11.31. Explicitly, the linear system $|D|$ inducing the second fibration is given as follows. In the affine Weierstrass form (12.11) choose the elliptic parameter $u = x/t^2$. This has poles at the zero section and deep into the II^* fibres at 0 and ∞. Substituting for x and simplifying by way of $y = vt^2$ (as in the resolution of the singularity at the affine origin), one obtains the affine equation

$$X: \quad v^2 = t^2 u^3 - 3\,A\,t^2 u + t\,(t^2 - 2\,B\,t + 1).$$

As a cubic in coordinates t, v over $k(u)$, up to scaling already in Weierstrass form, with rational points at $(0, 0)$ and O, the above equation describes an elliptic surface over $\mathbb{P}^1_u$ with an I_{12}^* fibre at $u = \infty$ and 2-torsion section $(0, 0)$.

12.3.5.2 Second Example

Along similar lines, consider complex elliptic curves E, E' which are isomorphic, but without CM. Through the Shioda–Inose structure, they give rise to a K3 surface X with

$$\mathrm{NS}(X) = U \oplus 2E_8^- \oplus A_1^-, \quad T(X) = U \oplus \langle 2 \rangle.$$

Here one can choose $M = E_7^-$ as partner lattice. Since E_r admits no embedding into (orthogonal sums of) A_n and D_n (cf. [Nis96], for instance), there are exactly four Niemeier lattices that admit an embedding of E_7. The embeddings into E_8^3 and $E_8 \oplus D_{16}'$ correspond to degenerations of the two elliptic fibrations of the above non-isogenous case. The other two embeddings where the root types of the Niemeier lattices are $A_{17} \oplus E_7$ resp. $D_{10} \oplus E_7^2$, have not been detected on the previous K3 surfaces.

As for explicit equations, the K3 surfaces arise from Inose's pencil (12.11) by setting $j(E) = j(E')$ (which causes an additional reducible fibre as required, see

Observation 12.6). In order to connect the fibrations, it is convenient to start from the one corresponding to the root type A_{17}. These fibrations arise from the rational elliptic surface in Example 7.15 by quadratic base change ramified at ∞ as this sees the I_9 fibre replaced by a I_{18} fibre while preserving the 3-torsion section. What's convenient about this model is that in the diagram displaying the fibre components and the three torsion sections, we can identify reducible singular fibres of all three other elliptic fibrations. Hence Corollary 11.28 applies directly, and it remains to work out the linear systems. We leave the details to the reader.

Remark 12.27 The above two examples of K3 surfaces are rather special because they have finite automorphism group and they contain only finitely many smooth rational curves by [Kon89]. In fact, even the configuration of rational curves is known—for the first example, it is exactly displayed in Fig. 12.1. Essentially, all of this can be proved using lattices, see [Nik79, Nik83, Nik85].

12.4 Supersingular K3 Surfaces

Supersingular K3 surfaces form a very particular subject among K3 surfaces that has gained considerable attention in recent years. Notably it displays numerous features special to positive characteristic, yet it shares a big portion of the rich theory over $\mathbb{C}$. Here we will only lightly touch this subject, mostly with a view towards elliptic fibrations and Mordell–Weil lattices.

12.4.1 Two Notions of Supersingularity

Classically, there have been two notions of a supersingular K3 surface, necessarily over a field of positive characteristic (because the second notion would visibly contradict the Lefschetz bound (4.15)):

1. Artin [Art74a] defined a K3 surface X to be supersingular if its height h, i.e. the height of the formal Brauer group $\widehat{\mathrm{Br}}(X)$, is infinite: $h = \infty$ (see Sect. 11.3.7).
2. Shioda defined a K3 surface X (or any smooth projective surface) to be supersingular if the Picard number is maximal: $\rho(X) = b_2(X)$.

It was long known that Shioda's notion implies Artin's notion. The converse direction was subject to the Tate conjecture (Theorem 12.2). Thus when talking about supersingular K3 surfaces, there is no difference between the two above notions. Hence it is now justified to assume that supersingularity is expressed by

$$\rho(X) = b_2(X),$$

which is what we will do in the following. Throughout we will let k denote an algebraically closed field of characteristic $p > 0$.

12.4.2 Artin Invariant

Artin showed in [Art74a] that for a supersingular K3 surface X over k, one has

$$\operatorname{disc}(\mathrm{NS}(X)) = -p^{2\sigma} \quad \text{for some } \sigma \in \{1, 2, \ldots, 10\}.$$

Definition 12.28 One calls σ the *Artin invariant* of X.

Here's the most important feature of the Artin invariant: the moduli spaces of supersingular polarized K3 surfaces can be stratified according to σ into subvarieties of dimension $\sigma - 1$ ([Art74a]; see also [Ogus79, Ogus83]). Note that strata need not be irreducible (see e.g. [Shima04]), and the stratification can be extended to the full moduli spaces of polarized K3 surfaces by means of the height as shown in [Art74a] (see also [vGK00]).

More precisely, Artin also proved that the discriminant group of $\mathrm{NS}(X)$ is p-elementary, i.e. this abelian group is p-torsion:

$$A_{\mathrm{NS}(X)} = \mathrm{NS}(X)^\vee/\mathrm{NS}(X) \cong (\mathbb{Z}/p\mathbb{Z})^{2\sigma}. \tag{12.25}$$

Subsequently, Rudakov and Shafarevich proved that the isometry class of $\mathrm{NS}(X)$ only depends on p and σ [RS83]. We can thus denote it by $\Lambda_{p,\sigma}$.

Throughout this section, we will pay particular attention to the case of Artin invariant $\sigma = 1$. Such a supersingular K3 surface is unique up to isomorphism (over $\bar{k}$) [Ogus79]. For later use, or reference to earlier considerations, we mention two canonical ways to construct supersingular K3 surfaces with Artin invariant $\sigma = 1$ (for one of the first instances cf. [Shio73, Shio75]).

Example 12.29 Assume that p is odd. Let E, E' be supersingular elliptic curves (which are automatically isogenous). Then the Kummer surface of $E \times E'$ as in Sect. 11.2.1 is the supersingular K3 surface of Artin invariant $\sigma = 1$, and so is the K3 surface arising from the Shioda–Inose structure Sect. 12.2.2.

There are several ways to determine the Artin invariant in Example 12.29. For instance, outside characteristic 2, we can use the fact that presently, E and E' are either isomorphic, or they admit an isogeny of degree d relatively prime to p. Together with the classes of the fibres of the projections to either factor, the graph of the isogeny (resp. of the isomorphism if $d = 1$) generates a sublattice

$$N \cong U \oplus \langle -2d \rangle \subset \mathrm{NS}(E \times E')$$

(for a very detailed account of generators of $\mathrm{NS}(E \times E')$, see [Kat15]). Since the discriminant group of $\mathrm{NS}(E \times E')$ is only p-torsion, but A_N contains no p-torsion by assumption, it follows that $A_{\mathrm{NS}(E\times E')}$ embeds into $A_{N^\perp}$. This implies

$$2\sigma = \text{length}(A_{\text{NS}(E\times E')}) \leq p\text{-length}(A_{N^\perp}) \leq \text{rk}N^\perp = 3,$$

so $\sigma = 1$. Almost literally, this argument carries over to $\text{Km}(E \times E')$, where the graph of the isogeny gives an additional fibre component or section of a given elliptic fibration.

An alternative line of argument will appear momentarily: lift $\text{Km}(E \times E')$ to a singular K3 surface over $\mathbb{C}$, or $E \times E'$ to a singular abelian surface, and apply Shimada's argument leading to Example 12.30.

Thanks to Inose's fibration Sect. 12.2.3, Example 12.29 has nice explicit consequences. For instance, for any odd p one can use it to derive quite easily a model over $\mathbb{F}_p$ of the supersingular K3 surface in characteristic p of Artin invariant $\sigma = 1$; additionally, this can be twisted to have all but one generator of $\text{NS}(X)$ also defined over the prime field $\mathbb{F}_p$ (but never all of them), see [Schü12].

For $p = 2$, the Kummer construction involving a supersingular abelian surface degenerates (cf. [Shio74b, Kat78]), but interestingly it is possible to engineer a Kummer surface for the self-product of the cuspidal cubic by [Schr07].

Generally, one can also appeal to another canonical way to produce supersingular K3 surfaces with Artin invariant $\sigma = 1$, starting from singular K3 surfaces.

By [SI77], any singular K3 surface X admits a Shioda–Inose structure Sect. 12.2.2; more precisely, this relates X with a product of isogenous elliptic curves with complex multiplication. It follows from classical CM theory (see [Shimu71]) that X can be defined over some number field. Specifically, if we let d denote the discriminant of X, then, by [Schü07b, Proposition 4.1], X admits a model over the ring class field $H(d)$. This is an abelian extension of $\mathbb{Q}(\sqrt{d})$ with prescribed ramification whose Galois group is isomorphic to the class group $\text{Cl}(d)$ (see e.g. [Cox13]); in fact, by the main theorem of complex multiplication [Shimu71, Theorem 5.4], this extension is generated by the j-invariants of the elliptic curves from (12.11).

Example 12.30 Let X be a singular K3 surface of discriminant $d < 0$, defined over some number field K. Let $\mathfrak{p}$ denote a prime ideal of $\mathcal{O}_K$ of good reduction for X. Assume that $\mathfrak{p}$ has norm p^r, where the prime $p \in \mathbb{N}$ is inert in the quadratic extension $\mathbb{Q}(\sqrt{d})/\mathbb{Q}$. Then $X \otimes \bar{\mathbb{F}}_p$ is a supersingular K3 surface with Artin invariant $\sigma = 1$ by [Shima09] (compare [Schü16, Proposition 4.1]).

Constructing supersingular K3 surfaces of Artin invariant $\sigma > 1$ explicitly tends to be a delicate problem. In odd characteristic, K3 surfaces with $\sigma = 2$ can be given as Kummer surfaces of supersingular abelian surfaces which are not of product type, and there are also approaches using generalized Kummer constructions [Kat87]. Beyond that, there are isolated examples given by Delsarte surfaces ([Shio87b, Got96]). Another technique is related to the conjecture that algebraically simply connected supersingular surfaces are unirational. We have come across this notion when discussing Castelnuovo's criterion in Sect. 4.9. For supersingular K3 surfaces in characteristic greater than 3, a proof of unirationality was published by Liedtke [Lie15] (see also [BL18] for arguments to cover characteristic 3). Liedtke's approach uses genus one fibrations and inseparable torsors, leading to fibrations without sections (though the inseparability, and thus the validity of the proof of unirationality, has recently been challenged in [BL19]).

In more elementary terms, for small characteristic, one may use inseparable covers of rational surfaces; in the context of elliptic surfaces, these often occur as inseparable base change. We will use this technique in Sects. 12.4.3, 12.4.5; in a very instructive way it also features in [IL13]. Here we only illustrate this by a few results in characteristic $p = 2$:

Theorem 12.31 (Rudakov–Shafarevich [RS78], Shimada [Shima04]) *Any supersingular K3 surface in characteristic* 2 *can be given in terms of either:*

- *a quasi-elliptic fibration (generally without section), or*
- *a purely inseparable cover of* $\mathbb{P}^2$ *branched along a sextic.*

In addition, there are equations given in [RS78] which allow for reading off the Artin number.

In comparison, the approach starting from a rational elliptic surface and applying a purely inseparable base change fails to produce all supersingular K3 surfaces, even in characteristic 2. From the moduli viewpoint, this is explained by the fact that rational elliptic surfaces only have eight moduli by Theorem 5.48 and Proposition 5.51, so purely inseparable base changes cannot fill out the nine-dimensional moduli space of supersingular K3 surfaces. Explicitly, this can also be explained from the Weierstrass form

$$y^2 + a_1xy + a_3y = x^3 + a_2x^2 + a_4x + a_6. \tag{12.26}$$

Namely, after inseparable quadratic base change, all polynomials a_i have coefficients in t^2, say, so we can define a generic 2-torsion section by setting $\mathrm{x} = a_3/a_1$ and solving for $\mathrm{y}^2 = \mathrm{RHS}$. Generically, this results in a K3 surface with 12 fibres of type I_2, $\mathrm{MWL} = E_8(2)$ and a non-trivial 2-torsion section, so that the determinant formula (6.23) returns $\sigma = 9$ (compare [Lan88, Sect. 3]).

Proposition 12.32 *Any supersingular K3 surface X in characteristic $p = 2$ with Artin invariant $\sigma = 9$ arises from a rational elliptic surface by an inseparable quadratic base change.*

Proof Since $\mathrm{NS}(X) = \Lambda_{p,\sigma}$ is uniquely determined by $(p, \sigma) = (2, 9)$, we infer that

$$\mathrm{NS}(X) \cong U \oplus W^-, \tag{12.27}$$

where $W = (A_1^{12} \oplus E_8(2))'$ denotes the frame of the inseparable quadratic base change of a generic rational elliptic surface as above. By Proposition 11.33, this decomposition of $\mathrm{NS}(X)$ shows that X admits an elliptic fibration which, by Proposition 6.42, has 12 fibres of type I_2 and $\mathrm{MW}(X) \cong E_8(2) \times \mathbb{Z}/2\mathbb{Z}$. (This also rules out that the fibration is quasi-elliptic by Corollary 7.22.) Hence the discriminant Δ of any Weierstrass form of this fibration is a perfect square in $k(t)$, and so is the j-invariant $j = a_1^{12}/\Delta$. By Lemma 5.30 the Weierstrass form arises from

$$y^2 + xy = x^3 + Dx^2 + 1/j \quad \text{for some} \quad D \in k(t)$$

by way of minimalization. (Here the rational functions D account for the quadratic twists.) As in [RS18, Sect. 6.2] one can apply the variable transformation $y \mapsto y + D$ to detect the inseparable quadratic base change from some other elliptic surface $S \to \mathbb{P}^1$. To see that S is rational, one can regard it as a quotient $S = X/G$, where $G \subset \text{Aut}(X)$ is generated by the translation by the 2-torsion section P. By construction, P meets all I_2 fibres non-trivially, so $(P.O) = 1$, and $\text{Fix}(G)$ consists of the fibre where O and P intersect. (In the above construction starting from a rational elliptic surface S with (12.26), this would be the fibre at the root of $a_1(t^2)$.) Hence it follows from [DK01, Theorem 3.7] that S is indeed rational. (Alternatively, one can use the fact that purely inseparable base change preserves (the index of) wild ramification to deduce that S can only have 12 fibres of type I_1 by comparison of Table 5.2 and [SSc13, Proposition 5.1]. Then it follows from Proposition 5.51 that S is rational as claimed.) □

Strata for smaller Artin invariants can be realized by arranging for suitable configurations of singular fibres; for instance, a single fibre of Kodaira type I_2 leads to $\sigma = 8$, and generally we have the following relation:

Lemma 12.33 *Let S be a general semi-stable rational elliptic surface with n singular fibres (all multiplicative) over an algebraically closed field of characteristic* 2. *The inseparable quadratic base change of S is a supersingular K3 surface of Artin invariant $\sigma = n - 3$. Conversely, any supersingular K3 surface in characteristic* 2 *of Artin invariant $n - 3$ arises in this way.*

Proof The proof of the lemma follows the same lines as that of Proposition 12.32 and the construction leading to it. Let S be a rational elliptic surface with n multiplicative fibres, located above the subset $R \subset \mathbb{P}^1$ of cardinality n. Arguing as around (8.8), we find

$$r = \text{rank}\, \text{MW}(S) = n - 4 \tag{12.28}$$

(so in particular $n \geq 4$). The inseparable quadratic base change of S is a K3 surface X by Proposition 5.51, which is unirational by construction, hence supersingular. To obtain a suitable sublattice N of $\text{NS}(X)$ we consider the reducible fibres of type I_{2m_v} and the pull-back of $\text{MW}(S)$ which induces a sublattice

$$\text{MWL}(S)(2) \hookrightarrow \text{MWL}(X).$$

It remains to consider the torsion inside $\text{MW}(X)$ which clearly contains $\text{MW}(S)_{\text{tors}}$. Since $a_1 \equiv 0$ would yield additive fibres at the zeroes of a_3, we find the 2-torsion section $(a_3/a_1, \ldots) \in \text{MW}(X)$ as in the generic case. In contrast, $\text{MW}(S)$ does not contain a non-trivial 2-torsion section by Corollary 8.32, since all fibres of S are assumed to be multiplicative.

We now compute the determinant of the rank 22 sublattice $N \subseteq \mathrm{NS}(X)$, generated by fibre components, pull-back of $\mathrm{MW}(S)$ and the given 2-torsion section. The determinant formula (6.23), restricted to N, gives

$$\det N = (-1)^r \underbrace{\det \mathrm{Triv}(X)}_{=2^n \det \mathrm{Triv}(S)} \cdot \underbrace{\det \mathrm{MWL}(N)}_{=2^r \det \mathrm{MWL}(S)} / \underbrace{|\mathrm{MW}(N)_{\mathrm{tors}}|^2}_{\geq 2|\mathrm{MW}(S)_{\mathrm{tors}}|}$$
$$\leq 2^{n+r-2} \det \mathrm{NS}(S) \overset{(12.28)}{=} -2^{2n-6}.$$

Hence $\sigma \leq n-3$. On the other hand, rational elliptic surfaces with n singular fibres, all of which multiplicative, have $n-4$ dimensional moduli by Theorem 8.39. This implies $\sigma = n-3$ for the inseparable quadratic base change of a generic rational elliptic surface with n multiplicative fibres.

But then we can argue exactly as in the proof of Proposition 12.32 to see that *every* supersingular K3 surface in characteristic 2 with Artin invariant $n-3$ arises in the same way from a rational elliptic surface with the same fibre configuration.

To see that not only a generic (or very general) rational elliptic surface, but also a general one fits the lemma, it suffices to note that for σ to drop one has to enhance the sublattice $N \subseteq \mathrm{NS}(X)$ by an additional torsion section or by some section $P \in \mathrm{MW}(X)$ such that $2P \in \mathrm{MW}(N)$ (since the fibre configuration stays fixed by assumption). Each condition is closed in moduli. □

Remark 12.34 Incidentally, the above argument also proves that any stratum of rational elliptic surfaces with multiplicative fibres only, n in number, has dimension exactly $n-4$ in characteristic 2 (cf. Remark 8.40).

We continue by investigating two specific series of examples in detail before turning to the problem of elliptic fibrations on supersingular K3 surfaces starting in Sect. 12.4.4.

12.4.3 Explicit Generators of NS(X)

Even though we know the abstract shape of $\mathrm{NS}(X)$ for any supersingular K3 surface (only depending on p and σ),

$$\mathrm{NS}(X) \cong \Lambda_{p,\sigma},$$

by [RS83], it is another problem to determine explicit generators of $\mathrm{NS}(X)$. Here we focus on the basic case of Artin invariant $\sigma = 1$ which admits a straight answer due to [Shio07a] (see also [HS12]):

Proposition 12.35 *Let the supersingular K3 surface X be given by the Shioda–Inose structure for two supersingular elliptic curves E, E' as in Example 12.29. Then* $\mathrm{NS}(X)$ *is generated by the zero section and fibre components of Inose's pencil* (12.11) *as well as classes corresponding canonically to homomorphisms $E \to E'$.*

Regarding the proof of Proposition 12.35, the key input once again consists of a Shioda–Inose structure (cf. 12.2). Indeed, since the corresponding statement for the abelian surface $E \times E'$ almost obviously holds true (since the homomorphisms induce graphs on the product $E \times E'$), this mostly requires to carefully trace back divisors through the construction (and this can be accomplished over certain non-algebraically closed fields by [HS12]).

Even with the answer of Proposition 12.35 at hand, it remains an open problem to exhibit explicit equations of the generators of $\mathrm{NS}(X)$. In the sequel, we will answer this for two series of supersingular K3 surfaces using base change of isotrivial elliptic fibrations and the functorial properties of Mordell–Weil lattices (see Sect. 6.9). The supersingular K3 surfaces arise from what Vinberg refers to as the two most algebraic (complex) K3 surfaces [Vin83] by way of reduction as in Example 12.30.

12.4.3.1 First Series: $d = -3$

We start with the singular complex K3 surface X_3 with transcendental lattice $T(X_3) \cong A_2$, i.e. of discriminant $d = -3$. In terms of Inose's pencil (12.11), a model of X_3 can be given by setting $j(E) = j(E') = 0$. That is, (12.11) specialises as

$$X_3 : y^2 = x^3 + t^5(t-1)^2 \qquad (12.29)$$

with an additional reducible fibre of Kodaira type IV at $t = 1$. For rank reasons (due to (4.15)), the trivial lattice already equals the Néron–Severi lattice:

$$\mathrm{NS}(X_3) = U \oplus 2E_8^- \oplus A_2^-.$$

This also allows us to verify the claimed shape of $T(X_3)$ by Proposition 2.41 and elementary class group theory related to positive-definite even binary quadratic forms. (These forms admit a very efficient reduction algorithm which allows for writing down representatives for all equivalence classes easily, cf. [Cox13]. For determinant at most 11, this results in exactly one class (which also implies that the corresponding imaginary quadratic field has class number one).)

Now consider the reductions of X_3. By Example 12.30, $X_3 \otimes \bar{\mathbb{F}}_p$ is supersingular for any odd $p \equiv -1 \mod 3$ (in agreement with the fact that elliptic curves with $j = 0$ are supersingular for $p \equiv -1 \mod 3$). Hence the Picard number increases to 22 upon reduction. Since the singular fibres do not degenerate (except possibly at $p = 2$ which we have excluded from our considerations), the jump in the Picard number is accounted for by sections. By Corollary 6.7:

$$\mathrm{rk}\,\mathrm{MW}(X_3 \otimes \bar{\mathbb{F}}_p) = 2.$$

We shall now not only exhibit two independent sections for any $p \equiv -1 \mod 3$ generating $\mathrm{MW}(X_3 \otimes \bar{\mathbb{F}}_p)$, but also compute $\mathrm{MWL}(X_3)$. To this end, we note that the elliptic fibration (12.29) arises from the rational elliptic surface S in Example

6.55 by a purely inseparable base change of degree p. Thus the sections pull back from S to X_3. Write $p = 6m + 5$. After minimalizing we find the following section on X_3:

$$P^* = (\beta(t-1)^{4(m+1)}/t^{2m}, (t-1)(t+1)^p/(2t^{3m})).$$

Applying the Galois action or the automorphism from Example 6.59, we obtain explicit generators of the rank two sublattice $A_2^\vee(p)$ of $\mathrm{MWL}(X_3)$ by Theorem 6.53. By Corollary 6.7, this sublattice together with the trivial lattice gives a finite index sublattice N of $\mathrm{NS}(X_3 \otimes \bar{\mathbb{F}}_p)$ whose determinant is $-3 \cdot p^2/3 = -p^2$ by (6.23). Comparing determinants, we deduce $N = \mathrm{NS}(X_3 \otimes \bar{\mathbb{F}}_p)$ as claimed.

Along the same lines, one could argue with another isotrivial elliptic fibration of X_3, this time with three fibres of type IV*. This can be given in affine normal form,

$$X_3: \quad y^2 + t^2(t-1)^2 y = x^3, \tag{12.30}$$

which even works in characteristic $p = 2$ (see [Schü08]).

12.4.3.2 Second Series: $d = -4$

A similar reasoning applies to the singular complex K3 surface X_4 with $T(X_4) = A_1^2$. This corresponds to the case $j(E) = j(E') = 12^3$. Here Inose's fibration (12.11) also degenerates as it attains two additional reducible fibres of Kodaira type I_2. Thus

$$\mathrm{NS}(X_4) = U \oplus 2E_8^- \oplus 2A_1^-,$$

giving the claimed shape of the transcendental lattice $T(X)$ as before. Note, however, that the given fibration is not isotrivial, and the two fibres of type I_2 prevent an inseparable base change construction in odd characteristic as in Sect. 12.4.3.1. Instead, we consider an alternative fibration corresponding to the embedding of the partner lattice $M = D_6$ into the Niemeier lattice with root system $\mathcal{R}(N) = D_{10} \oplus E_7^2$ from Theorem 12.21. (All elliptic fibrations on X_4 were determined in terms of the root types of the singular fibres by Nishiyama in [Nis96].) Explicitly, this can be given in affine Weierstrass form

$$X_4: \quad y^2 = x^3 - t^3(t-1)^2 x, \tag{12.31}$$

with singular fibres of type III* at $t = 0, \infty$ and I_0^* at $t = 1$ as well as the 2-torsion section $(0, 0)$. By Example 12.30, the reduction $X_4 \otimes \bar{\mathbb{F}}_p$ is supersingular for any

$p \equiv -1 \mod 4$. Here the fibration (12.31) arises from the isotrivial rational elliptic surface

$$y^2 = x^3 - t(t-1)^2 x$$

by purely inseparable base change of degree p. As before, we can thus derive explicit generators of

$$\mathrm{MWL}(X_4 \otimes \bar{\mathbb{F}}_p) \cong A_1^\vee(p)^2$$

from the rational elliptic surface. We leave the details of the computation to the interested reader who might find an ansatz as in Example 8.14 useful (compare [Bran17]).

Remark 12.36 It is quite entertaining to see how these explicit descriptions allow us to derive results on certain aspects of dynamics on the supersingular K3 surfaces in question. For instance, they admit explicit automorphisms of Salem degree 22 by [Bran17, Schü16] (i.e. the characteristic polynomial of the induced action on NS is irreducible of degree 22, a so-called Salem polynomial). It follows from Hodge theory that these automorphisms cannot lift to any smooth proper characteristic zero model by [EO15].

12.4.4 The Supersingular K3 Surface with $(p, \sigma) = (2, 1)$

We now turn to the problem of classifying all elliptic fibrations on supersingular K3 surfaces (sometimes even without sections, see Sect. 12.4.5). In these paragraphs, we give a brief account of the situation for the supersingular K3 surface of Artin invariant $\sigma = 1$ in characteristic $p = 2$ following [ElS15] (which, just like a supersingular K3 surface of Artin invariant one in any other characteristic, is unique up to isomorphism by [Ogus79]). The supersingular K3 surface of Artin invariant $\sigma = 1$ in characteristic $p = 3$ has been treated along similar lines by Sengupta in [Sen17].

In order to determine all genus one fibrations on the supersingular K3 surface X of Artin invariant $\sigma = 1$ in characteristic $p = 2$ up to isomorphism, we pursue the method pioneered by Nishiyama for complex K3 surfaces (see Sect. 12.3.2). To this end, we start with a partner lattice of $\mathrm{NS}(X)$ as in Definition 12.22. We point out that for Artin invariant $\sigma \geq 3$, such a partner lattice M cannot exist in any characteristic because the length of M would equal that of NS, which is 2σ by (12.25), exceeding the rank bound of 4 for M from Definition 12.22.

Presently, however, $\mathrm{NS}(X)$ admits the partner lattice D_4. This root lattice possesses 18 inequivalent embeddings into Niemeier lattices, as one can check from the root types in Theoren 12.21. For instance, the elliptic fibration on X induced by $X_3 \otimes \bar{\mathbb{F}}_2$ in (12.30) corresponds to $\mathcal{R}(N) = E_6^4$. Quite remarkably, with Mordell–Weil rank 2, this already attains the maximum rank for all elliptic fibrations on X, as tabulated below:

Explicit equations for each genus one fibration on X, and connections between each of them are given in [ElS15] (but we have seen some of them in this book, too, such as #1 in Example 5.38). Here we point out several noteworthy facts about the above classification. For instance, each of the above fibrations is uniquely determined on X by the type $\sum T_v$, contrary to the situation on Kummer surfaces of product type sketched in Sect. 12.3.1.

12.4.4.1 Elliptic Fibrations Versus Quasi-elliptic Fibrations

Among the genus one fibrations on X, there are both elliptic and quasi-elliptic ones (as indicated in the last column of Table 12.2). Presently, this can be decided based purely on the configuration of root lattices associated to the singular fibres; in greater generality, there is a surprising criterion given by the shape of the Mordell–Weil group which we explain briefly. The criterion builds on the fact that quasi-elliptic fibrations are extremal, i.e. the trivial lattice equals $\mathrm{NS}(X)$ up to finite index. Equivalently, the Mordell–Weil group $\mathrm{MW}(X)$ is finite, and the surface is supersingular. Note that for an a quasi-elliptic surface S over $\mathbb{P}^1$, both conditions are automatic: S is unirational and thus supersingular, while all fibres being additive (especially the generic and the general fibre) implies that any section is annihilated by 2 (Corollary 7.22). Using

Table 12.2 Genus one fibrations on X

#	$\mathcal{R}(\mathcal{N})$	$\sum T_v$	rk(MW)	Torsion	Elliptic?
1	$D_4A_5^4$	A_5^4	0	$\mathbb{Z}/3\mathbb{Z}\times\mathbb{Z}/6\mathbb{Z}$	Yes
2	D_4^6	D_4^5	0	$(\mathbb{Z}/2\mathbb{Z})^4$	No
3	$D_5^2A_7^2$	$D_5A_7^2$	1	$\mathbb{Z}/8\mathbb{Z}$	Yes
4	$D_6A_9^2$	$A_1^2A_9^2$	0	$\mathbb{Z}/10\mathbb{Z}$	Yes
5	D_6^4	$A_1^2D_6^3$	0	$(\mathbb{Z}/2\mathbb{Z})^3$	No
6	$E_6D_7A_{11}$	D_7A_{11}	2	$\mathbb{Z}/4\mathbb{Z}$	Yes
7	$E_6D_7A_{11}$	$A_3E_6A_{11}$	0	$\mathbb{Z}/6\mathbb{Z}$	Yes
8	E_6^4	E_6^3	2	$\mathbb{Z}/3\mathbb{Z}$	Yes
9	D_8^3	$D_4D_8^2$	0	$(\mathbb{Z}/2\mathbb{Z})^2$	No
10	D_9A_{15}	D_5A_{15}	0	$\mathbb{Z}/4\mathbb{Z}$	Yes
11	E_7A_{17}	$A_1^3A_{17}$	0	$\mathbb{Z}/6\mathbb{Z}$	Yes
12	$E_7^2D_{10}$	$A_1^3E_7D_{10}$	0	$(\mathbb{Z}/2\mathbb{Z})^2$	No
13	$E_7^2D_{10}$	$D_6E_7^2$	0	$\mathbb{Z}/2\mathbb{Z}$	No
14	D_{12}^2	D_8D_{12}	0	$\mathbb{Z}/2\mathbb{Z}$	No
15	E_8D_{16}	D_4D_{16}	0	$\mathbb{Z}/2\mathbb{Z}$	No
16	E_8D_{16}	$D_{12}E_8$	0	0	No
17	E_8^3	$D_4E_8^2$	0	0	No
18	D_{24}	D_{20}	0	0	No

Ito's classification of extremal elliptic surfaces over $\mathbb{P}^1$ in [Ito02], one can show the following:

Proposition 12.37 ([ElS15, Proposition 5]) *Let X be a supersingular surface admitting a genus one fibration over $\mathbb{P}^1$ with section. Assume that X is not rational. Then X is quasi-elliptic if and only if* $\mathrm{MW}(X)$ *is 2-torsion.*

As an application, one derives the last column of Table 12.2 (whose entries often also follow from the presence of certain fibre types (impossible on quasi-elliptic surfaces in characteristic 2) or Euler–Poincaré characteristic-considerations (exceeding 24 in case of an elliptic fibration). Remarkably, the size of the 2-torsion subgroup of $\mathrm{MW}(X)$ can be as big as 16 for a quasi-elliptic fibration on X. In comparison, a proper elliptic fibration can only have one non-trivial 2-torsion section in characteristic 2 by Proposition 6.33 *(vi)*, but Proposition 12.37 provides the much stronger statement that there any elliptic fibration on X (or on any supersingular non-rational elliptic surface over $\mathbb{P}^1$) has a section of order greater than two (including ∞).

12.4.4.2 Peculiar Torsion Sections

The 2-torsion sections of the proper elliptic fibrations #1, 4 and 11 are peculiar: they intersect the zero section transversally in a single point, contrary to the case for torsion sections of order relatively prime to the characteristic (Proposition 6.33 *(v)*). This can be checked either with the height formula from Sect. 6.5, applied to the full Mordell–Weil group including the torsion sections, or with the explicit equations in [ElS15], or, for #1, using Example 5.38) as shall exploit below.

Example 12.38 (*Hesse pencil revisited*) Consider the inseparable quadratic base change of the Hesse pencil from Example 5.38 in characteristic 2, together with the section $P = [s, 1, -1]$. Clearly, (P) meets (O) for $O = [0, 1, -1]$ (transversally in the fibre at $s = 0$), and indeed this choice of zero section determines P to have order two. In comparison, one finds that (P) meets every I_6 fibre at the component opposite to the one met by (O), so also the height pairing returns zero for $h(P)$ as required.

We also point out that there are cases with additive fibres whose combination with torsion sections of the given order is otherwise prevented by Proposition 6.33 *(iv)*: type IV^* with a 2-torsion section on #7, and type I_1^* with an 8-torsion section on #3.

12.4.4.3 Jacobian Genus One Fibrations

The elliptic fibration #1 on X singles out 42 rational curves on X: the 24 components of the singular fibres and the 18 sections. It is easily verified that the configuration of these 42 rational curves gives the incidence graph of the points and lines of $\mathbb{P}^2(\mathbb{F}_4)$. This gives another explanation of a large finite subgroup of the automorphism group (see [DK03])

$$\mathrm{PGL}_3(\mathbb{F}_4) \rtimes \mathbb{Z}/2\mathbb{Z} \hookrightarrow \mathrm{Aut}(X).$$

From the classification of genus one fibrations on X we can extract information about specific subgraphs of the incidence graph such as the following:

Theorem 12.39 *The incidence graph of points and lines in* $\mathbb{P}^2(\mathbb{F}_4)$ *does not contain any cycle of length* 14 *or* $2n$ *with* $n \geq 10$.

The proof of Theorem 12.39 is based on Theorem 11.24 as any cycle in the incidence graph of $\mathbb{P}^2(\mathbb{F}_4)$ will induce, via the identification with rational curves on fibration #1, a genus one fibration on X. For details of the proof the reader is referred to [ElS15]. Here we only want to point out a central ingredient of the proof which is of great relevance for our issues: in order to apply the classification, we need to know that any cycle of rational curves on X does indeed give rise to a *jacobian* genus one fibration on X by Theorem 11.24, i.e. to a genus one fibration *with section*. Presently, this is guaranteed by the following result from [ElS15]:

Theorem 12.40 *Any genus one fibration on a supersingular K3 surface of Artin invariant 1 admits a section.*

Proof Let Y denote the supersingular K3 surface of Artin invariant 1 in characteristic p. Given a genus one fibration, we denote the class of a fibre by F and the multisection index by $m \in \mathbb{N}$. That is,

$$m\mathbb{Z} = \{D.F, \quad D \in \mathrm{NS}(Y)\}.$$

Then the fibration has a section if and only if $m = 1$. Assume $m > 1$. Then $F/m \in \mathrm{NS}(Y)^\vee$, and in fact

$$N := \langle \mathrm{NS}(X), F/m \rangle \quad \text{is an even integral lattice,}$$

since $F^2 = 0$. Presently, F is indivisible in $\mathrm{NS}(Y)$ since there cannot be any multiple fibres by the canonical bundle formula (Theorem 5.44). Hence $\mathrm{NS}(Y)$ has index m in N from which we infer

$$\mathrm{disc}(N) = \mathrm{disc}(\mathrm{NS}(X))/m^2.$$

Since the determinant of N is an integer, it follows at once that $m = p$ and N is an even unimodular lattice. Then the signature $(1, 21)$ of N gives the required contradiction by Theorem 2.9. □

Remark 12.41 The above argument may be applied to any elliptic surface without a section being known as long as the fibre class is indivisible in NS (so Enriques surfaces, for instance, are excluded). In detail, it works for any K3 surface Y where we know a finite index sublattice $L \subset \mathrm{NS}(Y)$ such that:

- L has determinant of absolute value p^2 for some prime p and $\rho(Y) \neq 2, 10, 18$ (since an even unimodular lattice of signature (s, t) satisfies $s \equiv t \mod 8$ by Theorem 2.9), or
- L has determinant of absolute value $4n$ for some odd square-free number $n \in \mathbb{N}$ and $\rho(Y)$ is odd (since an even lattice of odd rank has even determinant).

The reader may want to compare Keum's result for complex elliptic K3 surfaces Y [Keu99] which states in the analogous notation of the proof of Theorem 12.40 that the jacobian of Y has $\mathrm{NS} = N$.

12.4.5 Duality of Néron–Severi Lattices

Recently, Kondo and Shimada discovered an interesting duality for supersingular K3 surfaces. Based on the results of [RS83], they derived the following structure for the Néron–Severi lattices $\Lambda_{p,\sigma}$ of supersingular K3 surfaces of Artin invariant σ in characteristic p:

Lemma 12.42 *Suppose that $\sigma + \sigma' = 11$. Then there is the following duality between $\Lambda_{p,\sigma}$ and $\Lambda_{p,\sigma'}$:*

$$\Lambda_{p,\sigma'} \cong \Lambda_{p,\sigma}^{\vee}(p).$$

Kondo and Shimada draw surprising consequences from Lemma 12.42. Here we mention their main results concerning genus one fibrations. Recall from Theorem 11.24 how a divisor class $D \in \mathrm{NS}(X)$ on a K3 surface X with $D^2 = 0$ induces a genus one fibration $X \to \mathbb{P}^1$. More precisely, D is a fibre class if and only if it is nef and primitive. Using this characterization, one can try to determine the number of non-isomorphic genus one fibrations (as in Sect. 12.3.1), or purely on the lattice level, the number of fibrations modulo $O^+(\mathrm{NS}(X))$. This leads to the notion of *lattice equivalence classes of genus one fibrations* on X which are collected in the finite set $\mathbb{E}(X)$ (with finiteness following from Lemma 2.43). By what we have seen above, this set only depends on p and σ, so we shall simply write $\mathbb{E}_{p,\sigma}$. We are now ready to state some of the consequences of the duality in Lemma 12.42 found in [KonS14]:

Theorem 12.43 *Suppose that $\sigma + \sigma' = 11$. Then there is a canonical bijection between $\mathbb{E}_{p,\sigma}$ and $\mathbb{E}_{p,\sigma'}$.*

This bijective correspondence can be used to study the genus fibrations in more detail:

Theorem 12.44 *Suppose that $\sigma + \sigma' = 11$. A class of genus one fibrations in $\mathbb{E}_{p,\sigma}$ admits a section if and only if the corresponding class of genus one fibrations in $\mathbb{E}_{p,\sigma'}$ does not admit a section.*

Theorem 12.40 thus leads to the following surprising result:

Corollary 12.45 *A genus one fibration on a supersingular K3 surface of Artin invariant $\sigma = 10$ does not admit a section.*

In fact, the techniques from [KonS14] are explicit enough to classify the types of all genus one fibrations on a supersingular K3 surface of Artin invariant $\sigma = 10$ in characteristic 2 and 3 with input coming from Table 12.2 and the corresponding result in [Sen17]. The results can be found in [KonS14, Sect. 4]; they show in particular that a supersingular K3 surface of Artin invariant $\sigma = 10$ admits exactly one quasi-elliptic fibration in characteristic 2 and none in characteristic 3. Here we should emphasize that supersingular K3 surfaces with $\sigma = 10$ (or big Artin invariant in general) are not very accessible from the explicit viewpoint, and only in characteristic $p = 2$ they are really well understood (see Theorem 12.31).

Assuming a different viewpoint, there is a clear conceptual interpretation. Namely, thanks to recent work of [Lie15], the supersingular K3 surfaces of given Artin invariant form degree p torsors over the supersingular K3 surfaces with Artin invariant one less (which may be inseparable or not, cf. [BL19]). On the level of genus one fibrations, the torsors feature as multisections of degree p (and there are no proper sections).

Admittedly, carrying out these constructions explicitly may be a different business.

We conclude with an application to the problem of finding (supersingular) elliptic K3 surfaces with maximum possible Mordell–Weil rank (the positive characteristic analogue of Lemma 11.39).

Theorem 12.46 *Let k be an algebraically closed field of characteristic $p > 0$. Then there exists an elliptic K3 surfaces of Mordell–Weil rank 20 over k.*

Proof In characteristic $p > 3$, this is Corollary 12.15. In characteristics 2, 3, we infer from [KonS14, Tables 4.1, 4.2] that any supersingular K3 surface with Artin invariant 10 admits a genus one fibration (or, in fact, plenty) without any singular fibres. Using Corollary 12.45, the Jacobian is an elliptic supersingular K3 surface with Artin invariant 9 with the same singular fibres (i.e. no reducible fibres). Hence, by Corollary 6.7, the Jacobian has Mordell–Weil rank 20 as claimed. □

Historical Notes

The notion of isogeny of K3 surfaces from Definition 12.8 may be a bit misleading, since unlike for abelian varieties there usually need not be a dual isogeny (unless we are in a sandwich situation as in Sect. 12.2.1). On the other hand, it is quite remarkable that as little as a dominant rational map suffices to read off symmetries between the K3 surfaces. Note that this can simply be regarded as a consequence of basic Hodge theory, valid for any algebraic surfaces of geometric genus one.

The possibility to work with different elliptic fibrations on a given K3 surface can be very useful. For instance, in the classification of K3 surfaces with finite automorphism group, this implies that every elliptic fibration has finite Mordell–Weil rank; this was heavily exploited by Nikulin, see [Nik79, Nik85, Nik14]. Remarkably, analogous arguments carry over to Enriques surfaces (where they prove to be even more restrictive [Kon86]). Other applications include dynamics (where different fibrations can foster the construction of automorphisms of positive entropy) and rank problems; in the latter direction, instead of applying successive quadratic twists as in Sect. 12.2.7, one may switch to an alternative high rank fibration on the same surface, as we will discuss in Sect. 13.1.4.

With the Tate conjecture confirmed for K3 surfaces over finite fields, there fortunately is no need anymore to distinguish the two notions of supersingularity, due to Artin and one of us. Elliptic fibrations played a central role in several steps of the proof, notably in the initial paper by [ASD73], but also lately in Maulik's arguments [Mau14] which paved the way to the final solution.

For us, the starting point to consider this circle of ideas arose from elliptic modular surfaces (as studied in [Shio72]). For these, such as the surface $S(\Gamma(4))$ with full level-4 structure, i.e. for the full modular group $\Gamma(4)$, the Tate conjecture predicted that the Mordell–Weil rank would go up in certain characteristics. This was in stark contrast with the fact that the Mordell–Weil rank was always zero over $\mathbb{C}$ (so the elliptic fibrations are extremal as in Sect. 8.8). Rather surprisingly, this means that the underlying moduli problem assumes a substantially different flavour in certain characteristics (which is still a kind of mystery to us if we view this purely from the moduli viewpoint). Indeed, this left us rather sceptical about the Tate conjecture at first, and it was only gradually that we got convinced that it might in fact hold true (see [Shio73]). In this direction, it was instrumental to notice that the Tate conjecture indeed holds for $S(\Gamma(4))$ not only in characteristics $p \equiv 1 \mod 4$ (where there are no additional sections), but also first for $p = 3$ and subsequently for all $p \equiv 3 \mod 4$ (where the Mordell–Weil rank jumps to two, so that this K3 surface has $\rho = 22$, cf. [Shio75]).

Research on supersingular K3 surfaces started with Artin's seminal paper [Art74a] introducing the stratification of the moduli space. It has led to so many influential findings of Ogus and others, like Ogus [Ogus79, Ogus83, Shio79], only some of which we could reflect in this book. It should be clear from Sect. 12.4 that supersingular K3 surfaces remain a very active area offering many interesting new developments.

Chapter 13
Ranks and Sphere Packings

We conclude this book with another circle of constructions and applications which played a central role for the genesis of Mordell–Weil lattices: the rank problem for elliptic curves over a given field and the sphere packing problem. Before going into the details, we give a brief overview.

Concerning the rank problem, one fixes any field K—for us, this will usually be

$$\mathbb{Q},\ \mathbb{C}(t),\ \mathbb{Q}(t),\ \text{or } \mathbb{F}_q(t). \tag{13.1}$$

If the base field is uncountable, most instructively for $\mathbb{C}(t)$, then of course we have to rule out trivial cases such as a constant elliptic curve over k, considered as base change over $k(t)$, in order for rank statements and the like to make sense in the first instance (as we did before in Convention 5.10). Given this convention, it is natural to investigate the rank problem:

Problem 13.1 Given K, is the rank of all (non-constant) elliptic curves over K bounded? I.e.,

$$\#\{\text{rk } E(K);\ E \text{ elliptic curve over } K\} < \infty?$$

Of course, one can ask the same question for

$$\sup\{\text{rk } E(K);\ E \text{ elliptic curve over } K\}, \tag{13.2}$$

if this exists at all. In a way, this problem is even closer to what we have seen before in this book, since in case $K = k(t)$ or generally $K = k(C)$ for some curve C, it naturally leads to the quest for elliptic surfaces

$$S \to C \tag{13.3}$$

of high Mordell–Weil rank over k. The relevant background information here is the following: as soon as the base curve C is rational or elliptic of positive rank over k (and

M. Schütt and T. Shioda, *Mordell–Weil Lattices*, Ergebnisse der Mathematik und ihrer Grenzgebiete. 3. Folge / A Series of Modern Surveys in Mathematics 70,
https://doi.org/10.1007/978-981-32-9301-4_13

k is infinite, for instance $k = \mathbb{Q}$), there are specialization arguments originating from Néron and improved by Tate and Silverman which ensure that (13.3) has infinitely many fibres over k whose rank does not drop compared to the rank of the generic fibre over $k(C)$. In fact, once one has an elliptic surface of high rank, one may aim for the opposite effect and look for special fibres of (13.3) whose rank goes up, and ideally quite far. We will see in Sect. 13.1.4 how this enabled Elkies to push the current record for the rank Problem 13.1 over $\mathbb{Q}$ to 28.

For the first three fields from (13.1), $\mathbb{Q}$, $\mathbb{C}(t)$, $\mathbb{Q}(t)$, the problem is open as of today (despite some announcement of Lapin [Lap64], see [Scho90]), but it is still very instructive to study what is known. We will review the historical origins, notably Néron's pioneering work on the subject, and then explain how Mordell–Weil lattices, together with K3 surfaces and Delsarte surfaces, enter the picture.

On the contrary, for $\mathbb{F}_q(t)$, Problem 13.1 is known to have a negative answer; that is, the rank is not bounded! Already over $\mathbb{F}_p(t)$, for any odd prime p, this goes back to the ground-breaking work of Tate and Shafarevich [ShT67] (see Sect. 13.3.1). Mordell–Weil lattices (and bits of surface theory) allowed one of us to give an alternative, somewhat more explicit and geometric proof over $\mathbb{F}_{p^2}(t)$ in the early 1990s [Shio91a], again for any prime p (including $p = 2$). Subsequently, Ulmer extended this line of argument to $\mathbb{F}_p(t)$ [Ulm02]. We will sketch the details of this geometric approach in Sects. 13.3.2, 13.3.3.

What makes these developments even more remarkable is that, beyond being of their very own interest, they were instrumental for the definition of Mordell–Weil lattices. Namely, it was a simultaneous discovery due to Elkies and one of us in the late 1980s that the elliptic curves involved in the construction of Tate and Shafarevich (and similar elliptic curves) tend to give rise to dense lattices via their rational points—and these are what we call now Mordell–Weil lattices. In fact, around that time, the development of Mordell–Weil lattices formed a key technique to construct lattice packings of new record densities (see [CS99, p. xvi]). We will review a small portion of this development in Sect. 13.4.

13.1 Rank Problem over $\mathbb{Q}$ and $\mathbb{Q}(t)$

The most elementary incarnation of Problem 13.1 might concern the field $\mathbb{Q}$ (or, slightly more involved, any other fixed number field). As it stands, the problem is wide open, but it fits very naturally into the theory of Mordell–Weil lattices, first motivating the latter before it actually gained genuinely new input from the fruitful combination of K3 surface theory and Mordell–Weil lattices, as sketched in the previous two chapters. We start by reviewing the innovative, and now classical, ideas of Néron [Ner54], dating back more than 60 years. For an alternative account, see for instance Serre [Ser97, Sect. 11] or [Top87].

13.1.1 An Effective Version of Néron's Method

Let us start with eight distinct points

$$P_1, \ldots, P_8 \in \mathbb{P}^2_k$$

such that no three are on a line and no six on a conic (i.e. in slightly less than general position) for some field k (which we could for simplicity assume to be $\mathbb{Q}$). Then there is a pencil of cubics over k through these points which corresponds to an elliptic surface

$$S \to \mathbb{P}^1,$$

or equivalently to an elliptic curve over $K = k(t)$ with eight special K-rational points (the quasi-elliptic case in characteristic 2 and 3 can be ruled out as in the proof of Proposition 7.23). By assumption, the pencil has only irreducible members, even over $\bar{k}$, and correspondingly, S has no reducible fibres unless the ninth base point of the pencil, P_9, coincides with one of the other base points P_i $(i = 1, \ldots, 8)$. Hence, by Theorem 7.4, S has Mordell–Weil rank 8 over $\bar{k}$.

Presently, this can be improved by noting that the Mordell–Weil group is generated by sections over k. This holds because with the first eight base points $P_1, \ldots, P_8$ defined over k by construction, so is P_9. Taking any of these points as zero section of S for the group law, the remaining eight generate the Mordell–Weil group up to index three (compare Theorem 7.26). Note that due to the trivial Galois action on a finite index sublattice of the Mordell–Weil lattice, the remaining section needed to generate the full Mordell–Weil lattice does not require any base extension.

In summary, we have constructed an elliptic curve over K together with explicit generators for the Mordell–Weil group (up to finite index).

In order to push the Mordell–Weil rank further up, one can follow Néron and use specialization. Often, however, it is much more beneficial to endow the cubic pencil with additional structure. This approach was pursued by one of us in [Shio91e]; explicitly, we imposed that the pencil contain a cuspidal curve. Recall that the smooth locus of the cuspidal curve admits the structure of the additive group $\mathbb{G}_a$ with addition given by Lemma 9.8 *(i)*.

In practice, one may first fix the cuspidal curve C over k, given affinely by

$$y^2 = x^3.$$

Recall from Sect. 3.4.1 that the smooth K-rational points on C correspond to $K = \mathbb{G}_a(K)$ by way of the group bijection

$$\mathbb{G}_a \ni u \mapsto Q(u) = (u^{-2}, u^{-3}), \tag{13.4}$$

where $u = 0$ is mapped to the point O at infinity and three points $Q(u_i)$ for $u_i \in \mathbb{G}_a (i = 1, \ldots, 3)$ are indeed collinear if and only if

$$u_1 + u_2 + u_3 = 0.$$

We continue by fixing eight distinct k-rational smooth points $Q_0, \ldots, Q_7$ on C no three of which are collinear. Writing $Q_i = Q(u_i)$ by way of (13.4), we set

$$u_0 = 0$$

and require

$$\sum_{i=1}^{3} 1/u_i = 0. \tag{13.5}$$

The latter condition is equivalent to the following fact which can be checked by direct calculation: let L_i denote the unique line through Q_i which meets C tangentially in another point; then (13.5) is satisfied if and only if the lines $L_i (i = 1, 2, 3)$ intersect at a single point P_0 as we will use shortly. The eight points determine a unique pencil of cubics

$$S : \quad F_0 + t\,(y^2 - x^3) = 0,$$

including the cuspidal curve C (at $t = \infty$); here F_0 may be taken as any cubic polynomial through $Q_0, \ldots, Q_7$ which is not a scalar multiple of the equation of C. Let $Q_8 = Q(u_8)$ denote the ninth base point of the pencil. One easily checks that u_8 is determined by the equality

$$\sum_{i=0}^{8} u_i = 0.$$

We shall now assume the following conditions (open in the parameter space):

- $Q_i \neq Q_8$ for all $i = 0, \ldots, 7$ (so that all base points are distinct);
- no three base points are collinear.

From Lemma 7.17 we draw the central conclusion for the cubic pencil S, phrased in terms of the induced genus one fibration

$$S \to \mathbb{P}^1.$$

Conclusion 13.2 *S is an elliptic surface without reducible fibres.*

It is a direct consequence of Conclusion 13.2, and the rank formula (7.1) that S has Mordell–Weil rank 8. Indeed, once a base point is chosen as zero for the group law (say $Q_0 = O$), the other base points generate a subgroup of $\mathrm{MW}(S)$ of index 3 by Theorem 7.26.

It is almost obvious how this construction already gives rise to

$$\sup\{\mathrm{rk}\, E(k);\ E \text{ elliptic curve over } k\} \geq 9$$

over global fields k. Namely, every point $P \in \mathbb{P}^2_k$ (other than the fixed base points) lies on a unique member of the pencil, and usually the point P should be independent of the base points $P_1, \ldots, P_9$ on this elliptic curve.

Raising the rank (much) further, however, requires some new ideas. Here we continue to follow the approach pursued in [Shio91e] which relies on the tangents L_i to the base points Q_i for $i = 1, 2, 3$. Recall our assumption (13.5) ensuring that the three lines meet in a single point P_0. By construction, each line L_i defines a bisection of the elliptic fibration $S \to \mathbb{P}^1$ given by the cubic pencil; this leads to a base change with an additional section as discussed around (5.18). Concretely, the bisection is split on a quadratic extension $K_i = K(R_i)$ of K which is obtained by adjoining the coordinates of either of the conjugate intersection points R_i, R_i' with the generic fibre. By the choice of L_i, the extension K_i/K ramifies at $t = \infty$ and at a second value $t_i \in k$. Here is the summary what we have obtained so far:

Lemma 13.3 *Fix an $i \in \{1, 2, 3\}$. Attached to the elliptic curve C_t/K_i, there is an elliptic K3 surface $S_i \to \mathbb{P}^1$ with*

- IV*-fibre at $t = \infty$ and*
- *Mordell–Weil rank at least 9.*

Fixing $Q_0 = O$ as zero for the group law, there are nine independent sections given by $Q_1, \ldots, Q_8$ and R_i.

Proof All the statements should be clear except for the rank estimate. For the latter we compute the height pairing of the given sections. For the Q_j, it suffices to use Theorem 7.26 and pull-back to find

$$\langle Q_j, Q_l \rangle = \begin{cases} 4, & j = l, \\ 2, & j \neq l. \end{cases}$$

In $\mathbb{P}^2$, the line L_i meets Q_i, but none of the other Q_j. Hence R_i meets Q_i on S_i transversally in a single point, and we obtain

$$\langle R_i, Q_j \rangle = \begin{cases} 1, & i = j, \\ 2, & i \neq j. \end{cases}$$

Finally, we find the height of R_i, $h(R_i) = 2\chi(\mathcal{O}_S) = 4$, since R_i meets neither O nor the non-identity components of the singular fibres at $t = \infty$ or possibly t_i. Thus we find the determinant of the height pairing to be $2^9 \cdot 3^2$. □

Having found three elliptic curves of rank at least 9 over the fields K_i, we are going to throw them all together by considering C_t over the composite field

$$k(B) = K(R_1, R_2, R_3). \tag{13.6}$$

Here B is a curve of genus 1 determined by the ramification data of the K_i. For instance, this can be seen as follows. The intermediate field $K(R_1, R_2)$ is still the function field of $\mathbb{P}^1$, corresponding to the degree 4 morphism

$$\mathbb{P}^1 \to \mathbb{P}^1 \tag{13.7}$$

ramifying exactly at t_1, t_2, ∞ with ramification index 2 each. The final composition step $k(B)/K(R_1, R_2)$ then corresponds to the degree 2 map

$$B \to \mathbb{P}^1$$

ramifying exactly at the four pre-images of t_3 under (13.7), thus B has genus 1 as claimed. Note in this context that

$$\bar{k}(B) = \bar{k}(\sqrt{t - t_1}, \sqrt{t - t_2}, \sqrt{t - t_3});$$

the corresponding statement over k holds up to scaling with roots of k-constants, compare (13.11).

Lemma 13.4 *The elliptic curve C_t has rank at least* 11 *over* $k(B)$.

Proof We argue with the elliptic surface

$$X \to B$$

which is the base change of the rational elliptic surface $S \to \mathbb{P}^1$ by the composite degree 8 morphism $B \to \mathbb{P}^1$. Most entries in the height pairing can be obtained from the S_i in Lemma 13.3 and its proof by scaling according to Theorem 6.53. For the remaining three entries depending on the intersection numbers of R_i and $R_j (i \neq j)$, we can work on the elliptic surface $S_{ij} \to \mathbb{P}^1$ arising from S by the base change by (13.7), applied to the intermediate field $K(R_i, R_j)$. Since the lines L_i, L_j meet in $\mathbb{P}^2$, and each splits into two sections on S_{ij}, the sections R_i, R_j meet transversally in a single point on S_{ij}. Hence

$$\langle R_i, R_j \rangle_{S_{ij}} = 3,$$

and on S the height pairing returns $\langle R_i, R_j \rangle = 6$. The resulting 11×11 matrix has full rank and determinant $2^{33} \cdot 3^2$. □

Having established a family of elliptic curves of rank 11 over $k(B)$ for the genus 1 curve B, it remains to verify (or rather choose the Q_i resp. u_i in such a way) that this gives rise to infinitely many elliptic curves over k with rank at least 11. To this end, we assume that k is a number field and appeal to specialization results of Silverman [Sil83] and Tate [Tat83] which generalize Néron's original arguments from [Ner54] to the extent that it only requires that

$$B(k) \text{ is an infinite set.} \tag{13.8}$$

Obviously, this requirement will always be met over some small extension of k where B becomes an elliptic curve of positive rank. Presently, however, it is not hard at all to see that (13.8) should almost always hold. For this purpose, we interpret B as a fibre product

$$B = L_1 \times_{\mathbb{P}^1} L_2 \times_{\mathbb{P}^1} L_3$$

corresponding to the degree 8 field extension (13.6). We claim that generally this endows B with eight k-rational points. To see this, consider the unique member C_{t_0} of the cubic pencil containing the intersection point P_0 of the three lines L_1, L_2, L_3. The assumption

$$t_0 \neq t_1, t_2, t_3 \tag{13.9}$$

ensures that each line L_i intersects C_{t_0} in three distinct points:

$$L_i \cap C_{t_0} = \{Q_i, P_0, P_i\}.$$

Here, since both Q_i and P_0 are k-rational, so is P_i. Over t_0, the curve B is thus equipped with eight k-rational pre-images, given by

$$(P_0, P_0, P_0), (P_0, P_0, P_3), \ldots, (P_1, P_2, P_3) \in B = L_1 \times_{\mathbb{P}^1} L_2 \times_{\mathbb{P}^1} L_3.$$

To see (13.8), it suffices to verify that these eight points generate a subgroup of $B(k)$ of infinite order. In explicit examples, one could thus continue by determining directly whether the points only produce torsion points or not. However, it turns out to be much more practical to work right away with a Weierstrass equation for B. This is obtained by realizing that K contains as a subfield the function field $k(B')$ of another elliptic curve B' given in Legendre form

$$B': \quad s^2 = (a_1t + b_1)(a_2t + b_2)(a_3t + b_3). \tag{13.10}$$

Here the $a_i, b_i \in \mathbb{Z}$ are chosen in such a way that

$$K(R_i) = K(\sqrt{a_it + b_i}), \tag{13.11}$$

i.e. $t_i = -b_i/a_i$, but a_i, b_i are *not* necessarily coprime. By construction, there is an isogeny of degree 4 from B to B'. T. Ekedahl pointed out that due to the special shape of the field extensions in question, in particular since $k(B)/K$ is a Kummer extension of exponent 2, the isogeny cannot be cyclic. Hence it can only be multiplication by 2, and B and B' are in fact isomorphic. At any rate, B' contains a rational point at $t = t_0$, say

$$b'_0 = (t_0, s_0),$$

since at t_0 each factor on the RHS evaluates as a perfect square in k corresponding to the two intersection points P_0, P_i of C_{t_0} and L_i. Thus all we require is that

$$b_0' \in B(k) \text{ has infinite order.} \tag{13.12}$$

(Note that this implies (13.9) since otherwise b_0' would be 2-torsion.)

In summary, we have worked out a rather elementary way to find elliptic curves over number fields, and in particular over $\mathbb{Q}$, in abundance with rank at least 11. More precisely, the specialization results of Silverman [Sil83] and Tate [Tat83] show that the subset of points in $B(k)$ where the rank of the fibre of $X \to B$ drops is in fact finite, so as soon as $B(k)$ is infinite, for instance when b_0' is not torsion, then we derive infinitely many elliptic curves over k with rank at least 11.

For an explicit example, see Shioda [Shio91e, Sect. 7] where the above idea is worked out for the numerical data

$$u_0 = 0, \quad u_1 = 1, \quad u_2 = 2, \quad u_3 = -\frac{2}{3},$$

$$u_i = i - 1 \ \ (4 \leq i \leq 7), \quad u_8 = -\frac{61}{3}.$$

13.1.2 Mestre's Method

From the purely geometric viewpoint, this was the end of the rank story for a while. The next big progress was achieved when more algebraic considerations entered due to Mestre. The following paragraphs give a brief account of the ideas involved, and the progress on Problem 13.1 made, before we will turn to geometry again.

Mestre's main result in the 1991 paper [Mes91] was to push the rank record over $\mathbb{Q}(t)$ to 11 (compared to rank 10 over $\mathbb{Q}(t)$ and rank 11 over $\mathbb{Q}(B)$ using Néron's method). Roughly, his arguments went as follows: Given the input consisting of 12 district points $x_i \in \mathbb{Q}$ and polynomials $A_2, A_3 \in \mathbb{Q}[x]$ of degree at most 2 resp. 3, we will consider the plane cubic curve C, given affinely as

$$C: \quad y^3 + A_2(x)y + A_3(x) = 0.$$

Then one asks for another polynomial $R \in \mathbb{Q}[x]$, this time monic and of degree 4, such that its graph in $\mathbb{A}^2$,

$$y = R(x),$$

intersects C at the 12 points

$$P_i = (x_i, R(x_i)).$$

Fixing any of the 12 points as origin on the elliptic curve C, one would expect to obtain rank 11 (which actually works). It thus remains to work out when the above situation persists, that is

$$\prod_{i=1}^{12}(x - x_i) = R(x)^3 + A_2(x)R(x) + A_3(x). \tag{13.13}$$

Expanding and comparing coefficients starting from x^{12}, the x_i (or rather their elementary symmetric polynomials s_i) determine $R(x)$ uniquely. Likewise we can recover A_2 and A_3 from the coefficients starting from x^6, so the problem amounts to the coefficients of x^7 in (13.13) to match:

$$s_5 - \frac{2}{3}s_1s_4 + \frac{5}{9}s_1{s_2}^2 - \frac{40}{81}s_2s_1^3 + \frac{22}{243}s_1^5 + \frac{5}{9}{s_1}^2s_3 - \frac{2}{3}s_3s_2 = 0. \tag{13.14}$$

Thus one is led to search for positive-, ideally high-dimensional rational subvarieties of the zero locus of (13.14), so admittedly geometry does enter after all. Subsequently, the rational varieties should be checked to preserve the (conjectural/naive) independence of the $\mathbb{Q}$-rational points P_i on C. As it happens, the most obvious rational subvariety interferes with independence (but instead introduces a 2-torsion point which is thus of its own interest), but Mestre found a three-dimensional subvariety parametrizing elliptic curves over $\mathbb{Q}(t)$ with rank at least 11. In later years, this was also the starting point for the efforts of Nagao, Mestre and Kihara to push the record rank over $\mathbb{Q}(t)$ as far as 14, but we will not discuss this here.

Regarding high rank specializations over $\mathbb{Q}$, machine-aided computations enabled a variety of mathematicians starting from Mestre and Nagao to raise the rank up to 24, attained by a curve due to Martin and McMillen. It should be pointed out, though, that all these computations used Mestre's original three-dimensional family with its symmetries as opposed to the later record holders over $\mathbb{Q}(t)$.

13.1.3 Elkies' Use of Elliptic K3 Surfaces

This was the end of the rank story over $\mathbb{Q}$ until Elkies proposed to use specific features of elliptic K3 surfaces, thus replacing the rational elliptic surfaces which were the basis for Néron's construction. With the Mordell–Weil rank no longer bounded by 8, but by 18, this would give ample room for improvement. We first review some of the general results over $\mathbb{C}$ before concentrating on $\mathbb{Q}$ as the ground field.

13.1.3.1 Complex K3 Surfaces of Mordell–Weil Rank 18

It seems that Cox was the first to realize in [Cox82] that the Torelli theorem implies the existence of complex elliptic K3 surfaces attaining the maximum possible Mordell–Weil rank of 18 (cf. Lemma 11.39). While this result was not explicit, we have seen how such surfaces may arise as base changes from Inose's pencils in Sect. 12.2.6. The first explicit examples, however, seem to be due to Nishiyama who used embeddings into Niemeier lattices à la Sect. 12.3.2 to prove the following result in [Nis95]:

Proposition 13.5 *If the complex K3 surface X has the transcendental lattice*

$$T(X) = \begin{pmatrix} 2 & 1 \\ 1 & 10n(17n+3) \end{pmatrix}$$

for some $n \in \mathbb{N}$, then X admits an elliptic fibration of Mordell–Weil rank 18.

For those who still do not find this explicit enough, we refer to a paper of Chahal, Meijer and Top [CMT00] originating from work of Kuwata (cf. [Kuw00]) which produces 18 independent sections in the Mordell–Weil group of the elliptic K3 surface given by the affine Weierstrass form

$$y^2 = x^3 - 27(t^6 - 11 - 1/t^6).$$

Later on, one of us worked out another example in detail in [Shio07b], this time with a full set of generators of the Mordell–Weil group (arising from two rational elliptic surfaces), which we shall briefly put into context.

Example 13.6 Start with an elliptic K3 surface X given by a similar Weierstrass form

$$X: \quad y^2 = x^3 + t^5 - 11 - 1/t^5.$$

Note how this arises by a cyclic base change of degree 5 from the K3 surface associated by means of the Shioda–Inose structure with the two isogenous CM-elliptic curves with j-invariant $j = 0$ and $j' = 2^4 3^3 5^3$ (see [Sil94, Appendix A.3]) for a table of all CM-elliptic curves over $\mathbb{Q}$). That is,

$$X = X^{(5)}(0, 2^4 3^3 5^3)$$

in the notation from Sect. 12.2.6, and the transcendental lattice of X can be computed as

$$T(X) = \begin{pmatrix} 20 & 10 \\ 10 & 20 \end{pmatrix}.$$

As an illustration of the ideas behind Proposition 13.5, we explain how the maximal rank elliptic fibration arises lattice-theoretically from the Kneser–Nishiyama technique. We start with the partner lattice L which can be taken as the unique (up to isometry) even integral index 3 overlattice of

$$L_0 = A_2 \oplus A_2 \oplus A_2(10).$$

The lattice L_0 embeds into the Niemeier lattice N with root lattice $R(N) = A_2^{12}$ by picking two orthogonal summands of the root lattice and adding up generators of the remaining 10 for $A_2(10)$. Explicitly, if e_i, f_i denote the standard generators of the ith summand A_2 in $R(N)$, then the embedding of L_0 into $R(N)$ can be given as

$$L_0 = \langle e_1, f_1 \rangle \oplus \langle e_2, f_2 \rangle \oplus \left\langle \sum_{i=3}^{12} e_i, \sum_{i=3}^{12} f_i \right\rangle.$$

Consider the auxiliary vector

$$v = \sum_{i=1}^{12} e_i - \sum_{i=1}^{12} f_i.$$

By making compatible choices for e_i and f_i in defining the embedding $L_0 \hookrightarrow R(N)$, we may assume that this vector not only generates L as an overlattice of L_0, but also that v is 3-divisible in N (since N is unimodular). Hence we can extend the given embedding $L_0 \hookrightarrow R(N)$ to a primitive embedding $L \hookrightarrow N$. By construction, there are no roots in $L^\perp \subset N$.

Returning to the general theme of (complex) K3 surfaces with an elliptic fibration of Mordell–Weil rank 18, we note that such a surface X necessarily is singular in the terminology of Definition 11.3, i.e. its Picard number attains the maximum

$$\rho(X) = h^{1,1}(X) = 20$$

by Corollary 6.7. It was first noted in [SI77], as a consequence of Shioda–Inose structures and classical CM theory of elliptic curves, that singular K3 surfaces always admit models over some number field (see (12.11)). Based on the work by Inose [Inos78], this was much later improved in [Schü07b] to models over (a certain "diagonal" subfield of) the ring class field $H(d)$ (where d is the determinant of $\mathrm{NS}(X)$ as explained in Example 12.30; see [Cox13] for relevant background on class fields). Remarkably, there even is a model of X with a set of generators of $\mathrm{NS}(X)$ defined over $H(d)$ (again, the proof of this result in [HS12] essentially is an application of Shioda–Inose structures and Mordell–Weil lattices). These results are far from being sharp, as we shall explore in the next section by specializing the problem to the ground field $\mathbb{Q}$.

13.1.3.2 Mordell–Weil Rank 18 Over $\mathbb{Q}$?

Since we care about maximal Mordell–Weil rank, the following question comes very naturally:

Problem 13.7 Is there an elliptic K3 surface X with Mordell–Weil rank 18 over $\mathbb{Q}$ (i.e. with $\mathrm{MW}(X)$ of rank 18 generated by sections defined over $\mathbb{Q}$)?

Skipping a little bit of the history of this problem between the authors (which can be extracted from [Shio05a, ST06]), we directly jump to Elkies' announcement at a conference in Nagoya in 2007 that Problem 13.7 has a negative answer. For lack of better reference, we briefly elaborate on an alternative argument from [Schü10]:

Theorem 13.8 *Let X be a singular K3 surface of discriminant d. If* $\mathrm{NS}(X)$ *has generators defined over some field K, then $K(\sqrt{d})$ contains the ring class field $H(d)$.*

The main idea from [Schü10] is to employ the Artin–Tate conjecture, a refinement of the Tate conjecture [Tat66b] to be discussed around (13.32), in order to find restrictions on the discriminant. Paired with some classical class group theory, Theorem 13.8 can be deduced using the modularity of X (whose zeta function is associated with some Hecke character by [SI77], see also [Liv95] for the case over $\mathbb{Q}$). For instance, applied to the singular K3 surface X from Example 13.6, this implies that a full set of generators of $\mathrm{MW}(X)$ always requires a degree $6 = h(-300)$ extension of $\mathbb{Q}$ (where $h(d)$ denotes the class number of $d < 0$, i.e. the degree of the abelian extension $H(d)/\mathbb{Q}(\sqrt{d})$, or equivalently the number of inequivalent positive-definite even binary quadratic forms of discriminant d, see [Cox13]); indeed, in [Shio07b], generators are explicitly worked out over the (larger) field $\mathbb{Q}(\zeta_3, \zeta_5, \sqrt[3]{10})$.

On the other hand, for singular K3 surfaces over $\mathbb{Q}$, the following conclusion is immediate:

Corollary 13.9 *If a singular K3 surface has* NS *of determinant d generated by divisors over $\mathbb{Q}$, then $h(d) = 1$.*

By Stark's class number one theorem [Sta66], there are thus only 13 possible values for d in Corollary 13.9:

$$d \in \{-3, -4, -7, -8, -11, -12, -16, -19, -27, -28, -43, -67, -163\}. \tag{13.15}$$

In fact, one can moreover show under the assumption of Corollary 13.9 that the quadratic form on the transcendental lattice of X is primitive, i.e. not divisible as an even binary quadratic form. In particular, by (12.8), X cannot be a Kummer surface (as was already noted in [Shio05a]). Even further, these lattices come with the special feature that both orientations (given by choosing the order of a basis) are equivalent (under $\mathrm{SL}(2, \mathbb{Z})$, not only under $\mathrm{GL}(2, \mathbb{Z})$). By the Torelli theorem for singular K3 surfaces (cf. Sect. 12.2.2), there are thus, up to complex automorphisms, exactly 13 singular K3 surfaces X which might a priori admit an elliptic fibration of Mordell–Weil rank 18 over $\mathbb{Q}$. By class group theory, their transcendental lattices are determined as

$$T(X) = \begin{cases} [2, 0, -d/2] = \operatorname{diag}(2, -d/2) & \text{if } d \text{ is even}, \\ [2, 1, (1-d)/2] = \begin{pmatrix} 2 & 1 \\ 1 & (1-d)/2 \end{pmatrix} & \text{if } d \text{ is odd}. \end{cases} \tag{13.16}$$

There is the following explicit description of the Néron–Severi lattice:

$$\mathrm{NS}(X) \cong U \oplus (E_8^-)^2 \oplus T(X)^-. \tag{13.17}$$

The main problem now is to decide whether the above lattice (for any of the 13 singular K3 surfaces of class number one) admits an alternative decomposition

$$\mathrm{NS}(X) \stackrel{?}{=} U \oplus M$$

where the even integral lattice M of rank 18 is negative definite without admitting any roots. Such a lattice M, if it were to exist, would break the density record for sphere packings in dimension 18 (which stands at $\sqrt{3}/8$ according to [NS]), but, at least for $d = -163$, it would not contradict the known sphere packing bounds (cf. [CS99, Chaps. 1, 9]). However, there are gluing techniques (in the vein of the Kneser–Nishiyama method, cf. [CS99, 4.3]) to prove that the possible determinants have too small an absolute value to admit a definite rank 18 lattice without roots (compare with the determinant $d = -300$ in Example 13.6 where this is barely possible). Hence Problem 13.7 has a negative answer.

13.1.4 Mordell–Weil Rank 17 over $\mathbb{Q}$

Since Problem 13.7 has a negative answer, we now turn to the problem of determining elliptic K3 surfaces with Mordell–Weil rank 17 over $\mathbb{Q}$. Given a K3 surface X (in any characteristic), this amounts to a primitive embedding

$$U \oplus M \hookrightarrow \mathrm{NS}(X) \tag{13.18}$$

where M denotes an even integral negative-definite lattice of rank 17 without roots. Disregarding the demand that M should correspond to (the narrow part, or a sublattice of) the Mordell–Weil lattice $\mathrm{MWL}(X)$ generated by sections over $\mathbb{Q}$, we infer from Proposition 11.15 that, at least over $\mathbb{C}$, such a K3 surface X comes in a one-dimensional family of lattice polarized K3 surfaces. Such families (with the full Néron–Severi group generically generated over the ground field) are parametrized by either modular curves or Shimura curves. For instance, if NS generically admits a decomposition as in (13.17), but with $T(X)^-$ replaced by an orthogonal summand

$\langle -2d \rangle$ of rank one ($d > 1$),

$$\mathrm{NS} = U \oplus (E_8^-)^2 \oplus \langle -2d \rangle \quad (d > 1), \tag{13.19}$$

then these K3 surfaces are parametrized by the (non-CM points of) the modular curve $X_0(d)/w_d$, where w_d denotes the Fricke involution (compare, for instance, [Dol96]). Even more concretely, this is illustrated by the following example for $d = 1$.

Example 13.10 Consider the complex K3 surfaces with

$$\mathrm{NS} = U \oplus (E_8^-)^2 \oplus A_1^-.$$

We know from the second example in Sect. 12.3.5.2 that these K3 surfaces arise from the Shioda–Inose structure for $E \times E$ for some elliptic curve E (without CM, if $\rho = 19$ as above), so they are parametrized by the j-line, i.e. by the classical modular curve $X_0(1)$.

In comparison, Shimura curves parametrize abelian surfaces with quaternionic multiplication. This translates into K3 surface language most naturally using the work of Dolgachev [Dol04] and Kumar [Kum08]. In fact, in the converse direction, the moduli interpretation through lattice polarized K3 surfaces can be very useful in determining explicit equations for the Shimura curves, cf. [Elk08].

Focussing again on the problem of finding an elliptic K3 surface of Mordell–Weil rank (at least) 17 with 17 independent sections defined over $\mathbb{Q}$, we are facing two constraining objectives:

- the parametrizing curve must have a rational point (expected to be neither a cusp nor a CM point), and the point can be verified to correspond to a K3 surface over $\mathbb{Q}$ (with the desired properties);
- the generic Néron–Severi lattice must have determinant large enough to admit a decomposition (or primitive embedding) as in (13.18) with rootless M of rank 17 (instead of leading to a lattice with density too large, as happened in rank 18).

Note that, as the determinant of the generic NS grows, so does the genus of the parametrizing curves eventually. Hence rational points tend to be increasingly rare (and finite in number anyway by Faltings' theorem [Fal83]). For the modular curves $X_0(d)/w_d$, no suitable rational points are known (neither are they expected to exist). The Shimura curves, in contrast, have genus not increasing as rapidly with the determinant as it does for modular curves. Indeed, let $N = 6 \cdot 79$ and consider abelian surfaces with endomorphism ring the maximal order in the unique indefinite quaternion algebra over $\mathbb{Q}$ whose finite ramified primes are precisely $2, 3, 79$ (the prime divisors of N). These abelian surfaces are parametrized by a Shimura curve, say $\mathcal{C}$; using the methods of [GR04], Elkies computes the following model:

$$\mathcal{C}: \quad y^2 = 16x^6 - 19x^4 + 88x^2 - 48. \tag{13.20}$$

The curve $\mathcal{C}$ is endowed with three involutions giving a Klein four-group V_4 generated by the sign changes in y and in x. Elkies locates a $\mathbb{Q}$-rational CM point of discriminant -43 at the orbit of $(x, y) = (2, 32)$ under V_4, and a non-CM point at the orbit of $(14/13, 2^6 \cdot 251/13^3)$. By [Dol04, Kum08], there is a family of K3 surfaces parametrized by $\mathcal{C}$ with NS generically of rank 19 and determinant $2N$. On the lattice level, we can build up this family from the CM member X_0. A priori, X has discriminant some square multiple of -43. Presently, however, X_0 is endowed with a rank 19 sublattice of $\text{NS}(X_0)$ which, by construction, is defined over $\mathbb{Q}$. Hence $\text{NS}(X_0)$ is either defined over $\mathbb{Q}$, too, or over a quadratic extension of $\mathbb{Q}$. But then there is no order in the ring of integers of $\mathbb{Q}(\sqrt{-43})$ of class number 2 by [Cox13, Theorem 7.24], so Theorem 13.8 rules out the second alternative, and X_0 has discriminant -43 and transcendental lattice $T(X_0) = [2, 1, 22]$ from (13.16). Therefore, we can construct the desired rank 19 lattice of determinant $2N$ uniquely by gluing $T(X_0)$ to a vector v of square $v^2 = -12 \cdot 43 \cdot 79$:

$$T = \begin{pmatrix} 2 & 1 & 0 \\ 1 & 22 & 1 \\ 0 & 1 & -22 \end{pmatrix} \supset \begin{pmatrix} 2 & 1 \\ 1 & 22 \end{pmatrix} \oplus \mathbb{Z}v. \tag{13.21}$$

In search for a suitable orthogonal complement of T inside the K3 lattice Λ, we find the lattice

$$L = U \oplus D_9^- \oplus A_2^- \oplus \langle A_5^-, w \rangle, \tag{13.22}$$

where w is a vector of square $w^2 = -14$ meeting A_5^- only in one exterior vertex (with multiplicity one). Indeed, it is easy to check that the discriminant forms satisfy

$$q_L = -q_T,$$

so the discussion around Proposition 2.41 applies to prove that $T = L^\perp \subset \Lambda$. Moreover, up to isometries, L is the unique such lattice by Theorem 2.40. In conclusion, we have found the generic Néron–Severi lattice of the K3 surfaces parametrized by the Shimura curve $\mathcal{C}$. We now justify why these K3 surfaces are relevant to the rank problem:

Lemma 13.11 *A K3 surface X with* $\text{NS}(X) \cong L$ *admits an elliptic fibration with Mordell–Weil rank* 17.

Proof We seek a decomposition (13.18) of L where the lattice M contains no roots. Following Sect. 12.3.3, we work with a partner lattice M_0. Presently, one can take

$$M_0 = E_6 \oplus \langle 4 \cdot 79 \rangle,$$

since $q_{M_0} = q_L$. Given this, the existence of an elliptic fibration with maximal Mordell–Weil rank is equivalent to an embedding of M_0 into some Niemeier lattice N such that the resulting frame $M_0^\perp$ has no roots. For this purpose, pick N from

Theorem 12.21 with
$$\mathcal{R}(N) = E_6 \oplus D_7 \oplus A_{11}.$$

Of course, E_6 embeds (only) into the orthogonal summand of the same type, so it remains to embed the vector u of square $4 \cdot 79$ into $D_7 \oplus A_{11}$ (or into the index 2 overlattice contained in N). We first identify a simple vector u_1 of small square inside A_{11} such that $u_1^\perp$ contains no roots. To this end, we number the vertices of A_{11} as in Fig. 2.2. Note that, up to sign, the roots A_{11} take the shape

$$\alpha_{i,j} = \alpha_i + \ldots + \alpha_j \quad 1 \leq i \leq j \leq 11.$$

We work with a sparse vector

$$u_1 = \sum_{k=1}^{6} a_k \alpha_{2k-1}, \quad a_k \in \mathbb{Z}$$

supported on the odd-indexed vertices of A_{11}. For u_1 to satisfy $\langle u_1, \alpha_{i,j} \rangle \neq 0$ for all i, j translates as
$$a_k \neq 0, \quad a_k \neq \pm a_l \quad \forall k \neq l.$$

This is readily achieved by taking $a_k = k$ so that $u_1^2 = 182$. It remains to find a vector $u_2 \in D_7$ such that $u_2^2 = 134$ and $u_2^\perp$ contains no roots. By enumerating all roots in D_7, one can easily find such a vector; in the numbering from Fig. 2.2, one may take, for instance,
$$u_2 = \alpha_1 + 4\alpha_2 + 2\alpha_4 + 5(\alpha_6 + \alpha_7).$$

Hence $u \mapsto u_1 + u_2$ defines the required embedding $M_0 \hookrightarrow N$ which, by construction, has rootless orthogonal complement, and which is primitive, since M_0 admits no non-trivial even integral finite index overlattice. □

We continue by working out explicitly the family of K3 surfaces with NS $= L$ generically. To this end, we fix the elliptic fibration induced by L. By Sect. 11.4, this will have reducible fibres of type I_5^*, I_6, I_3 (or, a priori, IV, but this special case will not be relevant to our issues). Moreover, the vector w translates into a section P of height $79/6$ which meets the I_6 fibre in one of the fibre components adjacent to the identity component, say in Θ_1, while P meets all other fibres in the identity component and $(P.O) = 5$ (by Theorem 6.24). Note that the K3 surfaces parametrized by $\mathcal{C}$ will have all these divisors defined over the ground field, but it will be much easier to first consider the geometric situation by allowing for a Galois action on the singular fibres. Indeed, these K3 surfaces will be parametrized by the quotient $\mathcal{C}_0 = \mathcal{C}/V_4$ which is rational, and $\mathcal{C}$ can be recovered easily from $\mathcal{C}_0$ (as we will see). Still, the height of the section P is much too large to parametrize these K3 surfaces directly, so we follow the approach indicated by Elkies in [Elk07, Elk08] to circumvent these computational difficulties.

We start by determining a large rank sublattice $L_0 \subset L$ such that K3 surfaces with generic NS $= L_0$ are easy enough to parametrize. Presently, the lattice

$$L_0 = U \oplus D_9^- \oplus A_2^- \oplus A_5^-$$

suggests itself since K3 surfaces with this Néron–Severi lattice admit an elliptic fibration with reducible fibre I_5^*, I_3, I_6. The choice draws on the considerations from Sect. 8.9 that fibre types (or degenerations) are encoded in the (vanishing orders of the) discriminant Δ of the generic fibre and thus relatively easy to control (at least with few reducible fibres). As in Sect. 8.10, we work with an extended Weierstrass form which allows us to move the nodes of the Weierstrass model to (0, 0) and ensure certain fibre types from the very beginning. Locating I_6 at $t = 0$, I_3 at $t = 1$ and I_5^* at $t = \infty$, we can start with the following extended Weierstrass form:

$$y^2 = x^3 + (t^3 + a_2t^2 + a_1t + u)x^2 + 2(t-1)t^2(rt^2 + bt + su)x + (t-1)^2t^4(r^2t + s^2u)$$

with $a_1, a_2, r, b \in \mathbb{Q}$ and $u, r, s \in \mathbb{Q}^\times$. Here we have already chosen the constant and top degree coefficients of the entries so that the fibre types are at least I_5, I_2, I_3^*. They are promoted to I_6, I_2, I_4^* by solving for the bottom and top degree coefficients of Δ to vanish, which gives easy expressions for a_1, a_2. Then

$$\Delta = t^6(t-1)^2\Delta' \quad \deg(\Delta') = 6.$$

The two remaining promotions of fibre types require

$$\Delta'(1) = 0 \quad \text{and} \quad \Delta'(\infty) = 0,$$

where $\Delta'(\infty)$ denotes the top coefficient of Δ'. Here $\Delta'(1)$ factors into $4r^6s^6$ times a simple factor h and a square factor where the fibre type becomes additive. We simplify by considering

$$\Delta'(\infty) - 4h = -r^3s^2(4br - r^3 - 8s^2u),$$

which gives an easy expression for b. The remaining equation is a quadric in u whose discriminant $64r^5s^7(r + s)$ allows for an easy parametrization. Thus we obtain the required family of elliptic K3 surfaces with reducible fibres I_6, I_3, I_5^*. The extended Weierstrass form simplifies greatly by allowing the I_3 fibre to move around, and we obtain

$$\begin{aligned} y^2 = x^3 &+ \left(t^3 + 3(\lambda + 3\mu)t^2 - (4\lambda + 1 + 6\mu)t + \lambda + \mu\right)x^2 \\ &+ 2(4\mu t - (\lambda+1)^2 - 4\mu)t^2\left(t^2 + (2\lambda + 3\mu)t - \lambda - \mu\right)x \\ &+ (4\mu t - (\lambda+1)^2 - 4\mu)^2t^4\,(t + \lambda + \mu)\,. \end{aligned} \quad (13.23)$$

Instead of trying to solve directly for the subfamily $\mathcal{X}$ of K3 surfaces admitting a section P of height $79/6$, we first compute the special member X_0, a singular K3 surface of discriminant -43, and then deform to obtain the whole one-dimensional family $\mathcal{X}$.

Proposition 13.12 *The singular K3 surface X_0 is located inside the two-dimensional family* (13.23) *at* $\lambda_0 = 15/49, \ \mu_0 = -16/49$.

Proof Lattice theoretically, it is easy to see how X_0 sits inside $\mathcal{X}$: this happens exactly at those points where the vector $v \in T$ from (13.21) becomes algebraic. Hence $\mathrm{NS}(X_0)$ is obtained from L by gluing to v. More precisely, one glues $v/2N$ to generators of the discriminant groups of each orthogonal summand of L in (13.22). By construction, this gives a vector $v' \in \mathrm{NS}(X_0)$ such that

$$\mathrm{NS}(X_0) = \langle L, v' \rangle.$$

Moreover, we can predict precisely which fibre components v' intersects. In terms of the orthogonal projection

$$\varphi : \mathrm{NS}(X_0)_{\mathbb{Q}} \to (\mathrm{Triv}(X_0))^{\perp}$$

from Sect. 6.6.1, we read off that

$$\varphi(v') = \varphi(P')$$

for a section $P' \in \mathrm{MW}(X_0)$ of height $145/4$ which meets I_3 at a non-identity component, I_5^* at a far component and I_6 at Θ_5, while $(P'.O) = 18$ and $(P'.P) = 3$. For the Mordell–Weil lattice of X_0, this returns

$$\mathrm{MWL}(X_0) = \langle P', P \rangle = \begin{pmatrix} \frac{145}{4} & \frac{131}{6} \\ \frac{131}{6} & \frac{79}{6} \end{pmatrix} \cong \begin{pmatrix} \frac{3}{2} & -\frac{1}{6} \\ -\frac{1}{6} & \frac{5}{12} \end{pmatrix} = \langle Q', Q \rangle$$

of determinant $43/72$, affirming that P, P' generate $\mathrm{MW}(X_0)$. Here the second Gram matrix is obtained from the first by the standard reduction technique for (even integral) binary quadratic forms, with basis $Q' = 2P' - 3P$ and $Q = 5P - 3P'$. Since we know which fibre components P and P' meet, we can easily compute the fibre components met by Q, Q': Q meets I_3 trivially, I_5^* at a far component and I_6 at Θ_2, while Q' meets I_3 at the non-identity component not met by P', I_5^* at the near component and I_6 at Θ_1. It also follows from Theorem 6.24 that both Q and Q' are integral, i.e. $(Q.O) = (Q'.O) = 0$, and $(Q.Q') = 1$. These sections have height small enough to implement directly. We start by computing the members of the two-dimensional family of elliptic K3 surfaces in (13.23) which admit the section Q. By integrality,

$$Q = (\mathrm{x}(Q), \mathrm{y}(Q)) \quad \text{with polynomials x, y of degree 4 \ resp. 6.}$$

For Q to meet the I_6 and I_5^* fibres as stated, $\mathrm{x}(Q)$ has to vanish quadratically at both fibres; it thus takes the shape $\mathrm{x}(Q) = -4\mu t^2$. But then the substitution into the RHS of (13.23) yields a perfect square (over $\bar{\mathbb{Q}}(t)$) if and only if $\lambda = -4\mu - 1$. In the second step, we impose the section Q' as well—with simple vanishing of $\mathrm{x}(Q')$ at all reducible fibres and leading coefficient determined as

$$\mathrm{x}(Q') = -t(t - 1 - 4\mu)(t + \gamma) \qquad (\gamma \in \bar{\mathbb{Q}}).$$

Again, substitution gives a perfect square in $\bar{\mathbb{Q}}[t]$ without degenerations if and only if

$$\mu = -16/49, \quad \gamma = 9/7.$$

Hence the singular K3 surface X_0 is located at $\lambda = 15/49, \ \mu = -16/49$ as claimed. □

Remark 13.13 For sections of greater height (such as P, P'), it is often not advisable to try to solve directly for the members admitting the given sections. However, for special surfaces such as X_0 this can often be helped by p-adic techniques (after finding a section modulo p by extensive search for a suitable small prime p) and/or by point counting over finite fields (using the modularity of singular K3 surfaces over $\mathbb{Q}$ due to Livné [Liv95]). To get an idea of this, see [Elk08] or [ElS13].

With the special surface X_0 at our disposal, we shall now try to deform it together with the section P inside the two-dimensional family (13.23). For this purpose, we perform a quadratic twist over $\mathbb{Q}(\sqrt{\lambda + \mu})$ to ensure that all components of the I_6 fibre, and thus also the section P, are rational. After this twist, the section $P \in \mathrm{MW}(X_0)$ has x-coordinate $\mathrm{x}(P) = tU(t)/(7^5 V(t)^2)$, where

$$\begin{aligned}
U = {} & 930522055948828\, t^{13} + 10544334113583403\, t^{12} + 40506969914916932\, t^{11} \\
& + 61068316736953526\, t^{10} + 30845805199623456\, t^9 + 9825189556364581\, t^8 \\
& + 2889385414308008\, t^7 + 637406008338932\, t^6 + 88199953375996\, t^5 \\
& + 7891808371621\, t^4 + 472180640148\, t^3 + 18584532918\, t^2 \\
& + 449047368\, t + 5250987, \\
V = {} & 3411821\, t^5 + 10600415\, t^4 + 2519874\, t^3 + 236734\, t^2 + 8225\, t + 147.
\end{aligned}$$

We want to deform X_0 together with P to find the image C of $\mathcal{C}_0 = \mathcal{C}/V_4$ inside the (λ, μ)-plane. We thus look for a polynomial $F \in \mathbb{Q}[\lambda, \mu]$ such that $C = Z(F)$. For a general choice of $\lambda \in \mathbb{Q}$, the curve C will always contain the same number d of $\bar{\mathbb{Q}}$-points, given by the polynomial F of degree d in μ over $\mathbb{Q}$ evaluated at λ. We would not know how to effectively compute this polynomial over the complex numbers, but there is a p-adic technique which applies despite the large height if λ is close to λ_0 in the p-adic topology. Namely, for λ fixed, we can use p-adic Newton iteration to lift the original solution comprising λ_0, μ_0, P to any given p-adic accuracy (fixing λ).

Once the accuracy is big enough, one can use lattice reduction to obtain an integer relation between the first d powers of μ. Up to a scalar, this gives a candidate for the equation F of C evaluated at the fixed value of λ.

Example 13.14 Fix $\lambda = 3$, which is close to λ_0 for $p = 11$. Lifting the solution to accuracy p^{2^9}, lattice reduction returns the first relation in degree $d = 7$:

$$\mu^7 + 117948\mu^6 - 1204907952\mu^5 + 15566815026880\mu^4 - 17995763072256\mu^3 + 5112068944896\mu^2 + 366401660350464\mu + 415815392968704 = 0.$$

In the same manner, one can collect as many data for relations in μ as seems appropriate. Then one can try to solve for F as a degree d' polynomial in λ matching all these relations when evaluating at the respective value of λ. This amounts to solving linear equations in the purported coefficients of F. Presently, we find that

$$\begin{aligned} F = {} & 59072816401\,\lambda^7 + 339603374887\,\lambda^6\mu + 614692809669\,\lambda^5\mu^2 \\ & + 336797807843\,\lambda^4\mu^3 + 2641922963\,\lambda^3\mu^4 + 6370869\,\lambda^2\mu^5 + 4567\,\lambda\mu^6 + \mu^7 \\ & + 190736834583\,\lambda^6 - 506323793046\,\lambda^5\mu - 5207391374343\,\lambda^4\mu^2 \\ & - 6815955079284\,\lambda^3\mu^3 + 445321455513\,\lambda^2\mu^4 - 936962070\,\lambda\mu^5 + 104247\,\mu^6 \\ & + 394036643109\,\lambda^5 + 1386652259529\,\lambda^4\mu + 6535344775218\,\lambda^3\mu^2 \\ & + 13068521207634\,\lambda^2\mu^3 + 2883538372473\,\lambda\mu^4 + 1548640437\,\mu^5 \\ & + 487221201331\,\lambda^4 + 3952018825228\,\lambda^3\mu + 10263102266994\,\lambda^2\mu^2 \\ & + 6405579518860\,\lambda\mu^3 + 2836974889843\,\mu^4 + 391875388371\,\lambda^3 \\ & + 2394649028169\,\lambda^2\mu + 2754665927001\,\lambda\mu^2 + 1920972207843\,\mu^3 \\ & + 189766192581\,\lambda^2 + 415436020074\,\lambda\mu + 454190402277\,\mu^2 \\ & + 23594416695\,\lambda + 45991151271\,\mu + 855036081. \end{aligned}$$

fits a data set of some 50+ relations (where $d' = 7$).

Lemma 13.15 *The curve $C = Z(F) \subset \mathbb{A}^2$ is rational over $\mathbb{Q}$.*

Proof One directly verifies that C is irreducible and contains 15 nodes $P_1, \ldots, P_{15}$, so it is rational over $\bar{\mathbb{Q}}$ by [Har77, IV.3.11.1]. We can make this explicit by considering a pencil of quartics through 13 of the nodes, say $P_3, \ldots, P_{15}$. For instance, one may take $P_1 = (-19/68, -1/4)$ and $P_2 = (19/45, -16/45)$. Then the pencil of quartics is $G_1 + sG_2$ for

$$\begin{aligned} G_1 = {} & 15557108\,\lambda^4 + 74228153\,\lambda^3\mu + 92890515\,\lambda^2\mu^2 - 1611061\,\lambda\mu^3 + 1277\,\mu^4 \\ & - 101117536\,\lambda^3 - 457392645\,\lambda^2\mu - 423975786\,\lambda\mu^2 - 10033861\,\mu^3 \\ & - 150770136\,\lambda^2 - 290742837\,\lambda\mu - 157818621\,\mu^2 \\ & - 41017536\,\lambda - 63983223\,\mu - 5152572, \\ G_2 = {} & 14762131\,\lambda^4 + 55734139\,\lambda^3\mu + 57602967\,\lambda^2\mu^2 - 1283831\,\lambda\mu^3 + 1114\,\mu^4 \\ & - 19029956\,\lambda^3 - 288544479\,\lambda^2\mu - 396516354\,\lambda\mu^2 - 10175207\,\mu^3 \\ & - 24543678\,\lambda^2 + 39540945\,\lambda\mu + 216667239\,\mu^2 \\ & + 4746780\,\lambda + 70840107\,\mu + 9654147. \end{aligned}$$

By construction, C intersects the pencil at the 13 given nodes with multiplicity two each, and at two further points Q_1, Q_2 depending on s which we do not write down explicitly for space reasons (but we will give a parametrization below). Thus we obtain a degree two map

$$C \to \mathbb{P}^1,$$

and the genus of the desingularization $\tilde{C}$ can be computed using Hurwitz' formula. We find that the map $C \to \mathbb{P}^1$ branches at the roots of $20s^2 + 28s - 51$ and at 0 and ∞. But the latter points have pre-images the nodes P_1, P_2 by the choice of G_1, G_2, which are replaced by two points of $\tilde{C}$ each. Hence the morphism $\tilde{C} \to \mathbb{P}^1$ has exactly two ramification points, so $\tilde{C}$ and C are rational as claimed. □

The above argument has the benefit that it can be used to find a rational parametrization of C. Namely, we can replace the parameter s of the pencil by

$$s = -3\frac{4u^2 + u - 1}{5u^2 + 3}$$

to make the two intersection points Q_1, Q_2 of C and the pencil $\mathbb{Q}(u)$-rational; e.g.

$$Q_1 = Q_1(u) = \left(-\frac{(16u^2 + 27u + 9)(11u^2 + 10u - 5)(8u^3 + u^2 - 2u + 9)}{(3+u)(5u+7)^2(4u^2 + 3u - 3)^2}, \right.$$
$$\left. 16\frac{(-3+7u)(u^3 - u^2 + 5u + 3)^2}{(3+u)(5u+7)^2(4u^2 + 3u - 3)^2} \right).$$

This gives the desired parametrization of C. Recall that C (and thus $\tilde{C}$) is supposed to parametrize elliptic K3 surfaces with geometric Néron–Severi lattice L from (13.22), but for our purposes we need all fibre components (and the section P) to be defined over $\mathbb{Q}$. Presently, all components of the I_6 fibre and the section P are defined over $\mathbb{Q}$ (by virtue of the quadratic twist applied), but the other fibres generically involve the following quadratic extensions:

$$I_3/\mathbb{Q}(\sqrt{\mu(\lambda+\mu)}), \qquad I_5^*/\mathbb{Q}(\sqrt{-\mu(\lambda+1)(3\lambda-1)(\lambda+\mu)}).$$

At $Q_1 \in C$, parametrized as above, the first square root is easily made rational by replacing

$$u = -6\frac{\nu + 3}{\nu^2 + 7}.$$

Then the second square root leads to the genus two curve

$$C': \quad z^2 = \nu^6 - 2\nu^5 + 63\nu^4 - 28\nu^3 - 1761\nu^2 - 5858\nu - 4703.$$

This is converted to the exact shape of C from (13.20) by the Möbius transformation

$$\nu \mapsto (7\nu + 2)/(2 - \nu).$$

Conversely, we can recover the CM-43 point $Q_1(0) = (15/49, -16/49)$ on C as well as the (presumably non-CM) point

$$Q_1\left(-\frac{45}{64}\right) = \left(-\frac{2^3 \cdot 3 \cdot 5 \cdot 11 \cdot 17 \cdot 37 \cdot 47 \cdot 59 \cdot 491}{7^2\, 223^2\, 1069^2}, -\frac{13^2\, 97^2\, 1223^2}{7^2\, 223^2\, 1069^2}\right) = (\lambda_1, \mu_1),$$

where, indeed, all fibre components of the corresponding elliptic K3 surface X_1 are defined over $\mathbb{Q}$. In summary, this provides ample evidence that $\tilde{C} = C_0$ and $C' = C$. A complete proof of these equalities would amount to computing the section P with coefficients in $\mathbb{Q}(u)$ deforming from X_0 all over C. This can be achieved in essentially the same manner as the computation of F: compute the section P over $\mathbb{Q}$ for enough fixed values of u which are p-adically close to $u_0 = 0$ using p-adic Newton iterations and lattice reduction, and then determine the rational functions in u which are the coefficients of P by solving linear equations. As a confirmation, the resulting section can be validated by substituting into the extended Weierstrass form of the generic K3 surface over C.

Here, however, we will content ourselves with a weaker result which admits a less involved argument, but which will be perfectly sufficient for our purposes.

Theorem 13.16 *The two-dimensional family of K3 surfaces* (13.23) *contains a member X_1 with* $\mathrm{NS}(X_1) = L$, *generated by divisors classes defined over* $\mathbb{Q}$, *at* (λ_1, μ_1).

Proof We first prove that $L \subseteq \mathrm{NS}(X_1)$. To this end, it suffices to compute the section $P_1 \in \mathrm{MW}(X_1)$ of height $79/6$. We proceed by first deforming (X_0, P_0) at $u = 0$ p-adically to X' at $u' = 73/64$ for $p = 73$ until the accuracy at $p^{2^{11}}$ is high enough to read off the $\mathbb{Q}$-rational coefficients of P' (which can be verified a posteriori symbolically to be a section of X'). Then we deform (X', P') p'-adically to X_1 at $u_1 = -45/64$, but now for $p' = 59$. The same accuracy as above allows for computing and verifying the anticipated section $P_1 \in \mathrm{MW}(X_1)$ of height $79/6$. For space reasons, we do not display all of its coefficients, but only the leading coefficient of the numerator of $\mathrm{x}(P_1)$ (after normalizing the denominator, cf. the set-up introduced after Remark 13.13):

$$19^2\, 41^2\, 71^2\, 97^2\, 127^2\, 1223^2\, 3769^2\, 9901^2/(7^8\, 31^2\, 257^2\, 3522399323903^2).$$

This proves that $L \subseteq \mathrm{NS}(X_1)$ as claimed. If this were not an equality, then X_1 would be a singular K3 surface, i.e. $\mathrm{NS}(X) = L_1$ of rank 20, say of discriminant $d_1 < 0$. At a prime p of good reduction, there are therefore only two alternatives for the Picard number:

$$\rho(X_1 \otimes \bar{\mathbb{F}}_p) \in \{20, 22\}.$$

In the latter case, $X_1 \otimes \mathbb{F}_p$ would be supersingular, while in the former case $\mathrm{NS}(X_1 \otimes \bar{\mathbb{F}}_p) = L_1$. Both cases can be checked using the formula from the Artin–Tate conjecture [Tat66b] which we will review around (13.32) (without the argument requiring its validity, since otherwise we would have $\rho(X_1) = 19$ anyway). In detail, the $\mathbb{Q}$-rational generators of $L \subset \mathrm{NS}(X_1)$ provide enough eigenvalues of Frobenius to read off the full characteristic polynomial on $H^2_{\text{ét}}(X_1 \otimes \bar{\mathbb{F}}_p, \mathbb{Q}_\ell)$, up to one potential ambiguity, from a simple point count over $\mathbb{F}_p$. For instance, the trace of Frobenius reads

$$\mathrm{tr}\,\mathrm{Frob}_p^*(H^2_{\text{ét}}(X_1 \otimes \bar{\mathbb{F}}_p, \mathbb{Q}_\ell)) = 19p \pm p + a_p = \#X_1(\mathbb{F}_p) - 1 - p^2.$$

Here a_p would account for the trace of Frob_p in the Galois representation associated with the conjectural rank two transcendental lattice $T(X_1)$, and the sign depends on the Galois action on the remaining divisor class generating $\mathrm{NS}(X_1)$ together with L. Clearly, $X_1 \otimes \mathbb{F}_p$ is supersingular if and only if $p \mid a_p$. Otherwise, one infers as in [Schü10] that $a_p^2 - 4p^2$ lies in the same square class inside $\mathbb{Q}$ as the determinant of $\mathrm{NS}(X_1 \otimes \bar{\mathbb{F}}_p)$. By assumption, this is just d_1, so we can establish a contradiction (in the spirit of [vLui07]) by counting points over $\mathbb{F}_p$ for $p = 59, 73$ and comparing the resulting square classes of $a_p^2 - 4p^2$:

$$\begin{aligned} p = 59: &\quad a_p^2 - 4p^2 = -7^2 \cdot 11 \cdot 17 \quad \text{or} \quad -3^2 \cdot 23 \cdot 167, \\ p = 73: &\quad a_p^2 - 4p^2 = -2^8 \cdot 3 \cdot 19 \quad \text{or} \quad -2^2 \cdot 3 \cdot 5 \cdot 7 \cdot 41. \end{aligned}$$

Since the resulting square classes are distinct, we infer that $\rho(X_1) = 19$ as claimed. □

Remark 13.17 The surface X_1 has a large number of bad primes where the Weierstrass equation or the fibre configuration degenerates. Therefore one of the most delicate parts in the above proof is the choice of the appropriate surface X' as an intermediate deformation between X_0 and X_1.

Given an elliptic K3 surface, we have seen in Sect. 12.3 how one can switch to other elliptic fibrations. Often this can be achieved by considering lattices which are 2- or 3-neighbours in Kneser's terminology, i.e. which share a common index 2 or 3 sublattice, see [EK14, Sect. 5]. A typical example of this was treated in Sect. 12.3.5.1. In this way, Elkies finds an elliptic fibration on the K3 surface X_1 of Mordell–Weil rank 17 over $\mathbb{Q}$. Base change by a bisection (just as in Néron's original construction of high rank elliptic curves, compare Sect. 13.1.1 and the discussion around (5.18) enables him to exhibit (infinitely many) honestly elliptic surfaces over $\mathbb{P}^1$ with $\chi = 4$ and Mordell–Weil rank 18. Another base change by a bisection induces an elliptic curve B as base, but it can be defined in such a way that B has positive rank over $\mathbb{Q}$ (as in the setting originating from Néron's method in Sect. 13.1.1 which culminates in (13.12)); in particular, specialization gives infinitely many elliptic curves over $\mathbb{Q}$ with rank at least 19 by Silverman's theorem [Sil83].

Elkies proceeds by identifying particular specializations where the rank of the elliptic curve over $\mathbb{Q}$ increases even further, in fact up to 28, the current rank record for an elliptic curve over $\mathbb{Q}$. Experimentally, this can be achieved using a heuristicdue to Mestre which is based on the Birch–Swinnerton-Dyer conjecture. This heuristic indicates that the following elliptic curve ought to have very high rank over $\mathbb{Q}$:

$$E: \quad y^2 + xy + y = x^3 - x^2 - a_4 x + a_6,$$

where

$$\begin{aligned} a_4 &= 20067762415575526585033208209338542750930230312178956502, \\ a_6 &= 3448161179503055646703298569039072037485 5\\ &\quad\ 94435931918036126600829629193944873224342 9. \end{aligned}$$

Theorem 13.18 (Elkies, Klagsbrun–Sherman–Weigandt) *The Mordell–Weil group $E(\mathbb{Q})$ has rank at least* 28. *Subject to GRH (for number fields), the rank equals* 28.

Proof Even with the rational points specializing from the elliptic surfaces, verifying that the curve admits indeed 28 independent $\mathbb{Q}$-rational points is non-trivial; it is achieved by Elkies using once more the structure coming from the elliptic K3 surface X of Mordell–Weil rank 17 as one can search for rational points near half-lattice holes of $\mathrm{MWL}(X)$ (as in Remark 2.20; the holes are also instrumental for the set-up of the bisections used for the base changes).

It remains to show (conditionally) that the rank does not exceed 28. This was only recently established in [KSW19] to which we refer for the details. The proof uses the classical method of 2-descent which presently, under GRH, implies that the 2-Selmer group $\mathrm{Sel}_2(E/\mathbb{Q})$ has dimension exactly 28. One of the key steps in the computation (conditional on GRH for number fields) shows that the 2-torsion field of E has class group with 2-rank 20 (which only seems to be surpassed by the 2-rank 22 of the 2-torsion field of a rank 27 elliptic curve over $\mathbb{Q}$, also due to Elkies). □

13.2 Rank Problem over $\mathbb{C}(t)$

The rank problem for arbitrary elliptic curves over $\mathbb{C}(t)$ takes a different flavour since this will lead to elliptic surfaces over $\mathbb{P}^1$ with Kodaira dimension $\kappa = 1$ (sometimes called honestly elliptic surfaces). Hence, there are no structures available whose understanding comes anywhere close to the moduli theory of lattice-polarized K3 surfaces (as exploited heavily in Sect. 13.1.3). In consequence, much less can be said about this general situation, although there are some strong results. As a sample, we mention the current record holder (cf. [Shio92a, p. 110]):

Theorem 13.19 (Shioda) *For any $n \in \mathbb{N}$, the elliptic curve*

$$E: \ y^2 = x^3 + t^{360n} + 1$$

has rank 68 over $\mathbb{C}(t)$.

The elliptic curves above can be studied in a rather uniform way using Delsarte surfaces. We shall review these surfaces in the next section before returning to the proof of Theorem 13.19 in Sect. 13.2.2. We will also comment on other approaches towards constructing elliptic curves with high rank over $\mathbb{C}(t)$.

Remark 13.20 The argument in favour of unbounded ranks by Lapin, given in [Lap64], is believed to be wrong, see [Scho90].

13.2.1 Delsarte Surfaces

Usually it is hard, if not impossible, to compute the Picard number of a given algebraic surface (although there has been substantial progress on the theoretical, yet not effective side, see [Cha14, PTvL15]). A notable exception, both over $\mathbb{C}$ and in positive characteristic, consists of Delsarte surfaces as introduced in [Shio86]. In brief, this is because they are covered by Fermat surfaces. Since the cohomology of the latter is very well understood, even to the extent of deciding about algebraicity of cohomology classes (as we outline below—the key ingredient is a large abelian group of automorphisms whose eigenspaces in cohomology are studied), we can obtain precise knowledge about the quotient surfaces through the invariant (transcendental) cycles. In this section, we briefly recall the relevant theory.

13.2.1.1 Fermat Surfaces

Fermat surfaces have been demonstrating the interplay of arithmetic and geometry, as manifested in the Weil conjectures, ever since André Weil's groundbreaking paper concerning their zeta functions [Wei49]. Here we quickly outline the basics on Fermat surfaces relevant for our purposes.

The Fermat surface of degree $m \in \mathbb{N}$ is the hypersurface in $\mathbb{P}^3$ given by

$$S_m = \{x^m + y^m + z^m + w^m = 0\} \subset \mathbb{P}^3,$$

considered over some algebraically closed field k of characteristic $p \geq 0$. Here usually one restricts to the reduced (and smooth) case where $p \nmid m$. The automorphism group of S_m does not only contain the symmetric group on four letters, but also three independent scalings of coordinates by the group μ_m of m-th roots of unity. Of course, in the K3 case of $m = 4$, there are many additional non-linear automorphisms (for instance, originating from elliptic fibrations where we have translations by sections and the inversion with respect to the group law), but otherwise (if also $m \neq 1, 2$) the above linear automorphisms generate the full automorphism group of S_m (see [Shio87c], but we will not need this property); that is, in fact, unless we are in a very special characteristic as we shall highlight in Sect. 13.3, see in particular the discussion following (13.34).

Following ideas of many mathematicians including Weil, Katz, Ogus and one of us, one can analyse the induced action of the above linear automorphisms on the cohomology of S_m. Since the hyperplane section is fixed by linear automorphisms, we work with primitive cohomology $H^2_{\text{prim}}(S_m)$, i.e. the orthogonal complement of the hyperplane section H, understood in singular or ℓ-adic étale cohomology, after extending scalars by the m-th roots of unity, if necessary (or choosing $\ell \equiv 1 \mod m$ right away). It turns out that we can decompose $H^2_{\text{prim}}(S_m)$ into one-dimensional eigenspaces $V(\alpha)$ with character α for the induced action of scaling coordinates by $\mu_m^3 = \mu_m^4/\text{diagonal}$:

$$H^2_{\text{prim}}(S_m) \cong \bigoplus_{\alpha \in \mathfrak{A}_m} V(\alpha).$$

Here α runs through the following subset of the character group of μ_m^3 (identified as a subset of the character group of μ^4 compatible with the above quotient by the diagonal):

$$\mathfrak{A}_m := \left\{ \alpha = (a_0, a_1, a_2, a_3) \in (\mathbb{Z}/m\mathbb{Z})^4 \mid a_i \neq 0, \sum_{i=0}^{3} a_i = 0 \right\}.$$

What makes this situation even more special is the following fortunate property which allows one to decide which eigenspaces $V(\alpha)$ are algebraic (i.e. contained in the image of $\text{NS}(S_m)$ under the cycle class map, tensored with $\mathbb{Q}(\zeta_m)$ resp. $\mathbb{Q}_\ell(\zeta_m)$)). Over $\mathbb{C}$, this is related to the Galois action of the unit group $\text{Gal}(\mathbb{Q}(\mu_m)/\mathbb{Q}) \cong (\mathbb{Z}/m\mathbb{Z})^\times$ permuting the eigenspaces $V(\alpha)$ (given by multiplication of entries of α):

$$(\mathbb{Z}/m\mathbb{Z})^\times \ni u : \quad V(\alpha) \mapsto V(u \cdot \alpha).$$

In particular, interpreted inside $H^2(S_m, \mathbb{C})$, the complex conjugate eigenspace is given by

$$\overline{V(\alpha)} = V(-\alpha). \tag{13.24}$$

The algebraicity of the eigenspace $V(\alpha)$ can be checked in different ways, for instance by writing down some divisor in $\text{NS}(S_m) \otimes \mathbb{Q}(\zeta_m)$ which is an eigenvector for the action of μ_m^3 with character α. This has been achieved for several α, notably in [Shio82] for those in the subset $\mathfrak{D}_m \subset \mathfrak{A}_m$ to be defined in (13.26), but the most canonical approach builds on the Hodge type of $V(\alpha)$. For this purpose, we write $\alpha \in \mathfrak{A}_m$ in canonical representatives $0 < b_0, b_1, b_2, b_3 < m$ (i.e. $\alpha = (b_0, b_1, b_2, b_3)$) and define

$$|\alpha| = \sum_{i=0}^{3} \frac{b_i}{m} - 1 \in \{0, 1, 2\}. \tag{13.25}$$

The key point now is that we can determine the induced action of μ_m^3 on $H^{2,0}(S_m)$ and, in fact, decompose the latter space into eigenspaces (for $m \geq 4$, of course). To this end, consider the regular 2-form $\omega = dx_1 \wedge dx_2/x_3^{m-1}$ and the affine representation of $H^{2,0}(S_m)$ from (4.6) in the chart $x_0 = 1$:

$$H^{2,0}(S_m) = \{f\omega;\ \ f \in k[x_1, x_2, x_3],\ \ \deg(f) \leq m-4\}.$$

A basis is obtained by varying f over the monomials

$$x_1^i x_2^j x_3^l \quad \text{where} \quad i, j, l \geq 0, \ \ i+j+k \leq m-4.$$

Let $g \in \mu_m^3$ act affinely on S_m as

$$[x_1, x_2, x_3] \mapsto [\zeta_1 x_1, \zeta_2 x_2, \zeta_3 x_3], \qquad \zeta_i \in \mu_m \ (i = 1, 2, 3).$$

Then

$$g^*(x_1^i x_2^j x_3^l \omega) = \zeta_1^{i+1} \zeta_2^{j+1} \zeta_3^{l+1} x_1^i x_2^j x_3^l \omega,$$

so $x_1^i x_2^j x_3^l \omega$ generates the eigenspace $V(\alpha)$ for $\alpha = (m-3-i-j-l, i+1, j+1, k+1)$. Note that this works for any $\alpha \in \mathfrak{A}_m$ with $|\alpha| = 0$ (or equivalently, $\sum b_i = m$), so all these spaces $V(\alpha)$ have Hodge type $(2-|\alpha|, |\alpha|)$. By complex conjugation (13.24), the same formula holds true for the eigenspaces $V(\alpha) \subset H^{0,2}(S_m)$ (with $|\alpha| = 2$). It follows that $V(\alpha)$ has Hodge type $(1, 1)$ if and only if $|\alpha| = 1$ (i.e. $\sum b_i = 2m$). We thus deduce the following algebraicity criterion:

Criterion 13.21 *The eigenspace $V(\alpha)$ is algebraic over $\mathbb{C}$ if and only if every member of its orbit under* $\mathrm{Gal}(\mathbb{Q}(\mu_m)/\mathbb{Q})$ *has Hodge type* $(1, 1)$.

To be explicit, $V(\alpha)$ is algebraic if and only if $|t \cdot \alpha| = 1$ for each $t \in (\mathbb{Z}/m\mathbb{Z})^\times$. In particular, we can compute the Picard number combinatorially by singling out all $\alpha \in \mathfrak{A}_m$ whose Galois orbit does not leave the Hodge type $(1, 1)$:

$$\rho(S_m) = 1 + \#\mathfrak{B}_m, \quad \mathfrak{B}_m = \{\alpha \in \mathfrak{A}_m;\ |t \cdot \alpha| = 1 \ \ \forall t \in (\mathbb{Z}/m\mathbb{Z})^\times\}.$$

Often these $\alpha \in \mathfrak{B}_m$ are referred to as the *Hodge classes* in $\mathfrak{A}_m$. The computations become especially transparent in special cases. For example, if the degree m is relatively prime to 6, then the structures are uniform enough to prove, for instance, that $\mathrm{NS}(S_m)$ is generated over $\mathbb{Z}$ by lines by [Deg15] (see also [Shio82] where this was initially proved up to finite index, and [SSvL10] which introduced a supersingular reduction technique to cover all m up to 100).

For later reference, we point out how the span of lines relates to the eigenspaces within the primitive second cohomology group $H^2_{\text{prim}}(S_m)$, i.e. the orthogonal complement of the subspace spanned by the hyperplane class (which is invariant under any linear transformation anyway). To this end, we introduce the set $\mathfrak{D}_m$ of *decomposable elements* in $\mathfrak{B}_m$:

$$\mathfrak{D}_m = \{\alpha = (a_0, a_1, a_2, a_3) \in \mathfrak{B}_m;\ \exists i \neq j : a_i = -a_j\}. \tag{13.26}$$

Then the main result from [Shio82] describes the span of lines inside $H^2(S_m, \mathbb{C})$ (or projected orthogonally to primitive cohomology) in terms of decomposable elements:

$$\mathbb{C}\langle\text{classes of lines}\rangle = \bigoplus_{\alpha\in\mathfrak{D}_m} V(\alpha) \oplus \mathbb{C}H \subset H^2(S_m, \mathbb{C}).$$

What might be surprising at first is that, although the degree m of S_m can be arbitrarily large, one can also give a fairly concrete description of the other algebraic eigenspaces. In detail, they fall into essentially four cases. There are three kinds of standard elements, given in terms of $a \in \mathbb{Z}/m\mathbb{Z}$ by the following Hodge classes $\alpha \in \mathfrak{B}_m$ and their coordinate permutations:

$$\begin{aligned}
\alpha &= (a, a + m/2, m/2, m - 2a) \quad 2 \mid m,\ 2a \neq 0 \quad (\nu = 1),\\
\alpha &= (a, a + m/3, a + 2m/3, m - 3a) \quad 3 \mid m,\ 3a \neq 0 \quad (\nu = 2),\\
\alpha &= (a, a + m/2, 2a + m/2, m - 4a) \quad 2 \mid m,\ 4a \neq 0 \quad (\nu = 3).
\end{aligned}$$

The above Hodge classes together with all their permutations are usually called *standard classes*. By [AS83] (see also [AS10]), each of the standard classes is represented by an explicit algebraic curve on S_m, obtained as an irreducible component of the intersection with certain hypersurfaces such as $w^{\nu+1} = \gamma_\nu xyz^{\nu-1}(\nu = 1, 2, 3,\ \gamma_\nu \in \bar{\mathbb{Q}}$ given).

With all decomposable and standard Hodge classes realized geometrically, here comes the key insight conjectured in [Shio82] needed to produce a closed formula for the Picard number of S_m: all other Hodge classes $\alpha \in \mathfrak{A}_m$, called *exceptional*, arise from a finite number of *primitive exceptional classes* where $\gcd(a_0, \dots, a_3) = 1$, as proven by [Aok83]; in fact, there are exactly 22080 primitive exceptional elements in number, and all occur on the Fermat surfaces of degree $m \leq 180$ not relatively prime to 6 (although by far not all of them have been realized geometrically yet, see Sect. 13.2.3 for an instructive example using rational elliptic surfaces and Mordell–Weil lattices). As an application and illustration, the reader is invited to inspect the closed formula for the Picard number $\rho(S_m)$ from [Aok83] (confirming the conjecture from [MN81, Shio82]).

For our purposes, an alternative approach concentrating on transcendental cycles will be of practical importance. Namely, one can compute the *Lefschetz number* over $\mathbb{C}$,

$$\lambda(S) = b_2(S) - \rho(S), \tag{13.27}$$

by summing up the sizes of the $\mathrm{Gal}(\mathbb{Q}(\mu_m)/\mathbb{Q})$-orbits of all $\alpha \in \mathfrak{A}_m$ of Hodge type $(2, 0)$, i.e. with $\sum_i b_i = m$. Note that the Lefschetz number is a birational invariant (as indicated in Remark 12.10) which over $\mathbb{C}$ is bounded from below by twice the geometric genus p_g (by Lefschetz $(1, 1)$-theorem, cf. 4.15).

Example 13.22 ($m = 4$) On the complex Fermat quartic surface S_4, there is only one character $\alpha \in \mathfrak{A}_4$ of Hodge type $(2, 0)$:

$$\alpha = (1, 1, 1, 1).$$

Since its orbit has length 2, we find $\lambda = 2$, $\rho = 20$. i.e. S_4 is a singular K3 surface, and in fact $\mathrm{NS}(S_4)$ is generated by lines.

Example 13.23 ($m = 5$) The complex Fermat quintic S_5 admits four characters $\alpha \in \mathfrak{A}_5$ of Hodge type $(2, 0)$:

$$\alpha = (1, 1, 1, 2) \text{ and coordinate permutations,}$$

in agreement with $p_g(S_5) = 4$. Since each character has orbit of length 4 under $(\mathbb{Z}/5\mathbb{Z})^\times$, and since the orbits are disjoint, we find $\lambda = 16$ and thus $\rho = 37$.

13.2.1.2 Delsarte Surfaces

A Delsarte surface X is a minimal smooth surface birational to a surface X_0 in $\mathbb{P}^3$ which can be defined by an irreducible homogeneous polynomial of degree $d \in \mathbb{N}$ consisting of four monomials:

$$X_0 = \left\{ \sum_{i=1}^{4} a_i x_1^{e_{i,1}} x_2^{e_{i,2}} x_3^{e_{i,3}} x_4^{e_{i,4}} = 0 \right\} \subset \mathbb{P}^3,$$

where $0 \leq e_{i,j} \leq d$ for any i, j and $\sum_j e_{i,j} = d$ for any i. Since $a_i \neq 0$ for all i, we can rescale the coordinates (at least if $k = \bar{k}$) to achieve the normalization $a_i = 1$ which we will retain (up to sign) throughout this section.

Regarding the Picard number, the instrumental property of Delsarte surfaces is that each Delsarte surface is birational to the quotient S_m/G of a Fermat surface S_m of degree m by a finite subgroup $G \subset \mu_m^3$. By [Shio86], there is an algorithm which determines a choice of m and G in terms of the cofactor matrix associated with the exponent matrix

$$(e_{i,j})_{1 \leq i,j \leq 4}$$

of the defining polynomial of X_0. Note that the quotient S_m/G is Galois (i.e. the extension of functions fields $k(S_m)/k(S_m/G)$ is Galois), since it sits at an intermediate level inside the abelian Galois quotient

$$S_m \to S_m/(\mu_m)^3 = S_1.$$

The Delsarte surface X arises from S_m/G by resolution of singularities and minimalization (if necessary). Hence it may be a non-trivial task to keep track exactly of the algebraic part of $H^2(X)$ in terms of S_m and the quotient map. However, the transcendental cycles, comprising the orthogonal complement $T(X)$ of $\mathrm{NS}(X)$ inside $H^2(X)$ (which we previously alluded to as the transcendental lattice, but now considered as a $\mathbb{Q}$-Hodge structure in singular cohomology), are governed completely by S_m, since the Lefschetz number is a birational invariant. In detail, $T(X)$ decomposes exactly into the non-algebraic eigenspaces $V(\alpha) \subset H^2(S_m)$ which are invariant under the induced G-action. Explicitly, if $g \in G$ acts on S_m as

$$\begin{array}{llll} g: & S_m & \to & S_m \\ & [x_0, x_1, x_2, x_3] & \mapsto & [x_0, \zeta_1 x_1, \zeta_2 x_2, \zeta_3 x_3], \qquad \zeta_i \in \mu_m \ (i = 1, 2, 3), \end{array}$$

then $V(\alpha)$ is invariant under g^* if and only if

$$\alpha(g) = \prod_{i=1}^{3} \zeta_i^{a_i} = 1 \quad (\alpha = (a_0, a_1, a_2, a_3)).$$

As in Sect. 13.2.1.1 this implies that we can compute the Lefschetz number of X purely combinatorially (see [Shio86]); we shall use the suggestive, though slightly abusing notation

$$\lambda(S) = \lambda(S_m/G) = \lambda(S_m)^G. \tag{13.28}$$

Eventually, the Picard number of X can be computed a posteriori using (13.27) (with $b_2(S)$ computed as in Example 4.22). Because $\lambda(X)$ is a birational invariant, the above technique applies to any affine or birational model of X, in particular to surfaces in weighted projective space and thus to elliptic surfaces (as soon as they admit a model defined by four monomials). To illustrate this approach, we shall discuss two K3 examples in detail. They are related to the family from Theorem 13.19 and give a good idea of the arguments to be used in the next section (compare the proof of Theorem 13.26).

Example 13.24 Consider the elliptic K3 surface X given in affine Weierstrass form

$$X: \quad y^2 = x^3 + t^7 + 1.$$

It has a singular fibre of type II* at ∞, so $U \oplus E_8^- \subseteq \mathrm{NS}(X)$ and $\rho(X) \geq 10$. There is an obvious cover by S_{42} such that X is birational to the quotient of S_{42} by the abelian group $G = \mu_{21} \times \mu_{14} \times \mu_6$ (although starting from the isomorphic model

$y^2 + y = x^3 + t^7$, one can even do with S_{21}, but this would make the shape of G and thus the computations to follow more complicated (but has benefit of working in characteristic two as well)). The unique G-invariant eigenspace $V(\alpha)$ of Hodge type $(2, 0)$ is given by $\alpha = (21, 14, 6, 1)$. Since this has G-orbit of length 12, we infer

$$\lambda(X) = \lambda(S_{42})^G = 12.$$

Hence $\rho(X) = 10$ and $\mathrm{NS}(X) = U \oplus E_8^-$.

Example 13.25 Consider the elliptic K3 surface $X = X^{(6)}(0, 12^3)$ in the notation of Sect. 12.2.6, i.e. the elliptic fibration arising from the Kummer surface of the elliptic curves on j-invariants 0 and 12^3 by a cyclic base change of degree 3 (compare Sect. 12.2.6). Then $\rho(X) = 18$ by (12.13) and Theorem 12.13. Here we pursue an alternative line of argument starting from the Delsarte Weierstrass form

$$X: \quad y^2 = x^3 + t^{12} + 1.$$

There is an obvious cover by S_{12} with Galois group $G = \mu_6 \times \mu_4$. The only G-invariant transcendental eigenspaces $V(\alpha)$ occur within a single $(\mathbb{Z}/12\mathbb{Z})^\times$-orbit,

$$\alpha = (6, 4, 1, 1), (6, 8, 5, 5), (6, 4, 7, 7), (6, 8, 11, 11),$$

with Hodge types $(2, 0)$, $(1, 1)$, $(1, 1)$, $(0, 2)$ respectively. Hence $\lambda(X) = 4$ in agreement with our previous calculations.

We point out that both examples could have been treated without any reference to Delsarte surfaces by using purely non-symplectic automorphisms of order 21 resp. 12 (i.e. with induced action of the same order on $H^{2,0}(X)$), or using Inose's isogeny notion from Sect. 12.2.5. Indeed, at least the first example admits a non-symplectic automorphism of maximal order, see [Kon92].

13.2.2 Elliptic Delsarte Surfaces and Rank Record over $\mathbb{C}(t)$

We now return to the rank record over $\mathbb{C}(t)$. To this end, we start by giving a proof of Theorem 13.19. Recall that for fixed $n \in \mathbb{N}$, we deal with the complex elliptic surface given in Weierstrass form

$$X: \quad y^2 = x^3 + t^{360n} + 1. \tag{13.29}$$

Note that this elliptic fibration is isotrivial without reducible fibres; indeed the only singular fibres are cuspidal curves (Kodaira type II) and they are located at the zeroes of $t^{360n} + 1$. In particular, X has Euler–Poincaré characteristic $e(X) = 720n$ and $\chi(\mathcal{O}_X) = 60n$. In order to prove Theorem 13.19, it suffices to prove that

$$\rho(X) = 70$$

regardless of the choice of n. In what follows, we will sketch the proof of a somewhat stronger result which will imply Theorem 13.19. For this, we consider a slightly more general series of elliptic surfaces, given by the Weierstrass form

$$X_m : \quad y^2 = x^3 + t^m + 1 \quad (m \in \mathbb{N}). \tag{13.30}$$

Theorem 13.26 (Shioda) *For any $m \in \mathbb{N}$, the complex elliptic surface X_m has Mordell–Weil rank $r \leq 68$. Equality is attained if and only if $360 \mid m$.*

Proof If the MW rank for X_m is denoted by r_m, then it is obvious that we have $r_m \leq r_{m'}$ if $m \mid m'$. Thus, for the inequality bit of Theorem 13.26, it suffices to consider the case where m is a multiple of 6.

Let $m = 6d$. Then all the singular fibres of X_m are of type II at the zeroes of $t^m + 1$ (there is a regular fibre at $t = \infty$), and thus we have

$$e(X_m) = 2m \quad \text{and} \quad b_2(X_m) = 2m - 2.$$

Consider the finite subgroup

$$G = \mu_{2d} \times \mu_{3d} \times \{1\} \subset \mu_m^3$$

such that S_m/G is birational to X_m. Let $L = \mu_3 \times \mu_2 \times \mu_m \subset \mu_m^3$ denote the group of characters of μ_m^3 which are trivial on G (in particular $L \cong \mu_m^3/G$). Then (similar to Example 13.25) the set $L \cap \mathfrak{A}_m$ consists of $2m - 4$ elements,

$$\pm\, \alpha_i = (3d, 2d, i, d - i) \in (\mathbb{Z}/m\mathbb{Z})^4; \quad i \in \mathbb{Z}/m\mathbb{Z}, \;\; i \neq 0, d. \tag{13.31}$$

It follows that $H^2(X_m)$ is spanned by the $2m - 4$ invariant eigenspaces $V(\alpha)$ together with the hyperplane section of (13.30) and the zero section O (at ∞ in terms of the affine model (13.30)). By (13.28), we have $\lambda = |L \cap (\mathfrak{A}_m \setminus \mathfrak{B}_m)|$. Hence Corollary 6.7 gives

$$r_m = \rho(S_m) - 2 = b_2(S_m) - \lambda - 2 = 2m - 4 - \lambda = |L \cap \mathfrak{B}_m|.$$

We now investigate the different kinds of Hodge classes contained in L. To this end, take $\alpha = \alpha_i \in L \cap \mathfrak{B}_m$. Then α is *decomposable*, i.e. $\alpha \in \mathfrak{D}_m$ if and only if either $i = 3d$ or $d - i = 3d$; that is,

$$\alpha = d \cdot (3, 2, 3, 4) \;\text{ or }\; \alpha = d \cdot (3, 2, 4, 3),$$

so there are exactly four decomposable elements $\alpha \in \mathfrak{D}_m$ invariant under G.

Next assume that $\pm\alpha$ is a standard element, as listed in Sect. 13.2.1.1. An elementary case-by-case analysis shows that $\pm\alpha$ falls in the first case ($\nu = 1$), i.e. $\pm\alpha$ equals some $(a, a + m/2, m/2, m - 2a) = (a, a + 3d, 3d, 6d - 2a)$ with $a \neq 0, 3d$ up to permutation. Indeed, this occurs if and only if

$$\alpha = d \cdot (3, 2, 2, 5) \text{ or } \alpha = d \cdot (3, 2, 5, 2).$$

Thus there are four standard elements $\alpha \in \mathfrak{B}_m$ invariant under G. Note that all the above eight invariant Hodge classes are divisible by d, i.e. induced from the Fermat sextic surface (case $d = 1$, $m = 6$). We shall now see, and exploit, that this behaviour follows a general pattern. To this end, let $\alpha \in \mathfrak{B}_m$ be a G-invariant Hodge class which is not primitive, i.e. where the a_i generate a proper subgroup $I \subset \mathbb{Z}/m\mathbb{Z}$. This means that α is induced from a Hodge class α' on the Fermat surface $S_{m'}$ where $m' = m/n$ for the minimal generator $n \geq 2$ of I; here $\alpha = n \cdot \alpha'$.

Lemma 13.27 *In the above notation, the Hodge class α' descends to the elliptic surface $X_{m'}$.*

Proof Since α is G-invariant, it takes the shape of (13.31). Hence $n \mid 2d$ and $n \mid 3d$ which implies $n \mid d$. It follows that there is a commutative diagram

$$\begin{array}{ccc} S_m & \dashrightarrow & S_{m'} \\ \downarrow & & \downarrow \\ X_m \sim S_m/G & \dashrightarrow & S_{m'}/G' \sim X_{m'} \end{array}$$

with the subgroup $G' = \mu_{3d/n} \times \mu_{2d/n} \times \{1\} \subset \mu_{m'}^3$. In particular, α' is G'-invariant if and only if $\alpha = n \cdot \alpha'$ is G-invariant. □

Note that since $n \mid d$, we still have $6 \mid m'$, i.e. m' fulfils the starting assumption of these arguments and $X_{m'}$ is among the surfaces under consideration. Thus we have seen:

Fact 13.28 *In order to compute all Hodge classes on S_m descending to X_m, it suffices to compute the primitive Hodge classes on $S_{m'}$ descending to $X_{m'}$ for all $6 \mid m' \mid m$.*

Since we have already computed all G-invariant decomposable and standard elements, it remains to deal with the exceptional elements in $\mathfrak{B}_m$. By Fact 13.28, this problem reduces to computing all *primitive* exceptional elements in $\mathfrak{B}_{m'}$, and among them the G'-invariants for the respective subgroup G' as above. By checking the list of primitive exceptional elements in $\mathfrak{B}_{m'}$ (which is a finite set of size 22080, with the largest $m' \leq 180$ (see [MN81, Aok83])), we find that, up to the action of the unit group $(\mathbb{Z}/m'\mathbb{Z})^\times$, the G'-invariant exceptional elements α are obtained as the $(\mathbb{Z}/m'\mathbb{Z})^\times$-orbits of the following elements (with $d' = d/n = m'/6$):

d'	m'	α
2	12	(6, 8, 9, 1), (6, 8, 1, 9)
3	18	(9, 12, 14, 1), (9, 12, 1, 14)
4	24	(12, 16, 19, 1)
5	30	(15, 20, 24, 1), (15, 20, 1, 24)
10	60	(30, 40, 49, 1)

By inspection, each α has $(\mathbb{Z}/m'\mathbb{Z})^\times$-orbit of size $\phi(m')$, with ϕ the Euler totient function. In particular, we obtain a closed formula for the MW-rank r_m of X_m by summing up 8 (for the contribution from decomposable and standard elements) and all the orbit lengths for those m' from the table which actually divide the given m. In particular, the maximum MW-rank is attained exactly by the X_m with $360 \mid m$:

$$r_{\max} = 8 + 2\phi(12) + 2\phi(18) + \phi(24) + 2\phi(30) + \phi(60) = 68.$$

This completes the proof of the inequality in Theorem 13.26. To check the equivalence statement, it remains to note that if $d = m/6$ were to attain the rank maximum at 68, then we would still have $60 \mid d$, so d would also be 6-divisible. Thus d would be covered by the above analysis, and $360 \mid d$ after all. □

Note that by specialization, we obtain infinitely many elliptic curves over some number field with rank at least 68!

Remark 13.29 The record rank for a non-constant elliptic curve over $\mathbb{C}(t)$ which is not isotrivial, seems to be 56, attained for instance by the following curve from [Shio86] (also to be found in [Sti87]):

$$y^2 = x^3 + x + t^{1260}.$$

It is unknown whether these rank records, for non-constant resp. non-isotrivial elliptic fibrations over $\mathbb{P}^1$, persist if we consider all (non-constant) elliptic curves over $\mathbb{C}(t)$, and actually we are hesitant to formulate any kind of conjecture, in particular whether the ranks may be bounded or not. To say the least, recently it was proven by Heijne in [Hei12] that the above records hold for Delsarte elliptic surfaces, i.e. those elliptic surfaces which can be defined by an irreducible polynomial of four monomials taking one of the following shapes:

- an affine cubic curve over $\mathbb{C}(t)$ including
- a Weierstrass form over $\mathbb{C}(t)$;
- a double cover of $\mathbb{P}^1$ ramified in four points over $\mathbb{C}(t)$.

Theorem 13.30 (Heijne) *Let X be a Delsarte elliptic surface over $\mathbb{C}(t)$. Then*

$$\textit{rank}\ \mathrm{MW}(X) \leq 68,$$

with equality if and only if X is isomorphic to one of the surfaces from (13.29).

Heijne's first main tool consists of the Newton polygon which is associated with the defining polynomial with four monomials only. This is shown to have exactly four vertices, but only one interior point. The elementary classification of these polygons leads to 12 cases which are analysed independently. Refining the arguments from [Shio86], Heijne then proceeds to derive Theorem 13.30 by delicate combinatorial computations.

Remark 13.31 It is possible, but unclear whether the above result applies to all Delsarte surfaces which admit an elliptic fibration. As an illustration, consider the K3 surface defined as a double sextic by the affine equation

$$X: \quad w^2 = x^5 + y^5 + 1.$$

Then, of course, X admits elliptic fibrations, for instance, by Theorem 11.24, or by a linear change of variables which takes the double covering to the third shape above (see [Bran19]). We are not sure, however, whether these fibrations admit representations of either of the above shapes, thus falling into Heijne's classification. At any rate, the Mordell–Weil ranks will not exceed 18 as long as this only happens for K3 surfaces.

Lattice structure of MWL(X_m)

It is quite instructive to see how the lattice structures of MWL(X_m) for all $m \in \mathbb{N}$ can be given. These were determined by Usui in a series of papers [Usu00+] which we shall quickly comment on.

For $m = 1, \ldots, 6$, the elliptic surface X_m is rational, so MWL(X_m) is given by Classification Theorem 8.8. The key result from the first paper of [Usu00+] was the insight that any MWL(X_m) can be computed from the following building blocks (which should look familiar, compare the degrees m' appearing in the proof of Theorem 13.26):

$$\mathrm{MWL}(X_m), \quad m = 2, 3, 4, 5, 6, 9, 12, 18, 24, 30, 60,$$

using base change (see Theorem 6.53) and also the involution on X_m induced by the symmetry $t \mapsto 1/t$. Incidentally, also the elliptic K3 surfaces $X^{(n)}(0, 12^3)$ from Sect. 12.2.6 arising from the Inose pencil for the elliptic curves with j-invariants 0 and 12^3 play an important role (see also Example 13.25).

As an application, Usui gives in the fourth paper of [Usu00+], for instance, the number of minimal sections for all m (which are exactly the integral sections if $m \equiv 0$ or $-1 \mod 6$). As a sample, for $m = 72N$, there are 2472 minimal sections (of height $24N$).

13.2.3 Exceptional Hodge Classes

In the proof of Theorem 13.26 (and Theorem 13.19) given above, we have identified several exceptional Hodge classes which correspond to the characters α occurring in the computation of the rank of the Mordell-Weil lattices. See, in particular, the table on Sect. 13.2.2.

Reversing this approach, we can construct some explicit curves on Fermat surfaces which will represent these exceptional characters. Roughly speaking, these curves are induced by some sections of an elliptic Delsarte surface X, under pull-back by a map from the Fermat surface S to X.

As an illustration, we borrow an instructive example of a Delsarte rational elliptic surface, the so-called (E_8, q_0)-model, from [Shio99a]:

$$X: \quad y^2 = x^3 + t^5 + q_0,$$

where $q_0 \in k$ can, of course, be taken to be 1 over an algebraically closed field k. Recall that this surface has appeared in Examples 5.8, 6.49 and also prominently in the context of polynomial sections in Example 9.37. We have seen that

$$\mathrm{MWL}(X) \cong E_8,$$

so let us fix any section $P \in \mathrm{MW}(X)$ of height $h(P) = 2$. Previously, we used the cyclic subgroup $G \subset \mathrm{Aut}(X)$ of order $m = 30$ generated by the automorphism

$$g_1: \quad (x, y, t) \mapsto (\zeta_3 x, -y, \zeta_5 t),$$

where ζ_j denotes a primitive j-th root of unity. Of course, G acts on $\mathrm{NS}(X)$; in fact, it preserves the class of a fibre F and the zero section O, so we will concentrate on the induced action on

$$M = \langle (O), F \rangle^{\perp} = \mathrm{MWL}(X)^{-}.$$

Here G acts on $\mathrm{MWL}(X)$ by isometries, and in fact, the action is compatible with the specialization map at ∞. Writing $G \cong \mu_m$, we have $G^{\vee} \cong \mathbb{Z}/m\mathbb{Z}$ and obtain the orthogonal decomposition

$$M \otimes \mathbb{C} = \oplus_{\chi \in G^{\vee}} V(\chi)$$

into the eigen-subspaces $V(\chi)$. Let $\chi_n : G \to \mathbb{C}^{\times}$ denote the character $\chi_n(g_1) = (\zeta_m)^n$ $(n \in \mathbb{Z}/m\mathbb{Z})$.

Theorem 13.32 *In the above notation, we have*

$$V(\chi_n) \neq 0 \Longleftrightarrow n \in (\mathbb{Z}/m\mathbb{Z})^{\times}.$$

Moreover, $\dim V(\chi_n) = 1$ *for each* $n \in (\mathbb{Z}/m\mathbb{Z})^{\times}$.

Proof (Outline). For ease of notation, we write $V(n) = V(\chi_n)$ and $C = (P)$ for any section $P \in \mathrm{MW}(X)$ with $h(P) = 2$. It is enough to prove $V(1) \neq 0$. For we have then $V(n) \neq 0$ also for any $n \in (\mathbb{Z}/m\mathbb{Z})^{\times}$ by considering the action of $Gal(\mathbb{Q}(\zeta_m))/\mathbb{Q})$. Hence the assertion results from the following inequalities:

$$8 = \mathrm{rk}(M) = \sum_{n \leq m} \dim V(n) \geq \sum_{n \in (\mathbb{Z}/m\mathbb{Z})^{\times}} \dim V(n) \geq |(\mathbb{Z}/m\mathbb{Z})^{\times}| = 8.$$

Fix $\chi = \chi_1$ and let

$$\omega_C := \sum_{g \in G} \chi_1(g^{-1}) g^*(C) \in \mathrm{NS}(X) \otimes \mathbb{C},$$

the image of C under the projector $V \to V(\chi_1)$ (up to constant factor). It is easy to see that ω_C belongs to $V(1) = V(\chi_1)$. We prove that $V(1) \neq 0$ by following the standard procedure from [Shio82, AS83], namely by showing that the intersection number of ω_C and its complex conjugate is non-zero (for a suitable choice of P). Explicitly, we have $(\omega_C . \bar{\omega_C}) = m(C.\omega_C)$ and

$$(C.\omega_C) = \sum_{g \in G} \chi_1(g^{-1})(C.g^*(C)) = \sum_i \zeta_m^{-i}(C.(g_1^i)^*(C)).$$

Hence the proof is reduced to the computation of the intersection numbers of C and $C_i := (g_1^i)^*(C)$ for all $i = 0, \dots, m-1 = 29$. Note that $C_i = (P_i)$ for the section $P_i = (g_1^i)^*(P)$.

To verify that $(C.\omega_C) \neq 0$, it is convenient to pass to the reduction X_p of X at a good split prime p. The terminology *split* means that the universal polynomial Φ of degree $N = 240$ for our Mordell–Weil lattice isomorphic to E_8 which is given in Theorem 9.21 (or, in the present context, in [Shio99a]) splits into linear factors over the prime field $\mathbb{F}_p$.

In the case under consideration, we can choose $p = 6271$, for instance, with a fixed primitive m-th root of unity, say, $593 \in \mathbb{F}_p$.

Lemma 13.33 *In the above notation with $p = 6271$, let $P \in \mathrm{MW}(X)$ correspond to the root $u_0 = 576$ of Φ. The following table lists, for $i = 0, \dots, 15$, the intersection number $(C.C_i)$ and the height pairing $\langle P, P_i \rangle$:*

i	0	1	2	3	4	5	6	7	8	9	10	11	12	13	14	15
$(C.C_i)$	−1	1	0	0	1	0	1	1	1	1	2	1	2	2	1	3
$\langle P, P_i \rangle$	2	0	1	1	0	1	0	0	0	0	−1	0	−1	−1	0	−2

Completion of proof of Theorem 13.32. From the data given in Lemma 13.33, the height pairings of P with $P_i = -P_{i-15}$ for $i = 16, \dots, 29$ are obtained by linearity which in turn yields $(C.C_i)$. We obtain from the reduction modulo p that

$$(C.\omega_C) = -2(3 + \zeta_m^2 + \zeta_m^3 + \zeta_m^{27} + \zeta_m^{28}) \neq 0.$$

In particular, $(\omega_C . \bar{\omega_C}) \neq 0$ as claimed. □

Proof of Lemma 13.33. We apply the addition theorem on X_p to compute the (x, y)-coordinates (x, y) of the section $P - P_i$ as a pair of rational functions in $\mathbb{F}_p(t)$ (which, of course, can also be done over $\mathbb{C}$, but then the functions are much harder to control). The intersection number $(C.C_i)$ (and thus the height pairing $\langle P, P_i \rangle$) is easily determined by this data, in particular because no distinct sections intersect on the singular fibre at ∞ as we have seen in Example 9.37. It follows that $(C.C_i)$ equals exactly half the degree of the pole divisor of x (after cancelling common factors), or equivalently, one third of the degree of the pole divisor of y. □

To conclude this section, we return to our original problem of determining explicit curves on Fermat surfaces corresponding to exceptional Hodge classes. We point out that in the way sketched above, we find some explicit curve on the Fermat surface $S = S_m$ which can be used to represent the exceptional class (character) $\alpha = (15, 20, 24, 1) \mod m = 30$. The curve arises by pulling back the curve $C = (P)$ from the elliptic surface X by the rational quotient map

$$\psi : S \to X,$$

raising affine coordinates to suitable powers (corresponding to the Galois group $\mu_{15} \times \mu_{10} \times \mu_6$). Presently, α descends to the character χ_{-1} on $G = \mu_m$, as g_1 is induced by the automorphism $[\zeta_m, \zeta_m, \zeta_m, 1] \in \mathrm{Aut}(S)$ and $\alpha([\zeta_m, \zeta_m, \zeta_m, 1]) = (\zeta_m)^{-1}$.

Remark 13.34 There are several other cases in [Shio99a] which can be treated in a similar way. This gives a few further explicit representations of exceptional Hodge classes. In particular, all classes α listed in the table on Sect. 13.2.2 can be related to rational elliptic surfaces, except for the last one, $\alpha = (30, 40, 49, 1) \mod m = 60$.

Finally, this α comes from the Delsarte elliptic K3 surface $y^2 = x^3 + t(t^{10} + 1)$ (cf. [Usu00+]), and the analogous result to Theorem 13.32 holds with $m = 60$.

13.3 Rank Problem in Positive Characteristic

Let k denote a field of positive characteristic p. Upon considering the rank problem for elliptic curves over $k(t)$, or equivalently for an elliptic surface

$$X \to \mathbb{P}^1,$$

the whole set-up changes drastically compared to characteristic zero; notably, this is due to the absence of Lefschetz' bound (4.15). As explained around (4.14), there is only the weaker bound

$$\rho(X \otimes \bar{k}) \leq b_2(X)$$

due to Igusa. Indeed, the algebraic part in $H^2(X \otimes \bar{k}, \mathbb{Q}_\ell)$ (the image of $\mathrm{NS}(X \otimes \bar{k})$ under the ℓ-adic cycle class map) can be much larger than in characteristic zero, even to the extent that all of $H^2(X \otimes \bar{k}, \mathbb{Q}_\ell)$ may be algebraic. Recall that surfaces with this property (i.e. $\rho(X \otimes \bar{k}) = b_2(X)$) are called (Shioda-) supersingular; they can be seen as the main reason why the rank problem has a positive answer in positive characteristic:

Theorem 13.35 *For any field k of positive characteristic, elliptic curves over $k(t)$ have unbounded rank.*

In this section, we sketch some of the ideas which lead to Theorem 13.35. We first treat the ground-breaking joint work of Tate and Shafarevich [ShT67] in Sect. 13.3.1 which settled Theorem 13.35 in the first instance. Then we put this into the context of the previous section by extending the techniques for Fermat surfaces and Delsarte surfaces to positive characteristic, first over an algebraically closed field in Sect. 13.3.2 and then over the prime field in Sect. 13.3.3.

13.3.1 Tate–Shafarevich Approach

In order to review the work of Tate and Shafarevich [ShT67], we start by explaining the main ideas over an algebraically closed field $k = \bar{k}$ of characteristic $p > 0$. Consider two abelian varieties A_1, A_2. If both are supersingular (i.e. $\rho(A_i) = b_2(A_i)$ for $i = 1, 2$ outside dimension one), then so is the product $A_1 \times A_2$. Naturally, the same holds true for any variety X of the same dimension as $A_1 \times A_2$ which is dominated by $A_1 \times A_2$.

Applying this to a supersingular elliptic curve E and an algebraic curve C whose Jacobian variety $\mathrm{Jac}(C)$ is supersingular, we find that any surface birational to a quotient of $E \times C$ will be supersingular, too. Assuming $p \neq 2$, we apply this to the situation where C is hyperelliptic:

$$C: \quad u^2 = f(t), \quad f \in k[t].$$

Here we may assume f to be square-free, and write $g = g(C) = (d-2)/2$, where $d = \deg(f)$ if this is even, or $d = \deg(f) + 1$ otherwise (i.e. d is the degree of f thought homogeneously by interpreting C as a double cover of $\mathbb{P}^1$). We take E to be given in extended Weierstrass form

$$E: \quad y^2 = x^3 + a_2x^2 + a_4x + a_6, \quad a_i \in k \ \ (i = 0, 1, 2).$$

Lemma 13.36 *The following quadratic twist of E has rank* $2d - 4$ *over* $k(t)$*:*

$$E': \quad f(t)\, y^2 = x^3 + a_2x^2 + a_4x + a_6.$$

Proof By construction, E' gives an isotrivial elliptic surface X'. The base change of E' to $k(C)$ is birational to the product $E \times C$. Since the latter product is supersingular by assumption, so is X'. It remains to compute the rank r of E', i.e. the Mordell–Weil rank r of X'. Note that X' has exactly d fibres of type I_0^* and no other singular fibres, so $e(X') = 6d$ and $b_2(X') = 6d - 2$. By Corollary 6.7 it follows from the supersingularity that

$$6d - 2 = \rho(X') = 2 + 4d + r \implies r = 2d - 4.$$

This exactly is the claim of the lemma. □

Assuming for the moment that there are supersingular hyperelliptic curves of arbitrary genus in any characteristic (see the discussion around (13.33)), we derive the following consequence:

Corollary 13.37 *For a prime p, let k be an algebraically closed field of characteristic p. The ranks of non-constant elliptic curves over $k(t)$ are unbounded.*

It remains to transfer this geometric result to the arithmetic setting of non-closed ground fields, and notably over $\mathbb{F}_p$. This is achieved by means of the Tate conjecture 12.1 which we recall for the reader's convenience in a different formulation:

Conjecture 13.38 (Tate conjecture) Let X be a smooth algebraic surface defined over some finite field $\mathbb{F}_q$. Then the Picard number of X over $\mathbb{F}_q$ equals exactly the order of the pole of its zeta function at $s = 1$.

Put differently, consider the reciprocal characteristic polynomial P_2 of the induced action of Frob_q^* on $H^2_{\text{ét}}(X \otimes \bar{\mathbb{F}}_q, \mathbb{Q}_\ell)$ for any prime $\ell \neq p$ as in Sect. 10.5.2. Then $P_2 \in 1 + T\mathbb{Z}[T]$ is independent of the choice of $\ell \neq p$, and the validity of the Tate conjecture for X over $\mathbb{F}_q$ is equivalent to a factorization 0

$$P_2(T) = (1 - qT)^{\rho(X)} \hat{P}(T),$$

where $\hat{P}(1/q) \neq 0$ (see (12.2)). For later use, we emphasize that the special value of the factor $\hat{P}$ at $T = 1/q$ still carries a lot of useful information about $\text{NS}(X)$. This is measured in the Artin–Tate conjecture [Tat66b] stating that

$$\hat{P}(1/q) = \frac{|\operatorname{Br}(X)| \cdot |\det(\operatorname{Num}(X))|}{q^{\alpha(X)} (\#\operatorname{NS}(X)_{\text{tors}})^2}. \tag{13.32}$$

Here $\text{Br}(X)$ denotes the Brauer group of X over $\mathbb{F}_q$, and $\alpha = \chi(\mathcal{O}_X) - 1 + \dim \text{PicVar}(X)$. The Artin–Tate conjecture is equivalent to the Tate conjecture by [Mil75] (for characteristic 2, this requires input from [LLR05] which proves that not only in odd characteristic the size of the Brauer group is a square, so that Milne's original argument also goes through in characteristic two).

In general, the Tate conjecture seems to be wide open, but fortunately it is known for central cases:

1. abelian varieties [Tat66a],
2. products of curves [Mil75],
3. and very recently also for K3 surfaces (see Theorem 12.2).

Specializing to the prime field $\mathbb{F}_p$, we know since Deuring's work that there exists a supersingular elliptic curve E over $\mathbb{F}_p$. Automatically (outside characteristics 2, 3) the characteristic polynomial χ_E of Frob_p^* on $H^1_{\text{ét}}(E \otimes \bar{\mathbb{F}}_p, \mathbb{Q}_\ell)$ is

$$\chi_E(T) = \det(T - \text{Frob}_p^*) = T^2 + p.$$

One can then continue to determine a suitable supersingular hyperelliptic curve over $\mathbb{F}_p$. Geometrically, this can be achieved much in the spirit of Fermat surfaces and Delsarte surfaces: Picking a hyperelliptic curve C given by a 3-term equation, C is automatically covered by some Fermat curve C_d of degree d. Then C_d is supersingular if there exists some p-power p^ν such that

$$p^\nu \equiv -1 \mod d, \tag{13.33}$$

and of course, the same holds for C. Geometrically, i.e. over $\bar{\mathbb{F}}_p$, this argument justifies Corollary 13.37, so it solely remains to control the characteristic polynomial of Frob_p^* on $H^1_{\text{ét}}(C \otimes \bar{\mathbb{F}}_p, \mathbb{Q}_\ell)$. This can be achieved using Hecke Größencharaktere, or Jacobi sums, as pioneered by Weil in [Wei49] (which will reappear in Sect. 13.3.3).

13.3.2 Approach Through Delsarte Surfaces

We shall now discuss a slightly different approach to the rank problem, namely through Delsarte surfaces. Although this is in spirit quite similar to the Shafarevich–Tate approach, the idea is more geometric and much simpler, especially in the supersingular case. This is because we have:

1. a short direct proof of the unirationality of the Fermat surface of degree m in char p such that

$$p^\nu \equiv -1 \mod m \quad \text{for some} \quad \nu \in \mathbb{N} \tag{13.34}$$

(see the proof in [Shio74a] for $p \neq 2$, and [KatS79] for any p including 2);
2. the general fact that, if $X \dashrightarrow Y$ is a dominant rational map of surfaces, then $\lambda(X) \geq \lambda(Y)$, as remarked in [Shio74a, Sect. 2]. Here $\lambda = b_2 - \rho$ is the Lefschetz number from (13.27), and the inequality follows from the existence of a good cohomology theory such as complex or ℓ-adic cohomology as exploited earlier in this book (see e.g. Remark 4.5).

The similarity with Shafarevich–Tate approach from Sect. 13.3.1 lies in the fact that the Fermat surface S_m is covered by the self-product C_m^2 of the Fermat curve of the same degree (which fits into the general scheme of the inductive structure of Fermat varieties introduced in [KatS79]).

The first proof that Fermat surfaces satisfying (13.34) are supersingular goes back to Tate and Thompson (see [Tat66a, p. 102]), and it can also be generalized to higher dimension. They used that for $m' = p^\nu + 1$, the automorphism group of $S_{m'}$ contains the unitary group $U(4)$ over $\mathbb{F}_{p^{2\nu}}$. This can be seen to act on $H^2_{\text{ét}}(S_{m'}, \mathbb{Q}_\ell)$ through two irreducible representations, one of dimension 1 (fixing the hyperplane class) and one of dimension $b_2(S_{m'}) - 1$ (acting irreducibly on primitive cohomology). Since there are other algebraic classes in $H^2_{\text{ét}}(S_{m'}, \mathbb{Q}_\ell)$ than just the multiples of the hyperplane class, notably already $3(m')^2$ lines obtained by reduction from characteristic zero, this implies that $H^2_{ét}(S_{m'}, \mathbb{Q}_\ell)$ is spanned by algebraic classes (which can be defined over $\mathbb{F}_{p^{2\nu}}$), i.e. $S_{m'}$ is supersingular. Through the dominant morphism

$$S_{m'} \to S_m \quad \text{for} \quad m \mid m',$$

this also proves the supersingularity of S_m verifying (13.34).

Hence, by considering elliptic Delsarte surfaces which are covered by suitable (i.e. supersingular) Fermat surfaces, we can give another geometric proof of Corollary 13.37, the arguments being very close to the complex case in Sect. 13.2.2. In fact, combining the $U(4)$ action and the lines on $S_{m'}$ in the notation of the preceding paragraphs, we see that the construction can already be carried out over $\mathbb{F}_{p^{2\nu}}$.

Example 13.39 Let $m \in \mathbb{N}$ and consider the elliptic surface given in affine Weierstrass form

$$X: \quad y^2 + y = x^3 + t^{3m}.$$

Outside characteristics $p \mid 3m$, the Jacobi criterion shows that X is smooth outside $t = \infty$. If p is odd, then there are $3m$ singular fibres of Kodaira type II at the roots of $4t^{3m} + 1$, and if m is also odd, I_0^* at ∞. If $p = 2$ (and m is odd), then Tate's algorithm yields that the fibre at $t = \infty$ has type I_0^* (with index $6m$ of wild ramification). In either case, X is minimal with no further singular fibres.

The Delsarte surface X is covered by the Fermat surface S_{3m}. Indeed, writing affinely $S_{3m} = \{u^{3m} + v^{3m} + w^{3m} + 1 = 0\} \subset \mathbb{A}^3$, there is a dominant rational map $S_{3m} \dashrightarrow X$ given by

$$(u, v, w) \mapsto (x, y, t) = (-v^m u^m, v^{3m}, -wu).$$

Hence, for any m such that (13.34) holds with m replaced by $3m$, X will be supersingular over $\bar{\mathbb{F}}_p$. In particular, for $p \equiv -1 \mod 3$ fixed, we can let $3m$ run through all of the odd p-powers after adding one:

$$3m = p^{2\nu-1} + 1, \quad \nu \in \mathbb{N}$$

to derive infinitely many supersingular elliptic surfaces over $\bar{\mathbb{F}}_p$. Using the standard formulas, we find

$$e(X) = \begin{cases} 6m, \\ 6m+6, \end{cases} \qquad \rho(X) = \begin{cases} 6m-2, \\ 6m+4, \end{cases} \qquad \text{rk}\, MW(X) = \begin{cases} 6m-4, & \text{if } 2 \mid m, \\ 6m-2, & \text{if } 2 \nmid m. \end{cases}$$

Thus Mordell–Weil ranks are unbounded over $\bar{\mathbb{F}}_p$ for any $p \equiv -1 \mod 3$. We point out that this also proves Corollary 13.37 in characteristic 2.

For completeness, we quickly explain how one can let go the supersingularity condition for the covering Fermat surface S_m while retaining complete control about the algebraic cycles even in characteristic $p \nmid m$. To this end, consider the order o of

p in $(\mathbb{Z}/m\mathbb{Z})^\times$. Then, for $\alpha \in \mathfrak{A}_m$, the eigenspace

$$V(\alpha) \subset H^2_{\text{ét}}(S_m \otimes \bar{\mathbb{F}}_p, \bar{\mathbb{Q}}_\ell)$$

is algebraic over $\bar{\mathbb{F}}_p$ (i.e. spanned by some algebraic cycle) if and only if its Galois orbit is algebraic *on average*. That is, since Frobenius acts as multiplication by p,

$$\sum_{j=0}^{o-1} |p^j \cdot t \cdot \alpha| = o \quad \forall t \in (\mathbb{Z}/m\mathbb{Z})^\times, \tag{13.35}$$

where the summands are defined as in (13.25). (For a proof, see [KatS79, Sect. 3], esp. (3.9), building on Stickelberger's theorem on Jacobi sums, see [Wei49, Wei52].) As before, we can thus compute the Lefschetz number and the Picard number of S_m and the quotient Delsarte surface in a purely combinatorial way. Note that (13.35) is in perfect agreement with supersingularity of X_m being implied by (13.34).

Example 13.40 The Fermat quartic in $\mathbb{P}^3$ is supersingular over $\bar{\mathbb{F}}_p$ if and only if

$$p \equiv -1 \mod 4.$$

For the Fermat octic S_8 with $\rho(S_8 \otimes \mathbb{C}) = 176$ and $b_2(S_8) = 302$, for instance, the problem is a little more subtle. In view of the characters $(2, 2, 2, 2)$ induced from the Fermat quartic, and $(1, 1, 3, 3)$ one finds:

$$\rho(S_8 \otimes \bar{\mathbb{F}}_p) = \begin{cases} 176 & p \equiv 1 \mod 8, \\ 178 & p \equiv 3 \mod 8, \\ 188 & p \equiv 5 \mod 8, \\ 302 & p \equiv 7 \mod 8. \end{cases}$$

Hence S_8 is supersingular if and only if $p \equiv -1 \mod 8$.

13.3.3 Delsarte Approach over $\mathbb{F}_p(t)$

We conclude this section with a quick discussion of Ulmer's work [Ulm02] which blends the geometric approach through Delsarte surfaces from Sect. 13.3.2 with the computations based on products of supersingular curves and Jacobi sums from Sect. 13.3.1. Ultimately, this will allow us to produce non-isotrivial elliptic curves over $\mathbb{F}_p(t)$ of arbitrarily large rank.

For starters, let k be any field and $d \in \mathbb{N}$. Consider the elliptic curve over $k(t)$ given in affine Weierstrass form

$$E: \quad y^2 + xy = x^3 - t^d.$$

The corresponding elliptic surface has a fibre of Kodaira type I_d at $t = 0$, while the fibre at infinity is given by the congruence class $\bar{d}$ of d modulo 6:

$\bar{d}$	0	1	2	3	4	5
fibre type	I_0	II^*	IV^*	I_0^*	IV	II

Moreover, outside characteristics 2, 3, there are d fibres of type I_1 at the zeroes of $432t^d - 1$ (some of which may be merged if $\text{char}(k) \mid d$).

Let X denote the corresponding elliptic surface. It is easy to verify that X is dominated by the Fermat surface S_d of degree d; for instance, with signs adjusted, one may take

$$x = x_1^d x_2^d, \quad y = x_1^d x_2^{2d}, \quad t = x_1^2 x_2^3 x_3. \tag{13.36}$$

That is, X is birational to the Galois quotient S_d/G, where G is the subgroup of $\mu_d^3 \subset \text{Aut}(S_d)$ leaving the functions in (13.36) invariant. Here is what Ulmer proves in [Ulm02]:

Theorem 13.41 *Let p be prime, $\nu \in \mathbb{N}$ and $d = p^\nu + 1$. Then we have the following* MW*-ranks:*

$$\text{rank}\, E(\bar{\mathbb{F}}_p(t)) = \text{rank}(E(\mathbb{F}_{p^{2\nu}})) = \begin{cases} p^\nu & \text{if } 6 \nmid d, \\ p^\nu - 2 & \text{if } 6 \mid d; \end{cases}$$

$$\text{rank}\, E(\mathbb{F}_p(t)) \geq (p^\nu - 1)/2\nu.$$

As in Sect. 13.3.2, the proof of Theorem 13.41 relies on controlling the algebraic cycles in $\text{NS}(S_d)$ (which presently span all the second cohomology of S_d by (13.34)) and singling out those which are invariant under G. (In addition, those cycles have to be taken into account which are affected (upstairs and downstairs) by the birational map $S_d \dashrightarrow X$, but this is a relatively minor issue.) Then Euler-number considerations and the rank formula (Corollary 6.7) allow us to deduce the first claim of Theorem 13.41 almost immediately, just as for Example 13.39. The key improvement consists in appealing to the Tate conjecture not just geometrically, or over the field $\mathbb{F}_{p^{2\nu}}$ where generators of $\text{NS}(S_d \otimes \bar{\mathbb{F}}_p)$ are defined, compare the discussion after (13.34), but over the prime field $\mathbb{F}_p$. This can be achieved by analysing the corresponding Jacobi sums in detail, following [Wei49, Shio87a]. Previously, this had mostly been done over those finite fields $\mathbb{F}_q$ containing μ_d, i.e. for $q \equiv 1 \mod d$, but in the supersingular case, Ulmer adapts these calculations to the prime field, thus finding the stated lower bound for the Mordell–Weil rank over $\mathbb{F}_p$ (which is derived from a precise formula for the Mordell–Weil rank over $\mathbb{F}_p$).

More recently, Ulmer succeeded in producing explicit points on certain cyclic base changes of the Legendre curve

$$E: \quad y^2 = x(x-1)(x-t),$$

i.e. over the field $\mathbb{F}_p(\mu_d)(t^{1/d})$ for certain d—in fact enough points to prove in a rather elementary way that the Mordell–Weil rank is unbounded for given p when varying d [Ulm14] (this was lately extended further by Joshi in [Jos18a, Jos18b]). In a similar direction, there is interesting work by Berger [Ber08] on elliptic surfaces dominated by products of curves which enables her to give a somewhat different proof of the unboundedness of the ranks of elliptic curves over $\bar{\mathbb{F}}_p(t)$ for all p, based on a concept resembling the inductive structure of Fermat varieties, as exploited in [KatS79].

13.4 Application to Sphere Packings

In this final section we sketch an important application of Mordell–Weil lattices, namely to the sphere packing problem. Indeed, this was initially one of the key features which indicated that Mordell–Weil lattices might be of relevance. To this end, they were used independently by Elkies and one of us in order to construct lattices with higher density than previously known (compare [CS99] or [Oes90]). As it happens, some of these records are still standing (see (13.40) and compare the tables in loc. cit. or [Shio91a] with [NS]).

In what follows, we will describe one particular result from [Shio91a] in detail; this will give a finishing touch to our previous considerations as it genuinely builds on the theory of elliptic fibrations and Mordell–Weil lattices as laid out in the preceding chapters. In contrast, Elkies' beautiful results in the three subsequent papers [Elk94, Elk97, Elk01] follow a Shafarevich-style approach based on curve theory along the lines of Sect. 13.3.1. We include a brief recap in Sect. 13.4.3.

13.4.1 Exemplary Result

Let us state the main theorem of this section:

Theorem 13.42 *Let p be a prime number such that $p \equiv -1 \mod 6$, and let k be any field of characteristic p containing $\mathbb{F}_{p^2}$. Consider the elliptic curve E over $K = k(t)$ defined by the following Weierstrass form:*

$$E: \quad y^2 = x^3 + 1 + t^{p+1}. \tag{13.37}$$

Then the Mordell–Weil lattice $E(K)$ is a positive-definite even integral lattice with

$$\text{rank } E(K) = 2p - 2 \quad \text{and} \quad \det E(K) = p^{2p_g}, \qquad p_g = (p-5)/6.$$

Moreover, $E(K)$ has minimal norm $2\chi = (p+1)/3$, so its center density is

$$\delta = \frac{\left(\frac{p+1}{12}\right)^{p-1}}{p^{(p-5)/6}}. \tag{13.38}$$

Furthermore, the kissing number τ (the number of minimal vectors) has a lower bound:

$$\tau \geq 6p(p-1). \tag{13.39}$$

As the reader can infer from the reasoning in Sects. 13.2.2, 13.3.2, there are similar results for other Delsarte-type elliptic surfaces. For the most canonical case, the reader is referred to [Shio91a, Theorem 2]. As it happens, the densities from Theorem 13.42 at $p = 29$, $p = 41$ and $p = 53$ still seem to form the current records at ranks 56, 80 resp. 104 with

$$\log_2 \delta = 17.58, \quad 40.14 \quad \text{resp.} \quad 67.01. \tag{13.40}$$

The remainder of this section will be spent on explaining the proof of Theorem 13.42.

13.4.2 Invariants of the Elliptic Surface

Consider the elliptic surface X over k (or $\bar{k}$, thought geometrically) attached to E given by (13.37). Recall from Sect. 13.2.2 that X has the invariants

$$e(X) = 2p + 2, \quad b_2(X) = 2p. \tag{13.41}$$

We will also need the Euler characteristic and the geometric genus which we can compute by Noether's formula (or by the arguments from Sect. 5.13) as

$$\chi(\mathcal{O}_X) = (p+1)/6, \quad p_g(X) = (p-5)/6.$$

Since all singular fibres of the elliptic fibration $X \to \mathbb{P}^1$ have Kodaira type II, Corollary 6.7 gives

$$\text{rk}\, E(K) = \rho(X) - 2. \tag{13.42}$$

Moreover, $E(K)$ is visibly torsion-free since torsion sections are supported on the saturation of the root lattices corresponding to the singular fibres, of which there are none (alternatively due to the fibres of type II, compare Sect. 5.6.1). Hence the Mordell–Weil lattice $E(K)$ equals the narrow Mordell–Weil lattice which by definition is even and positive definite. This proves the first claim of Theorem 13.42. The rank of $E(K)$ follows from a comparison of (13.41) and (13.42) because X is not only supersingular by (13.34), but even has $\text{NS}(X)$ generated by divisor classes defined over $\mathbb{F}_{p^2}$—this is a consequence of the discussion directly after (13.34) which shows that the covering Fermat surface has NS generated by lines over $\mathbb{F}_{p^2}$.

We now turn to the computation of the determinant of $E(K)$, where we claim the following:

$$\det E(K) = p^{2p_g}. \tag{13.43}$$

This requires a little extra input as it can be proven along the lines of [Shio87b] where the Artin invariants of supersingular K3 surfaces with Delsarte model are calculated. Here the argument is only a little more involved since we do not have $p_g = 1$ anymore, but the techniques are the same. We learned them originally from T. Ekedahl (compare [Eke95]). Formally, the arguments are phrased in terms of F-crystals, crystalline cohomology and de Rham cohomology over Witt rings, but for the sake of brevity, we will be rather sketchy and refer the interested reader to [BO78, Ogus78, Ogus79]. For an alternative argument showing a slightly weaker statement (and pointing towards the approach in Sect. 13.4.3), see Remark 13.43.

First we invoke the crystalline viewpoint where

$$H = H^2(X/W)$$

is the F-crystal of X, i.e. a free module of rank $b_2(X)$ over the Witt ring W of $\mathbb{F}_p$, endowed with an action of the Frobenius morphism F (cf. [Ogus79]). Consider the maximal subspace $H' \subset H$ where F is divisible by p. Clearly, the image of NS(X) in H is contained in H'; in fact, by [Ill79, II, 5.8.5 & 5.20] (sometimes also called Nygaard's theorem) we know that

$$\mathrm{NS}(X) \otimes W = H'.$$

It follows by Poincaré duality that

$$|\det \mathrm{NS}(X)| = [H : H']^2, \tag{13.44}$$

so we shall proceed by computing the determinant of NS(X) through the latter index. To this end, we switch to de Rham cohomology for the obvious lift $\tilde{X}$ of X over W which preserves the group action used to split the primitive cohomology (ℓ-adic, it was back then) into one-dimensional eigenspaces. Recall from the proof of Theorem 13.26 (with $m = p + 1$) that the non-zero sections of X (which fill out all of NS(X) together with zero section and general fibre by supersingularity) are in correspondence with the $2p - 2$ characters α in the set

$$\mathfrak{A} = \{\pm\alpha_i = (3d, 2d, i, d - i) \in (\mathbb{Z}/m\mathbb{Z})^4; \quad i \in \mathbb{Z}/m\mathbb{Z}, \ i \neq 0, d\} \quad (d = (p + 1)/6)$$

by way of their one-dimensional eigenspaces $V(\alpha)$. Accordingly we consider

$$H^2_{DR}(\tilde{X}/W) = \bigoplus_{\alpha \in \mathfrak{A}} V(\alpha).$$

By the standard comparison theorem (cf. [BO78, Ogus78]), there is an isomorphism

$$H^2_{DR}(\tilde{X}/W) \cong H,$$

so it remains to compute H' inside $H^2_{DR}(\tilde{X}/W)$. The Frobenius morphism F acts by

$$V(\alpha) \stackrel{F^*}{\longrightarrow} V(p \cdot \alpha).$$

Since the product is taken componentwise modulo $m = p + 1$, our assumption on p in fact gives

$$V(\alpha) \stackrel{F^*}{\longrightarrow} V(-\alpha).$$

By Mazur's theorem (see [Maz72]), Frobenius is exactly divisible by $p^{|\alpha|}$ on $V(\alpha)$ (where $|\alpha|$ is again given by (13.25)). In order to calculate the determinant, it thus remains to consider the Frobenius orbits of the eigenspaces of Hodge type $(2, 0)$. These correspond bijectively to the p_g characters

$$\alpha_i = (3d, 2d, i, d - i) \quad \text{for} \quad 0 < i < d.$$

For each α_i, the orbit of $V(\alpha_i)$ under F is given by $V(\alpha_i) \oplus V(-\alpha_i)$. The subspace where F is p-divisible is thus exactly $pV(\alpha_i) \oplus V(-\alpha_i)$. Since this has index p in $V(\alpha_i) \oplus V(-\alpha_i)$ for each i, we find

$$\det \mathrm{NS}(X) = -p^{2p_g}$$

by (13.44) as claimed; this is equivalent to the given determinant of $E(K)$.

To complete the proof of Theorem 13.42, it remains to verify the statements about the minimal sections and kissing number. Note that since there are no reducible singular fibres, any section has height at least $2\chi = 2\chi(\mathcal{O}_X)$ by (6.15). Hence we have to show that there are indeed sections attaining this minimum, i.e. integral sections (not meeting the zero section), and compute a lower bound for their quantity. Presently, these are easily derived from the lines on the covering Fermat surface S_{p+1}. For instance, one can take the following section:

$$P = (-1, t^{(q+1)/2}).$$

This section is visibly disjoint from the zero section, hence integral and of height 2χ as stated. More generally, any unitary matrix

$$\begin{pmatrix} a & b \\ c & d \end{pmatrix} \in U(2, \mathbb{F}_{p^2})$$

gives rise to the integral section

$$P = (-(at+b)^{(q+1)/3}, (ct+d)^{(q+1)/2}).$$

One computes that there are $6p(p-1)$ distinct integral sections of this form. Hence we obtain the lower bound for the kissing number

$$\tau \geq 6p(p-1).$$

This concludes the proof of Theorem 13.42. □

Remark 13.43 Alternatively, one may argue with the Artin–Tate conjecture from (13.32). Presently, this uses as input that we not only know that the elliptic Delsarte surface X is supersingular in the given characteristic p, but also that $\mathrm{NS}(X \otimes \mathbb{F}_{p^2})$ is the full Néron–Severi group as explained above. It thus follows directly from (13.32) that

$$\det(\mathrm{NS}(X)) \mid p^{2p_g}.$$

13.4.3 Elkies' Results

From his results on Mordell–Weil lattices and sphere packings, Elkies published three papers on a certain series of elliptic curves in characteristic 2 in [Elk94, Elk97, Elk01]. We give a brief overview of the results.

Let $n > 1$ be some integer, $q = 2^n$ and $k = \mathbb{F}_{q^2}$. Then there is a (supersingular) elliptic curve E_0 over k whose number of k-rational points attains the maximum $(1+q)^2$ allowed by the Weil conjectures (see Sect. 10.5.2). Indeed, if n is odd, then E_0 can be taken as the elliptic curve with an automorphism of order 3,

$$E_0: \quad y^2 + y = x^3,$$

while even n requires a quadratic twist

$$E_0: \quad y^2 + y = x^3 + a \tag{13.45}$$

for some $a \in k$ with $k/\mathbb{F}_2$-trace 1. In favour of a uniform treatment, we set $a = 0$ if n is odd and define an elliptic curve over $k(t)$ as another quadratic twist of E, this time by t^{q+1}:

$$E: \quad y^2 + y = x^3 + t^{q+1} + a. \tag{13.46}$$

The corresponding elliptic surface S has only one singular fibre which is located at $t = \infty$ and of Kodaira type II if n is even, or I_0^* if n is odd (with non-identity components distinguished by third roots of unity). To see that S is supersingular with NS(S) generated by divisor classes defined over k (i.e. by sections), one can

- either interpret S as a Delsarte surface, at least if n is odd, and apply the techniques from Sect. 13.2.1,
- or pursue the Tate–Shafarevich style approach as in Sect. 13.3.1, building on the fact that the fibration $S \to \mathbb{P}^1$ is isotrivial.

Indeed, the fibration is trivialized by the base change to the following hyperelliptic curve C of genus $g = q/2$:

$$C: \quad u^2 + u = t^{q+1}. \tag{13.47}$$

Then it suffices to show that the Jacobian of C is isogenous to E_0^g, for instance by verifying that $\#C(k) = 1 + 2gq + q^2$ attains the maximum possible number of points over k (again!), so the claim follows from Tate's theorem [Tat66a].

We are now ready to look into the Mordell–Weil lattice $M = \mathrm{MWL}(S)$. By inspection of (13.46), S has Euler–Poincaré characteristic

$$e(S) = \begin{cases} 2(q+4) & \text{if } n \text{ is odd,} \\ 2(q+2) & \text{if } n \text{ is even.} \end{cases} \tag{13.48}$$

With trivial lattice $\mathrm{Triv}(S) = U \oplus D_4^-$ resp. $\mathrm{Triv}(S) = U$, supersingularity implies

$$\mathrm{rk}(M) = 2q.$$

The height pairing shows that M has minimal norm

$$h(P) \geq \begin{cases} 2\chi - 1 & \text{if } n \text{ is odd,} \\ 2\chi & \text{if } n \text{ is even,} \end{cases} \tag{13.49}$$

where $\chi = \chi(\mathcal{O}_S) = e(S)/12$ is given by (13.48). (This was originally proved in [Elk94] using the naïve and the canonical height following Tate.) At least in the case of odd n, the minimum is attained, as can be seen from the sections

$$P = (\omega t^{(q+1)/3}, 0) \quad \text{with} \quad \omega^3 = 1$$

(which meet distinct non-identity components of the I_0^* fibre). Note the consequence that, for odd n, the narrow Mordell–Weil lattice $M^0 = \mathrm{MWL}(S)^0$ has index 4 inside M. Of course, M^0 comes with the advantage that the minimal norm satisfies the standard inequality

$$h(P) \geq 2\chi$$

as in (13.49). For density considerations, this will outbalance the determinant contribution from the finite index inclusion, so we will mostly restrict to M^0 in what follows.

It remains to compute the determinant of M. Since $\mathrm{NS}(S)$ is generated by divisors defined over k, we can apply the Artin–Tate conjecture from (13.32) to obtain an upper bound for $\det(\mathrm{NS}(S))$, and subsequently for $\det(M)$ (as indicated in Remark 13.43). Indeed, using that $\mathrm{NS}(S)$ is torsion-free and $\alpha(S) = \chi - 1$, supersingularity implies that the left-hand side in (13.32) is trivial, so

$$\det(\mathrm{NS}(S)) \mid q^{\alpha(S)} = \begin{cases} q^{(q-2)/6} & \text{if } n \text{ is odd,} \\ q^{(q-4)/6} & \text{if } n \text{ is even.} \end{cases}$$

This equals the determinant of M^0 if n is even, or $\det(M^0)/4$ if n is odd. Hence we can determine a lower bound for the density of the sphere packing resulting from M^0:

Theorem 13.44 (Elkies) *The narrow Mordell–Weil lattice M^0 has center density*

$$\delta(M^0) \geq \begin{cases} \frac{1}{2} q^{-(q-2)/6} \left(\frac{q+4}{12}\right)^q & \textit{if } n \textit{ is odd,} \\ q^{-(q-4)/6} \left(\frac{q+2}{12}\right)^q & \textit{if } n \textit{ is even.} \end{cases}$$

At the time of publication, these narrow Mordell–Weil lattices provided record densities in dimensions 64, 128, 256, 512, 1024 (some of which have been improved in the meantime, see [Neb98, NS]).

Remark 13.45 The results about determinants and densities can be improved when the Brauer group in (13.32) is non-trivial. The Brauer group was computed by Elkies for small n, and later by Dummigan in [Dum95] for all n using the symmetries of the lattices, a group of size at least $48nq^2(1+q)$.

Remark 13.46 The same line of arguments applies to quotients of the hyperelliptic curve C from (13.47) (since these will again have Jacobian isogenous to a power of E_0). In particular, this leads to the isotrivial supersingular elliptic surfaces given by

$$S' : \quad y^2 + y = x^3 + t^m + a, \quad m \mid (q+1).$$

For $q = 64$ and $m = 13$, this has $\mathrm{MWL}(S')$ homothetic to the Leech lattice by [Elk97].

Historical Notes

We have presented the classical solutions and claims concerning the rank problem for elliptic curves due to Shafarevich–Tate [ShT67] and Lapin [Lap64] which seemed to settle the problem for a long time. The problem came to our attention through our study of Delsarte surfaces (following

[Del51]) and their Picard numbers and Lefschetz numbers [Shio86]. This was to enlarge the class of (hyper)surfaces for which analogous ideas as for Fermat surfaces could be brought to work. Originally, these considerations were geared towards complex projective surfaces in $\mathbb{P}^3$, but they immediately applied to elliptic surfaces over arbitrary fields, producing elliptic curves of rank 56 over $\mathbb{C}(t)$, and of unbounded rank over $\bar{\mathbb{F}}_p(t)$. Subsequently, the rank record for complex elliptic surfaces was pushed to 68 (see [Shio92a]). We note that, at that time, all these results only concerned ranks (or Picard numbers), but no lattice structures (so no Mordell–Weil lattices yet).

By the early 1980s, it was known that algebraic geometry (and number theory) can give rise to interesting codes and dense sphere packings (see [TV91]). In [Tsa92, Sect. 9], Tsfasman asks explicitly "what are the densities of Mordell–Weil lattices on abelian varieties", and this question (posed at the Journées Arithmétiques at Luminy in 1989) prompted Elkies to revisit Mordell–Weil lattices of certain elliptic curves [Elk94, pp. 360/361]. To us it appeared to be a lucky coincidence that the precise definition of Mordell–Weil lattices for elliptic curves over function fields brought with it some applications to sphere packings which enabled us to contribute to this area of research. It is instructive to note how complementary the two approaches sketched in this chapter are: Elkies' method in characteristics 2 and 3, building on the work of Tate–Shafarevich coming from curve theory, and our approach using surface theory which we mostly applied in characteristics different from 2 and 3.

As a final comment, we follow Elkies to mention that while Mordell–Weil lattices have proven very useful to construct lattices of record density in a certain range of ranks (roughly ranging from 44 to 1000, compare [CS99, p. xviii] and [NS]), the known constructions fail to beat the record densities in higher dimension, attained by lattices of large packing density whose existence was proved by Minkowski and Hlawka (cf. [CS99, pp. 14–16]). Explicit constructions for these record lattices are, however, only known in small dimensions, so that also in this respect Mordell–Weil lattices are a valuable complement.

Notation

$\mathbb{N}$	$= \{1, 2, \ldots\}$ the set of natural numbers
$\mathbb{Z}$	The ring of integers
$\mathbb{Q}$	The field of rational numbers
$\mathbb{R}$	The field of real numbers
$\mathbb{C}$	the field of complex numbers
$\mathbb{F}_q$	Finite fields with $q = p^r$ elements for some prime p and $r \in \mathbb{N}$
$\mathbb{P}^n$	Projective n-space over a given field
$E(K)$	The group of K-rational points of an elliptic curve E defined over K
$k(X)$	Function field of a variety X over a field k
K_X	Canonical divisor of a variety X
$b_i(X)$	i-th Betti number of a projective variety X
$e(X)$	Euler–Poincaré characteristic of a projective variety X
$\kappa(X)$	Kodaira dimension of a smooth projective variety
$\chi(\mathcal{F})$	Euler characteristic of a sheaf $\mathcal{F}$ (on a given variety)
$\chi(X)$	$= \chi(\mathcal{O}_X)$ Euler characteristic of a projective variety X
$g(C)$	Genus of a smooth projective curve C
$\mathrm{NS}(S)$	Néron–Severi group of an algebraic surface S
$\rho(S)$	$= \mathrm{rank}(\mathrm{NS}(S))$ Picard number of an algebraic surface S
$\mathrm{Num}(S)$	$= \mathrm{NS}(S)/(\mathrm{torsion})$ Néron–Severi lattice of an algebraic surface S
$\lambda(S)$	Lefschetz number of a projective surface S
$\mathrm{MW}(S)$	Mordell–Weil group of an elliptic surface S
$\mathrm{MWL}(S)$	Mordell–Weil lattice of an elliptic surface S
$\mathrm{Triv}(S)$	Trivial lattice of an elliptic surface S
(P)	Curve associated with a section P of an elliptic surface S
$v(f)$	Vanishing order of a rational function f
$(C.C')$	Intersection number of two projective curves C, C'
$(D.D')$	Intersection number of two divisors $D.D'$ (on some surface)
$\langle \cdot, \cdot \rangle$	Bilinear form, usually on a lattice
I	Gram matrix of a lattice with respect to a given basis
$L(n)$	Lattice L with bilinear form scaled by n

M. Schütt and T. Shioda, *Mordell–Weil Lattices*, Ergebnisse der Mathematik und ihrer Grenzgebiete. 3. Folge / A Series of Modern Surveys in Mathematics 70,
https://doi.org/10.1007/978-981-32-9301-4

$L^{\vee}$	Dual lattice of an integral lattice L
A_n	Root lattice of rank $n \geq 1$ and type (A)
D_m	Root lattice of rank $m \geq 4$ and type (D)
E_r	Root lattice of rank $r \in \{6, 7, 8\}$ and type (E)
A_L	Discriminant group of a lattice L
q_L	Discriminant form of a lattice L

References

[Ahl78] Ahlfors, L.: Complex Analysis. An Introduction to the Theory of Analytic Functions of One Complex Variable. International Series in Pure and Applied Mathkonematics, 3rd edn. McGraw-Hill Book Co., New York (1978)

[AK+01] An, S.Y., Kim, S.Y., Marshall, D.C., Marshall, S.H., McCallum, W.G., Perlis, A.R.: Jacobians of genus one curves. J. Number Theory **90**, 304–315 (2001)

[And57] Andreotti, A.: On the complex structure of a class of simply connected manifolds. Algebraic Geometry and Topology, pp. 58–77. Princeton University Press, Princeton (1957)

[Aok83] Aoki, N.: On some arithmetic problems related to the Hodge cycles on the Fermat varieties. Math. Ann. **266**, 23–54 (1983). Erratum: Math. Ann. **267**, 572 (1984)

[AS83] Aoki, N., Shioda, T.: Generators of the Néron-Severi group of a Fermat surface. In: "Arithmetic and Geometry, vol. I". Progress in Mathematics, vol. 35, pp. 1–12 (1983)

[AS10] Aoki, N., Shioda, T.: Correction and supplement to the paper "Generators of the Néron-Severi group of a Fermat surface". Comment. Math. Univ. St. Pauli **59**, 65–75 (2010)

[Art24] Artin, E.: Quadratische Körper im Gebiete der hoheren Kongruenzen. II. Analytischer Teil, Math. Z. **19**, 207–246 (1924)

[Art62] Artin, M.: Some numerical criteria for contractibility of curves on algebraic surfaces. Am. J. Math. **84**, 485–496 (1962)

[Art66] Artin, M.: On isolated rational singularities of surfaces. Am. J. Math. **88**, 129–136 (1966)

[Art74a] Artin, M.: Supersingular $K3$ surfaces. Ann. Scient. Éc. Norm. Sup. **7**(4), 543–568 (1974)

[Art74b] Artin, M.: Algebraic constructions of Brieskorn's resolutions. J. Algebra **29**, 330–348 (1974)

[Art77] Artin, M.: Coverings of the rational double points in characteristic p. Complex Analysis and Algebraic Geometry, pp. 11–22. Iwanami Shoten, Tokyo and Cambridge University Press, Cambridge (1977)

[AM77] Artin, M., Mazur, B.: Formal groups arising from algebraic varieties. Ann. Scient. Éc. Norm. Sup. **10**(4), 87–131 (1977)

[ARVT05] Artin, M., Rodriguez-Villegas, F., Tate, J.: On the Jacobians of plane cubics. Adv. Math. **198**, 366–382 (2005)

[ASD73] Artin, M., Swinnerton-Dyer, P.: The Shafarevich-Tate conjecture for pencils of elliptic curves on $K3$ surfaces. Invent. Math. **20**, 249–266 (1973)

M. Schütt and T. Shioda, *Mordell–Weil Lattices*, Ergebnisse der Mathematik und ihrer Grenzgebiete. 3. Folge / A Series of Modern Surveys in Mathematics 70,
https://doi.org/10.1007/978-981-32-9301-4

[Bad01] Badescu, L.: Algebraic Surfaces. Universitext, Springer, New York (2001)

[Bak66/67] Baker, A.: Linear forms in the logarithms of algebraic numbers. I, II, III, Mathematika. J. Pure Appl. Math. **13**, 204–216 (1966); 14, 102–107, 220–228 (1967)

[BFL19] Banwait, B., Fité, F., Loughran, D.: Del Pezzo surfaces over finite fields and their Frobenius traces. Math. Proc. Camb. Philos. Soc. **167**, 35–60 (2019)

[BHPV04] Barth, W., Hulek, K., Peters, C., van de Ven, A.: Compact Complex Surfaces. Erg. der Math. und ihrer Grenzgebiete, 3. Folge, Band, vol. 4, 2nd edn. Springer, Berlin (2004)

[Bea82] Beauville, A.: Les familles stables de courbes elliptiques sur $\mathbb{P}^1$ admettant quatre fibres singuliéres. C.R. Acad. Sci. Paris **294**, 657–660 (1982)

[Bea96] Beauville, A.: Complex Algebraic Surfaces. LMS Student Texts, vol. 34, 2nd edn. Cambridge University Press, Cambridge (1996)

[B+85] Beauville, A., et al.: Géométrie des surfaces K3: modules et périodes - Séminaire Palaiseau, Asterisque 126. Societe mathematique de France, Paris (1985)

[Ber08] Berger, L.: Towers of surfaces dominated by products of curves and elliptic curves of large rank over function fields. J. Number Theory **128**, 3013–3030 (2008)

[BL13] Bertin, M.J., Lecacheux, O.: Elliptic fibrations on the modular surface associated to $\Gamma_1(8)$. Arithmetic and Geometry of K3 Surfaces and Calabi-Yau Threefolds, Fields Institute Communications, vol. 67, pp. 1–38, Springer, New York (2013)

[BO78] Berthelot, P., Ogus, A.: Notes on Crystalline Cohomology. Princeton University Press, Princeton (1978)

[BS-D65] Birch, B., Swinnerton-Dyer, P.: Notes on elliptic curves (II). J. Reine Angew. Math. **218**, 79–108 (1965)

[BM08] Beukers, F., Montanus, H.: Explicit calculation of elliptic fibrations of K3-surfaces and their Belyi-maps. Number Theory and Polynomials, London Mathematical Society Lecture Note Series, vol. 352, pp. 33–51. Cambridge University Press, Cambridge (2008)

[BS15] Bhargava, M., Shankar, A.: Ternary cubic forms having bounded invariants, and the existence of a positive proportion of elliptic curves having rank 0. Ann. Math. **181**, 587–621 (2015)

[BKV17] Bloch, S., Kerr, M., Vanhove, P.: Local mirror symmetry and the sunset Feynman integral. Adv. Theo. Math. Phys. **21**, 1373–1453 (2017)

[BHT11] Bogomolov, F.A., Hassett, B., Tschinkel, Y.: Constructing rational curves on K3 surfaces. Duke Math. J. **157**, 535–550 (2011)

[BG06] Bombieri, E., Gubler, W.: Heights in Diophantine Geometry, New Mathematical Monographs 4. Cambridge University Press, Cambridge (2006)

[BM76/77] Bombieri, E., Mumford, D.: Enriques' classification of surfaces in char. p, II-III. Complex Analysis and Algebraic Geometry, pp. 23–42. Iwanami and Cambridge University Press, Tokyo, Cambridge (1977); Invent. Math. **35** 197–232 (1976)

[Bor87] Borcherds, R.: Automorphism groups of Lorentzian lattices. J. Algebra **111**, 133–153 (1987)

[Bor98] Borcherds, R.: Coxeter groups, Lorentzian lattices and K3 surfaces. Internat. Math. Res. Not. **19**, 1011–1031 (1998)

[BLR90] Bosch, S., Lütkebohmert, W., Raynaud, M.: Néron Models. Springer, Berlin (1990)

[Bou68] Bourbaki, N.: Groupes et algèbres de Lie, Chap. 4, 5 et 6, Hermann, Paris (1968)

[BL18] Bragg, D., Lieblich, M.: Twistor Spaces for Supersingular K3 Surfaces, preprint (2018). arXiv:1804.07282v3

[BL19] Bragg, D., Lieblich, M.: Perfect Points on Genus one Curves and Consequences for Supersingular K3 Surfaces, preprint (2019). arXiv:1904.04803

[Bram18] Bramble, C.C.: A collineation group isomorphic with the group of the double tangents of the plane quartic. Am. J. Math. **40**, 351–365 (1918)

[Bran17] Brandhorst, S.: Dynamics on supersingular K3 surfaces and automorphisms of Salem degree 22. Nagoya Math. J. **227**, 1–15 (2017)

[Bran19] Brandhorst, S.: The classification of purely non-symplectic automorphisms of high order on K3 surfaces. J. Algebra **533**, 229–265 (2019)

[BKW13] Braun, A.P., Kimura, Y., Watari, T.: On the Classification of Elliptic Fibrations Modulo Isomorphism on K3 Surfaces with Large Picard Number, preprint (2013). arXiv:1312.4421

[BCDT01] Breuil, C., Conrad, B., Diamond, F., Taylor, R.: On the modularity of elliptic curves over $\mathbb{Q}$: wild 3-adic exercises. J. AMS **14**, 843–939 (2001)

[Bri68] Brieskorn, E.: Die Auflösung der rationalen Singularitäten holomorphen Abbildungen. Math. Ann. **178**, 255–270 (1968)

[Bri71] Brieskorn, E.: Singular elements of semi-simple algebraic groups. In: Proceedings of ICM (Nice, 1970), pp. 279–284. Gauthier-Villars, Paris (1971)

[BW79] Bruce, J.W., Wall, C.T.C.: On the classification of cubic surfaces. J. Lond. Math. Soc. **19**, 245–256 (1979)

[BR75] Burns, D., Rapoport, M.: On the Torelli problem for Kählerian K3 surfaces. Ann. Scient. Ec. Norm. Sup. **8**, 235–274 (1975)

[Cas66] Cassels, J.W.S.: Diophantine equations with special reference to elliptic curves. J. Lond. Math. Soc. **41**, 193–291 (1966)

[Cas91] Cassels, J.W.S.: Lectures on elliptic curves. London Mathematics Society Student Texts, vol. 24. Cambridge University Press, Cambridge (1991)

[Cay49] Cayley, A.: On the triple tangent planes of surfaces of the third order. Camb. Dublin Math. J. **4**, 118–138 (1849)

[CMT00] Chahal, J., Meijer, M., Top, J.: Sections on certain $j = 0$ elliptic surfaces. Comment. Math. Univ. St. Paul. **49**, 79–89 (2000)

[Cha13] Charles, F.: The Tate conjecture for K3 surfaces over finite fields. Invent. Math. 194, 119–145 (2013). Erratum: Invent. Math. **202**, 481–485 (2015)

[Cha14] Charles, F.: On the Picard number of K3 surfaces over number fields. Algebra & Number Theory **8**, 1–17 (2014)

[CDH14] Cheng, M.C.N., Duncan, J.F.R., Harvey, J.A.: Umbral moonshine. Commun. Number Theory Phys. **8**, 101–242 (2014)

[CK09] Cohn, H., Kumar, A.: Optimality and uniqueness of the Leech lattice among lattices. Ann. Math. **170**(3), 1003–1050 (2009)

[CK+17] Cohn, H., Kumar, A., Miller, S.D., Radchenko, D., Viazovska, M.: The sphere packing problem in dimension 24. Ann. Math. **185**, 1017–1033 (2017)

[CvGL09] Colombo, E., van Geemen, B., Looijenga, E.: Del Pezzo moduli via root systems. Algebra, Arithmetic, and Geometry: in Honor of Yu. I. Manin, vol. I, 269, pp. 291–337. Progress in Mathematics, Birkhäuser, Boston, MA (2009)

[Con85] Conway, J., et al.: Atlas of Finite Groups. Clarendon Press, Oxford (1985)

[CPS82] Conway, J.H., Parker, R.A., Sloane, N.J.A.: The covering radius of the Leech lattice. Proc. R. Soc. Lond. A **380**, 261–290 (1982)

[CS99] Conway, J.H., Sloane, N.J.A.: Sphere Packings, Lattices and Groups, 3rd edn. Springer, New York (1999)

[CP08] Cossart, V., Piltant, O.: Resolution of singularities of threefolds in positive characteristic. I. Reduction to local uniformization on Artin-Schreier and purely inseparable coverings. J. Algebra **320**, 1051–1082 (2008)

[CP09] Cossart, V., Piltant, O.: Resolution of singularities of threefolds in positive characteristic II. J. Algebra **321**, 1836–1976 (2009)

[CD89] Cossec, F.R., Dolgachev, I.V.: Enriques surfaces. I. Progress in Mathematics, vol. 76. Birkhäuser, Boston (1989)

[CT14] Costa, E., Tschinkel, Y.: Variation of Néron-Severi ranks of reductions of K3 surfaces. Exp. Math. **23**, 475–481 (2014)

[Cox79] Cox, D.A.: Solutions of Weierstrass equations, Algebraic geometry (Copenhagen 1978). Lect. Notes Math. **732**, 43–59 (1979)

[Cox82] Cox, D.A.: Mordell-Weil groups of elliptic curves over $\mathbf{C}(t)$ with $p_g = 0$ or 1. Duke Math. J. **49**, 677–689 (1982)

[Cox13] Cox, D.A.: Primes of the form $x^2 + ny^2$: Fermat, Class Field Theory, and Complex Multiplication, 2nd edn. Wiley Interscience, New York (2013)

[CZ79] Cox, D., Zucker, S.: Intersection numbers of sections of elliptic surfaces. Invent. Math. **53**, 1–44 (1979)

[Dav65] Davenport, H.: On $f^3(t) - g^2(t)$. Norske Vid. Selsk. Forh. (Trondheim) **38**, 86–87 (1965)

[Deg15] Degtyarev, A.: Lines generate the Picard group of a Fermat surface. J. Number Theory **147**, 454–477 (2015)

[Deg16] Degtyarev, A.: Lines in Supersingular Quartics, preprint (2016). arXiv:1604.05836

[Del71] Deligne, P.: Formes modulaires et représentations ℓ-adiques, Sém. Bourbaki 1968/69, no. Lecture Notes in Mathematics, vol. 179, 355, pp. 139–172. Springer (1971)

[Del72] Deligne, P.: La conjecture de Weil pour les surfaces K3. Invent. Math. **15**, 206–226 (1972)

[Del73] Deligne, P.: La formule de Milnor, Groupes de Monodromie en Géometrie Algébrique (SGA 7 II). Lecture Notes in Mathematics, vol. 340, pp. 197–211. Springer (1973)

[Del74] Deligne, P.: La conjecture de Weil I. Publ. Math. IHÉS **43**, 273–308 (1974)

[Del75] Deligne, P.: Courbes Elliptiques: Formulaire (d'aprés J. Tate), Modular Functions in One Varibale, IV. Lecture Notes in Mathematics, vol. 476, pp. 53–74. Springer (1975)

[Del80] Deligne, P.: La conjecture de Weil II. Publ. Math. IHÉS **52**, 137–252 (1980)

[Del81] Deligne, P.: Relèvement des Surfaces K3 en caractéristique nulle. Algebraic Surfaces (Orsay, 1976–78). Lecture Notes in Mathematics, vol. 868, 58–79 (1981)

[DM69] Deligne, P., Mumford, D.: The irreducibility of the space of curves of given genus. Publ. Math. IHES **36**, 75–109 (1969)

[Del51] Delsarte, J.: Nombres de solutions des équations polynomiales sur un corps fini. Séminaire Bourbaki **39**, 1–9 (1951)

[DP80] Demazure, M., Pinkham, H., Teissier, B.: Séminaire sur les singularités des surfaces. Lecture Notes in Mathematics, vol. 777. Springer (1980)

[Deu41] Deuring, M.: Die Typen der Multiplikatorenringe elliptischer Funktionenkörper. Abh. Math. Sem. Univ. Hamburg **14**, 197–272 (1941)

[DD13] Dokchitser, T., Dokchitser, V.: A remark on Tate's algorithm and Kodaira types. Acta Arith. **160**, 95–100 (2013)

[Dol72] Dolgachev, I.: Euler characteristic of a family of algebraic varieties. Math. USSR-Sb. **18**, 303–318 (1972)

[Dol96] Dolgachev, I.: Mirror symmetry for lattice polarized K3 surfaces. J. Math. Sci. **81**, 2599–2630 (1996)

[Dol04] Dolgachev, I., appendix to: Galluzzi, F., Lombardi, G.: Correspondences between K3 surfaces. Mich. Math. J. **52**, 267–277 (2004)

[DK01] Dolgachev, I., Keum, J.H.: Wild p-cyclic actions on K3 surfaces. J. Algebraic Geom. **10**, 101–131 (2001)

[DK03] Dolgachev, I., Kondō, S.: A supersingular $K3$ surface in characteristic 2 and the Leech lattice. Int. Math. Res. Not. **1**, 1–23 (2003)

[DuV33] Du Val, P.: On isolated singularities of surfaces which do not affect the conditions of adjunction (Part I). Proc. Camb. Philos. Soc. **30**, 483–491 (1933–1934)

[Dum95] Dummigan, N.: The determinants of certain Mordell-Weil lattices. Am. J. Math. **117**, 1409–1429 (1995)

[DGO15] Duncan, J.F.R., Griffin, M.J., Ono, K.: Proof of the umbral moonshine conjecture. Res. Math. Sci. **2**, Art. 26, 47pp (2015)

[Dwo60] Dwork, B.: On the rationality of the zeta function of an algebraic variety. Am. J. Math. **82**, 631–648 (1960)

[Dyn57] Dynkin, E.: Semi-simple subalgebras of semi-simple Lie algebras. AMS Transl. **6**, 111–244 (1957)

[Ebe13] Ebeling, W.: Lattices and Codes. Advanced Lectures in Mathematics, 3rd edn. Springer Spektrum, Wiesbaden (2013)

[EOT11] Eguchi, T., Ooguri, H., Tachikawa, Y.: Notes on the K3 surface and the Mathieu group M_{24}. Exp. Math. **20**, 91–96 (2011)

[EgS03] Eguchi, T., Sakai, K.: Seiberg-Witten curve for E-string theory revisited. Adv. Theor. Math. Phys. **7**(3), 419–455 (2003)

[Eke90] Ekedahl, T.: An effective version of Hilbert's irreducibility theorem. Séminaire de Théorie des Nombres, Paris, pp. 1988–1989, 241–249. Birkhäuser, Boston (1990)

[Eke95] Ekedahl, T.: Varieties of CM Type, preprint (1995). arXiv:alg-geom/9512004

[Elk90] Elkies, N.D.: On Mordell-Weil Lattices, Arbeitstagung Bonn (1990)

[Elk94] Elkies, N.D.: Mordell-Weil lattices in characteristic 2, I. Construction and first properties. Int. Math. Res. Not. **8**, 343–361 (1994)

[Elk97] Elkies, N.D.: Mordell-Weil lattices in characteristic 2, II. The Leech lattice as a Mordell-Weil lattice. Inv. Math. **128**, 1–8 (1997)

[Elk01] Elkies, N.D.: Mordell-Weil lattices in characteristic 2, III. A Mordell-Weil lattice of rank 128. Exp. Math. **10**, 467–473 (2001)

[Elk07] Elkies, N.D.: Three Lectures on Elliptic Surfaces and Curves of High Rank, preprint (2007). arXiv:0709.2908

[Elk08] Elkies, N., D.: Shimura Curve Computations via K3 Surfaces of NS-Rank at Least 19. Lecture Notes in Computer Science 5011 (ANTS VIII, 2008), vol. 2008, pp. 196–211. Springer, Berlin (2008)

[EK14] Elkies, N.D., Kumar, A.: K3 surfaces and equations for Hilbert modular surfaces. Algebra Number Theory **8**, 2297–2411 (2014)

[ElS08] Elkies, N.D., Schütt, M.: K3 Families of High Picard Rank, preprint (2008)

[ElS13] Elkies, N.D., Schütt, M.: Modular forms and K3 surfaces. Adv. Math. **240**, 106–131 (2013)

[ElS15] Elkies, N.D., Schütt, M.: Genus 1 fibrations on the supersingular K3 surface in characteristic 2 with Artin invariant 1. Asian J. Math. **19**, 555–582 (2015)

[EJ11] Elsenhans, A.-S., Jahnel, J.: The Picard group of a K3 surface and its reduction modulo p. Algebra Number Theory **5**, 1027–1040 (2011)

[Enr14] Enriques, F.: Sulla classificazione delle superficie algebriche e particolarmente sulle superficie di genere $p^1 = 1$. Atti. Acc. Lincei V Ser. **23** (1914)

[Enr49] Enriques, F.: Le Superficie Algebriche. Societa tipografica editrice bolognese, Bologna (1949)

[Ern94] Erné, R.: Construction of a del Pezzo surface with maximal Galois action on its Picard group. J. Pure. Appl. Alg. **97**, 15–27 (1994)

[EO15] Esnault, H., Oguiso, K.: Non-liftability of automorphism groups of a K3 surface in positive characteristic. Math. Ann. **363**, 1187–1206 (2015)

[Fal83] Faltings, G.: Endlichkeitssätze für abelsche Varietäten über Zahlkörpern. Invent. Math. **73**, 349–366 (1983)

[Fri84] Friedman, R.: A new proof of the global Torelli theorem for K3 surfaces. Ann. Math. **120**, 237–269 (1984)

[Ful98] Fulton, W.: Intersection Theory, 2nd edn. Springer, New York (1998)

[Fus06] Fusi, D.: Construction of linear pencils of cubic curves with Mordell-Weil rank six and seven. Comment. Math. Univ. St. Pauli **55**, 195–205 (2006)

[GS12] Garbagnati, A., Schütt, M.: Enriques surfaces-Brauer groups and Kummer structures. Mich. Math. J. **61**, 297–330 (2012)

[GR04] Gonzáles, J., Rotger, V.: Equations of Shimura curves of genus two. Internat. Math. Res. Not. **14**, 661–674 (2004)

[Got96] Goto, Y.: The Artin invariant of supersingular weighted Delsarte K3 surfaces. J. Math. Kyoto Univ. **36**, 359–363 (1996)

[GM91] Gouvêa, F., Mazur, B.: The square-free sieve and the rank of elliptic curves. J. Am. Math. Soc. **4**, 1–23 (1991)

[GH78] Griffiths, P., Harris, J.: Principles of Algebraic Geometry. Wiley, New York (1978)

[GHS07] Gritsenko, V., Hulek, K., Sankaran, G.K.: The Kodaira dimension of the moduli of K3 surfaces. Invent. Math. **169**, 519–567 (2007)

[GZ86] Gross, B.H., Zagier, D.B.: Heegner points and derivatives of L-series. Invent. Math. **84**, 225–320 (1986)

[Gro58] Grothendieck, A.: The cohomology theory of abstract algebraic varieties. Edinb. Proc. Int. Cong. Math. **103–118** (1958)

[Hal05] Hales, T.C.: A proof of the Kepler conjecture. Ann. Math. **162**, 1065–1185 (2005)

[Har77] Hartshorne, R.: Algebraic Geometry. GTM, vol. 52. Springer, New York-Heidelberg (1977)

[Has36] Hasse, H.: Zur Theorie der abstrakten elliptischen Funktionenkörper I-III. J. Reine Angew. Math. **175**, 55–62, 69–88, 193–208 (1936)

[Hei12] Heijne, B.: The maximal rank of elliptic Delsarte surfaces. Math. Comput. **81**, 1111–1130 (2012)

[HS88] Hindry, M., Silverman, J.H.: The canonical height and integral points on elliptic curves. Invent. Math. **93**, 419–450 (1988)

[HS00] Hindry, M., Silverman, J.H.: Diophantine Geometry: An Introduction, GTM 201. Springer, New York (2000)

[Hir64] Hironaka, H.: Resolution of singularities of an algebraic variety over a field of characteristic 0. Ann. Math. **79**, 109–203 (1964)

[Hir78] Hirzebruch, F.: Topological Methods in Algebraic Geometry. Grundlehren der mathematischen Wissenschaften, 3rd edn. Springer, Heidelberg (1978)

[Hod52] Hodge, W.V.D.: The Theory and Applications of Harmonic Integrals, 2nd edn. Cambridge University Press, Cambridge (1952)

[HS12] Hulek, K., Schütt, M.: Arithmetic of singular Enriques surfaces. Algebra & Number Theory **6**, 195–230 (2012)

[Hum92] Humphreys, J.E.: Reflection Groups and Coxeter Groups. Cambridge Studies in Advanced Mathematics, vol. 29. Cambridge University Press, Cambridge (1992)

[Hus14] Husemöller, D.: Elliptic Curves, GTM 111, 2nd edn. Springer, New York (2014)

[Huy99] Huybrechts, D.: Compact hyperkähler manifolds: Basic results. Invent. Math. 135, 63–113 (1999). Erratum: Invent. Math. **152**, 209–212 (2003)

[Huy16] Huybrechts, D.: K3 Surfaces. Cambridge University Press, Cambridge (2016)

[Igu59] Igusa, J.-I.: Kroneckerian model of fields of elliptic modular functions. Am. J. Math. **81**, 561–577 (1959)

[Igu60] Igusa, J.-I.: Betti and Picard numbers of abstract algebraic surfaces. Proc. Nat. Acad. Sci. USA **46**, 724–726 (1960)

[Igu68] Igusa, J.-I.: On the algebraic theory of elliptic modular functions. J. Math. Soc. Jpn. **20**, 96–106 (1968)

[Iit71] Iitaka, S.: On D-dimensions of algebraic varieties. J. Math. Soc. Jpn. **23**, 356–373 (1971)

[Ill79] Illusie, L.: Complexe de de Rham-Witt et cohomologie cristalline. Ann. Scient. Éc. Norm. Sup. **12**, 501–661 (1979)

[Inos78] Inose, H.: Defining equations of singular $K3$ surfaces and a notion of isogeny. In: Proceedings of the International Symposium on Algebraic Geometry (Kyoto University, Kyoto, 1977), pp. 495–502. Kinokuniya Book Store, Tokyo (1978)

[IR82] Ireland, K., Rosen, M.: A Classical Introduction to Modern Number Theory. Graduate Texts in Mathematics, vol. 84. Springer, New York (1982)

[Ito94] Ito, H.: Mordell-Weil groups of unirational quasi-elliptic surfaces in characteristic 2. Tohoku Math. J. **46**, 221–251 (1994)

[Ito02] Ito, H.: On extremal elliptic surfaces in characteristic 2 and 3. Hiroshima Math. J. **32**, 179–188 (2002)

[IL13] Ito, H., Liedtke, C.: Elliptic K3 surfaces with p^n-torsion sections. J. Algebraic Geom. **22**, 105–139 (2013)

[JLR12] Jarvis, T.J., Lang, W.E., Ricks, J.R.: Integral models of extremal rational elliptic surfaces. Commun. Algebra **40**, 3867–3883 (2012)

[Jor70] Jordan, C.: Traité des substitutions et des équations algébriques. Gauthier-Villars, Paris (1870)

[Jos18a] Joshi, K.: A Method for Construction of Rational Points Over Elliptic Curves, preprint (2018). arXiv:1705.07514v2

[Jos18b] Joshi, K.: A Method for Construction of Rational Points Over Elliptic Curves II: Points Over Solvable Extensions, preprint (2018). arXiv:1801.06245

[JKZ08] Jouve, F., Kowalski, E., Zywina, D.: An explicit integral polynomial whose splitting field has Galois group $W(E_8)$. J. Théor. Nombres Bordeaux **20**, 761–782 (2008)

[Kat78] Katsura, T.: On Kummer surfaces in characteristic 2. In: Nagata, M. (ed.) Proceedings of the International Symposium on Algebraic Geometry, pp. 525–542. Kinokuniya Book Store, Tokyo (1978)

[Kat87] Katsura, T.: Generalized Kummer surfaces and their unirationality in characteristic p. J. Fac. Sci. Univ. Tokyo Sect. IA Math. **34**, 1–41 (1987)

[Kat15] Katsura, T.: The Chern class map on abelian surfaces. J. Algebra **425**, 85–106 (2015)

[KatS79] Katsura, T., Shioda, T.: On Fermat varieties. Tohoku Math. J. **31**, 97–115 (1979)

[KM85] Katz, N.M., Mazur, B.: Arithmetic moduli of elliptic curves. Annals of Mathematics Studies, vol. 108. Princeton University Press, Princeton, NJ (1985)

[Kaw08] Kawamata, Y.: Flops connect minimal models. Publ. Res. Inst. Math. Sci. **44**, 419–423 (2008)

[Keu99] Keum, J.: A note on elliptic K3 surfaces. Trans. AMS **352**, 2077–2086 (1999)

[KM16] Kim, W., Madapusi Pera, K.: 2-adic integral canonical models. Forum Math. Sigma **4**, e28, 34pp (2016)

[KSW19] Klagsbrun, Z., Sherman, T., Weigandt, J.: The Elkies curve has rank 28 subject only to GRH. Math. Comput. **88**, 837–846 (2019)

[Kle14] Kleiman, S.L.: The Picard scheme. L. Schneps (ed.) Alexandre Grothendieck: A Mathematical Portrait, pp. 35–74. International Press, Somerville, MA (2014)

[Klo05] Kloosterman, R.: Extremal elliptic surfaces and Infinitesimal Torelli. Mich. Math. J. **52**, 141–160 (2005)

[Klo07a] Kloosterman, R.: Higher Noether-Lefschetz loci of elliptic surfaces. J. Differ. Geom. **76**, 293–316 (2007)

[Klo07b] Kloosterman, R.: Elliptic K3 surfaces with geometric Mordell-Weil rank 15. Canad. Math. Bull. **50**(2), 215–226 (2007)

[Kne56] Kneser, M.: Klassenzahlen indefiniter quadratischer Formen in drei oder mehr Veränderlichen. Arch. Math. **7**, 323–332 (1956)

[Kne02] Kneser, M.: Quadratische Formen. Springer, Berlin-Heidelberg (2002)

[Kod63] Kodaira, K.: On compact analytic surfaces, II-III. Ann. Math. **77**, 563–626 (1963); **78**, 1–40(1963); Collected Works, vol. III, pp. 1269–1372. Iwanami and Princeton University Press (1975)

[Kod64+] Kodaira, K.: On the structure of compact complex analytic surfaces I-IV. Am. J. Math. 86, 751–798 (1964); 88, 682–721 (1966); 90, 55–83 (1968); 90, 1048–1066 (1968). Collected Works, vol. III, pp. 1389–1436, 1471–1510, 1525–1553, 1577–1595. Iwanami and Princeton University Press (1975)

[Koi12] Koike, K.: Elliptic K3 surfaces admitting a Shioda-Inose structure. Comment. Math. Univ. St. Pauli **61**, 77–86 (2012)

[KM98] Kollár, J., Mori, S.: Birational Geometry of Algebraic Varieties. Cambridge University Press, Cambridge (1998)

[Kol89] Kolyvagin, V.: Finiteness of $E(\mathbb{Q})$ and $Ш(E, \mathbb{Q})$ for a class of Weil curves. Math. USSR Izv. **32**, 523–541 (1989)

[Kon86] Kondō, S.: Enriques surfaces with finite automorphism groups. Jpn. J. Math. (N.S.) **12**, 191–282 (1986)

[Kon89] Kondō, S.: Algebraic K3 surfaces with finite automorphism groups. Nagoya Math. J. **116**, 1–15 (1989)

[Kon92] Kondō, S.: Automorphisms of algebraic K3 surfaces which act trivially on Picard groups. J. Math. Soc. Jpn. **44**(1), 75–98 (1992)

[Kon93] Kondō, S.: On the Kodaira dimension of the moduli space of K3 surfaces. Compos. Math. **89**, 251–299 (1993)

[Kon98a] Kondō, S.: Niemeier lattices, Mathieu groups, and finite groups of symplectic automorphisms of K3 surfaces (with an appendix by Shigeru Mukai). Duke Math. J. **92**, 593–603 (1998)

[Kon98b] Kondō, S.: The automorphism group of a generic Jacobian Kummer surface. J. Algebr. Geom. **7**, 589–609 (1998)

[Kon99] Kondō, S.: On the Kodaira dimension of the moduli space of K3 surfaces II. Compos. Math. **116**, 111–117 (1999)

[KonS14] Kondō, S., Shimada, I.: On certain duality of Néron-Severi lattices of supersingular K3 surfaces and its application to generic supersingular K3 surfaces. Algebr. Geom. **1**, 311–333 (2014)

[Kov94] Kovács, S.: The cone of curves of a K3 surface. Math. Ann. **300**, 681–691 (1994)

[Kub76] Kubert, D.S.: Universal bounds on the torsion of elliptic curves. Proc. Lond. Math. Soc. **33**, 193–237 (1976)

[Kul77] Kulikov, V.: Degenerations of K3-surfaces and Enriques surfaces. Izv. Akad. Nauk. SSSR, Ser. Math. **41**, 1008–1042 (1977)

[Kum08] Kumar, A.: K3 surfaces associated with curves of genus two. Internat. Math. Res. Not. 26pp (2008)

[Kum14] Kumar, A.: Elliptic fibrations on a generic Jacobian Kummer surface. J. Algebr. Geom. **23**, 599–667 (2014)

[KumK18] Kumar, A., Kuwata, M.: Inose's construction and elliptic K3 surfaces with Mordell-Weil rank 15 revisited. High. Genus Curves Math. Phys. Arith. Geom. Contemp. Math. **703**, 131–141 (2018)

[KumS13] Kumar, A., Shioda, T.: Multiplicative excellent families of elliptic surfaces of type E_7 or E_8. Algebra & Number Theory **7**, 1613–1641 (2013)

[Kuw00] Kuwata, M.: Elliptic $K3$ surfaces with given Mordell-Weil rank. Comment. Math. Univ. St. Pauli **49**(1), 91–100 (2000)

[KuwS08] Kuwata, M., Shioda, T.: Elliptic parameters and defining equations for elliptic fibrations on a Kummer surface. In: Algebraic Geometry in East Asia (Hanoi 2005), pp. 177–215 (2008). Adv. Stud. Pure Math. **50**, Math. Soc. Jpn. Tokyo (2008)

[Lan59] Lang, S.: Abelian Varieties. Interscience, New York (1959)

[Lan83] Lang, S.: Fundamentals of Diophantine Geometry. Springer, New York (1983)

[Lan87] Lang, S.: Elliptic Functions, GTM 112, 2nd edn. Springer, New York (1987)

[Lan88] Lang, W.E.: On Enriques surfaces in characteristic p. II. Math. Ann. **281**, 671–685 (1988)

[Lan91] Lang, W.E.: Extremal rational elliptic surfaces in characteristic p. I: Beauville surfaces. Math. Z. **207**, 429–438 (1991)

[Lan94] Lang, W.E.: Extremal rational elliptic surfaces in characteristic p. II. Surfaces with three or fewer singular fibres. Ark. Mat. **32**, 423–448 (1994)

[Lap64] Lapin, A.I.: Subfields of hyperelliptic fields. I. Izv. Akad. Nauk SSSR Ser. Mat. **28**, 953–988 (1964)

[Laz55] Lazard, M.: Sur les groupes de Lie formels à un paramètre. Bull. Soc. Math. Fr. **83**, 251–274 (1955)

[Lef24] Lefschetz, S.: L'analysis situs et la géométrie algébrique. Gauthier-Villars (1924)

[Lev68] Levin, M.: On the group of rational points on elliptic curves over function fields. Am. J. Math. **90**, 456–462 (1968)

[LL12] Li, J., Liedtke, C.: Rational curves on K3 surfaces. Invent. Math. **188**, 713–727 (2012)

[Lic68] Lichtenbaum, S.: Curves over discrete valuation rings. Am. J. Math. **85**, 380–405 (1968)

[LM18] Lieblich, M., Maulik, D.: A note on the cone conjecture for K3 surfaces in positive characteristic. Math. Res. Lett. **25**, 1879–1891 (2018)

[Lie15] Liedtke, C.: Supersingular K3 surfaces are unirational. Invent. Math. **200**, 979–1014 (2015)

[Lip75] Lipman, J.: Introduction to resolution of singularities. Algebr. Geom. (Arcata 1975), Proc. Symp. Pure Math. **24**, 187–230 (1975)

[Liu06] Liu, Q.: Algebraic Geometry and Arithmetic Curves, Oxford Graduate Texts in Mathematics 6. Oxford University Press, Oxford (2006)

[LLR05] Liu, Q., Lorenzini, D., Raynaud, M.: On the Brauer group of a surface. Invent. Math. **159**, 673–676 (2005). Corrigendum: Invent. Math. **214**, 593–604 (2018)

[Liv87] Livné, R.: Cubic exponential sums and Galois representations, Current trends in arithmetical algebraic geometry (Arcata, 1985) : Contemp. Math. **67**. Am. Math. Soc. 247–261 (1987)

[Liv95] Livné, R.: Motivic orthogonal two-dimensional representations of Gal($\bar{\mathbb{Q}}/\mathbb{Q}$). Israel J. Math. **92**, 149–156 (1995)

[Loo81] Looijenga, E.: A Torelli theorem for Kähler-Einstein K3-surfaces. Geom. Sympos. (Utrecht 1980), Lect. Notes Math. **894**, 107–112 (1981)

[LP81] Looijenga, E., Peters, C.: Torelli theorems for Kähler K3-surfaces. Compos. Math. **42**, 145–186 (1981)

[Ma12] Ma, S.: On K3 surfaces which dominate Kummer surfaces. Proc. AMS **141**, 131–137 (2012)

[Mad15] Madapusi Pera, K.: The Tate conjecture for K3 surfaces in odd characteristic. Invent. Math. **201**, 625–668 (2015)

[MM18] Malle, G., Matzat, B.H.: Inverse Galois Theory, 2nd edn. Springer, Berlin-Heidelberg (2018)

[Man64] Manin, Y.: The Tate height of points on an Abelian variety, its variants and applications. Izv. Akad. Nauk SSSR Ser. Mat. **28**, 1363–1390 (1964); A.M.S. Transl. **2**(59), 82–110 (1966)

[Man86] Manin, J.: Cubic Forms, 2nd edn. North-Holland (1986)

[Mar19] Martin, G.: Enriques surfaces with finite automorphism group in positive characteristic. **6**, 592–649 (2019)

[Mar03] Martinet, J.: Perfect Lattices in Euclidean Spaces. Grundlehren der Mathematischen Wissenschaften, vol. 327. Springer, Berlin (2003)

[Mau14] Maulik, D.: Supersingular K3 surfaces for large primes. Duke Math. J. **163**, 2357–2425 (2014)

[Maz72] Mazur, B.: Frobenius and the Hodge filtration. Bull. AMS **78**, 653–667 (1972)

[Maz77] Mazur, B.: Modular curves and the Eisenstein ideal. Publ. Math. IHES **47**, 33–186 (1977)

[Mes91] Mestre, J.-F.: Courbes elliptique de rang ≥ 11 *sur* $\mathbb{Q}(t)$. C.R. Acad. Sci. Paris, Sér. I (Math.) **313**, 139–142 (1991)

[MN81] Meyer, W., Neutsch, W.: Fermatquadrupel. Math. Ann. **256**, 51–62 (1981)

[Mil75] Milne, J.: On a conjecture of Artin and Tate. Ann. Math. **102**, 517–533 (1975)

[Mil80] Milne, J.: Étale Cohomology. Princeton University Press, Princeton (1980)

[Mir81] Miranda, R.: The moduli of Weierstrass fibrations over $\mathbb{P}^1$. Math. Ann. **255**, 379–394 (1981)

[Mir89] Miranda, R.: The Basic Theory of Elliptic Surfaces. Dottorato di ricerca in matematica, ETS Editrice, Pisa (1989)

[Mir90] Miranda, R.: Persson's list of singular fibers for a rational elliptic surface. Math. Z. **205**, 191–211 (1990)

[MP86] Miranda, R., Persson, U.: On extremal rational elliptic surfaces. Math. Z. **193**, 537–558 (1986)

[MT05] Miyamoto, R., Top, J.: Reduction of elliptic curves in equal characteristic. Canad. Math. Bull. **48**, 428–444 (2005)

[Miy76] Miyanishi, M.: Unirational quasi-elliptic surfaces in characteristic 3. Oaska Math. J. **13**, 513–522 (1976)

[Mord22] Mordell, L.: On the rational solutions of the indeterminate equation of the third and fourth degree. Proc. Camb. Phil. Soc. **21**, 179–192 (1922)

[Mord69] Mordell, L.: Diophantine Equations. Academic Press (1969)

[MM83] Mori, S., Mukai, S.: The uniruledness of the moduli space of curves of genus 11. Lect. Notes Math. **1016**, 335–353 (1983)

[Morr84] Morrison, D.R.: On K3 surfaces with large Picard number. Invent. Math. **75**(1), 105–121 (1984)

[Muk88] Mukai, S.: Finite groups of automorphisms of K3 surfaces and the Mathieu group. Invent. Math. **94**, 183–221 (1988)

[Muk06] Mukai, S.: Polarized K3 surfaces of genus thirteen, Moduli spaces and arithmetic geometry. Adv. Stud. Pure Math. **45**, 315–326 (2006)

[Muk16] Mukai, S.: K3 surfaces of genus sixteen. In: Kollár, J. et al. (ed.) Minimal Models and Extremal Rays (Kyoto 2011). Advanced Studies in Pure Mathematics, vol. 70, pp. 379–396 (2016)

[Mum66] Mumford, D.: Lectures on curves on an algebraic surface. Annals of Mathematics Studies, vol. 59. Princeton University Press (1966)

[Mum69] Mumford, D.: Enriques' classification of surfaces in char p. I. 1969 Global Analysis (Papers in Honor of K. Kodaira), pp. 325–339. University of Tokyo Press, Tokyo (1969)

[Mum74] Mumford, D.: Abelian Varieties. With Appendices by C.P. Ramanujam and Y. Manin. Corrected Reprint of the Second (1974) Edition. Tata Institute of Fundamental Research Studies in Mathematics, vol. 5. Hindustan Book Agency, New Delhi (2008)

[Mum76] Mumford, D.: Algebraic Geometry I: Complex Projective Varieties, Grundlehren der mathematischen Wissenschaften 221. Springer, Berlin (1976)

[MFK94] Mumford, D., Fogarty, J., Kirwan, F.: Geometric Invariant Theory. Ergebnisse der Mathematik und ihrer Grenzgebiete. 2. Folge, vol. 34, 3rd edn. Springer, Berlin–Heidelberg (1994)

[MS72] Mumford, D., Suominen, K.: Introduction to the theory of moduli. Algebraic Geometry (Oslo 1970), pp. 171–222. Wolters–Noordhoff Publishing (1972)

[Nar87] Naruki, I.: Configurations related to maximal rational elliptic surfaces. Complex Analytic Singularities, Advanced Studies in Pure Mathematics vol. 8, pp. 315–347. North-Holland, Amsterdam (1987)

[Neb98] Nebe, G.: Some cyclo-quaternionic lattices. J. Algebra **199**, 472–498 (1998)

[NS] Nebe, G., Sloane, N.J.A.: Catalogue of Lattices. http://www.math.rwth-aachen.de/~Gabriele.Nebe/LATTICES/index.html

[Ner54] Néron, A.: Propriétés arithmétiques de certaines familles de courbes algébriques. In: Proceedings of ICM Amsterdam III, vol. 481–488 (1954)

[Ner64] Néron, A.: Modèles minimaux des variétés abéliennes sur les corps locaux et globaux. Publ. Math. IHES **21**, 5–128 (1964)

[Ner65] Néron, A.: Quasi-fonctions et hauteurs sur les variétés abéliennes. Ann. Math. **82**, 249–331 (1965)

[Nie73] Niemeier, H.-V.: Definite quadratische Formen der Dimension 24 und Diskriminante 1. J. Number Theory **5**, 142–178 (1973)

[Nik75] Nikulin, V.V.: On Kummer surfaces. Math. USSR Izv. **9**, 261–275 (1975)

[Nik79] Nikulin, V.V.: Finite automorphism groups of Kähler surfaces of type K3. Proc. Moscow Math. Soc. **38**, 75–137 (1979)

[Nik80] Nikulin, V.V.: Integral symmetric bilinear forms and some of their applications. Math. USSR Izv. **14**(1), 103–167 (1980)

[Nik83] Nikulin, V.V.: On the quotient groups of the automorphism group of hyperbolic forms by the subgroup generated by 2-reflections. J. Sov. Math. **22**, 1401–1476 (1983)

[Nik85] Nikulin, V.V.: Surfaces of type K3 with a finite automorphism group and a Picard group of rank three. Proc. Steklov Inst. Math. **3**, 131–155 (1985)

[Nik91] Nikulin, V.V.: On rational maps between K3 surfaces. Constantin Carathéodory: An International Tribute, pp. 964–995. World Scientific Publishing (1991)

[Nik14] Nikulin, V.V.: Elliptic fibrations on K3 surfaces Proc. Edinb. Math. Soc. **57**, 253–267 (2014)

[Nis95] Nishiyama, K.-I.: Examples of Jacobian fibrations on some K3 surfaces whose Mordell-Weil lattices have the maximal rank 18. Comment. Math. Univ. St. Paul. **44**, 219–223 (1995)

[Nis96] Nishiyama, K.-I.: The Jacobian fibrations on some K3 surfaces and their Mordell-Weil groups. Jpn. J. Math. **22**, 293–347 (1996)

[Nis97] Nishiyama, K.-I.: A remark on Jacobian fibrations on K3 surfaces. Saitama Math. J. **15**, 67–71 (1997)

[Nor50] Northcott, D.G.: Periodic points on an algebraic variety. Ann. Math. **51**, 167–177 (1950)

[Nyg83a] Nygaard, N.O.: The Tate conjecture for ordinary K3 surfaces over finite fields. Invent. Math. **74**(2), 213–237 (1983)

[Nyg83b] Nygaard, N.O.: The Torelli theorem for ordinary K3 surfaces over finite fields. In: Arithmetic and Geometry, Progress in Mathematics, vol. I, 35, pp. 267–276. Birkhäuser Boston, Boston, MA (1983)

[NO85] Nygaard, N.: Ogus, A: Tate's conjecture for K3 surfaces of finite height. Ann. Math. **122**, 461–507 (1985)

[Oes90] Oesterlé, J.: Empilements de sphères, Sém. Bourbaki n^o **727**, 375–397 (1989–1990)

[Ogg67] Ogg, A.: Elliptic curves and wild ramification. Am. J. Math. **89**, 1–21 (1967)

[Ogui89] Oguiso, K.: On Jacobian fibrations on the Kummer surfaces of the product of non-isogenous elliptic curves. J. Math. Soc. Jpn. **41**(4), 651–680 (1989)

[Ogui90] Oguiso, K.: An elementary proof of the topological Euler characteristic formula for an elliptic surface. Comment. Math. Univ. St. Pauli **39**, 81–86 (1990)

[Ogui03] Oguiso, K.: Local families of K3 surfaces and applications. J. Alg. Geom. **12**, 405–433 (2003)

[OS90] Oguiso, K., Shioda, T.: The Mordell-Weil lattice of a rational elliptic surface. Comment. Math. Univ. St. Pauli **40**, 83–99 (1990)

[Ogus78] Ogus, A.: Griffiths transversality in crystalline cohomology. Ann. Math. **108**, 395–419 (1978)

[Ogus79] Ogus, A.: Supersingular K3 crystals, Journées de Géométrie Algébrique de Rennes (Rennes 1978), II, pp. 3–86, : Astérisque 64. Soc. Math, France, Paris (1979)

[Ogus83] Ogus, A.: A crystalline Torelli theorem for supersingular K3 surfaces. In: Arithmetic and Geometry, vol. II, 36, pp. 361–394. Progress in Mathematics, Birkhäuser Boston, Boston, MA (1983)

[Pas12] Pastro, V.: Construction of rational elliptic surfaces with Mordell-Weil rank four. Comment. Math. Univ. St. Pauli **61**, 29–42 (2012)

[Per85] Persson, U.: Double sextics and singular K3 surfaces. Algebraic Geometry (Sitges 1983). Lecture Notes in Mathematics, vol. 1124, pp. 262–328. Springer, Berlin-Heidelberg (1985)

[Per90] Persson, U.: Configurations of Kodaira fibers on rational elliptic surfaces. Math. Z. **205**, 1–47 (1990)

[PP81] Persson, U., Pinkham, H.: Degeneration of surfaces with trivial canonical bundle. Ann. Math. **113**, 45–66 (1981)

[Pet15] Peterson, A.: Modular Forms on the Moduli Space of Polarised K3 Surfaces, preprint (2015). arXiv:1511.06887

[PSS71] Piatetski-Shapiro, I.I., Shafarevich, I.R.: Torelli's theorem for algebraic surfaces of type K3. Izv. Akad. Nauk SSSR Ser. Mat. **35**, 530–572 (1971)

[PBT08] Polo Bianco, I., Top, J.: Explicit real cubic surfaces. Canad. Math. Bull. **51**, 125–133 (2008)

[PTvL15] Poonen, B., Testa, D., van Luijk, R.: Computing Néron-Severi groups and cycle class groups. Compos. Math. **151**, 713–734 (2015)

[Ray79] Raynaud, M.: "p-torsion" du schéma de Picard. Astérisque **64**, 87–148 (1979)

[RS15a] Rams, S., Schütt, M.: 64 lines on smooth quartic surfaces. Math. Ann. **362**, 679–698 (2015)

[RS15b] Rams, S., Schütt, M.: 112 lines on smooth quartic surfaces (characteristic 3). Quart. J. Math. **66**, 941–951 (2015)

[RS18] Rams, S., Schütt, M.: At most 64 lines on smooth quartic surfaces (characteristic 2). Nagoya Math. J. **232**, 76–95 (2018)

[RS78] Rudakov, A.N., Shafarevich, I.R.: Supersingular K3 surfaces over fields of characteristic 2. Izv. Akad. Nauk SSSR, Ser. Mat. **42**, 848–869 (1978)

[RS83] Rudakov, A.N., Shafarevich, I.R.: Surfaces of type K3 over fields of finite characteristic. J. Sov. Math. **22**, 1476–1533 (1983)

[S-D74] Saint-Donat, B.: Projective models of K3 surfaces. Am. J. Math. **96**, 602–639 (1974)

[Sai88] Saito, T.: Conductor, discriminant, and the Noether formula of arithmetic surfaces. Duke Math. J. **57**, 151–173 (1988)

[Salg09] Salgado, C.: Construction of linear pencils of cubics with Mordell-Weil rank five. Comment. Math. Univ. St. Pauli **58**, 95–104 (2009)

[Salm49] Salmon, G.: On the triple tangent planes to a surface of the third order. Camb. Dublin Math. J. **4**, 252–260 (1849)

[Schl63] Schläfli, L.: On the distribution of surfaces of the third order into species, in reference to the absence or presence of singular points, and the reality of their lines. Philos. Trans. R. Soc. Lond. R. Soc. **153**, 193–241 (1863)

[Scho88] Schoen, C.: On fiber products of rational elliptic surfaces with section. Math. Z. **197**, 177–199 (1988)

[Scho90] Schoen, C.: Bounds for rational points on twists of constant hyperelliptic curves. J. Reine Angew. Math. **411**, 196–204 (1990)

[Scho74] Schoeneberg, B.: Elliptic Modular Functions, Grundlehren der math. Wiss, vol. 203. Springer, Berlin–Heidelberg (1974)

[Schr07] Schröer, S.: Kummer surfaces for the selfproduct of the cuspidal rational curve. J. Algebr. Geom. **16**, 305–346 (2007)

[Schü04] Schütt, M.: New examples of modular rigid Calabi-Yau threefolds. Collect. Math. **55**, 219–228 (2004)

[Schü06] Schütt, M.: The maximal singular fibres of elliptic K3 surfaces. Arch. Math. **87**, 309–319 (2006)

[Schü07a] Schütt, M.: Elliptic fibrations of some extremal K3 surfaces. Rocky Mountain J. Math. **37**, 609–652 (2007)

[Schü07b] Schütt, M.: Fields of definition of singular K3 surfaces. Commun. Number Theory Phys. **1**, 307–321 (2007)

[Schü08] Schütt, M.: Arithmetic of a singular K3 surface. Mich. Math. J. **56**, 513–527 (2008)

[Schü10] Schütt, M.: K3 surfaces of Picard rank 20 over $\mathbb{Q}$. Algebra & Number Theory **4**(3), 335–356 (2010)

[Schü12] Schütt, M.: A note on the supersingular K3 surface of Artin invariant 1. J. Pure Appl. Algebra **216**, 1438–1441 (2012)

[Schü13] Schütt, M.: Sandwich theorems for Shioda-Inose structures. Izvestiya Mat. **77**, 211–222 (2013)

[Schü15] Schütt, M.: Picard numbers of quintic surfaces. Proc. LMS **110**, 428–476 (2015)

[Schü16] Schütt, M.: Dynamics on supersingular K3 surfaces. Comment. Math. Helv. **91**, 705–719 (2016)

[Schü17] Schütt, M.: $\mathbb{Q}_\ell$-cohomology Projective Planes from Enriques Surfaces in Odd Characteristic. arXiv:1611.03847v3

[Schü18] Schütt, M.: Divisibilities among nodal curves. Math. Res. Lett. **25**, 1359–1368 (2018)

[Schü19] Schütt, M.: $\mathbb{Q}_\ell$-cohomology Projective Planes and Enriques Surfaces in Characteristic Two; Épijournal de Géométrie Algébrique **3**, 24 (2019)

[SSc08] Schütt, M., Schweizer, A.: Davenport-Stothers inequalities and elliptic surfaces in positive characteristic. Q. J. Math. **59**, 499–522 (2008)

[SSc13] Schütt, M., Schweizer, A.: On the uniqueness of elliptic K3 surfaces with maximal singular fibre. Ann. Inst. Fourier **63**, 689–713 (2013)

[SSh10] Schütt, M., Shioda, T.: Elliptic surfaces, Algebraic geometry in East Asia (Seoul 2008). Adv. Stud. Pure Math. **60**, 51–160 (2010)

[SSvL10] Schütt, M., Shioda, T., van Luijk, R.: Lines on Fermat surfaces. J. Number Theory **130**, 1939–1963 (2010)

[ST06] Schütt, M., Top, J.: Arithmetic of the [19,1,1,1,1,1] fibration. Comment. Math. Univ. St. Pauli **55**, 9–16 (2006)

[SGA2] Séminaire de Géométrie Algébrique, Grothendieck, A.: Cohomologie locale des faisceaux cohérents et théorèmes de Lefschetz locaux et globaux, édition recomposée et annotée, Doc. Math. Soc. Math. Fr. **4** (2005)

[SGA4] Séminaire de géométrie algébrique, Artin, M., Grothendieck, A., Verdier, J.L.: Théorie des topos et cohomologie étale des schémas. Lecture Notes in Mathematics, vol. 269, 270, 305. Springer, Heidelberg (1972–1973)

[SGA7.2] Séminaire de géométrie algébrique, Deligne, P., Katz, N.M.: Groupes de monodromie en géométrie algébrique. Lecture Notes in Mathematics, vol. 340. Springer, Heidelberg (1973)

[Sen17] Sengupta, T.: Elliptic Fibrations on Supersingular K3 Surface with Artin Invariant 1 in Characteristic 3, In Analytic and algebraic geometry, 239–292. Hindustan Book Agency, New Delhi (2017)

[Ser55] Serre, J.-P.: Faisceaux algébriques cohérents. Ann. Math. **61**, 197–278 (1955)

[Ser55/56] Serre, J.-P.: Géométrie algébrique et géométrie analytique. Ann. Inst. Fourier **6**, 1–42 (1955–1956)

[Ser68] Serre, J.-P.: Abelian l-adic Representations and Elliptic Curves. Benjamin Inc, W. A (1968)

[Ser70] Serre, J.-P.: Cours d'Arithmétique. Presses Uuniversitaires de France (1970)

[Ser72] Serre, J.-P.: Propriétés galoisiennes des points d' ordre fini des courbes elliptiques. Invent. Math. **15**, 259–331 (1972)

[Ser85] Serre, J.-P.: Résumé des cours de 1984–1985. Annuaire du Collége de France **85–90** (1985)

[Ser97] Serre, J.-P.: Lectures on the Mordell–Weil Theorem, 3rd edn. Vieweg+Teubner (1997)

[Sha+65] Shafarevich, I.R., Averbukh, B.G., Vainberg, J.R., Zhizhchenko, A.B., Manin, Y., Moishezon, B.G., Tyurina, G.N., Tyurin, A.N.: Algebraic surfaces. Tr. Mat. Inst. Steklova **75**, 1–215 (1965)

[Sha66] Shafarevich, I.R.: Lectures on minimal models and birational transformations of two-dimensional schemes. Tata Inst. Fund. Res. Bombay **37** (1966)

[Sha94] Shafarevich, I.: R: Basic Algebraic Geometry, 2nd edn. Springer, Berlin (1994)

[ShT67] Shafarevich, I.R., Tate, J.: The rank of elliptic curves. Dokl. Akad. Nauk SSSR **175**, 770–773; Sov. Math. Dokl. **8**, 917–920 (1967)

[S-B19] Shepherd-Barron, N.: Weyl Group Covers for Brieskorn's Resolutions in all Characteristics and the Integral Cohomology of G/P, preprint (2019). arXiv:1711.10439v5

[Shima00] Shimada, I: On elliptic K3 surfaces. Mich. Math. J. **47**, 423–446 (2000) (Extended version on arXiv:ath/0505140)

[Shima04] Shimada, I.: Supersingular K3 surfaces in characteristic 2 as double covers of a projective plane. Asian J. Math. **8**, 531–586 (2004)

[Shima09] Shimada, I.: Transcendental lattices and supersingular reduction lattices of a singular $K3$ surface. Trans. Am. Math. Soc. **361**, 909–949 (2009)

[SZ01] Shimada, I., Zhang, D.Q.: Classification of extremal elliptic $K3$ surfaces and fundamental groups of open $K3$ surfaces. Nagoya Math. J. **161**, 23–54 (2001)

[Shimu55] Shimura, G.: Reduction of algebraic varieties with respect to a discrete valuation. Am. J. Math **77**, 134–176 (1955)

[Shimu71] Shimura, G.: Introduction to the Arithmetic Theory of Automorphic Forms. Iwanami and Princeton Univ, Press (1971)

[Shio72] Shioda, T.: On elliptic modular surfaces. J. Math. Soc. Jpn. **24**, 20–59 (1972)

[Shio73] Shioda, T.: On rational points of the generic elliptic curve with level N structure over the field of modular functions of level N. J. Math. Soc. Jpn. **25**, 144–157 (1973)

[Shio74a] Shioda, T.: An example of unirational surfaces in characteristic p. Math. Ann. **211**, 233–236 (1974)

[Shio74b] Shioda, T.: Kummer surfaces in characteristic 2. Proc. Jpn. Acad. **50**, 718–722 (1974)

[Shio75] Shioda, T.: Algebraic cycles on certain K3 surfaces in characteristic p. In: Proceedings of International Conference on Manifolds (Tokyo 1973), pp. 357–364. University of Tokyo Press, Tokyo (1975)

[Shio77] Shioda, T.: Some results on unirationality of algebraic surfaces. Math. Ann. **230**, 153–168 (1977)

[Shio78] Shioda, T.: The period map of abelian surfaces. J. Fac. Sci. Univ. Tokyo **25**, 47–59 (1978)

[Shio79] Shioda, T.: Supersingular K3 surfaces. Lect. Notes Math. **732**, 564–591 (1979)

[Shio82] Shioda, T.: On the Picard number of a Fermat surface. J. Fac. Sci. Univ. Tokyo Sect. IA Math. **28**(3), 725–734 (1982)

[Shio86] Shioda, T.: An explicit algorithm for computing the Picard number of certain algebraic surfaces. Am. J. Math. **108**, 415–432 (1986)

[Shio87a] Shioda, T.: Some observations on Jacobi sums. Galois Representations and Arithmetic Algebraic Geometry (Kyoto, 1985/Tokyo, 1986). Advanced Studies in Pure Mathematics, vol. 12, pp. 119–135. North-Holland, Amsterdam (1987)

[Shio87b] Shioda, T.: Supersingular K3 surfaces with big Artin invariants. J. Reine Angew. Math. **381**, 205–210 (1987)

[Shio87c] Shioda, T.: Arithmetic and geometry of Fermat curves. Algebraic Geometry Seminar **1988**, 95–102 (1987)

[Shio89a] Shioda, T.: The Galois representation of type E_8 arising from certain Mordell-Weil groups. Proc. Jpn. Acad. **65A**, 195–197 (1989)

[Shio89b] Shioda, T.: Mordell–Weil lattices and Galois representation, I, II, III. Proc. Jpn. Acad. **65A**, 268–271; 296–299; 300–303 (1989)

[Shio90a] Shioda, T.: On the Mordell-Weil lattices. Comment. Math. Univ. St. Pauli **39**, 211–240 (1990)

[Shio91a] Shioda, T.: Mordell-Weil lattices and sphere packings. Am. J. Math. **113**, 931–948 (1991)

[Shio91b] Shioda, T.: Mordell-Weil lattices of type E_8 and deformation of singularities. SLN **1468**, 177–202 (1991)

[Shio91c] Shioda, T.: Construction of elliptic curves with high rank via the invariants of the Weyl groups. J. Math. Soc. Jpn. **43**, 673–719 (1991)

[Shio91d] Shioda, T.: Theory of Mordell–Weil lattices. Proc. ICM Kyoto-1990, **I**, 473–489 (1991)

[Shio91e] Shioda, T.: An infinite family of elliptic curves over Q with large rank via Néron's method. Invent. Math. **106**, 109–119 (1991)

[Shio92a] Shioda, T.: Some remarks on elliptic curves over function fields. Proc. Journée Arithmetique **91**, Genève, Asterisque, **209**, 99–114 (1992)

[Shio92b] Shioda, T.: Existence of a rational elliptic surface with a given Mordell-Weil lattice. Proc. Jpn. Acad. **68A**, 251–255 (1992)

[Shio92c] Shioda, T.: Mordell-Weil lattices for higher genus fibration. Proc. Jpn. Acad. **68A**, 247–250 (1992)

[Shio95a] Shioda, T.: Weierstrass transformations and cubic surfaces. Comment. Math. Univ. St. Pauli **44**, 109–128 (1995)

[Shio95b] Shioda, T.: A uniform construction of the root lattices E_6, E_7, E_8 and their dual lattices. Proc. Jpn. Acad. **71A**, 140–143 (1995)

[Shio99a] Shioda, T.: Cyclotomic anologue in the theory of algebraic equations of type E_6, E_7, E_8. Integral Quadratic Forms Lattices, Contemp. Math. **249**, 87–96 (1999)

[Shio99b] Shioda, T.: Mordell-Weil lattices for higher genus fibration over a curve. New Trends in Algebraic Geometry, pp. 359–373. Cambridge University Press, Cambridge (1999)

[Shio03] Shioda, T.: The elliptic K3 surfaces with a maximal singular fibre. C.R. Acad. Sci. Paris, Ser. I, **337**, 461–466 (2003)

[Shio05a] Shioda, T.: On K3 surfaces defined over $\mathbb{Q}$—Correction to the paper in Vol. 43 (1994). Comment. Math. Univ. St. Pauli **54**, 87–89 (2005)

[Shio05b] Shioda, T.: Elliptic surfaces and Davenport-Stothers triples. Comment. Math. Univ. St. Pauli **54**, 49–68 (2005)

[Shio06] Shioda, T.: Kummer sandwich theorem of certain elliptic K3 surfaces. Proc. Jpn. Acad. Ser. A **82**, 137–140 (2006)

[Shio07a] Shioda, T.: Correspondence of elliptic curves and Mordell–Weil lattices of certain elliptic K3 surfaces. In: Algebraic Cycles and Motives, vol. 2, 344, pp. 319–339. LMS Lecture Notes Series, Cambridge University Press (2007)

[Shio07b] Shioda, T.: The Mordell-Weil lattice of $y^2 = x^3 + t^5 - 1/t^5 - 11$. Comment. Math. Univ. St. Pauli **56**, 45–70 (2007)

[Shio08] Shioda, T.: K3 surfaces and sphere packings. J. Math. Soc. Jpn. **60**, 1083–1105 (2008)

[Shio09] Shioda, T.: Some explicit integral polynomials with Galois group $W(E_8)$. Proc. Jpn. Acad. **85A**, 118–121 (2009)

[Shio10] Shioda, T.: Gröbner basis, Mordell-Weil lattices and deformation of singularities, I. II. Proc. Jpn. Acad. **86A**, 21–26, 27–32 (2010)

[Shio12] Shioda, T.: Multiplicative excellent family of type E_6. Proc. Jpn. Acad. **88A**, 46–51 (2012)

[Shio13] Shioda, T.: Elliptic fibrations of maximal rank on a supersingular K3 surface. Izvestia RAS. Ser. Math. **77**, 139–148 (2013)

[Shio15a] Shioda, T.: Mordell-Weil lattice of higher genus fibration on a Fermat surface. J. Math. Sci. Univ. Tokyo **22**, 443–468 (2015)

[Shio15b] Shioda, T.: The MWL-algorithms for constructing cubic surfaces with preassigned 27 lines. Comment. Math. Univ. St. Pauli **64**, 157–186 (2015); II, **67** (2019, to appear)

[SM74] Shioda, T., Mitani, N.: Singular abelian surfaces and binary quadratic forms. In: Classification of Algebraic Varieties and Compact Complex Manifolds. Lecture Notes in Mathematics, vol. 412, pp. 259–287 (1974)

[SI77] Shioda, T., Inose, H.: On singular $K3$ surfaces. In: Baily Jr., W.L., Shioda, T. (eds.) Complex Analysis and Algebraic Geometry, pp. 119–136. Tokyo, Cambridge University Press, Cambridge, Iwanami Shoten (1977)

[SU92] Shioda, T., Usui, H.: Fundamental invariants of Weyl groups and excellent families of elliptic curves. Comment. Math. Univ. St. Pauli **41**, 169–217 (1992)

[Sie29] Siegel, C.L.: Über einige Anwendungen diophantischer Approximationen. Abhandlungen Akad. Berlin, Physikalisch-mathematische Klasse, Nr. **1**, 70pp (1929)

[Sil83] Silverman, J.: Heights and the specialization map for families of abelian varieties. J. Reine Angew. Math. **342**, 197–211 (1983)

[Sil88] Silverman, J.H.: Computing heights on elliptic curves. Math. Comput. **51**, 339–358 (1988)

[Sil94] Silverman, J.: Advanced Topics in the Arithmetic of Elliptic Curves, GTM 151. Springer, New York (1994)

[Sil09] Silverman, J.: The Arithmetic of Elliptic Curves, vol. 106, 2nd edn. GTM, Springer, New York (2009)

[SiT15] Silverman, J., Tate, J.: Rational Points on Elliptic Curves, 2nd edn. Springer, New York (2015)

[Siu83] Siu, Y.T.: Every K3-surface is Kähler. Invent. Math. **73**, 139–150 (1983)

[Slo80] Slodowy, P.: Simple Singularities and Simple Algebraic Groups, SLN 815. Springer, Berlin (1980)

[Sta66] Stark, H.: On complex quadratic fields with class number equal to one. Trans. Am. Math. Soc. **122**, 112–119 (1966)

[Ste85] Sterk, H.: Finiteness results for algebraic $K3$ surfaces. Math. Z. **189**(4), 507–513 (1985)

[Sti87] Stiller, P.: The Picard numbers of elliptic surfaces with many symmetries. Pacific J. Math. **128**, 157–189 (1987)

[SV87] Stöhr, K.-O., Voloch, J.F.: A formula for the Cartier operator on plane algebraic curves. J. Reine Angew. Math. **377**, 49–64 (1987)

[Swi67] Swinnerton-Dyer, H.P.F.: The zeta function of a cubic surface over a finite field. Proc. Cambr. Phil. Soc. **63**, 55–71 (1967)

[Szy04] Szydlo, M.: Elliptic fibers over non-perfect residue fields. J. Number Theory **104**(1), 75–99 (2004)

[Tat52] Tate, J.: Genus change in purely inseparable extensions of function fields. Proc. AMS **3**, 400–406 (1952)

[Tat65] Tate, J.: Algebraic cycles and the pole of zeta functions. In: Arithmetical Algebraic Geometry, pp. 93–110. Harper and Row, New York (1965)

[Tat66a] Tate, J.: Endomorphisms of abelian varieties over finite fields. Invent. Math. **2**, 134–144 (1966)

[Tat66b] Tate, J.: On the conjectures of Birch and Swinnerton-Dyer and a geometric analog. Sém. Bourbaki, n^o **306**, 415–440 (1965/1966)

[Tat74] Tate, J.: The arithmetic of elliptic curves. Invent. Math. **23**, 179–206 (1974)

[Tat75] Tate, J.: Algorithm for determining the type of a singular fiber in an elliptic pencil. SLN **476**, 33–52 (1975)

[Tat83] Tate, J.: Variation of the canonical height of a point depending on a parameter. Am. J. Math. **105**, 287–294 (1983)

[TW95] Taylor, R., Wiles, A.: Ring-theoretic properties of certain Hecke algebras. Ann. Math. **141**, 553–572 (1995)

[Tju70] Tjurina, G.N.: Resolutions of singularities for flat deformations of isolated rational double points, Funkcional Anal. i. Prilozen **4**, 77–83 (1970). Translation: Functional Anal. Appl. **4**, 68–73 (1970)

[Top87] Top, J.: Néron's proof of the existence of elliptic curves over $\mathbb{Q}$ with rank at least 11. Master thesis, Universiteit Utrecht (1987)

[Top93] Top, J.: Descent by 3-isogeny and 3-rank of Quadratic Fields, Advances in Number Theory (Kingston, ON, 1991), pp. 303–317. Oxford University Press, New York (1993)

[TdZ09] Top, J., De Zeeuw, F.: Explicit elliptic $K3$ surfaces with rank 15. Rocky Mt. J. Math. **39**(5), 1689–1697 (2009)

[Tot43] Toth, L.F.: Über die dichteste Kugellagerung. Math. Z. **48**, 676–684 (1943)

[Tsa92] Tsfasman, M.A.: Global fields, codes and sphere packings. Journées Arithmétiques 1989, Astérisque **198-199-200**, 373–396 (1992)

[TV91] Tsfasman, M.A., Vladuts, S.G.: Algebraic-Geometric Codes. Kluwer Academic Publishers (1991)

[Ulm02] Ulmer, D.: Elliptic curves with large rank over function fields. Ann. Math. **155**, 295–315 (2002)

[Ulm14] Ulmer, D.: Explicit points on the Legendre curve. J. Number Theory **136**, 165–194 (2014)

[Usu93] Usui, H.: Mordell-Weil lattices of type D_5. Math. Nachr. **161**, 219–232 (1993)

[Usu00+] Usui, H.: On the Mordell–Weil lattice of the elliptic curve $y^2 = x^3 + t^m + 1$. I–IV. Comment. Math. Univ. St. Pauli **49**, 71–78 (2000); **50**, 65–87 (2001); **55**, 173–194 (2006); **57**, 23–63 (2008)

[vGK00] van der Geer, G., Katsura, T.: On a stratification of the moduli of K3 surfaces. JEMS **2**, 259–290 (2000)

[vLui07] van Luijk, R.: K3 surfaces with Picard number one and infinitely many rational points. Algebra Number Theory **1**(1), 1–15 (2007)

[VAZ09] Várilly-Alvarado, A., Zywina, D.: Arithmetic E_8 lattices with maximal Galois action. LMS J. Comput. Math. **12**, 144–165 (2009)

[Via17] Viazovska, M.: The sphere packing problem in dimension 8. Ann. Math. **185**, 991–1015 (2017)

[Vin83] Vinberg, È.B.: The two most algebraic $K3$ surfaces. Math. Ann. **265**(1), 1–21 (1983)

[Was08] Washington, L.C.: Elliptic Curves. Number Theory and Cryptography, 2nd edn. Chapman & Hall, Boca Raton (2008)

[Wei28] Weil, A.: L'arithmetique sur les courbes algébriques. Acta Math. **52**, 281–315 (1928); Collected Works, vol. I, pp. 11–45. Springer (1980)

[Wei46] Weil, A.: Foundations of Algebraic Geometry. AMS, New York (1946)

[Wei48a] Weil, A.: Sur les courbes algébriques et les variétés qui s'en déduisent. Actualités Sci. Ind. **1041**, Publ. Inst. Math. Univ. Strasbg. **7**, Hermann et Cie., Paris (1948)

[Wei48b] Weil, A.: Variétés abéliennes et courbes algébriques. Actualités Sci. Ind. **1064**, Publ. Inst. Math. Univ. Strasbg. **8**, Hermann et Cie., Paris (1948)

[Wei49] Weil, A.: Numbers of solutions of equations in finite fields. Bull. AMS **55**, 497–508 (1949)

[Wei50] Weil, A.: Number theory and algebraic geometry. In: Proceedings of ICM, vol. 2, pp. 90–100. Cambridge, MA (1950)

[Wei51] Weil, A.: Arithmetic on algebraic varieties. Ann. Math. **53**, 412–444 (1951)

[Wei52] Weil, A.: Jacobi sums as "Grössencharaktere". Trans. Am. Math. Soc. **73**, 487–495 (1952)

[Wei54] Weil, A.: Abstract versus classical algebraic geometry. In: Proceedings of ICM Amsterdam, vol. III, pp. 550–558 (1954); Collected Works, vol. II, pp. 180–188. Springer (1980)

[Wei58] Weil, A.: Final report on contract AF 18(603)-57, Scientific works. Collected Papers, II, pp. 391–395. Springer, Berlin–New York (2009)

[Wei62] Weil, A.: Foundations of Algebraic Geometry, 2nd edn. AMS (1962)

[Wil95] Wiles, A.: Modular elliptic curves and Fermat's Last Theorem. Ann. Math. **141**(2), 443–551 (1995)

[Wit37] Witt, E.: Theorie der quadratischen Formen in beliebigen Körpern. J. Reine Angew. Math. **176**, 31–44 (1937)

[Zar50] Zariski, O.: The fundamental ideas of abstract algebraic geometry. In: Proceedings of ICM, vol. 2, pp. 77–89. Cambridge, MA (1950)

[Zar58] Zariski, O.: On Castelnuovo's criterion of rationality $p_a = P_2 = 0$ of an algebraic surface. Ill. J. Math. **2**, 303–315 (1958)

[Zar71] Zariski, O.: Algebraic surfaces. Second supplemented edition. Ergebnisse der Mathematik und ihrer Grenzgebiete, vol. 61. Springer, New York–Heidelberg (1971)

Index

M. Schütt and T. Shioda, *Mordell–Weil Lattices*, Ergebnisse der Mathematik und ihrer Grenzgebiete. 3. Folge / A Series of Modern Surveys in Mathematics 70,
https://doi.org/10.1007/978-981-32-9301-4

The manufacturer's authorised representative in the EU is Springer Nature Customer Service Centre GmbH, Europaplatz 3, 69115 Heidelberg, Germany. If you have any concerns regarding our products, please contact ProductSafety@springernature.com

Printed and bound by CPI Group (UK) Ltd, Croydon, CR0 4YY
25/09/2026
02222882-0005